T0184323

# Graduate Texts in Physics

**Series Editors**

Kurt H. Becker, NYU Polytechnic School of Engineering, Brooklyn, NY, USA

Jean-Marc Di Meglio, Matière et Systèmes Complexes, Bâtiment Condorcet, Université Paris Diderot, Paris, France

Sadri Hassani, Department of Physics, Illinois State University, Normal, IL, USA

Morten Hjorth-Jensen, Department of Physics, Blindern, University of Oslo, Oslo, Norway

Bill Munro, NTT Basic Research Laboratories, Atsugi, Japan

Richard Needs, Cavendish Laboratory, University of Cambridge, Cambridge, UK

William T. Rhodes, Department of Computer and Electrical Engineering and Computer Science, Florida Atlantic University, Boca Raton, FL, USA

Susan Scott, Australian National University, Acton, Australia

H. Eugene Stanley, Center for Polymer Studies, Physics Department, Boston University, Boston, MA, USA

Martin Stutzmann, Walter Schottky Institute, Technical University of Munich, Garching, Germany

Andreas Wipf, Institute of Theoretical Physics, Friedrich-Schiller-University Jena, Jena, Germany

*Graduate Texts in Physics* publishes core learning/teaching material for graduate- and advanced-level undergraduate courses on topics of current and emerging fields within physics, both pure and applied. These textbooks serve students at the MS- or PhD-level and their instructors as comprehensive sources of principles, definitions, derivations, experiments and applications (as relevant) for their mastery and teaching, respectively. International in scope and relevance, the textbooks correspond to course syllabi sufficiently to serve as required reading. Their didactic style, comprehensiveness and coverage of fundamental material also make them suitable as introductions or references for scientists entering, or requiring timely knowledge of, a research field.

Manuel García-León

# Detecting Environmental Radioactivity

 Springer

Manuel García-León
Department of Atomic, Molecular
and Nuclear Physics
University of Seville
Sevilla, Spain

ISSN 1868-4513          ISSN 1868-4521  (electronic)
Graduate Texts in Physics
ISBN 978-3-031-09972-4          ISBN 978-3-031-09970-0  (eBook)
https://doi.org/10.1007/978-3-031-09970-0

© The Editor(s) (if applicable) and The Author(s), under exclusive license to Springer Nature
Switzerland AG 2022
This work is subject to copyright. All rights are solely and exclusively licensed by the Publisher, whether
the whole or part of the material is concerned, specifically the rights of translation, reprinting, reuse
of illustrations, recitation, broadcasting, reproduction on microfilms or in any other physical way, and
transmission or information storage and retrieval, electronic adaptation, computer software, or by similar
or dissimilar methodology now known or hereafter developed.
The use of general descriptive names, registered names, trademarks, service marks, etc. in this publication
does not imply, even in the absence of a specific statement, that such names are exempt from the relevant
protective laws and regulations and therefore free for general use.
The publisher, the authors, and the editors are safe to assume that the advice and information in this book
are believed to be true and accurate at the date of publication. Neither the publisher nor the authors or
the editors give a warranty, expressed or implied, with respect to the material contained herein or for any
errors or omissions that may have been made. The publisher remains neutral with regard to jurisdictional
claims in published maps and institutional affiliations.

Cover image: Olga_Kostrova

This Springer imprint is published by the registered company Springer Nature Switzerland AG
The registered company address is: Gewerbestrasse 11, 6330 Cham, Switzerland

*To María Dolores, center of my Universe.*
*To Alicia and Ernesto, its brightest stars.*

# Preface

In this book, an account of principles and methods for the measurement of radioactivity in the environment is given. The approach to this subject is very specific since it is a problem of measuring low or very low levels of radioactivity. Consequently, traditional methods for radiation counting and spectrometry do not apply. In contrast, specific low-level radiation counting and spectrometry or mass spectrometry techniques are needed. In the book, an academic and compact approach to these techniques is provided, together with an account of the presence of radioactivity in nature, including sources, distribution, levels, and dynamics. The text also includes a description of the fundamental concepts needed for a solid explanation of the main objective of the book. Thus, the phenomenology and laws of radioactivity as well as of the radiation interaction with matter are described. Once these bases have been established, the description of the different types of detectors and mass spectrometers is afforded. The measurement of radioactivity in the environment starts with sampling and follows with sample transport, storage, and preparation. All these aspects are incorporated in the text, including radiochemistry, which plays a very important role in this problem. For that, relevant concepts about chemical separations are also introduced in the book.

The measurement of low levels of radioactivity is a very specific problem in which many disciplines are involved. For that, the information is dispersed in different books and journals in which it is treated from different point of views. The need to bridge the pertinent information in a solid and compact text book is very clear. This is the most important aim of the book, where the reader, on the basis of well-structured and scientifically based information, can find a rigorous view of the topic with strong academic support.

This book intends to be a textbook for graduate or advanced undergraduate students in physics, chemistry, or engineering oriented to environmental sciences and in many other disciplines in which monitoring of the environment and its management is of interest. From another point of view, the determination of environmental radioactivity is always included in governmental protection programs because governments are required by national or international regulations to preserve the quality of the environment. These programs include the determination of radioactivity in the environment, which requires the techniques presented in the book. For that, this book will very likely also be of interest to the involved professionals.

This text is the product mainly of the long teaching experience the author has accumulated over the last 40 years but also of the generous contributions from the members of the Applied Nuclear Physics Research Group of the University of Sevilla. I am very grateful for the assistance of my group colleagues who have provided me with suggestions, data, figures, and, of course, warm support and friendship. The assistance of Alicia García-Roldán in the preparation of the figures and the constant encouragement of my family are gratefully acknowledged.

Sevilla, Spain                                                                              Manuel García-León

# Contents

# Radioactivity: History and Phenomenology

**1**

**ABSTRACT**

The main objectives of the chapter are the phenomenological description of the atomic nucleus and of radioactivity. The nucleus components are introduced, and the forces acting inside are presented. The stability conditions of nuclei against radioactivity are defined. Finally, a phenomenological description of the different types of radioactivity is given. Information is also provided on a few historical aspects of the discovery of radioactivity and the nomenclature used within this field presented.

## 1.1 Basic Description of the Atomic Nucleus. Nuclear Stability

### 1.1.1 Simple Nuclear Models

Radioactivity occurs in the atomic nucleus. The description of the atom is complex. However, as far as this book is concerned, a simple planetary model, as that of Rutherford [1], is enough to describe the atom and understand the phenomenon of radioactivity. According to such a model, we can assume that a cloud of $Z$ electrons orbits around a nucleus that contains $Z$ protons and $N$ neutrons. In this way, the nucleus contains $A = Z + N$ nucleons, according to the Heisenberg proposal of 1932 [2]. In Fig. 1.1, we give a representation of this picture.

The electron, $e^-$, is an elementary particle and has a negative electric charge equal to that of the proton but with opposite sign ($q_p = -q_e$), and a value of

$$q_e = 1.60217733 \times 10^{-19} \ C = 4.8032068 \times 10^{-10} \text{e.s.u.}$$

© The Author(s), under exclusive license to Springer Nature Switzerland AG 2022
M. García-León, *Detecting Environmental Radioactivity*, Graduate Texts in Physics, https://doi.org/10.1007/978-3-031-09970-0_1

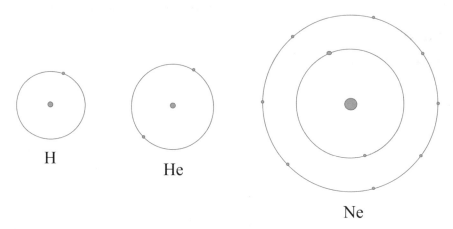

**Fig. 1.1** A pictorial representation of the Rutherford planetary model for the atom. The H, He, and Ne atoms are presented

Consequently, the nuclear electric charge is compensated for by that of the electron cloud of the atom, provided that the neutron has 0 electric charge. As a consequence, the atom is electrically neutral, which fully coincides with the experimental observations.

Furthermore, the mass of the electron, $m_e$, is approximately 1840 times smaller than that of the proton and neutron, which have similar masses, $m_p$ and $m_n$. Indeed,

$$m_e = 5.48579903 \times 10^{-4} \quad u = 0.511 \, \text{MeV}/\text{c}^2$$

$$m_p = 1.007276470 \quad u = 938.3 \, \text{MeV}/\text{c}^2$$

$$m_n = 1.008664904 \quad u = 939.6 \, \text{MeV}/\text{c}^2$$

Therefore, 99.975% of the atomic mass is concentrated in the nucleus. However, the dimensions of the atom are $\sim 10^{-10}$ m $= 1$ Å (Armstrong), while those of the nucleus are $\sim 10^{-15}$ m $= 1$ F (Fermi).

The long distance between the electronic cloud and the nucleus leads to a *quasi-*independent behavior of both systems. Thus, for example, the chemical properties of the atom are exclusively determined by the electronic cloud and not by the nucleus. In any case, there is an influence on the atomic energy levels of the electric, magnetic, and quadrupole moments, as well as the size and mass of the nucleus. This gives rise to the so-called hyperfine structure of the atom. Despite this, the hypothesis of independent behavior of the electron shell and the nucleus, even if it is not strictly true, can be accepted without risk of straying too far from the real thing.

The electromagnetic interaction makes the nucleus and the electron cloud, both with equal electric charge, although of opposite sign, attract each other. Within our picture, it is the rotation of the electrons that compensates for this attraction by means of centrifugal forces, avoiding the collapse of the electronic cloud toward the nucleus and ensuring the stability of the atom. However, the protons in the nucleus undergo electrostatic repulsion because they have an electric charge of the same sign. The key to nucleus stability lies in the strong interaction, which, broadly speaking, can be described as attractive and of short range, ~1 F, roughly the size of the nucleus. The strong interaction among nucleons compensates for electrostatic repulsion, ensuring the stability of the nucleus. Radioactivity breaks nuclear stability and leads to nuclei disintegration. Before describing radioactivity, let us review the nomenclature necessary to describe nuclear phenomena.

A more realistic description of the nucleus, in which protons and neutron are not elementary particles, can be found in many texts on Nuclear Physics as, for example, that cited in [3].

## 1.1.2 Atomic and Mass Numbers. Isobars, Isotopes, and Isotone Nuclei

$Z$ is the atomic number, $N$ is the number of neutrons, and $A = N + Z$ is the mass number. The nuclei are represented as follows.

$$_Z^A X$$

where $X$ is the symbol of the chemical element. It is true that $X$ and $Z$ give the same information; so the nucleus can be identified simply as:

$$^A X$$

A nuclei group with the same $Z$ is called isotopes of element $X$. Isotopes have the same chemical behavior, since they are in fact the same chemical element. For example, 13 isotopes are known of the chemical element C, $Z = 6$, of which only two of them, $^{12}$C and $^{13}$C, are stable. The rest are radioactive isotopes, among which $^{14}$C stands out for its great geochemical, geological, and archeological significance. The three mentioned C isotopes are found in nature, where they behave almost identically. This is the basis for the well-known $^{14}$C dating method. The abundance of a given isotope within an isotope group is known as isotopic abundance, which is usually expressed as a percentage. The isotopic abundance gives the ratio between the number of atoms of a given isotope and the total number of atoms of the element studied found in a natural sample. For example, the isotopic abundances of $^{12}$C and $^{13}$C are 98.90% and 1.10%, respectively. (The $^{14}$C isotopic abundance is negligible compared to the other two isotopes). Another example: for the element U, $Z = 92$, 19 isotopes are known, all radioactive. Only three of them are found in Nature, $^{238}$U, $^{235}$U, and $^{234}$U, whose isotopic abundances are 99.2745%, 0.720%, and 0.0055%, respectively.

Nuclei with equal $N$ are called isotones, while nuclei with equal $A$ are called isobars. In both cases, the relationships among isotones and isobars are important only from the nuclear point of view, not from the chemical point of view, as happens with isotopes, which have the same chemical behavior.

The nuclear properties of a given nuclear species depend on its energy state. Usually, nuclei are found in their fundamental, ground, state. However, there are some excited states of some longevity, metastable states, which give rise to the phenomenon of nuclear isomerism. We will discuss this phenomenon in the next sections. For now, we will just introduce the appropriate nomenclature to describe them. An isomeric state of the nucleus $^A X$ is represented by:

$$_Z^{Am} X$$

or simply by

$$^{Am} X$$

The chemical behavior of the isomers is identical to that of element $X$. However, the nuclear behavior is different from that of the nucleus ground state. Therefore, on some occasions, it is preferred to represent the nuclei as follows:

$$_Z^{Ag} X$$

or simply

$$^{Ag} X$$

where the use of $g$ highlights that we are representing the ground state. It is not strictly necessary to use this nomenclature. In fact, its use is not common. Normally, ground states are represented by the simple nomenclature used at the beginning of the paragraph. When the nucleus is in an excited state, of a nonmetastable type, it can be represented as:

$$^{A*} X$$

The word isotope is often used incorrectly to designate nuclear species. The correct way is to use the word nuclide. If the nucleus is radioactive, radionuclides can be used, although the word radioisotope is also acceptable.

## 1.1.3  Unstable Nuclides

There are approximately 3000 known nuclei, in addition to approximately 600 isomeric states of these nuclei [4]. They are isotopes of the 118 elements known to date. Approximately 280 of them appear to be stable with respect to radioactive decay. It is very interesting to observe that almost 60% of the stable nuclei have $Z$ and $N$ even, while the remaining 40% are distributed in an almost identical way between the group of nuclei with odd $Z$ and $N$ even and those with $Z$ even and $N$ odd. There are, in fact, only five stable nuclei with odd $Z$ and $N$ [3]. It appears that the proton–proton or neutron–neutron pairing situation, but not the proton–neutron, confers stability on the nucleus. This is a property of the strong interaction, known as the pairing effect, whose basic explanation goes beyond the limits of this text. However, the pairing effects are not sufficient to explain nuclear stability: A certain $\frac{N}{Z}$ ratio is necessary. Indeed, if we plot $Z$ against $N$ for the known stable nuclei, see Fig. 1.2, we observe that $Z = N$ for light nuclei. However, as $Z$ increases, stable nuclei have $\frac{N}{Z} \approx 1.5$. From $Z > 83$, all known nuclei are radioactive, essentially, although not only, by $\alpha$ decay, which we will study later.

Roughly speaking nuclear stability is a consequence of Coulomb repulsion and the nuclear or strong attraction balance. For that reason, the nucleus $\frac{N}{Z}$ ratio is important. Indeed, as $Z$ increases, we could say that more neutrons are needed to compensate for Coulomb repulsion. Stability is achieved by increasing $N$, which compensates for the repulsion that occurs among protons. However, when $\frac{N}{Z}$ is too high, we have neutron-rich nuclei that are candidates to disintegrate by $\beta^-$ decay (see the following Sections). During $\beta^-$ decay, N decreases, and the resulting nucleus has a more stable $\frac{N}{Z}$ ratio. If $\frac{N}{Z}$ is too low, we have proton-rich nuclei that very likely disintegrate by $\beta^+$ decay, whereby $Z$ decreases. In both cases, the resulting nucleus has a $\frac{N}{Z}$ value closer to nuclear stability within its $Z$ region.

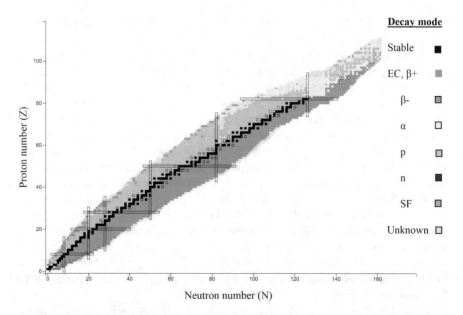

**Fig. 1.2** Stability diagram of nuclides. Plot obtained from the National Nuclear Data Center NuDat web application, www.nndc.bnl.gov/nudat/

## 1.2 Discovery of Radioactivity

### 1.2.1 Some Historic Data

Radioactivity was discovered by chance, as many other discoveries in Science. This occurred in 1896 as a result of the work of Henri Becquerel. This fact, on the other hand, has caused a historical treatment, perhaps somewhat unfair, of the scientific work that led Becquerel to establish this new property of matter [5].

H. Becquerel was studying another phenomenon, the fluorescence of some materials, when he came across radioactivity. However, in the scientific context of the time, the path that linked this fortuitous discovery with the identification of a new phenomenon of matter was not easy to travel, and required methodical, serious, and tenacious work that qualified himself to his discoverer. The discovery of X-rays by Röentgen in 1895 [6] concentrated many of the scientific efforts of the time toward the new field of work. This makes H. Becquerel, after 20 years of a previous scientific career, start studying the fluorescence of some materials. The reason for their choice is clear, considering the close relationship between this phenomenon and recently discovered X-rays. In fact, they excited the fluorescence of some substances. The question was whether the mechanism that explained both phenomena could be the same.

He was studying the fluorescence of double uranium and potassium salts. Specifically, their fluorescent excitation by sunlight was used in hopes that X-rays were also produced. To do this, he coated the fluorescent substance with a photographic film, following the suggestions of H. Poincaré. If sunlight could also excite the X-ray emission, the film should be blackened after exposure. On 24 February 1896, he observed the expected blackening. In addition, announced the discovery. However, he continued his work, trying to reproduce the results. The bad weather dissuaded him from continuing, and on February 26 he kept the uranium and potassium salts, covered by the photographic film, in a drawer, waiting for a sunny day. Four days later, he revealed the films and discovered that they presented an even more intense blackening than that found on February 24. Obviously, the Sun could not be responsible for the blackening. It was not possible to excite the light fluorescence and, presumably, the X-ray emission of the uranium and potassium chemical compounds. After that finding, he undertook a whole series of systematic experiments that allowed him to claim that a new invisible radiation, similar to X-rays, was emitted by all uranium compounds, including nonfluorescent ones. Furthermore, the new radiation did not appear in other nonuranic fluorescent substances. Approximately two months later, Becquerel concluded that the phenomenon discovered was specific to uranium.

The discovery of uranic rays aroused the interest of many scientists, who began the search for new radioactive elements. Among others, but very prominently, Marie Sklodowska, Marie Curie, for her marriage, very recent at that time, with the eminent physicist Pierre Curie. The result of the dedication of Marie Curie first and of Pierre Curie later, but together with his wife, to this new field was very fruitful.

In 1898, they coined the name radioactivity for the new phenomenon. Independent of Schmidt, they identified Thorium in early 1898. Later, but during the same year, Polonium and Radium [7].

In 1899, Debierne isolated actinium, and in 1900, Dorn isolated a radioactive gas produced by radiation. He calls it Radon [8]. In addition, during that year, Rutherford discovered an emanation of Thorium [9]. Finally, in 1925, around 25 radioactive species had already been identified [10].

## 1.2.2   Phenomenology of Radioactivity

The nature of the radiation emitted by the few radioactive elements discovered around 1900 was, in fact, very little known. The subsequent and simultaneous works of Becquerel himself, Marie and Pierre Curie, as well as Ernest Rutherford managed to establish several facts.

Two components of the radiation emitted by Radium were distinguished: one not very penetrating, a sheet of paper stopping it, and another more penetrating, in such a way that aluminum sheets were necessary to stop the radiation. Both were electrically charged, as they ionized the air and deviated oppositely in the presence of electric and magnetic fields [11]. Rutherford called them $\alpha$, positively charged, and $\beta$, negatively charged. On the other hand, the most penetrating radiation, the so-called $\beta$, had characteristics similar to those known as cathode rays at that time. They are identified as electrons.

Becquerel himself identified a third component of radiation that could pass through lead. Villard, later in 1900 [12], demonstrated that such radiation is not deflected in the presence of electric or magnetic fields. It had characteristics analogous to X-rays, although with a shorter wavelength. Debierne called it radiation $\gamma$. Villard and Becquerel proposed that it was electromagnetic radiation, a fact that was experimentally shown later in 1914 by Rutherford and Andrade. It also seemed clear that known radioactive elements do not emit all these radiations. Some emitted $\alpha$ and others emitted $\beta$, while $\gamma$ radiation frequently accompanied the previous ones.

From the analysis of the deviation that the $\alpha$ radiation undergoes in the presence of electric and magnetic fields, it was discovered that it is formed by particles whose mass is at least twice that of the hydrogen ion, the smallest positive charge known at that time [11]. However, it is not until 1903 that the riddle of the $\alpha$ particles is solved. It is Rutherford who does it, identifying it with helium nuclei or, better, ionized helium atoms.

Previously, Soddy together with Rutherford in 1902 [13] proposed that radioactive decay consists of the transformation of atoms that gives rise to the appearance of other radioactive elements and formulates the laws of radioactive displacement or transmutation that, expressed following the knowledge of those times, can be written as:

1. A radioactive transformation, accompanied by α radiation, lowers the atomic number by two units and moves the element back two places on the periodic table.
2. A transformation, accompanied by β radiation, increases the atomic number by one unit and advances the element one place in the periodic table.

According to the model that we have adopted for the atomic nucleus, these displacement laws can be translated to the following.

$$Radioactivity\ \alpha : \quad {}_{Z}^{A}X \rightarrow {}_{Z-2}^{A-4}Y + {}_{2}^{4}\mathrm{He}(\alpha)$$
$$Radioactivity\ \beta : \quad {}_{Z}^{A}X \rightarrow {}_{Z+1}^{A}Y + \beta(e)$$

As we shall see later, this description of β radioactivity is **wrong**. For now, it is enough, as it responds to the knowledge of those times.

On the other hand, we know today that γ radioactivity is the result of the energy deexcitation of the nucleus toward a lower energy state, and it consists of electromagnetic radiation. This result was inaccessible at the time. We represent the process as follows.

$$_{Z}^{A*}X \rightarrow {}_{Z}^{A}X + \gamma$$

## 1.3   Types of Radioactivity

### 1.3.1   Alpha Radioactivity

Through α decay, the radioactive nucleus spontaneously emits $^{4}$He nuclei, which we call α particles. They are charged particles, $Z = 2$ and $A = 4$, with a mass

$$m_{\alpha} = 4.002603\ \mathrm{amu} = 3728.4\ \mathrm{MeV/c}^{-2}$$

Some 7300 times the mass of the electron.

Due to the emission of α particles, a nucleus appears with $A$ four units smaller and $Z$ two units smaller, according to the following process.

$$_{Z}^{A}X \rightarrow {}_{Z-2}^{A-4}Y + \alpha$$

A few examples follow

$$^{235}\mathrm{U} \rightarrow {}^{231}\mathrm{Th} + \alpha$$
$$^{226}\mathrm{Ra} \rightarrow {}^{222}\mathrm{Rn} + \alpha$$
$$^{210}\mathrm{Po} \rightarrow {}^{206}\mathrm{Pb} + \alpha$$

It is a common decay mechanism of heavy nuclei, $A \geq 200$, where the electrostatic repulsion of nuclear protons is very intense because of the high $Z$. Less frequently, $\alpha$-decay occurs in lighter nuclei. Examples are

$$^{148}\text{Gd} \rightarrow {}^{144}\text{Sm} + \alpha$$
$$^{188\text{m}}\text{Bi} \rightarrow {}^{184}\text{Tl} + \alpha$$
$$^{174}\text{Au} \rightarrow {}^{170}\text{Ir} + \alpha$$
$$^{170}\text{Pt} \rightarrow {}^{166}\text{Os} + \alpha$$

Exceptionally, it also occurs in very light nuclei. In fact, $\alpha$ decay has been found in $^5$He, $^5$Li, $^6$Be, $^8$B, $^8$C and others. The available energy in $\alpha$ decay, $Q_\alpha$, corresponds to the difference between the masses of the original nucleus and those of the decay products. Thus:

$$Q_\alpha = (m_X - m_Y - m_\alpha)c^2 \tag{1.1}$$

where $m_x$ and $m_y$ are the masses of the parent and daughter nuclei in the corresponding nuclear state.

(It should be noted that the balance of energy is made based on atomic masses, which is the information available in a Table of Nuclear Masses. This means that the electrons must be considered in the balance that leads to $Q$. In this case, the result is the same as that obtained if we directly used nuclear masses. This will not happen when we describe the energy balance for $\beta$ radiation.)

$Q_\alpha$ is shared between the $\alpha$ particle and the recoil nucleus (the daughter nucleus) in a well-established way that derives from the principles of conservation of energy and momentum. Thus, the kinetic energy of the $\alpha$ particle is

$$E_\alpha = \frac{A - 4}{A} Q_\alpha \tag{1.2}$$

while that of the recoil nucleus is

$$E_N = \frac{4}{A} Q_\alpha \tag{1.3}$$

In other words, the $\alpha$ particle is always emitted with the same energy. That is, the $\alpha$ energy spectrum is discrete. However, a radionuclide can emit several $\alpha$ particles, each with a different energy. As we will see in the next chapter, radionuclides can take different decay pathways, each connecting well-defined parent and daughter nuclei energy states, i.e., with a well-defined $Q_\alpha$. Thus, according to (1.2) and (1.3), the emitted $\alpha$ particle will always have the same energy for each decay pathway. In the early days of radioactivity, this effect came to be called the fine structure of $\alpha$ emission. Figure 1.3 shows the $\alpha$ particles observed from the decay of a radionuclide. In the next chapter, we will describe the concept of the decay scheme, a diagram connecting the different radiation emitted by a radionuclide. It will help clarify how radioactive nuclei can emit different radiation.

**Fig. 1.3** Decay scheme of a radionuclide emitting several α particles. See the next chapter for more details

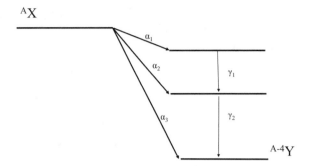

## 1.3.2 Beta Radioactivity: Electrons, Positrons, and Electron Capture

Three decay processes can be included in this section: electron emission, positron emission, and electron capture by the nucleus. In fact, they are described by the same theory of β decay. As in the case of α decay, the three processes can be found in Nature. However, at the beginning, β radioactivity was identified only by electron emission.

*Beta negative decay* (β⁻). *Electron emission*

It consists of the spontaneous emission of electrons by the nuclei. As a consequence, an isobar nucleus appears with Z one unit higher, according to the following process.

$$\begin{matrix} A \\ Z \end{matrix} X \rightarrow \begin{matrix} A \\ Z+1 \end{matrix} Y + e\left(\beta^{-}\right) + \overline{v}$$

where $\overline{v}$ is the antineutrino, a particle with a mass close to 0 and 0 electric charge, which is the antiparticle of the neutrino, $v$. Some examples are:

$$^{14}C \rightarrow {}^{14}N + \beta^{-} + \overline{v}$$
$$^{40}K \rightarrow {}^{40}Ca + \beta^{-} + \overline{v}$$
$$^{212}Bi \rightarrow {}^{212}Po + \beta^{-} + \overline{v}$$

β⁻ decay is possible for any $A$ value. However, it is typical for nuclei with an excess of neutrons, that is, with $N \geq Z$. In fact, it is the way the nuclei approach the line of stability by getting rid of excess neutrons.

At first, the emission of the antineutrino was not known. In fact, the history of β⁻ decay posed several challenges to the well-established principles of energy and momentum conservation. Indeed, the β⁻ energy spectrum was found to be a continuum (see Fig. 1.4), i.e., β⁻ was not a two-body problem in which the available energy is distributed between two physical bodies in a fixed way. It was only when Pauli in 1931 [14] postulated the existence of the neutrino and its corresponding antiparticle, the antineutrino, both without mass and electric charge, that

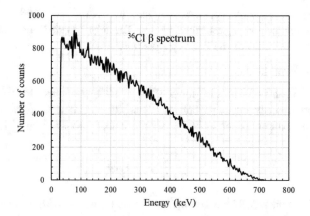

**Fig. 1.4** $\beta^-$ energy spectrum of $^{36}$Cl ($E_{max} = 710$ keV). The number of events recorded in a detector (count number) is plotted as a function of the energy deposited by the $\beta$ particle. The continuum nature of the spectrum was a difficult problem to resolve at the beginning

the puzzle started to be solved. The existence of these particles was experimentally demonstrated much later by Reines and Cowan in 1956 [15].

Whether antineutrinos are emitted in $\beta^-$ decay and neutrinos in $\beta^+$, as we will see below, is a question resolved in the theory of beta decay due to Fermi in 1934 [16]. We do not go into detail. According to Pauli, the mechanism of production of $e^-$ in the nucleus is governed by a fundamental force, the weak interaction, which causes the following process:

$$n \rightarrow p + e(\beta^-) + \overline{\nu}$$

The neutron disintegrates into one proton, one electron, and one antineutrino. This explains the emission of electrons from the nucleus, and the resulting daughter nucleus is an isobar of the parent with an atomic number one unit higher. Likewise, it explains that the energy spectrum of the emitted electrons is continuous, since the energy available in the reaction must be shared among three bodies, the electron, the antineutrino, and the recoil nuclei. The available energy in the decay, $Q_{\beta-}$, is

$$Q_{\beta-} = (m_X - m_Y)c^2 \qquad (1.4)$$

$m_x$ and $m_y$ are the masses of the parent and daughter nuclei. (Note that both are atomic masses and, as commented on before, the involved electrons must be taken into account in the balance, and the mass of the antineutrino is taken as 0.)

$Q_{\beta-}$ can be distributed in many possible ways among the recoil nucleus, the $\beta^-$ particle and the $\overline{\nu}$. The energy spectrum of $\beta^-$ radiation is therefore continuous. In other words, for the same transition, the particle can be emitted with different possible values of kinetic energy, $E_{\beta-}$, continuously distributed within zero, and a maximum energy, $E_{max} = Q_{\beta-}$. Or

$$0 \leq E_{\beta-} \leq Q_{\beta-}$$

The shape of the energy spectrum depends on several parameters of the parent and daughter nuclei and is described by the Theory of Beta Decay. On the other hand, the average value of the $\beta$ energy is approximately $\frac{1}{3}Q_{\beta^-}$ [17].

*Beta positive decay* ($\beta^+$)

$\beta^+$ decay is described by the same theory of beta decay. In this case, the nucleus spontaneously emits positrons, $e^+$, which is the antiparticle of the electron. It has the same mass and electric charge but of opposite sign. The process is as follows.

$$_Z^A X \rightarrow {}_{Z-1}^A Y + e^+ \left(\beta^+\right) + \nu$$

where $\nu$ is the neutrino. The resulting nucleus is an isobar with an atomic number one unit less than that of the parent.

The basic process that explains the appearance of $e^+$ is also governed by the weak interaction as follows.

$$p \rightarrow n + e^+ \left(\beta^+\right) + \nu$$

$\beta^+$ emission is typical of proton-rich nuclei ($N < Z$) in which the proton disintegrates into a neutron looking for the stability line, $\frac{N}{Z} \approx 1.5$. $\beta^+$ emissions can take place for any $A$ value. Some examples follow:

$$^{30}P \rightarrow {}^{30}SI + \beta^+ + \nu$$
$$^{40}K \rightarrow {}^{40}Ar + \beta^+ + \nu$$
$$^{53}Fe \rightarrow {}^{53}Mn + \beta^+ + \nu$$

The available energy is:

$$Q_{\beta^+} = (m_X - m_Y - 2m_e)c^2 \tag{1.5}$$

which also must be shared among three bodies. Consequently, the energy spectrum of the positrons is continuous with an average energy of approximately $\frac{1}{2}Q_{\beta^+}$ [17]. The energies of the emitted positrons, $E_{\beta^+}$, are distributed as

$$0 \leq E_{\beta^+} \leq Q_{\beta^+}$$

*Electron capture* (*EC*)

This mechanism is an alternative to the $\beta^+$ emission for proton-rich nuclei. Now, the nucleus 'captures' one electron from the internal electronic shells, prominently from the $K$ shell. The microscopic reaction is governed by the weak interaction and can be described as

$$p + e^- \rightarrow n + \nu$$

The EC decay would consequently be

$$_Z^A X + e^- \rightarrow {}_{Z-1}^A Y + \nu$$

The daughter nucleus is identical to that produced during $\beta^+$ decay, although now only one particle, the neutrino, $\nu$, is emitted. The available energy is:

$$Q_{EC} = (m_X - m_Y)c^2 \qquad (1.6)$$

which is distributed between the recoil nucleus and the $\nu$. Consequently, the $\nu$ energy spectrum is discrete.

$\beta^+$ emitters usually decay by EC as well. They are competing processes since they solve the same $\frac{N}{Z}$ instability. From an experimental point of view, EC has to be characterized indirectly, since it is almost impossible to detect neutrinos by conventional procedures. Fortunately, when EC occurs, the generated atomic vacancies give rise to the emission of X-rays or Auger electrons, which we will describe later. They witness the existence of EC.

### 1.3.3 Gamma Radioactivity: Electromagnetic Radiation, Conversion Electrons, and Isomers

In $\gamma$ decay, the nucleus in an excited state spontaneously dissipates the excess energy by emitting electromagnetic radiation. Since nuclear states are involved in the transition, the energy of the radiation falls within the so-called $\gamma$ region of the electromagnetic radiation spectrum (see Fig. 1.5).

The process can be described as follows:

$$^A X^* \rightarrow {}^A X + \gamma$$

Insofar as the nucleus is a quantum system with well-defined energy states, the $\gamma$ radiation energy spectrum is discrete. Each $\gamma$ transition, therefore, is characterized by a very well-defined energy. The typical $\gamma$ radiation energies range from 10 keV to several MeV. The energy available in the transition is shared between the $\gamma$ radiation emitted and the recoil nucleus. The latter only takes approximately 0.1% of the available energy. Thus, it can be considered that the emitted $\gamma$ radiation carries all the available energy [17].

$\gamma$ radiation accompanies other decay modes. Indeed, when a nucleus emits $\alpha$ or $\beta$ radiation, the daughter nuclei are usually left in an excited state, thereby giving rise to $\gamma$ radiation. This is illustrated in Fig. 1.6, where a decay scheme (see Chap. 2) is presented. There, the emission of $\alpha$ or $\beta$ particles feeds excited states of the daughter nuclei producing $\gamma$ radiation. Nuclear excited states rapidly de-excite. Their mean lives or half-lives are very short (see Chap. 2), typically $10^{-9}$ s, $10^{-12}$ s, or even $10^{-15}$ s. For this reason, there are no pure $\gamma$ emitters in nature, although $\gamma$ radiation can be observed from $\alpha$ or $\beta$ natural radionuclides.

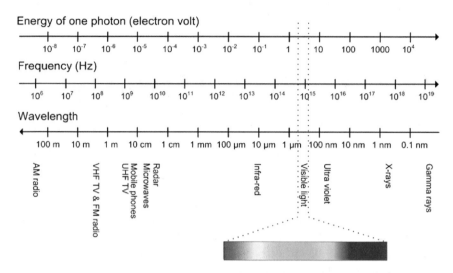

Fig. 1.5 Electromagnetic energy spectrum. The γ- and X-regions are displayed. Plot obtained from [18]

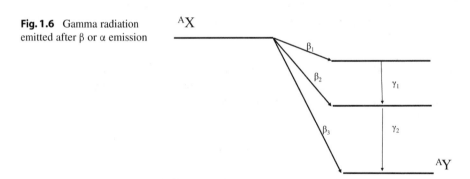

**Fig. 1.6** Gamma radiation emitted after β or α emission

However, there are unusually long-lived nuclear energy levels with mean or half-lives of min, h, or even years in extraordinary cases. They are the so-called metastable states and the corresponding γ decay known as isomeric transitions. Metastable (or isomeric) states are represented by $^{Am}X$, as we have already seen. Some examples of metastable states follow:

$$^{24m}\text{Na} \left( T_{1/2} = 20.20 \, \text{ms} \right)$$
$$^{60m}\text{Co} \left( T_{1/2} = 10.47 \, \text{min} \right)$$
$$^{99m}\text{Tc} \left( T_{1/2} = 6.02 \, \text{h} \right)$$
$$^{129m}\text{Xe} \left( T_{1/2} = 8.89 \, \text{d} \right)$$
$$^{242m}\text{Am} \left( T_{1/2} = 141 \, \text{a} \right)$$

$T_{1/2}$ is the half-life (see Chap. 2), which is defined as the time necessary to reduce the radioactivity by half of the original value. Metastable states can decay through alternative pathways to $\gamma$ decay, such as those studied so far.

Nuclear excited states can alternatively decay by the emission of conversion electrons (CE), which are produced during a process called internal conversion. In this case, the nucleus dissipates excess energy by transferring it to atomic electrons from the internal shells. The transformation can be represented as

$$^{A*}X \rightarrow {}^{A}X + e(CE)$$

The conversion electron kinetic energy $E_e$ can be found to be

$$E_e = E_\gamma - E_B \tag{1.7}$$

where $E_\gamma$ is the corresponding alternative $\gamma$ radiation energy and $E_B$ is the electron binding energy in the atomic shell. The CE energy spectrum is discrete as a consequence. This helps to experimentally differentiate them from $\beta^-$ radiation.

The so-called internal conversion coefficient, $\alpha$, serves to evaluate the relative probability of CE regarding $\gamma$ emission. It is defined as

$$\alpha_i = \frac{I_i}{I_\gamma} \tag{1.8}$$

Here, the subscript $i$ refers to the atomic shell (K, L, M, ...) of the emitted electron. I is the intensity of the transition (see Chap. 2), that is, the probability of decaying through $\gamma$ or CE. It can be said that in general,

$$\alpha_K > \alpha_L > \alpha_M$$

As we shall see below, CE decay is accompanied by the emission of X-rays because of the atomic vacancy left by the emitted electron. This also helps with their experimental characterization.

### 1.3.4 Other Radioactivity Types: Double Beta Decay, Proton and Neutron Emissions, Exotic Radioactivity, Fission

*β β double beta decay*

Double $\beta$ decay can be described as:

$$^{A}_{Z}X \rightarrow {}^{A}_{Z+2}Y + 2\beta^- + 2\bar{\nu}$$

and

$$^{A}_{Z}X \rightarrow {}^{A}_{Z-2}Y + 2\beta^+ + 2\nu$$

They are energetically possible processes but very unlikely. Despite this, it has been observed both in geochemical experiments and those based on direct observation of the emitted radiation. Some examples of double β emitters are $^{82}$Se, $^{100}$Mo, $^{128}$Te, $^{130}$Te, and $^{136}$Xe, among others, with half-lives (see Chap. 2) ranging from $10^{19}$ to $10^{24}$ y. In addition, there is an intense research effort aimed at the search for neutrinoless double β decay, for instance.

$$^{76}\text{Ge} \rightarrow {}^{76}\text{Se} + 2e$$

for which a $T_{1/2}$ lower limit is estimated to be around $10^{24}$ y. The existence of these processes has important consequences for the standard model in Particle Physics.

*Proton emission*
It is an alternative decay to β$^+$ or EC. It consists of the spontaneous emission of protons by proton-rich nuclei, i.e., low $\frac{N}{Z}$. Proton emission can be theoretically described similarly to the α decay. The energy spectrum of the emitted protons is discrete. Examples of *p*-emitting radionuclides are $^{11}$N, $^{19}$Na, $^{12}$O, and $^{53m}$Co. Some nuclei can emit more than one proton. This is the case for $^{6}$Be and $^{16}$Ne, which emit two protons in a single decay [19].

Delayed proton emission after β$^+$ or EC decay is more frequent. Examples of this process are observed in $^{9}$C, $^{13}$O, $^{17}$Ne, $^{22}$Al, ..., $^{151}$Yb, etc. Many of these nuclei also experience α delayed emission as an alternative to proton emission [19]

*Neutron emission*
In this case, only n delayed emission after β$^-$ decay has been observed thus far. It is a typical process of neutron-rich nuclei, i.e., high $\frac{N}{Z}$. It is an alternative decay to a single β$^-$ decay. From a physical viewpoint, it is analogous to the emission α and *p* and can be theoretically described in a similar way. The energy spectrum of the emitted neutrons is discrete. Examples of n-emitting radionuclides are $^{8}$He, $^{27}$Na, $^{49}$K, $^{210}$Tl, and others [19].

*Exotic radioactivity. A > 4 nuclei emission*

Since the first times of radioactivity, the possibility that very heavy nuclei emit *A* > 4 'particles' (nucleon clusters), i.e., nuclei heavier than α particles, is known. However, it was not until 1984 [20] that this possibility was experimentally confirmed. This process is typical of very heavy nuclei and competes with α decay. Some examples follow:

$$^{223}\text{Ra} \rightarrow {}^{209}\text{Pb} + {}^{14}\text{C}; \quad \frac{\lambda_{14_C}}{\lambda_\alpha} = (5.5 \pm 2.0) \times 10^{-10}$$
$$^{232}\text{U} \rightarrow {}^{208}\text{Pb} + {}^{24}\text{Ne}; \quad \frac{\lambda_{24_{Ne}}}{\lambda_\alpha} = (2.0 \pm 0.5) \times 10^{-12}$$
$$^{226}\text{Ra} \rightarrow {}^{212}\text{Pb} + {}^{14}\text{C}; \quad \frac{\lambda_{14_C}}{\lambda_\alpha} = (3.2 \pm 1.6) \times 10^{-11}$$

The branching ratios beside the decays give the relative decay probability with respect to the α emission of the parent nuclei. Their measured values justify the name of exotic radioactivity. From a theoretical point of view, this decay mechanism can be interpreted in a very similar way to α decay.

*Spontaneous fission (SF)*

SF occurs in heavy nuclei and competes with α emissions. It was discovered by Petrzak and Flerov in 1944 [21]. In SF, the parent nucleus splits into two other nuclei with similar masses, approximately half of the parent nucleus. Such nuclei, the fission products, are emitted along with a certain number of neutrons. The average neutrons emitted, $\eta$, are characteristic of the observed fission process, and as will be seen in Chap. 7, when studying the interaction of neutrons with matter, it has important technological significance. Fission is a highly symmetrical nuclear breaking mechanism, so it should be clearly distinguished from other processes in which the nucleus breaks into very different nucleon clusters. This is the case for spallation, in which the nucleus breaks down into some particles and two nuclei with very different masses [22].

SF can be described as follows:

$$\,^{A}_{Z}X \rightarrow \,^{A'}_{Z'}N + \,^{A''}_{Z''}N' + \eta n$$

with

$$A = A' + A'' + \eta n$$

and

$$Z = Z' + Z''$$

However,

$$m_N \approx m_{N'} \approx \frac{m_X}{2}$$

The available energy in fission is shared among the fission products, the neutrons, and other radiation emitted. Spontaneous fission in natural radionuclides is very unlikely compared to the α decay with which it competes in the same nuclei. Some nuclei that undergo spontaneous fission are $^{238}$U, $^{244}$Pu, $^{252}$Cf, $^{242}$Fm, $^{257}$Fm, and others.

## 1.4   X-rays. Auger Electrons

X-rays are electromagnetic radiation (see Fig. 1.5) of atomic origin. X-rays come from the deexcitation of bound atomic energy levels. As in the case of γ radiation, the X-ray energy spectrum is discrete. The excited atomic levels can alternatively decay by the emission of Auger electrons, which was discovered close in time by L. Meitner and P. Auger [23]. This process is equivalent to CE. Now, the excess of energy in the atom is communicated to electrons in outer energy shells instead of being emitted in the form of electromagnetic radiation. The Auger electron is emitted with a well-defined kinetic energy, which is

$$E_e = E_X - E_B \tag{1.9}$$

where $E_X$ is the alternative X-ray energy and $E_B$ is the electron binding energy to the atom.

### Exercises

1. Use https://www.nndc.bnl.gov/nudat2/ to identify isotopic and isotone families corresponding to $Z = 8$ and $N = 8$. Explain the decay modes found for each radionuclide with respect to the stability line.
2. Find (1.1). Note: as described in the text, the available data in Tables are atomic masses.
3. Find (1.5) and (1.6). Explain the differences.
4. Estimate the kinetic energy of the recoil nucleus for the main $\alpha$ decays of $^{239}$Pu and $^{210}$Po.
5. By using (1.5) and (1.6) estimate the maximum energy of the $^{90}$Sr $\beta$-spectrum. Make the same calculation for $^{90}$Y and $^{22}$Na. What is the average energy of the emitted $\beta^-$ or $\beta^+$?
6. Calculate the energy of conversion electrons emitted during the decay of $^{137m}$Ba. Consider that the binding energy of the Ba K-shell electrons is 37.4 keV.
7. In the figure, we have the electron spectrum from a $^{137}$Cs source. Describe the different parts of the spectrum. Identify the peaks. What is their origin?

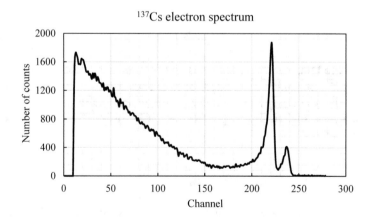

$^{137}$Cs electron spectrum

8. $^{11}$N, $^{19}$Na, and $^{12}$O are proton emitters. Calculate the energies of the emitted protons and the recoil nuclei.

9. Estimate the energies of $^{14}C$ and $^{209}Pb$ during the exotic decay of $^{223}Ra \rightarrow {}^{209}Pb + {}^{14}C$. Compare these energies with those of the recoil nuclei during the $^{223}Ra$ $\alpha$ decay. Explain the differences.

10. $^{235}U$ undergoes spontaneous fission and $\alpha$-decay. Estimate the $Q$ value for both decays and compare them. Use $^{235}U \rightarrow {}^{87}Br + {}^{145}La + 3n$ for fission calculations.

11. Calculate the Q value in the neutron-induced fission of $^{235}U$ through the specific fission channel: $n + {}^{235}U \rightarrow {}^{139}La + {}^{95}Mo$.

12. Compare the energy produced during:
    (a) Total combustion of 1 g C (6 eV per $CO_2$ formed molecule).
    (b) Total fission of 1 g of $^{235}U$ (200 meV per fission).
    (c) Total fusion of 1 g deuterium ($d + d \rightarrow He + n + 3.35$ meV).

## References

1. E. Rutherford, Phil. Mag. **21**, 669 (1911)
2. W. Heisenberg, Über den Bau der Atomkerne. II. Z. Phys. **78**(3–44), 156–164 (1932) and Über den Bau der Atomkerne. III. Z. Phys. **80**(9–10), 587–596 (1933)
3. K.S. Krane, *Introductory Nuclear Physics*, Chapter 3 (Wiley & Sons, 1988)
4. https://www.nndc.bnl.gov/wallet/wallet11.pdf
5. H. Becquerel, *Comptes Rendus*, **122**, 420, 501 (1896)
6. W. K. Roentgen: Sitzler. Wurgbcrger, Phys.-Med. Ges. (1895)
7. P. Curie, M. Curie, G. Bémont, Sur une nouvelle substance fortement radioactive contenue dans la pechblende. Comptes Rendus **127**, 1215–1217 (1898)
8. F.E. Dorn, Die von radioactiven Substanzen ausgesandte Emanation. Abhandlungen der Naturforschenden Gesellschaft zu Halle **23**, 1–15 (1900)
9. E. Rutherford, Phil. Mag. **xlix**, 1, 161 (1900)
10. N. E. Holden, 'History of the Origin of Chemical Elements and Their Discoverers', BNL-NCS-68350- 01/10-REV, prepared for the 41st IUPAC General Assembly in Brisbane, Australia, 29 June–8 July 2001
11. E. Rutherford, T. Royds, Phil. Mag. **17**, 281 (1909)
12. P. Villard, Sur la réflexion et la réfraction des rayons cathodiques et des rayons déviables du radium. Comptes rendus. **130**, 1010–1012 (1900) and Sur le rayonnement du radium. Comptes rendus. **130**, 1178–1179 (1900)
13. F. Soddy, *Chemistry of the Radio-Elements, Part II* (Longmans, Green & Co., Ltd., London, 1914)
14. The hypothesis of neutrino was proposed by W. Pauli in December 1930 in the famous letter addressed to Lise Meitner and Hans Geiger and a group of radioactive people at the Gauverein meeting in Tübingen, Dec. 4, 1930. The letter started with the famous sentence 'Dear Radioactive Ladies and Gentlemen'
15. F. Reines, C.L. Cowan, Phys. Rev. **92**, 830 (1953); F. Reines, C.L. Cowan, Nature **178**, 446 (1956); F. Reines, C.L. Cowan, Phys. Rev. **113**, 273 (1959)
16. K.S. Krane, *Introductory Nuclear Physics*, Chapter 9 (Wiley & Sons, 1988)
17. K.S. Krane, *Introductory Nuclear Physics*, Chapter 10 (Wiley & Sons, 1988)
18. Date accessed on October 13, 2021, at http://what-when-how.com/introduction-to-video-and image-processing/image-acquisition-introduction-to-video-and-image-processing-part-1/
19. https://www.nndc.bnl.gov/nudat3/
20. H.J. Rose, G.A. Jones, Nature **307**, 245 (1984); S. Gales, E. Hourani, M. Houssonnois, J.P. Schapira, L. Stab, M. Vergnes, Phys. Rev. **53**, 759 (1984)

21. K.A. Petrzhak, G.N. Flerov, Spontaneous fission of uranium. Proc. USSR Acad. Sci. **28**(6), 500 (1940) and Phys. Rev. **58**, 89 (1940)
22. B. Rossi, Über die Eigenschaften der durchdringenden Korpuskularstrahlung im Meeresniveau. Z. Phys. **82**, 151–178 (1933)
23. L. Meitner, Über die Entstehung der β-Strahl-Spektren radioaktiver Substanzen. Z. Phys. **9**, 131–144 (1922) and P. Auger, Sur les rayons β secondaires produits dans un gaz par des rayons X, C.R.A.S. **177**, 169–171 (1923)

# Radioactivity: Decay Law, Definitions, and Units

**2**

**ABSTRACT**

Once radioactivity has been described, some mathematical laws are derived and discussed in this chapter to quantify the new phenomenon. Thus, the exponential decay law together with the decay constant, half-life, and mean-life and associated parameters are introduced and studied from a physical point of view. The Bateman equations for radioactive series are also presented. A new magnitude, Activity, is introduced in addition to the adequate physical units. Conditions for transient and secular equilibrium among members of a radioactive series, of paramount importance for nuclear medicine, radiogeochronology or environmental radioactivity, are also studied.

## 2.1 Exponential Decay Law. Decay Constant, Half-Life and Mean-Life

Once the phenomenology of radioactivity is known, it is necessary to have some mathematical laws to study the behavior of radioactive samples. The question of how the radioactive sample evolves over time is the most immediate. In what follows, we will answer that question following a strict macroscopic approach.

Let us suppose a sample of radioactive nuclei consisting of $N$ nuclei. Since $N$ is $\sim N_A$, the Avogadro number, it can be considered large enough to take it as a continuous variable. Thus, the number of nuclei that decay per unit of time is proportional to $N$, that is,

$$\frac{dN}{dt} \propto N \tag{2.1}$$

In other words,

$$\frac{dN}{dt} = -\lambda N \tag{2.2}$$

© The Author(s), under exclusive license to Springer Nature Switzerland AG 2022
M. García-León, *Detecting Environmental Radioactivity*, Graduate Texts
in Physics, https://doi.org/10.1007/978-3-031-09970-0_2

The 'negative' sign appears because $N$ decreases over time. The proportionality factor, $\lambda$, is called the decay constant and can be interpreted as the decay probability per unit time. It is interesting to see that $\lambda$ contains information about the parent and daughter nuclei. In fact, it is a measurement of the probability of transition between two nuclear states. Thus, the knowledge of $\lambda$ provides much information about the structure of the parent and daughter nuclei. In fact, $\lambda$ can be calculated on the basis of a nuclear model that can be tested by comparing the calculations with its experimental determination. The integration of (2.2) is not trivial. Some hypotheses about $\lambda$, of considerable physical interest, are needed. Indeed:

1. Since $\lambda$ is a nuclear feature, it should not be affected by external agents. Neither the temperature nor the pressure nor even the chemical state in which the atom we are studying is found should influence the value of $\lambda$.
2. Since $\lambda$ characterizes a given nuclear structure, it is the same for all nuclei of the same species.
3. The decay constant is independent of time. The nuclear structure representing $\lambda$ does not change over time.

Under these conditions, the integration of (2.2) becomes:

$$N = N_0 e^{-\lambda t} \qquad (2.3)$$

$N_0$ is the number of nuclei at $t = 0$, and $N$ is the number of nuclei at time $t$, that is, the number of nuclei surviving in the sample after such a time period. Equation 2.3 is known as the exponential law of radioactive decay, and it is obvious that the number of nuclei $N$ that have decayed during $t$ is:

$$N' = N_0(1 - e^{-\lambda t}) \qquad (2.4)$$

It is most convenient to determine $\lambda$ by using two related magnitudes: the half-life, $T_{1/2}$, and the mean-life, $\tau$. Both have time dimensions and are easier to determine from an experimental point of view.

$T_{1/2}$ is the time taken by one-half of the radionuclide population to disintegrate, and it can be demonstrated that:

$$T_{\frac{1}{2}} = \frac{ln2}{\lambda} \qquad (2.5)$$

Additionally, it is very easy to see that:

$$\frac{N_0}{N} = 2^{\frac{t}{T_{\frac{1}{2}}}} \qquad (2.6)$$

Figure 2.1 illustrates the meaning of half-life, and a list of $T_{1/2}$ for different radionuclides is found in Table 2.1.

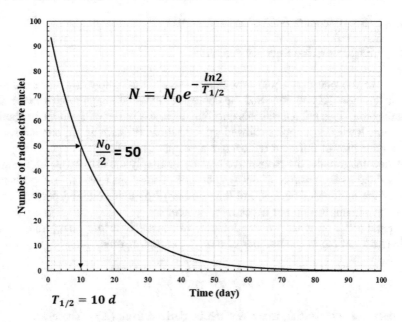

**Fig. 2.1** Activity exponential law with a representation of 10 day half-life

**Table 2.1** Some examples of half-lives of different radionuclides. More data can be found at https://www.nndc.bnl.gov/nudat2/ [3]

| Radionuclide | Half-life |
|---|---|
| $^3$H | 12.32 y |
| $^{14}$C | 5700 y |
| $^{15}$B | 10.18 ms |
| $^{36}$Na | <180 ns |
| $^{38}$P | 0.64 s |
| $^{41}$Ca | $9.94 \times 10^4$ y |
| $^{90}$Y | 64.053 h |
| $^{96}$Sr | 1.07 s |
| $^{99}$Tc | $2.111 \times 10^5$ y |
| $^{129}$I | $1.57 \times 10^7$ y |
| $^{147}$Pm | 2.6234 y |
| $^{238}$U | $4.468 \times 10^9$ y |

$T_{1/2}$ is a concept related to the population of nuclei rather than to individual nuclei. For them, the concept of mean life, $\tau$, is more useful. The mean life can be calculated as follows:

$$\tau = \frac{\int_0^\infty te^{-\lambda t}dt}{\int_0^\infty e^{-\lambda t}dt} = \frac{1}{\lambda} \tag{2.7}$$

## 2.2    Radioactive Activity and Units

### 2.2.1   Exponential Law of Activity

To answer the question about how 'radioactive' is a sample, we must define a new magnitude. It is obvious that the concept of mass, although related to the new phenomenon, does not answer this question, as it was observed from the beginning that some low-mass samples showed more 'radioactive activity' in comparison to heavier samples of different radioactive elements.

A new magnitude, the radioactive activity or simply the activity, $A$, of a sample was defined as the number of decays per unit of time observed. It is important to specify the number of decays, not the number of particles emitted per unit time. These two quantities do not necessarily coincide.

Equation 2.2 is what we are looking for, since it is the decay rate of the radioactive sample or the number of decays per unit of time. So:

$$A = -\frac{dN}{dt} = \lambda N \tag{2.8}$$

Hence, an exponential decay law can be derived from (2.3). Therefore,

$$A = A_0 e^{-\lambda t} \tag{2.9}$$

where $A_0$ is the activity of the sample at $t = 0$. Consequently, $T_{1/2}$ can be interpreted as the time needed to decrease the sample activity by half.

### 2.2.2   Becquerels and Curies

The dimensions of $A$ are $T^{-1}$. The units defined to quantify A are:

(a) 1 Curie $\equiv$ 1 Ci $= 3.7 \times 10^{10}$ disintegrations per second and corresponds to the activity of 1 g of $^{226}$ Ra. It is the historical unit, and its use is not recommended. Nevertheless, there are still texts in which Ci and its multiples, kCi, MCi, etc., and submultiples, mCi, $\mu$Ci, etc., appear.

(b) 1 Bq $\equiv$ 1 Bq $= 1$ disintegration per second. It is the unit currently used by international agreement. It is the natural unit and replaces Ci, which is certainly too large. It can be easily found that 1 Ci $= 37$ GBq with 1 GBq $= 10^9$ Bq.

Although not recommended, the units dpm, disintegrations per minute, or dps, disintegrations per second, are sometimes used in the laboratory at an informal level.

The relationship of $A$ with the sample mass, m, follows from (2.8), since $N = \frac{m}{P_m} N_A$, where $P_m$ is the molecular weight of the sample. The same mass of different radioactive elements will give different values of $A$, since $\lambda$ is different.

Being this true, it is also clear that an exponential law applies for all the quantities related to $N$ in the same sample: mass, volume, percentage of radioactive species in the sample, etc.

It is useful to define the concept of specific activity, which must not be confused with that of activity concentration. While the latter refers to the activity of a radionuclide per unit mass or volume of a given sample, the specific activity, $A_e$, is the activity of the radioactive isotope of an element per unit mass of the element in nature. It is clear that

$$A_e = \frac{\lambda N_A}{P_m} \qquad (2.10)$$

where $N_A$ is the Avogadro number and $P_m$ is the average molecular weight of the isotopic series of the element.

## 2.3 Radioactive Series

### 2.3.1 Bateman Equations

It is very common that the descendant of a radioactive nucleus is also radioactive. Some examples are given in Table 2.2. We will see in Chap. 3 that there exist natural radioactive series that have paramount geochronological importance.

The appearance of radioactive descendants poses a very interesting problem. How does the number of descendant nuclei or its activity evolve over time? The simplest problem, although not uncommon, is that of a series or family of two components (see Fig. 2.2).

Let us suppose that the parent nucleus, $A$, with a decay constant $\lambda_A$, decays into the daughter nucleus, $B$, which in turn decays with a constant $\lambda_B$ into $C$, which is stable. It is clear that

$$N_A(t) = N_A(0)e^{-\lambda_A t} \qquad (2.11)$$

as we found previously. It is also evident that

$$\frac{dN_B}{dt} = (\lambda_A N_A - \lambda_B N_B) \qquad (2.12)$$

**Table 2.2** Some examples of daughter nuclei that are also radioactive. More at https://www.nndc.bnl.gov/nudat2/

| Parent (half-life) | Daughter (half-life) |
| --- | --- |
| $^{32}$Si (153 y) | $^{32}$P (14.268 d) |
| $^{90}$Sr (28.90 y) | $^{90}$Y (64.053 h) |
| $^{99}$Mo (65.976 h) | $^{99m}$Tc (6.0072 h) |
| $^{137}$Cs (30.08 y) | $^{137m}$Ba (2.552 min) |
| $^{210}$Pb (22.20 y) | $^{210}$Bi (5.012 d) |
| $^{241}$Am (432.6 y) | $^{237}$Np ($2.344 \times 10^6$ y) |

**Fig. 2.2** Representation of a two-component family (or series) of radioactive nuclei

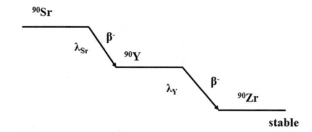

$$N_A \overset{\lambda_A}{\Rightarrow} N_B \overset{\lambda_B}{\Rightarrow} N_C(\textbf{\textit{stable}})$$

The equation can be very easily solved. Indeed, if $N_B = 0$ at $t = 0$, the following can be obtained:

$$N_B(t) = \frac{\lambda_A N_A(0)}{\lambda_B - \lambda_A}(e^{-\lambda_A t} - e^{-\lambda_B t}) = \frac{\lambda_A N_A(t)}{\lambda_B - \lambda_A}\left(1 - e^{-(\lambda_B - \lambda_A)t}\right) \quad (2.13)$$

and the corresponding equation for activities is simply:

$$A_B(t) = \lambda_B N_B(t) \quad (2.14)$$

Figure 2.3 shows the variation in $N_B(A_B)$ over time, according to (2.13) and (2.14).

It is easy, although laborious, to solve the corresponding differential equations for the case of radioactive families with more than two components. This was done by Bateman in 1910 [1], and the following recurrence relationship can be demonstrated for a family (series) of $n$ components, which includes the results obtained in (2.13):

$$N_n(t) = \lambda_1 \lambda_2 \ldots \lambda_{n-1} N_1(0) \sum_1^n C_i e^{-\lambda_i t} \quad (2.15)$$

$N_n(t)$ is the number of nuclei at time t of the radioactive species that occupies the place $n$ in the series, while $N_1(0)$ is the number of nuclei of the parent at $t = 0$. On the other hand, the coefficient $C_i$ is:

$$C_i = \prod_{j=1}^n (\lambda_j - \lambda_i)^{-1}; \quad j \neq i \quad (2.16)$$

These equations are valid if $N_i(0) = 0; \forall i > 1$ [5].

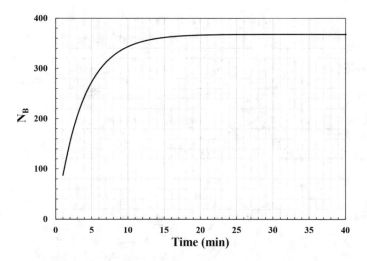

**Fig. 2.3** Time evolution of the number of descendant nuclei and its activity in the case of a two-component family. The half-life of the daughter nuclei is taken as 2.55 min, while that of the parent nuclei is 30 y. Within some 10 daughter half-lives, secular equilibrium is reached, and the daughter apparently decays with the parent half-life

The case of a series of three components

$$A \xrightarrow{\lambda_A} B \xrightarrow{\lambda_B} C \xrightarrow{\lambda_C} \ldots$$

is illustrated in Fig. 2.4, where the number of daughter and grand-daughter nuclei is given over time.

## 2.3.2 Transient and Secular Equilibria

The relationship among the decay constant of the radioactive series components determines the behavior of the Bateman equations. In natural radioactive series, as we will see in Chap. 3, such relationships lead to the concepts of transient and secular equilibria [2]. To understand both concepts, we can start studying the simplest case of a two-component radioactive series as presented in Fig. 2.2. It is quite usual to find in real cases that:

$$\lambda_B \gg \lambda_A \tag{2.17}$$

Under this condition, (2.13) becomes:

$$N_B(t) \approx \frac{\lambda_A N_A(t)}{\lambda_B}\left(1 - e^{-\lambda_B t}\right) \tag{2.18}$$

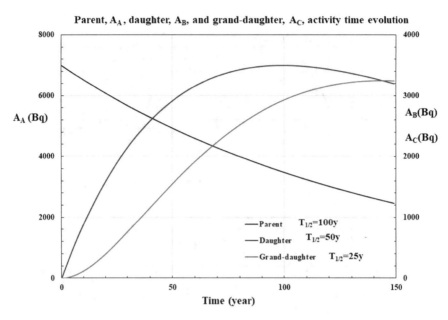

**Fig. 2.4** Time evolution of the parent, daughter, and grand-daughter activities in the case of a three-component family. The half-lives of the parent and descendants are given in the figure

Furthermore, if

$$t \gg T_{\frac{1}{2}}^{(B)} \rightarrow \lambda_B \gg 1 \tag{2.19}$$

It is very easy to find that

$$A_B(t) \approx A_A(t) \tag{2.20}$$

the daughter nucleus has approximately the same activity as the parent; i.e., it apparently decays with the parent half-life. This is the so-called secular equilibrium situation: For it to appear, both conditions expressed in (2.17) and (2.19) must occur simultaneously.

In Fig. 2.5, we can see the time variation of the number of daughter nuclei of a two-component series in which the parent and daughter half-lives are 20 h and 1 min, respectively. After a time, higher than several half-lives of the daughter (2.19), the latter apparently begins to decay with the half-life of the parent.

Using (2.15), it is easy to generalize the secular equilibrium situation for a series of n radioactive components. For this, the conditions in (2.17) and (2.19) must be met for each of the members of the series, i.e., the half-lives of the daughters, grand-daughters, etc., must be very small compared to that of the parent radionuclide. We will see that this happens in the natural radioactive series of $^{238}$U, $^{232}$Th, and $^{235}$U, but not yet for that of $^{237}$Np, in which (2.19) does not apply.

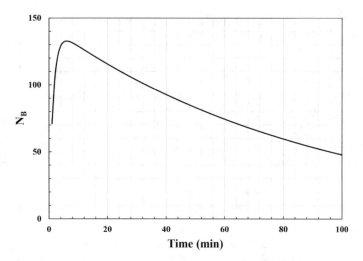

**Fig. 2.5** Same as Fig. 2.3 but for a daughter half-life of 1 min and a parent half-life of 1 h. As in Fig. 2.3, after a few half-lives of the daughter, the daughter starts decaying with the half-life of the parent

A less drastic situation is known as transient equilibrium. In this case, instead of (2.17) it happens to be:

$$(\lambda_B - \lambda_A)t \gg 1 \tag{2.21}$$

Now, the decay constant of the daughter nucleus is greater than that of the parent but not much greater. The result of this condition in (2.13) leads to:

$$\frac{N_B(t)}{N_A(t)} \approx \frac{\lambda_B}{\lambda_B - \lambda_A} \to \frac{A_B(t)}{A_A(t)} \approx \frac{\lambda_B}{\lambda_B - \lambda_A} = k > 1 \tag{2.22}$$

In other words, the daughter activity is always greater than that of the parent, but it apparently decays with the parent half-life. In Fig. 2.6, we observe the time evolution of the parent and daughter nuclei under transient equilibrium conditions.

## 2.4 Partial Activities. Branching Ratio and Intensity of Radiation

Very frequently, nuclei disintegrate through different mechanisms. For example, heavy nuclei can decay by α or β emission or through different α or β pathways [3]. To evaluate the relative probability of emission through the different decay modes, it makes sense to introduce the concepts of intensity, branching ratio, and partial activity.

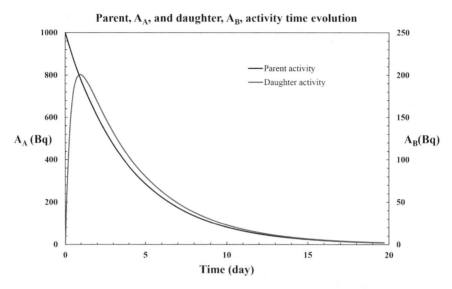

**Fig. 2.6** Transient equilibrium between daughter and parent with half-lives of 6.01 and 65.92 h, respectively. Time evolution of the parent and daughter activities, $A_A$ and $A_B$, respectively

Let us suppose a nucleus decays through n different alternative pathways, each with its own decay constant $\lambda_i$. Since the modes are alternative, it is obvious that the total probability of decay of the nucleus, or simply its decay constant, is:

$$\lambda = \sum_{i=1}^{n} \lambda_i \tag{2.23}$$

The activity of a sample with $N$ radioactive nuclei will be as follows:

$$A = \lambda N = N \sum_{i=1}^{n} \lambda_i \tag{2.24}$$

Each term of the summation

$$A_i = \lambda_i N = \lambda_i N_0 e^{-\lambda t} \tag{2.25}$$

is known as partial activity, representing the activity for each specific pathway. It is very interesting to observe that the decay constant present in the exponential is the total $\lambda$. This is consistent with the fact that the nucleus decays following all possible modes. Consequently, the temporal variation of the population of nuclei is governed by $\lambda$.

However, if we are interested in giving the number of decays per unit time occurring through the i pathway, we must multiply $N$ by $\lambda_i$, since this gives us the probability per unit time that the nucleus decays through the i pathway (and emits particles corresponding to such decay mode).

The case of $n = 2$ is frequent and very interesting. Moreover, trivially, the following can be found:

$$\lambda = \lambda_1 + \lambda_2 \tag{2.26}$$

$$\tau = \frac{\tau_1 \tau_2}{\tau_1 + \tau_2} \tag{2.27}$$

$$T_{1/2} = \frac{T_{1/2}^{1)} T_{1/2}^{2)}}{T_{1/2}^{1)} + T_{1/2}^{2)}} \tag{2.28}$$

and

$$r = \frac{\lambda_1}{\lambda_2} \tag{2.29}$$

is called the branching ratio.

For the general case of n decay modes,

$$I_i = \frac{A_i}{A} = \frac{\lambda_i}{\lambda} \tag{2.30}$$

is the so-called intensity of radiation and gives a measure of the relative probability that the nucleus decays through the $i$ pathway.

## 2.5 Decay Schemes

Unstable nuclei emit a variety of radiation. Daughter, grand-daughter nuclei and beyond can also emit radiation. All of this radiation is observed when measuring a radioactive sample, and it is necessary to know the different relationships among the emitted radiation. Decay schemes help to order all this information and facilitate interpreting the results obtained during the measurement of a radioactive sample. Some of them are presented in Fig. 2.7.

These diagrams include all the available information on the radiation emitted and its characteristics. Usually, the y-axis diagram corresponds to the energy of the nuclear state, and the x-axis corresponds to the atomic number Z of the nucleus. The different energy states of the parent and daughter nuclei and the radiation that links them are presented in a decay scheme. In this way, the interpretation is immediate. Other features of the nuclear levels, such as quantum numbers, mean-lives, magnetic or electric moments, are also included. There are several web pages where this information is available and continuously updated [4].

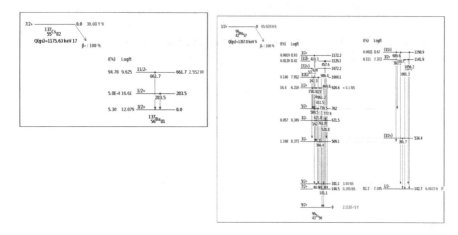

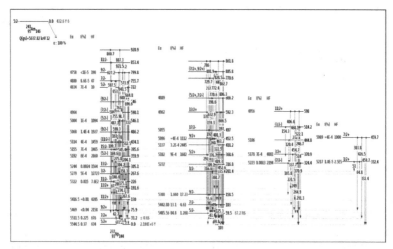

**Fig. 2.7** Decay schemes for several radionuclides [3]. The schemes gather all the information about the nuclide radiation emitted and their physical characteristics. Plots obtained from the NuDat web application of the National Nuclear Data Center, www.nndc.bnl.gov/nudat/

### Exercises

1. Find the relationship between the decay constant and the half-life of a radionuclide by using the exponential decay law.

2. The natural isotopic abundance of U is 99.27% $^{238}$U and 0.72% $^{235}$U. Suppose that both isotopes were initially produced in the same amounts. Estimate the age of the elements.

3. Assume the conditions of exercise 1. How long has it been since the abundance was 70% for $^{238}$U and 30% for $^{235}$U? How much time must elapse from now for the abundances to be 99.75% and 0.25%, respectively?

4.  Calculate the isotopic abundance of $^{234}$U in natural U. Suppose that it was not present when the $^{238}$U series was formed.
5.  How many $^{226}$Ra grams are there in a $^{238}$U t?
6.  Find the activity of 10 t of natural U.
7.  Find the activity of 10 t of 5% enriched U. (Enriched U contains an isotopic abundance of $^{235}$U higher than in natural samples. This is necessary to use U as fuel in nuclear reactors. Thus, 5% enriched U means that the $^{235}$U isotopic abundance is 5% and consequently that of $^{238}$U is 95%).
8.  Which natural U mass is necessary to have the activity corresponding to 1 t 5% enriched U?
9.  Show that the Ci definition is correct. Which $^{238}$U mass is necessary to have 1 Ci?. Make the same calculation for $^{99m}$Tc?
10. Natural U contains $^{234}$U with 0.0055% isotopic abundance. Calculate the natural U-specific activity in $Bqg^{-1}$. Make the same calculation considering now that no $^{234}$U is present. Compare the results.
11. A total of 1 cm$^3$ of a $^{99m}$Tc solution containing 0.24 kBq was injected into the blood of a patient. Two hours later, 10 cm$^3$ of blood was found to contain 0.29 Bq. Estimate the total blood volume of the patient.
12. A total of 1 cm$^3$ of a $^{24}$Na 200 Bq solution was injected into the blood of a patient. After 5 h, the activity of 10 cm$^3$ of the blood was found to be 16 dpm. Estimate the total blood volume of the patient.
13. $^{212}$Bi decays through α-emission (36% intensity) and through β-emission (64%) up to $^{208}$Pb.
    (a)  Find the α and β activities of 1 ng $^{212}$Bi sample.
    (b)  Find the α and β activities of the same sample as determined 6 h later.
14. Before the nuclear era, the $^{14}$C specific activity of carbonates dissolved in seawater was determined to be 16 dpm per C g. Calculate the total $^{14}$C activity of seawater in MCi and TBq if the amount of carbonates has been estimated to be $5 \times 10^{13}$ t.
15. Suppose we have $^{137}$Cs in secular equilibrium with its daughter $^{137m}$Ba. A chemical separation was carried out, and 80% of the Ba was extracted. As a result, we now have two samples: one containing 20% Ba at $t = 0$ and the other with 80% Ba at $t = 0$. How long does it take for the two samples to have the same gamma activity?
16. A $^{137m}$Ba sample is counted with a NaI(Tl) detector. Each 10 s, the sample activity is measured for 8 s. The following table gives the data series obtained. Calculate the half-life of $^{137m}$Ba.

| T | N |
|---|---|
|   | 30,000 |
| 1 | 28,700 |
| 2 | 27,300 |
| 3 | 26,150 |
| 4 | 24,890 |

| T  | N      |
|----|--------|
| 5  | 23,900 |
| 6  | 22,600 |
| 7  | 21,750 |
| 8  | 20,730 |
| 9  | 19,780 |
| 10 | 18,900 |

17. A 100 d half-life radioactive substance emits β-particles with $5 \times 10^{-7}$ erg average energy. Such a substance is used to activate a thermoelectric device. If the yield is 10%, calculate the amount of substance (in mol) needed to generate 5 W of electric power. Find the equivalent activity.

18. An emanation device contains 200 mg of $^{226}$Ra. $^{222}$Rn is extracted regularly every two days. What is the available $^{222}$Rn activity (mCi) per extraction? What will be the maximum available activity in the case where $^{226}$Ra–$^{222}$Rn are allowed to reach secular equilibrium?

19. A $^{99}$Mo–$^{99m}$Tc radioisotope generator from nuclear medicine was constructed three weeks ago. Are both radionuclides in secular equilibrium? Why?

20. $^{137m}$Ba is eluted from a $^{137}$Cs–$^{137m}$Ba based radioisotope generator constructed weeks ago. What is the half-life of $^{137m}$Ba in the eluted solution? Is it possible to restore the secular equilibrium in the generator? Describe the conditions needed to achieve equilibrium.

# References

1. H. Bateman, Proc. Cambridge Phil Soc. **15**, 423 (1910)
2. R.D. Evans, *The Atomic Nucleus*, Chapter 15 (McGraw-Hill Publishing Company, 1955)
3. See the following webpage https://www.nndc.bnl.gov/nudat3/
4. See the following webpages https://www.nndc.bnl.gov/nudat3/, http://nucleardata.nuclear.lu.se/toi/, https://www-nds.iaea.org/relnsd/vcharthtml/VChartHTML.html
5. E. Segre, *Nuclei and Particles: An Introduction to Nuclear and Subnuclear Physics*, Chapter V (Benjamin-Cummings Publishing Company, 1978)

# Natural and Artificial Radioactivity

**3**

**ABSTRACT**

The main objective of this chapter is to describe the origin of the known radionuclides. Thus, primordial, cosmogenic, and artificial radionuclides are described. A general equation for the production of artificial (man-made) radionuclides in accelerators or nuclear reactors is presented.

## 3.1 Primordial Radionuclides

There are radioactive elements in nature, and in fact, life has coexisted with radioactivity since its appearance on Earth.

Natural radionuclides can be classified into two groups: primordial and cosmogenic radionuclides. The half-lives of primordial radionuclides are higher than or comparable to the age of the Earth, $\sim 10^9$ years. Thus, they have survived on Earth since their formation. As we shall see later, this is not the case for cosmogenic radionuclides. However, they are continuously being produced in the Earth by the interaction of cosmic radiation with elements in the atmosphere. For that, they are present in Nature.

Primordial radionuclides can be grouped into two classes, as described in the following.

### 3.1.1 Long-Lived Radionuclides

As mentioned above, primordial radionuclides are those that have survived on Earth since their formation. The reason being that its $T_{1/2}$ is comparable to, or higher than, the age of Earth, which is approximately $10^9$ years. These radionuclides are supposed to have appeared during, or even before, the formation of the solar system [1].

© The Author(s), under exclusive license to Springer Nature Switzerland AG 2022
M. García-León, *Detecting Environmental Radioactivity*, Graduate Texts
in Physics, https://doi.org/10.1007/978-3-031-09970-0_3

**Table 3.1** Some primordial radionuclides are listed together with their half-lives

| Radionuclide | Half-life (year) | Decay mode | Isotopic abundance (%) |
|---|---|---|---|
| $^{40}$K | $1.248 \times 10^9$ | $\beta^-$, EC | 0.0117 |
| $^{50}$V | $2.1 \times 10^{17}$ | $\beta^-$, EC | 0.250 |
| $^{87}$Rb | $4.97 \times 10^{10}$ | $\beta^-$ | 27.83 |
| $^{115}$In | $4.41 \times 10^{14}$ | $\beta^-$ | 95.71 |
| $^{123}$Te | $>9.2 \times 10^{16}$ | EC | 0.89 |
| $^{138}$La | $1.02 \times 10^{11}$ | $\beta^-$, EC | 0.08881 |
| $^{144}$Nd | $2.29 \times 10^{15}$ | $\alpha$ | 23.798 |
| $^{147}$Sm | $1.06 \times 10^{11}$ | $\alpha$ | 14.99 |
| $^{148}$Sm | $7 \times 10^{15}$ | $\alpha$ | 11.24 |
| $^{176}$Lu | $3.76 \times 10^{10}$ | $\beta^-$ | 2.599 |
| $^{174}$Hf | $2 \times 10^{15}$ | $\alpha$ | 0.16 |
| $^{187}$Re | $4.33 \times 10^{10}$ | $\beta^-$ | 62.60 |
| $^{190}$Pt | $6.5 \times 10^{11}$ | $\alpha$ | 0.012 |

In Table 3.1, we present some of them. This list of radionuclides is not exhaustive, and in fact, the number of radionuclides increases as detection techniques allow the determination of lower concentrations of activity (see Chaps. 14 and 15). The reason is their extremely long half-lives, which make it very difficult to identify them as radioactive rather than stable elements. Primordial radionuclides populate the nature associated with specific geological systems, making them an invaluable chronological tool. In fact, rock and mineral dating can be performed with the help of many radionuclides in Table 3.1. Examples are $^{39}$Ar, $^{40}$K, $^{87}$Rb, $^{176}$Lu, $^{174}$Hf, etc. [2].

## 3.1.2  Natural Radioactive Series

Among the primordial radionuclides, the most interesting are $^{238}$U (99.3% isotopic abundance in U), $^{235}$U (0.7%) and $^{232}$Th (100% isotopic abundance in Th). Indeed, their $T_{1/2}$ are comparable to the age of the Earth (see Table 3.1), but in addition, they give rise to the appearance of radioactive series in nature. The characteristics of these series are summarized in Table 3.2, while in Fig. 3.1, we present each of these radioactive series in more detail. The structure of the series is very similar, all ending in a stable Pb isotope, except that of Np, which is a special case, as we will see later.

Radioactive series are defined by the mass number of their components. Thus, the $^{232}$Th family is called the 4n series because the A of the members is divisible by 4.

**Table 3.2** Natural radioactive series

| Mass number | Series | Parent | Half-life (year) | End nuclide |
|---|---|---|---|---|
| 4n | Thorium | $^{232}$Th | $1.41 \times 10^{10}$ | $^{208}$Pb |
| 4n + 1 | Neptunium | $^{237}$Np | $2.14 \times 10^{6}$ | $^{209}$Bi |
| 4n + 2 | Uranium–Radium | $^{238}$U | $4.51 \times 10^{9}$ | $^{206}$Pb |
| 4n + 3 | Uranium–Actinium | $^{235}$U | $7.18 \times 10^{8}$ | $^{207}$Pb |

The $^{238}$U family is the 4n + 2 series since it contains nuclei with A-2 divisible by 4. Finally, in the 4n + 3 series, the $^{235}$U family, members have A-3 divisible by 4.

The half-lives of all descendants are clearly shorter than that of the parent and, of course, shorter than the age of Earth. Despite this fact, radionuclides can be found in nature, as they are continuously produced by the disintegration of the series parent.

With the exception of the $^{237}$Np series, the rest meet the conditions of secular equilibrium (see Chap. 2). Indeed, the half-lives of the descendants are all smaller than those of the parent, and the time elapsed since the series formation, approximately $10^9$ years, is very large compared to their half-lives. In conclusion, all the descendants apparently decay with the half-life of the series parent. This result is certainly important, but more so is the fact that the secular equilibrium is very frequently broken in nature. Such breaking allows geological dating, for example, because the radionuclides produced associate with geological systems according to their geochemical behavior and begin to decay with their own half-life [2].

Indeed, the rupture of secular equilibrium releases a wide variety of radionuclides with a varied geochemical behavior, all decaying with their own half-life. Some of them are incorporated into the lithosphere, some into the atmosphere, some into the hydrosphere, etc. In this way, the system is specifically marked, and its time evolution can be followed by measuring the radionuclide content.

The rupture of secular equilibrium due to the emanation of $^{222}$Rn is an important example. $^{222}$Rn is a noble gas that belongs to the $^{238}$U series and can be released from Earth to the lower layers of the atmosphere. There, it decays with its half-life ($T_{1/2} = 3.8$ d), producing radionuclides until $^{210}$Pb ($T_{1/2} = 22.2$ y), which is useful for dating (see Fig. 3.2). The $^{210}$Pb dating method allows us to study the temporal evolution of sediments.

Radioactive disequilibria between $^{238}$U and $^{234}$U, as well as that of $^{230}$Th and $^{238}$U [3], make it possible to date some geological terrestrial and marine formations. In the first case, the disequilibrium consists of an excess of $^{234}$U in a geological sample. It is supposed that excess $^{234}$U decays with its own half-life ($2.455 \times 10^5$ y), while the rest of $^{234}$U apparently disintegrates with the $^{238}$U half-life ($4.468 \times 10^9$ y). The determination of $^{238}$U in the sample allows one to subtract $^{234}$U in equilibrium from the total $^{234}$U. By applying the corresponding Bateman equation, the age of the rock can be found. The method based on the disequilibrium between $^{230}$Th and $^{238}$U lies in the fact that certain systems do not

**Fig. 3.1** Natural radioactive
series

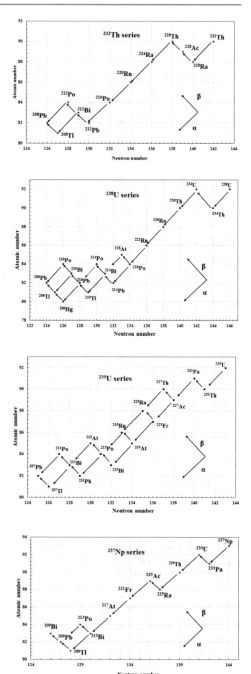

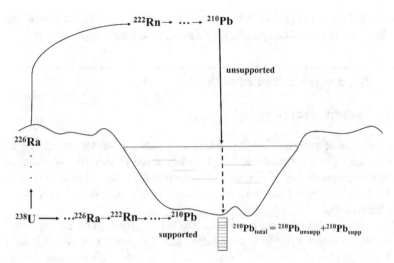

**Fig. 3.2** $^{210}$Pb dating method

incorporate Th during its formation. For that, all the $^{230}$Th found in the sample is supposed to have been formed by the decay of $^{238}$U.

The $^{235}$U series also provides some dating methods. Perhaps the most popular method is based on the $\frac{^{231}Pa}{^{235}U}$ activity ratio [4]. It is expected to be 1 due to secular equilibrium (see Fig. 3.1), but in some systems, it is <1 since $^{231}$Pa is still forming after the incorporation of $^{235}$U. In many cases, $^{235}$U is estimated from the measurement of $^{238}$U. Very often the $\frac{^{231}Pa}{^{235}U}$ activity ratio method is combined with that of $\frac{^{230}Th}{^{238}U}$ to gain more information about the system time evolution [4]

The U–Pb dating methods are probably the most experienced procedures for dating geological systems [2]. They are based on the measurement of the ratios between the contents of $^{238}$U and $^{206}$Pb or $^{235}$U and $^{207}$Pb in some specific rocks that do not incorporate Pb during its formation. The background of the method is similar to that based on the $\frac{^{230}Th}{^{238}U}$ activity ratio. Therefore, it is assumed that the corresponding Pb isotope was formed exclusively by the decay of $^{238}$U or $^{235}$U (see Table 3.2). By applying the corresponding Bateman equation, it is possible to calculate t, which is the age of the rock.

Because of the range of half-lives involved in these methods, they can be used to date geological objects aged $10^5$ to $10^9$ years.

The case of the $^{237}$Np series is somewhat different. Now, the parent half-life, $T_{1/2} = 2.14 \times 10^6$ y, is considerably shorter than the age of the Earth. Therefore, it can be considered an extinct series. However, its presence is observed in several natural compartments. The reason lies in its reintroduction to the Earth during the nuclear era (see Chap. 4), since $^{237}$Np or $^{241}$Am, its parent, are produced during the nuclear fuel cycle [5]. In this case, secular equilibrium has not been reached since the second condition has not yet been fulfilled (see Chap. 2).

## 3.2 Cosmogenic Radionuclides

### 3.2.1 Cosmic Radiation

Energetic cosmic rays irradiate the Earth at a flux that varies depending on the height, latitude, and longitude of the place where the observation is being carried out. Two components are distinguished in cosmic rays: the cosmic solar radiation produced during solar flares and the galactic cosmic radiation produced outside the solar system.

Both components of cosmic radiation are called primary cosmic rays and contain mainly protons with 10% $\alpha$ particles and 1% heavier nuclei [6]. In Table 3.3, the composition and other features of the cosmic radiation are given. The galactic component is more energetic, from 0.1 to 20 GeV, than the solar cosmic radiation, from 5 to 100 MeV. However, the flux of the solar component is clearly higher, $\approx$100 particles cm$^{-2}$ s$^{-1}$, than the galactic component, $\approx$3 particles cm$^{-2}$ s$^{-1}$. A global view of the cosmic-ray production in the atmosphere is depicted in Fig. 3.3.

Cosmic radiation causes nuclear reactions at the upper layers of the atmosphere, giving rise to the appearance of so-called secondary cosmic radiation. It consists essentially of neutrons, protons, hyperons, mesons, light nucleons, and others. These particles have enough energy to produce new nuclear reactions and generate additional particles that can decay and produce muons, electrons, positrons, photons, and others [7]. The so-called secondary cosmic radiation has three components: The hard component basically contains muons, the nucleonic component mostly contains neutrons (97%) and protons, and the soft component contains electrons, positrons, and photons.

**Table 3.3** Components, energy, and average flux of cosmic rays

| Cosmic-ray component | Composition | Energy (GeV) | Flux (m$^{-2}$ s$^{-1}$) |
|---|---|---|---|
| *Primary* | | | |
| | Protons (89%) $\alpha$ (10%) A>4 (1%) | | |
| Solar | | $10^{-3}$ to $10^{-1}$ | ~$10^6$ |
| Galactic | | $10^{-1}$ to 20 | ~$10^4$ |
| *Secondary* | | | |
| Muonic or hard | $\mu^{\pm}$, $\pi^{\pm}$, ... | | ~$10^2$ |
| Hadronic or nucleonic | Nucleons (97% neutrons) | | ~10 |
| Electromagnetic or Soft | $\pi^0$, e$^{\pm}$, $\gamma$ | | ~10 |

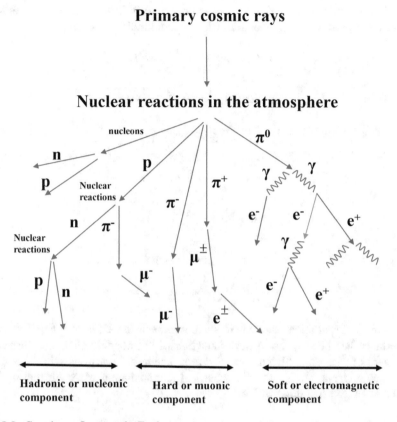

**Fig. 3.3** Cosmic ray flux over the Earth

The composition, energy, and flux of the different components of cosmic radiation are given in Table 3.3.

## 3.2.2 Production of Radionuclides by Cosmic Radiation

Secondary cosmic radiation can undergo additional nuclear reactions with the elements present in the atmosphere, giving rise to the appearance of cosmogenic radionuclides. Some of them have a small half-life even very small compared to the age of the Earth, yet they exist in nature as they are continuously being produced in the atmosphere. The production rate of cosmogenic radionuclides is not constant. In contrast, it varies over time depending on the history of irradiation of the Earth by cosmic radiation. This is highly dependent on solar activity and the magnitude of the Earth's magnetic field. Of special importance is the influence of the 11-year solar cycle that modulates the flux of cosmic rays over Earth [8].

A list of cosmogenic radionuclides is presented in Table 3.4. As an example, we give the production ways of some cosmogenic radionuclides:

$$n(fast) + {}^{14}N \rightarrow {}^{12}C + {}^{3}H$$

$$n(slow) + {}^{14}N \rightarrow {}^{14}C + {}^{3}H$$

$$p + {}^{18}O \rightarrow {}^{18}F + p$$

$$n + {}^{40}Ar \rightarrow {}^{39}Ar + 2n$$

$$n + {}^{35}Cl \rightarrow {}^{36}Cl + \gamma$$

$$n + {}^{40}Ca \rightarrow {}^{41}Ca + \gamma$$

$$n + {}^{80}Kr \rightarrow {}^{81}Kr + \gamma$$

The role of spallation reactions in nucleosynthesis is well known [9]. The same importance has been given to the production of cosmogenic nuclides. These are the cases of ${}^{7}Be$ and ${}^{10}Be$ produced during spallation reactions with N and O, the cases of ${}^{22}Na$, ${}^{26}Al$, and ${}^{32}Si$ caused by the spallation of Ar, or that of ${}^{129}I$ produced by the spallation of Xe.

**Table 3.4**  Some relevant cosmogenic radionuclides

| Radionuclide | Half-life | Decay mode | Production mode |
|---|---|---|---|
| ${}^{3}H$ | 12.32 y | $\beta^-$ | $(n, {}^{3}H)$, $(p, {}^{3}H)$, $(\mu, {}^{3}H)$ on ${}^{14}N$ or ${}^{16}O$ |
| ${}^{10}Be$ | $1.51 \times 10^6$ y | $\beta^-$ | p spallation on N or O isotopes |
| ${}^{14}C$ | 5730 y | $\beta^-$ | ${}^{14}N(n, p){}^{14}C$ |
| ${}^{26}Al$ | $7.17 \times 10^5$ y | EC | p spallation on Ar isotopes |
| ${}^{36}Cl$ | $3.01 \times 10^5$ y | $\beta^-$, EC | p spallation on K isotopes |
| ${}^{39}Ar$ | 269 y | $\beta^-$ | ${}^{40}Ar(n, 2n){}^{39}Ar$ |
| ${}^{81}Kr$ | $2.29 \times 10^5$ y | EC | ${}^{80}Kr(n, \gamma){}^{81}Kr$ |
| ${}^{85}Kr$ | 10.739 | $\beta^-$ | ${}^{84}Kr(n, \gamma){}^{85}Kr$ |
| ${}^{129}I$ | $1.57 \times 10^7$ y | $\beta^-$ | p spallation on Xe isotopes and n reactions with ${}^{128}Te$ and ${}^{130}Te$ |

## 3.3  Artificial Radionuclides

### 3.3.1  Some Historic Data

In 1934, during their experiences with radioactivity, F. Joliot and I. Curie irradiated a Geiger–Müller counter (see Chap. 9) with a $^{210}$Po sample. They discovered that the detector continued to count even though they removed the sample. They also found that the count rate decreased exponentially as a function of time, in the same way as that of natural radionuclides that had been recently discovered. It was clear that they had produced a radioactive isotope, most likely by the following reaction:

$$\alpha +^{27} \text{Al} \rightarrow^{30} \text{P} + \text{n}$$

since the Geiger–Müller counter had an Al window and $^{210}$Po emitted $\alpha$ particles. The reasoning leading to this hypothesis is well explained in the original paper [10]. $^{30}$P, a P isotope, does not exist in nature owing to its very short half-life, $T_{1/2}$ = 2.498 min. Of course, they did not discover a new phenomenon. They were able to artificially produce an extinct radionuclide.

Today, humans can produce approximately three thousand radioactive nuclei that are not present in nature and that have very important scientific, medical, and industrial applications. As we shall see later, there is a very abundant source of artificial radioactive elements: nuclear fission [11]. Nuclear fission (see Chap. 1) generates many radioactive elements and fission products, with a wide variety of half-lives and, of course, chemical properties.

### 3.3.2  Production of Radionuclides in Accelerators

However, the most important method is the production of radionuclides with accelerators or neutron sources. In this case, the radionuclide is produced by a nuclear reaction with charged particles or neutrons. The use of accelerators allows for controlled production of radionuclides, as there exists the possibility of selecting the target as well as the projectile and its energy.

It is possible to generate some equations for radionuclide production in accelerators that are similar to the Bateman equations. As we will see in the next section, they can be applied to the production of radionuclides by nuclear reactors as well.

We can start from the simpler case. Thus, let us suppose the following nuclear reaction:

$$x + T \rightarrow y + B(\lambda)$$

or

$$T(x, y)B$$

in which a flux $\phi$ (particles/cm$^2$ s) of particles x impinges a target containing $N_T$ nuclei. The nuclear reaction gives rise to nuclei of particles y and $N_B$ that are radioactive and disintegrate with a decay constant $\lambda$

The important problem in this case is to calculate the number $N_B$ that will be produced during an irradiation time $t$. This problem is identical to that solved by the Bateman equations for the case of a series of two components. In fact, this process can be represented by

$$T \xrightarrow{\sigma\phi} B \xrightarrow{\lambda} \ldots$$

Now, $\sigma\phi$ plays the role of the decay constant of $T$, since it is just the probability per unit time for the transformation of $T$ into $B$, following the probabilistic interpretation of the cross section. It seems apparent that the time evolution of the $N_T$ and $N_B$ nuclei during irradiation will be as follows.

$$N_T(t) = N_T(0)e^{-\sigma\phi t} \tag{3.1}$$

and

$$N_B(t) = \frac{\sigma\phi N_T(0)}{\lambda - \sigma\phi}(e^{-\sigma\phi t} - e^{-\lambda t}) \tag{3.2}$$

Note that both equations are simply (2.11) and (2.13). It is very common that $\sigma\phi t \ll 1$, and so $N_T(t) \approx N_T(0)$. In other words, the number of target nuclei does not vary appreciably during the irradiation time. On the other hand, if $\lambda \gg \sigma\phi$, which is also very frequent, we have:

$$\lambda N_B(t) \approx \sigma\phi N_T(0)(1 - e^{-\lambda t}) \tag{3.3}$$

Finally, when $\lambda t \gg 1$, we find the saturation condition, that is:

$$\lambda N_B(t) \approx \sigma\phi N_T(0) \tag{3.4}$$

Under these conditions, as many $B$ nuclei are created, as they are destroyed. This condition is similar to that expressed in (2.20) when describing secular equilibrium

Production processes can be more complicated. For instance, $T$ nuclei could be radioactive as well, the projectiles could also react with $B$, the produced y particles could react with the target, or we might be interested in the descendant of $B$, etc. Of course, the equations to be considered in these cases are more complicated, as they must include more processes of nuclei disintegration and production, but they are not difficult to solve. For example, in the first case.

$$dN_B = [(\sigma\emptyset + \lambda_T)N_T - \lambda N_B]dt \tag{3.5}$$

$\lambda_T$ is the decay constant of $T$. The solution of (3.5) is not complicated.

Or, in the second case,

$$dN_B = [\sigma\phi N_T - (\sigma_B\phi + \lambda)N_B]dt \qquad (3.6)$$

Now, $\sigma_B$ is the cross section for the reaction of $B$ with the projectiles. This equation is also readily solved. However, there is an assumption that is not completely correct. Indeed, it is supposed that $\phi$ affecting $T$ is the same as that interacting with $B$. Of course, this will depend on how the cross sections compare and therefore on the concentrations of $T$ and $B$ in the sample over time.

We will see some examples in Chap. 19 when dealing with the neutron activation problem applied to realistic cases.

### 3.3.3  Production of Radionuclides in Nuclear Reactors

The production of radionuclides in nuclear reactors is another way to obtain artificial or man-made radionuclides. The main reactions that take place inside a nuclear reactor are neutron-induced nuclear fission and neutron activation, both boosted by neutrons at different energies [11]. In Chap. 7, an account will be given of these reactions that can also be used in some special neutron detectors. In the first case, neutrons generate the fission (see previous Chapter) of the components of the fuel in a nuclear reactor, essentially $^{235}U$ and $^{239}Pu$ or $^{238}U$ in the case of fast neutrons (see Chap. 7).

These reactions produce large amounts of radionuclides, the so-called fission products, normally β-emitters, that at the same time disintegrate into some nuclides, many of which are radioactive. In Fig. 3.4, we give the fission yield for $^{235}U$ and $^{239}Pu$, i.e., the probability of producing nuclides of mass A during the nuclear fission of both nuclides.

Activation reactions also occur in nuclear fuel. These reactions will be studied more deeply in Chaps. 7 and 19, since they give rise to a powerful analytical tool: neutron activation. They are of the type

$$n +^A X \rightarrow {}^{A+1}X + \gamma$$

The $^{A+1}X$ isotope resulting in the reaction usually has an excess of neutrons with regard to the stability line and, therefore, is a $\beta^-$-emitter.

As is known, the neutron flux within the core of the nuclear reactor is high enough to produce a wide variety of nuclear reactions that lead to the appearance of many radioactive species, the so-called activation products [12]. A view of the different reactions that take place inside the core of a nuclear reactor is given in Fig. 3.5. Some relevant activation processes can be identified. In addition, neutron activation of fission products or components of the reactor vessel multiplies the production of such radionuclides. A list of activation products and their half-lives is given in Table 3.5.

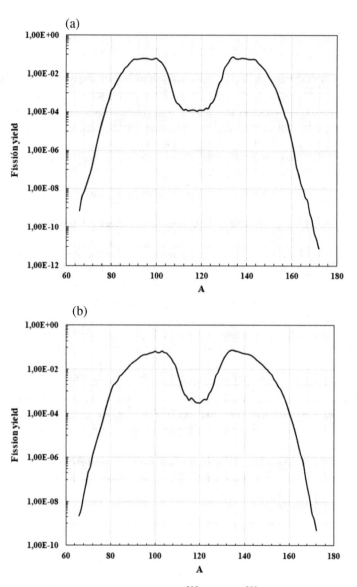

**Fig. 3.4** Radionuclides produced in the fission of $^{235}$U (**a**) and $^{239}$Pu (**b**)

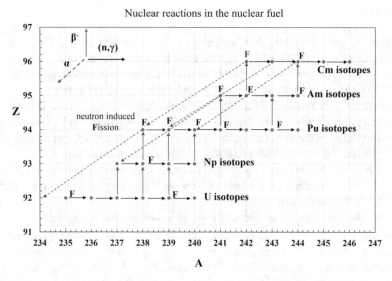

**Fig. 3.5** Nuclear reactions took place in the core of a nuclear reactor. F points to fissile nuclides

**Table 3.5** Some relevant activation products appearing during the operation of nuclear reactors

| Radionuclide | Half-life | Target | Production mode |
|---|---|---|---|
| $^3$H | 12.32 y | Li | (n, α) |
| $^{10}$Be | $1.51 \times 10^6$ y | Be | (n, γ) |
| $^{14}$C | 5730 y | N | (n, p) |
| $^{26}$Al | $7.17 \times 10^5$ y | Al | (n, 2n) |
| $^{36}$Cl | $3.01 \times 10^5$ y | Cl | (n, γ) |
| $^{41}$Ca | $9.94 \times 10^4$ y | Ca | (n, γ) |
| $^{53}$Mn | $3.74 \times 10^6$ y | Fe | (n, d) |
| $^{55}$Fe | 2.744 y | Fe | (n, γ) |
| $^{59}$Ni | $7.6 \times 10^4$ y | Ni | (n, γ) |
| $^{60}$Co | 1925.28 d | Co | (n, γ) |
| $^{63}$Ni | 101.2 y | Ni | (n, γ) |
| $^{81}$Kr | $2.29 \times 10^5$ y | Sr | (n, α) |
| $^{97}$Tc | $4.21 \times 10^6$ y | Ru | (n, γ) |
| $^{98}$Tc | $4.2 \times 10^6$ y | Ru | (n, p) |
| $^{99}$Tc | $2.111 \times 10^5$ y | Mo | (n, γ) |
| $^{107}$Pd | $6.5 \times 106$ y | Pd | (n, γ) |
| $^{108m}$Ag | 438 | Ag | (n, γ) |
| $^{129}$I | $1.57 \times 10^7$ y | Te | (n, γ) |
| $^{137}$La | $6 \times 10^4$ y | Ce | (n, γ) |
| $^{151}$Sm | 90 y | Sm | (n, γ) |
| $^{152}$Eu | 13.517 y | Eu | (n, γ) |
| $^{233}$U | $1.592 \times 10^5$ y | Th | (n, γ) |
| $^{236}$U | $2.342 \times 10^7$ y | U | (n, γ) |
| $^{239}$Pu | 24,110 y | U | (n,γ) |

**Exercises**

1. The $^{14}$C specific activity in living beings is $0.268 \pm 0.002$ Bq/g C. To find the age of a wood coffer, 20 g of C extracted from the wood is measured with a counter recording 100 cpm (counts per minute). If the detector counting efficiency is 25% and the background 30 cpm), calculate the age of the coffer.

2. Depleted U is known to be a sample in which the $^{235}$U abundance is below that expected in Nature. Consider a recent 0.2% depleted U sample. Calculate the sample mass needed to obtain the same activity as that of 1 kg of natural U.

3. Calculate the activity in Bq of 250 g of this depleted U sample.

4. $^{87}$Rb is a $\beta$-emitter decaying to $^{87}$Sr. Its isotopic abundance in minerals from Rb is 27.8%. A mineral sample with 0.85% de Rb was found to contain 0.0050% $^{87}$Sr. Estimate the age of the mineral.

5. $^{40}$K decays into $^{40}$Ar (https://www.nndc.bnl.gov/nudat2/). Thus, a dating method can be developed based on the $\frac{^{40}K}{^{40}Ar}$ atom ratio. Find an expression for the age, $t$, of a geological sample in terms of the K and Ar atom numbers found in the sample. Assume that all $^{40}$Ar comes from the $^{40}$K decay. The isotopic abundances of $^{40}$K and $^{40}$Ar are 0.0117% and 99.6%, respectively.

6. The isotope ratio $\frac{^{206}Pb}{^{207}Pb}$ can be used to date some rocks since it can be related to the $\frac{^{235}U}{^{238}U}$ ratio. Find the relationship between them from which the age of the rock, t, can be deduced.

7. The $^{238}$U radioactive series ends in $^{206}$Pb, while that of $^{232}$Th ends in $^{208}$Pb. A given mineral contains 55% Th, 38% U, and 2% Pb (atom percentages). Calculate the age and original composition of the mineral. Assume that all Pb originates from U and Th and that all the half-lives of the series members are sufficiently shorter than the age.

8. Which He mass has emanated from 10 t of such mineral along $10^6$ y?

9. In most natural systems, $^{230}$Th is in secular equilibrium with $^{238}$U. However, in coralline systems, for instance, they are not. In fact, corals can accumulate U but not Th. Assuming that all the $^{230}$Th comes from the $^{238}$U decay, show that the age of the coral sample can be calculated with (4.2) if $^{234}$U and $^{238}$U are in secular equilibrium.

10. Guess now that $^{234}$U and $^{238}$U are not in secular equilibrium with an activity ratio $R = \frac{A(^{234}U)}{A(^{238}U)} > 1$. Show that the coral age is given by (4.3).

11. Suppose that you find an activity ratio $\frac{^{228}Th}{^{232}Th} > 1$ in a natural sample. Is a dating method possible on the basis of this disequilibrium? Explain the method and deduce the necessary equations.

12. Marine crustaceans absorb Ca and Ra from surrounding seawater to form their exoskeletons. Could its age be determined by measuring the $\frac{^{228}Ra}{^{228}Th}$ activity ratio? Find the appropriate equations. Measurable ages with this method range from months to a few years. Could you estimate the age limit that we could give?

13. Approximately 4.5% of the neutron-induced $^{235}$U fissions lead to $^{133}$Sb through

$$^{133}\text{Sb} \rightarrow {}^{133}\text{Te} \rightarrow {}^{133}\text{I} \rightarrow {}^{133}\text{Xe} \rightarrow {}^{133}\text{Cs}$$

One gram of natural U is irradiated with a neutron flux of $10^{13}$ ncm$^{-2}$ s$^{-1}$ for 60 min. If $\sigma_f = 582$ barn, calculate the $^{133}$Sb and $^{133}$Te atoms in the sample at the time the irradiation stops.

14. Suppose, for the sake of simplicity, that in a thermal nuclear reactor with 8% enriched U, fission and neutron capture processes occur for thermal neutrons. At these energies, the possible processes are as follows:
    (a) $^{235}$U fission, $\sigma_f = 580$ barn;
    (b) $^{238}$U(n, $\gamma$)$^{239}$U $\to \cdots \to ^{239}$Pu, $\sigma_c = 2.7$ barn.
    (c) $^{235}$U(n, $\gamma$)$^{236}$U, $\sigma_c = 98$ barn.
    Which $^{238}$U proportion will have been transformed into $^{239}$Pu when 5% of $^{235}$U has been consumed? Assume that this will happen in a time much longer than the $^{239}$U half-life.

15. Suppose now that there is a mass of 100 t of enriched U in the reactor. What $^{239}$Pu mass has been produced?

16. A total of 0.669% of the $^{235}$U thermal n-induced fissions generate $^{129}$I, which decays into $^{129}$Xe. Suppose that in a 5% enriched U nuclear thermal reactor, fission processes occur for thermal neutrons at a constant neutron flux of $\phi = 2 \times 10^{12}$ cm$^{-2}$ s$^{-1}$.
    Calculate the abundances of $^{235}$U, $^{129}$I, and $^{129}$Xe when the reactor is shut down after 50 years.

17. Calculate the $^{235}$U and $^{129}$I activities present among the residues generated by the initial fuel load of 100 t in the reactor of exercise 17. In addition, calculate the $^{235}$U, $^{129}$I, and $^{129}$Xe masses produced.

18. $^{99}$Mo decays to $^{99m}$Tc. Let us have a pure $^{99}$Mo sample that starts to decay at $t = 0$.
    (a) Find the time for which both the $^{99m}$Tc and $^{99}$Mo activities are the same.
    (b) Find the time at which the $^{99m}$Tc activity will be maximum.
    (c) After 100 h, find the activity ratio of $^{99}$Mo to $^{99m}$Tc.

19. Radionuclides B (half-life $T_B$) is produced in an accelerator by irradiating radionuclides A ($T_A$) with a beam of a given particle. Find the irradiation time necessary to obtain the maximum B activity. Assume that all the necessary data are known and assume that the amount of radionuclide A remains practically constant during the irradiation time.

20. In Nuclear Medicine, Mo–Tc generators are frequently used. Among them, $^{99}$Mo is found to be in secular equilibrium with $^{99m}$Tc. The latter is removed and injected into the patients. Suppose that you perform an extraction and separate 70% of the Tc. Thus, you have two samples: one in which Mo has 30% Tc at $t = 0$ and the eluate that contains 70% Tc. What time must elapse for the two samples to have the same $\gamma$ activity?.

21. Sm is a rare earth with seven natural isotopes, of which $^{147}$Sm is radioactive and decays into $^{143}$Nd. A moon sample presents the following isotopic data normalized to the $^{144}$Nd abundance:

| Sample | $^{147}$Sm/$^{144}$Nd | $^{143}$Nd/$^{144}$Nd |
|--------|----------------------|----------------------|
| 1.     | 0.1727               | 0.512356             |

| Sample | $^{147}Sm/^{144}Nd$ | $^{143}Nd/^{144}Nd$ |
|--------|---------------------|---------------------|
| 2.     | 0.2434              | 0.513861            |

(a) Derive the isochron equation: the mathematical expression that relates the $^{147}Sm/^{144}Nd$ and $^{143}Nd/^{144}Nd$ ratios at time t to the $^{143}Nd/^{144}Nd$ ratio at $t = 0$.

(b) Calculate the age of the samples.

(c) Calculate the $^{143}Nd/^{144}Nd$ ratio at $t = 0$.

# References

1. White, *Isotope Geochemistry*, Chapter 1 (Wiley, 2015)
2. A.P. Dicking, *Radiogenic Isotope Geology* (Cambridge University Press, 2018)
3. M. Ivanovich, R.S. Harmon, *Uranium Series Disequilibrium: Applications to Environmental Problems in Earth Sciences* (Clarendon Press, 1982)
4. M. Ivanovich, Uranium series disequilibrium methods: applications to paleoclimatic studies, in *Low-Level Measurements and Their Applications to Environmental Radioactivity*, ed. by M. Garca-León, G. Madurga (World Scientific Pub. Co., 1988), pp. 135–170
5. G.R. Choppin, J.O. Liljenzin, J. Rydbeg, *Radiochemistry and Nuclear Chemistry*, Chapter 21 (Butterworth-Heinemann, 2002)
6. C. Castagnoli, D. Lal, Solar modulation effects in terrestrial $^{14}C$ production. Radiocarbon **22**, 133–158 (1980)
7. J.C. Gosse, F.M. Phillips, Terrestrial in situ cosmogenic nuclides: theory and application. Q. Sci. Rev. **20**, 1475–1560 (2001)
8. M.S. Potgieter, Solar modulation of cosmic rays. Living Rev. Solar Phys. **10**, 3 (2013)
9. B. Borchers, S. Marrero, G. Balco, M. Caffee, B. Goehring, N. Lifton, K. Nishiizumi. F. Phillips, J. Schaefer, J. Stone, Geological calibration of spallation production rates in the CRONUS-Earth project. Q. Geochronol. **31**, 188–198 (2016); S. Ivy-Ochs, F. Kober, Surface exposure dating with cosmogenic nuclides. Q. Sci. J. **57**, 179–209 (2008)
10. I. Curie, F. Joliot, 'Un nouveau type de radioactivité', Comptes rendus hebdomadaires des séances de l'Académie des Sciences **198**, 254–256 (1934); F. Joliot, I. Curie, Artificial production of a new kind of radio element. Nature **133**, 201–202 (1934)
11. G.R. Choppin, J.-O. Liljenzin, J. Rydbeg, *Radiochemistry and Nuclear Chemistry*, Chapter 15 (Butterworth-Heinemann 2002)
12. J.C. Evans, E.L. Lepel, R.W. Sanders, C.L. Wilkerson, W. Silker, C.W. Thomas, K.H. Abel, D.R. Robertson, *Long-Lived Activation Products in Reactor Materials*, NUREG/CR-3474 Report 1984

# Environmental Radioactivity

**4**

**ABSTRACT**

In this chapter, a description is given of the presence of primordial and cosmo-genic radionuclides in the different environmental compartments. Depending on their origin and geochemical behavior, natural radionuclides are preferentially associated with specific compartments. This fact allows us to develop dating tools that are also outlined in the chapter. On the other hand, special atten-tion has been given to the industrial use of some materials with relatively high natural radioactivity content (NORM). It poses in some cases a radiological problem for the workers and the population. The sources of artificial radioac-tivity in the environment are also described. The beginning of the nuclear era during the last 1940s reintroduced extinct radionuclides into the Earth. The subsequent development of the nuclear industry has caused the distribution of artificial radionuclides in the environment throughout the different steps of the nuclear fuel cycle.

## 4.1 Presence of Natural Radioactivity in the Environment

### 4.1.1 Primordial Radionuclides

As described in Chap. 3, primordial radionuclides are those having a half-life close to Earth age, i.e., some $10^9$ years. Tables 3.1 and 3.2, as well as Fig. 3.1, gather the most interesting information on them. Originally, uranium isotopes are present in soils and rocks at different concentrations depending on the geological sub-strate. Some examples are given in Table 4.1, where the concentration of $^{238}U$ in some rocks is presented [1]. Usually, concentrations range from 7 to 40 Bqkg$^{-1}$, although some specific minerals contain comparatively higher amounts of U. This is the case for phosphate rocks that some time ago were even suggested as a com-mercial U source. As we shall see later, phosphate rock is the basic raw material

© The Author(s), under exclusive license to Springer Nature Switzerland AG 2022       51
M. García-León, *Detecting Environmental Radioactivity*, Graduate Texts
in Physics, https://doi.org/10.1007/978-3-031-09970-0_4

**Table 4.1** $^{238}$U, $^{232}$Th, $^{226}$Ra, and $^{40}$K radioactivity concentrations (Bqkg$^{-1}$) ranges or averages in different rocks

| Rocks | $^{238}$U | $^{232}$Th | $^{226}$Ra | $^{40}$K |
|---|---|---|---|---|
| Basalt | 7–10 | 10–15 | 80 | 900 |
| Granite | 60 | 90 | 80 | 1100 |
| Limestones | 15 | 5 | 16 | 160 |
| Shale sandstones | 15 | 50 | 180 | 500 |
| Quartz | 10–40 | 10–30 | | 400 |
| Beach sands | 40 | 30 | | |
| Carbonates | 25 | 10 | 7–40 | 70 |

for the phosphoric acid industry, whose radiological environmental impact is not negligible.

The geochemical mobility of U makes it appear in different environmental compartments. Thus, it can be found in fresh or marine waters, sediments, and the atmosphere. Some examples are given in Table 4.2, where some average data on U isotope concentrations are presented in river water, sediment, and rainwater. In the absence of a local source of radioactive contamination, the levels of U in river waters depend on the geological substrate of the freshwater system. U concentrations range in the case presented in Table 4.2 from 1 to 3 $\mu$gl$^{-1}$ [2]. The differences along the same fluvial system depend on the dynamics of the river. For example, the influence of the sea on estuarine systems is very clear, contributing in many cases to enhanced U concentrations [2].

There are data in the literature on U concentrations in the atmosphere. Some of them are presented in Table 4.2, where U is given in rainwater [3]. The presence of U in the atmosphere is related to soil resuspension effects in the studied region and, of course, to the possible transport of these materials by the action of winds. Thus, specific concentrations are essentially affected by local weather conditions, such as rainwater precipitation patterns. For example, in the case presented in the mentioned Table, levels range from 0.3 to 1.8 $\mu$l$^{-1}$ with a time variation clearly connected with the precipitation pattern, as seen in Fig. 4.1. Dry periods are followed by high concentrations of U in rainwater, supporting the idea of its soil origin.

**Table 4.2** Typical concentrations or ranges of U isotopes in river water, sediments, and rainwater. In some cases, the U concentrations are given in $\mu$gl$^{-1}$ or $\mu$gg$^{-1}$. The conversion of mBql$^{-1}$ or mBqg$^{-1}$ can be carried out by multiplying the radioactivity concentration by 0.081

| Sample | $^{238}$U | $^{234}$U/$^{238}$U |
|---|---|---|
| River water (mBql$^{-1}$) | 12–38 | 1.03–1.71 |
| Sediment (mBqg$^{-1}$) | 6.2–10 | 1.00–1.25 |
| Rainwater (mBql$^{-1}$) | 1–22 | 1.00–1.68 |

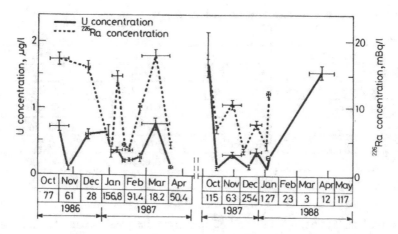

**Fig. 4.1** U and $^{226}$Ra in rainwater. Figure reprinted from [3] with permission from Springer Nature

Small U concentrations also appear in drinking water. The levels are very variable. Thus, in Table 4.3, concentrations below $10^{-1}$ $\mu gl^{-1}$ are found at most locations [1, 4]. However, there are some special places where the U content is very high, and this depends on the geological origin of the water. This is the case for waters collected in some places in Finland with concentrations of up to 30 $\mu gl^{-1}$. Water treatment plants are normally oriented to the depuration of conventional biological and chemical contaminants. The case of radionuclides must be included specifically in the treatment plans. In Table 4.3, some U concentrations are given for commercial mineral waters [4]. Only in a few cases are concentrations unacceptable for human consumption, as seen in Table 4.3.

The bioavailability of U has also been studied and will be discussed in Chap. 5. Some data in biological tissues and foodstuffs are given in Table 4.4 [1].

As we will see in Chap. 5, isotopic ratios are a good measure of radionuclide origin and dynamics in the environment and, for that, are very much used in environmental radioactivity studies.

$^{238}$U and $^{235}$U are the parents of two radioactive series (see Chap. 3 and Fig. 3.1) that populate the environment with varied radionuclides with different half-lives and geochemical behaviors. Since the conditions for secular equilibrium are

**Table 4.3** U isotopes and $^{226}$Ra in drinking and mineral waters (ground water). Estimative ranges found in the literature

| Sample | $^{238}$U (mBql$^{-1}$) | $^{234}$U (mBql$^{-1}$) | $^{226}$Ra (mBql$^{-1}$) |
|---|---|---|---|
| Drinking Waters | 0.4–260 | 0.12–350 | 20–50 |
| Mineral waters | 0.76–800 | 2–1000 | 3–4000 |
| Commercial waters | 0.6–200 | 1.1–700 | 4–1200 |

**Table 4.4** U and other natural isotopes in foods (mBqg$^{-1}$)

| Sample | $^{238}$U | $^{226}$Ra | $^{40}$K |
|---|---|---|---|
| Milk and milk products | $10^{-4}$–$5 \times 10^{-3}$ | $4 \times 10^{-4}$-200 | 1.7–100 |
| Cereals | $3 \times 10^{-3}$–0.4 | 0.04–4.3 | 0.4–200 |
| Root vegetables | $3 \times 10^{-3}$–3 | 0.02–1.3 | 72–180 |
| Leafy vegetables | 0.02–2 | 0.02–5 | 60–450 |
| Fruit | $5 \times 10^{-4}$–1.3 | 0.01–3 | 23–260 |
| Meat | $10^{-3}$–0.02 | 0.03–3 | 50–200 |
| Eggs | $10^{-3}$ | 0.14–3.2 | 60–100 |
| Fish | 0.5–7.4 | 0.02–7.8 | 60–600 |

accomplished, it is expected that all of them apparently decay with the half-life of the parent in a natural sample. However, disequilibrium is very often observed in environmental samples, as seen in Tables 4.2 and 4.3. In river waters, disequilibrium between $^{238}$U and $^{234}$U is found very frequently, with activity ratios ranging from 1.00 to 1.70 [2]. Ratios close to 1.7 are found in rainwater [3].

Nature is a complex system, and the conditions for secular equilibrium can easily be broken. The disequilibrium between $^{234}$U and $^{238}$U activities is very frequent and depends on the environmental conditions. For example, in river waters with acidic pH values, activity ratios above 2 can be observed [5]. As shown in Table 4.2, the $\frac{^{234}U}{^{238}U}$ activity ratio is usually lower, but in general, it is >1. The existence of unsupported $^{234}$U is attributed to various phenomena. The most immediate is related to the geochemical behavior of the radionuclides produced by $^{238}$U disintegration before $^{234}$U [6]. Furthermore, the α-particle emitted from $^{238}$U decay could damage the crystal lattice, which facilitates the migration of the descendants across the interface of the natural compartments in contact. As a result, an excess of $^{234}$U appears in the environment, which leads to an increase in the activity ratio. The isotopic ratio therefore characterizes the sample, its origin and the interactions that have taken place with other compartments.

As a byproduct of this disequilibrium, the possibility of developing dating methods appears since $^{234}$U in excess decays with its own half-life [6]. In fact, several specific geological systems, such as some speleothems, can be dated based on unsupported $^{234}$U [7]. For that, $^{234}$U and $^{238}$U are measured in the sample, and a simple application of the Bateman equations described in Chap. 2 leads to

$$R(t) - 1 = [R(0) - 1]e^{-\lambda_{234}t} \qquad (4.1)$$

$R(t)$ and $R(0)$ are the $\frac{^{234}U}{^{238}U}$ activity ratios at the time of observation, $t$, and at the beginning, respectively; $\lambda_{234}$ is the decay constant of $^{234}$U. Of course, $R(0)$ must be known, and in fact, it is known for some specific geological systems [7]. The equation can be readily solved, and with this method, it is possible to date systems up to $10^6$ years old.

Similarly, another disequilibrium can be used for dating purposes. Indeed, since corals are known to accumulate U but not Th from the surrounding marine environment, it seems clear that the $^{230}$Th found in fossil corals must have been produced by the decay of $^{238}$U [7]. This way, it turns out.

$$T(t) = \left(1 - e^{-\lambda_{230}t}\right). \tag{4.2}$$

T is the $\frac{^{230}Th}{^{238}U}$ activity ratio at the time of observation, $t$ and $\lambda_{230}$ the $^{230}$Th decay constant. Equation 4.2 applies when $R = 1$. In the case where there is a disequilibrium between $^{234}$U and $^{238}$U, the unsupported $^{234}$U independently produces $^{230}$Th as well, and then the dating equation results.

$$T(t) = \left(1 - e^{-\lambda_{230}t}\right) + \frac{\lambda_{230}}{(\lambda_{230} - \lambda_{234})}[R(t) - 1]\left[1 - e^{(\lambda_{234} - \lambda_{230})t}\right] \tag{4.3}$$

with an obvious meaning for the notation.

From a radiological point of view, the most interesting descendants of the $^{238}$U radioactive series are $^{226}$Ra and $^{222}$Rn. Specifically, $^{226}$Ra and its direct descendants contribute highly to the radiation dose received by the population [1], as it tends to concentrate in the bones. As in the case of $^{238}$U, $^{226}$Ra is present in several natural compartments. Thus, many studies on its concentrations in freshwater, marine waters, and sediments can be found in the current literature [1, 8]. In addition, even in drinking waters. Some of these data can be seen in Tables 4.3 and 4.5. $^{226}$Ra activity concentrations from 2 to 30 mBql$^{-1}$ can be found in river waters from the case presented in Table 4.3 [2].

On the other hand, they are very close to other freshwater systems all over the world, with the exception of those affected by local sources of radioactivity or exceptionally high radiation areas. As in the case of U, the presence of $^{226}$Ra in rainwaters is affected by surrounding soils, as well as wind transport and local weather conditions. Figure 4.1 shows the variation of $^{226}$Ra and U in rainwater in southwestern Spain [3]. The discussion about the origin and variations of U in rainwater also applies in this case.

In many cases, the analysis of the $\frac{^{226}Ra}{^{238}U}$ activity ratios gives very interesting information on the dynamics of the studied environment. For example, in Fig. 4.2, such an activity ratio is presented as a function of the sampling station along the Guadalquivir River [2]. A very apparent increase in the activity ratio observed for samples collected near the mouth of the river reveals the redissolution of $^{226}$Ra from the sediment as the salinity of the water increases.

**Table 4.5** $^{226}$Ra concentrations in several natural compartments. Additional information can be found in the previous tables

| Sample type | $^{226}$Ra |
|---|---|
| River waters (mBql$^{-1}$) | 2–30 |
| River sediments (mBqg$^{-1}$) | $\leq$60 |
| Marine waters (mBql$^{-1}$) | Average 1–2 |

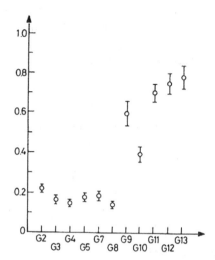

$^{226}$Ra/$^{234}$U activity ratios along the Guadalquivir river channel in samples taken in 1989

**Fig. 4.2** $^{226}$Ra/$^{234}$U activity ratio in the Guadalquivir River. The activity ratio increases as the sampling point approaches the river mouth. Figure reprinted from [2] with permission from Springer Nature

For radiological reasons, the presence of $^{226}$Ra in drinking and mineral waters is very important. The concentrations are very variable and depend on the origin of the water, as shown in Table 4.3. The $^{226}$Ra present in mineral or drinking waters in general can contribute significantly to the total annual radiation dose received by the population by ingestion (see Chap. 6). Knowledge of its concentration in mineral waters is a key issue as it is the development of efficient systems for the removal of Ra during water purification for human consumption.

$^{222}$Rn is a direct descendant of $^{226}$Ra and belongs to the $^{238}$U radioactive series. Rn is a noble gas and can emanate from soils or waters where $^{226}$Ra or $^{238}$U are present. Of course, the other Rn isotopes must be taken into account, but $^{222}$Rn is the most radiologically relevant. Its contribution to the total radiation absorbed dose by inhalation is very high, and in fact, several strategies have been implemented to decrease its concentration in air at houses and workplaces [1]. A more detailed account of its presence, levels, and distribution in the environment is given in Chap. 5, where a specific section on the 'Radon problem' is given.

The $^{222}$Rn emanation from the soils breaks secular equilibrium with $^{238}$U. This is very important and, as introduced in Chap. 3, makes it possible to develop a new dating method based on $^{210}$Pb. In fact, once the decay of $^{222}$Rn is not governed by that of $^{238}$U, it disintegrates with its own half-life ($T_{1/2} = 3.8$ d). Their immediate descendants are all very short-lived (see Tables 3.1 and 3.2 and Fig. 3.1). Thus, $^{210}$Pb is produced in the lower layers of the atmosphere, decaying with its own half-life ($T_{1/2} = 22.3$ y). After some weeks, this unsupported $^{210}$Pb is deposited on the Earth's crust or on a lake surface (see Fig. 3.2) and then on lacustrine sediment.

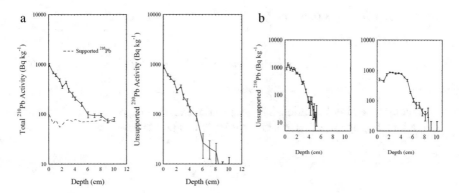

**Fig. 4.3** $^{210}$Pb activity depth profiles **a** following the CIC model and **b** following the CRS model. Figures reprinted from [9]

In this way, each sediment layer receives a certain amount of unsupported $^{210}$Pb. Under some circumstances, the measurement of unsupported $^{210}$Pb activity as a function of the sediment depth, the activity depth profile, gives information on the age of each layer or the lake sedimentation rate in the studied area. Some examples are given in Fig. 4.3 [9]. On some occasions, the analysis of the depth profile is simple. Indeed, as seen in Fig. 4.3a, the profile can be represented as an exponential that can be modeled as

$$C_z = C_0 e^{-\lambda t} = C_0 e^{-\lambda \frac{z}{r}} \tag{4.4}$$

$C_z$ and $C_0$ are, respectively, the activity of $^{210}$Pb per unit mass at depth $z$ and at the sediment surface, $t$ is the age of the sediment layer at depth $z$ and $r$ is the sedimentation rate. Finally, $\lambda$ is the decay constant of $^{210}$Pb. This approach implicitly assumes that $C_0$ is constant throughout the sediment column, which is the basis of the constant initial concentration (CIC) model. Of course, $r$ is also supposed to be constant over a reasonable time period. However, in other cases, the interpretation of the activity profile is not as simple as seen in Fig. 4.3b. It is very apparent that an exponential shape for the profile does not fit the experimental data.

It seems that $C_0$ and $r$ cannot be considered constant during the studied period of time. To solve this problem, a more complex mathematical model is needed. One possible alternative is the application of the so-called constant rate of supply (CRS) model, in which the $^{210}$Pb rate of supply, $P$, over the sediment surface is considered constant [10]. This is certainly a good hypothesis under steady geological conditions. A new quantity is defined to apply this model: the integral activity per unit area. This is the accumulated activity, $A_z$, per unit area from the sediment surface to depth $z$. It is obvious that

$$A_z = \int_0^t P e^{-\lambda t} dt = \frac{P}{\lambda}\left(1 - e^{-\lambda t}\right) \tag{4.5}$$

Of course, a direct relationship between $z$ and $t$ must be assumed to allow integration. The age, $t$, of each sediment layer can be obtained as

$$t = \frac{1}{\lambda}\ln\frac{A_\infty}{(A_\infty - A_z)} \tag{4.6}$$

where $A_\infty = \frac{P}{\lambda}$ is the integral activity per unit area over the whole sediment profile. $A_\infty$ is mathematically obtained by integrating (4.5) from 0 to $\infty$ and experimentally determined by adding the $^{210}$Pb activities along the sediment column.

Both the CIC and CRS models are simple mathematical approaches to analyze $^{210}$Pb sediment profiles but do not truly represent the behavior of the studied lacustrine systems. For that, more realistic models have been developed. A collection of them can be found in Table 4.6 [11]. With the $^{210}$Pb method, it is possible to date sediment changes within the last $10^2$ y [12].

$^{232}$Th is relatively insoluble and is mostly associated with rocks and soils. Typical concentrations in some rock types are given in Table 4.1. Compared to U isotopes, it is apparent that its concentration is typically higher. Although $^{232}$Th can be present in marine or riverine waters, concentrations in the corresponding sediments are higher. An example is provided in Table 4.7, where we present

**Table 4.6** List of $^{210}$Pb dating models. Adapted from [11]

| Name | Description |
|---|---|
| CIC | Constant initial concentration |
| CF–CS | Constant flux: constant sedimentation |
| CRS | Constant rate of supply |
| CMZ–CS | Complete mixing zone with constant SAR |
| CF–CS | Constant diffusion |
| CF–CS | Depth-dependent diffusion and/or translocational mixing |
| IMZ | Incomplete mixing zone |
| SIT | Sediment isotope tomography |
| NID–CSR | Nonideal deposition, constant sedimentation rate |
| CICCS | Constant initial concentration and constant sedimentation rate |
| IP–CRS | Initial penetration, constant rate of supply |
| TERESA | Time estimates from random entries of sediments and activities |

**Table 4.7** $^{232}$Th concentration ranges in waters and sediments in the Odiel River [13] and in waters and sediments in the Guadalquivir River [14]

| Sample type | Guadalquivir River | Odiel River |
|---|---|---|
| Water (mBql$^{-1}$) | 0.08–3.5 | 0.006–0.048 |
| Sediment (mBqg$^{-1}$) | 10.7–23.5 | 0.1–2.5 |

some results on the $^{232}$Th concentrations in waters, suspended matter, and sediments taken in two riverine systems. It can be clearly seen that the concentration of $^{232}$Th radioactivity is higher in the solid phases and that the concentration of $^{238}$U is systematically higher in the waters. This low environmental mobility makes $^{232}$Th appear in very small amounts in biological tissues.

$^{228}$Ra belongs to the $^{232}$Th radioactive series. It is a β-emitter and, like $^{226}$Ra, is of great radiological interest. $^{228}$Ra behaves as $^{226}$Ra, and thus, it can be found in waters, sediments, the atmosphere, and even in biological tissues. The data archive for $^{228}$Ra is not as large as that of $^{226}$Ra, but the known data demonstrate that radioactivity concentrations are very similar in several natural compartments, such as groundwater and river waters [1].

Among the other primordial radionuclides in Table 3.1, $^{40}$K is the only radionuclide relevant within the field of environmental radioactivity. It is known to contribute significantly to the absorbed dose by humans (see Chap. 6). In fact, the $^{40}$K contained in the human body, due to the natural presence of K, contributes almost to 10% of the annual allowed dose equivalent for the public [1]. Some $^{40}$K in the Earth's crust are given in Table 4.1.

In general, it can be said that primordial radionuclides are ubiquitous in nature, although the concentration distribution is very heterogeneous, with variations within two orders of magnitude [8]. Differences are associated with the geochemical behavior of radionuclides. As a consequence, there are specific rocks where the primordial radionuclides are concentrated (Table 4.8).

Thus, there are some sites with exceptionally high concentrations of natural radioactivity due to the abundance of some of the minerals and rocks in Table 4.8, where the concentrations of U isotopes and $^{232}$Th are especially high. This is the case for the deposits of monazite and ilmenite sands in Espirito Santo (Brazil) and Kerala and Tamil Nadu (India). Monazite is very rich in $^{232}$Th and, to a lesser extent, in U isotopes. This leads to an exposure to radiation higher than in normal areas [8]. Apatite is a U-rich rock that, as in the previous cases, is present in some special areas giving rise to an anomalous radiation dose increase. One example is the case of the Minas Gerais region, also in Brazil [8]. Alkaline intrusive rocks contain high amounts of U and Th. Morro do Ferro (Brazil) is a site of special interest with regard to this rock. Beach sands are in many cases unsuspected deposits of ilmenites and monazites, although in low concentrations. The black sand that is often found on some beaches is a compound of these rocks. In Fig. 4.4, the X-ray fluorescence spectrum taken from a black sand particle found on a beach shows the presence of Th and a shape compatible with that expected for monazite [17].

**Table 4.8** Minerals and rocks with high concentrations of U and Th, as well as other natural radionuclides [15, 16]

| Material | Radionuclides | $^{238}U$-$^{232}Th$ activity (mBqg$^{-1}$) | U-Th content | Composition |
|---|---|---|---|---|
| Monazite | Th- U-isotopes | 7300–4450 | 3–9% ThO$_2$, 0.3% U$_3$O$_8$ | Rare earth phosphates, (Ce, Nd, La)SmPO$_4$ |
| Zircon | Th- U-isotopes | 2300–500 | | ZrSiO$_4$ |
| Rutile | Th- U-isotopes | 1240–1600 | | TiO$_2$ |
| Ilmenite | Th- U-isotopes | 130–620 | | FeO, TiO$_2$ |
| Xenotime | Th- U-isotopes | | 0.8% U$_3$O$_8$, 0.8% ThO$_2$ | (Y, Yb, Er, Dy …)PO$_4$ |
| Bastnaesite | Th-isotopes | | 0.1–2% ThO$_2$ | (Ce, La, Y)CO$_3$F |
| Apatite | U-isotopes | | 5–300 ppm U | Ca$_5$(PO4)$_3$ |
| Phosphorites | Mostly U-isotopes | 150–4800 $^{238}U$ 16–78 $^{232}Th$ | | Other PO$_4$ compounds |
| Bauxite | Th- U-isotopes | 500–400 | | AlO |
| Pyrochlore | Th- U-isotopes | | 0.05% for both U and Th oxides | (Na, Ca)$_2$Nb$_2$O$_6$(OH, F) |
| Coal | Th- U-isotopes | 20–22 | | C-rich minerals |
| Natural Gas | Ra- Rn-isotopes | Basically, $^{226}Ra$-$^{222}Rn$. Up to 400 | | Hydrocarbon mixture |
| Granite | Th- U-isotopes | 60–90 and up to $10^3$ $^{40}K$ | | Quartz, feldspars, mica, and amphiboles |
| Limestone | Th- U-isotopes | 31–9 | | Calcite, aragonite (CaCO$_3$ forms) |

Several of the rocks described in Table 4.8 are of economic importance, as we will see in the next sections. Of course, the intrusion of radionuclides into the food chain has also been demonstrated in these natural high concentration areas [8].

## 4.1.2  Cosmogenic Radionuclides

Table 3.4 gives a list of cosmogenic radionuclides, including the production modes. As already explained in Chap. 3, the production of cosmogenic radionuclides depends on the cosmic-ray flux and therefore on the different processes that modulate it, i.e., the solar activity and the intensity and orientation of the geomagnetic field [18]. On the other hand, the cosmic-ray flux increases with altitude since their main components are less absorbed by the atmosphere. It also increases with

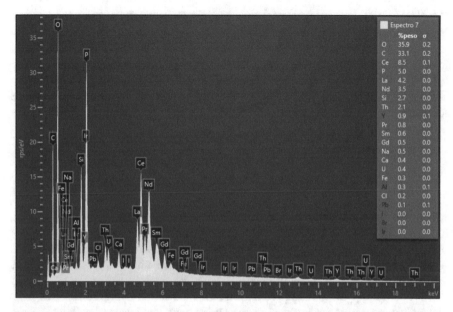

**Fig. 4.4** X-ray spectrum taken from a 'black sand' particle collected on a beach. Spectrum courtesy of J.-L. García-León

latitude, where the charged component of cosmic rays is less deflected by the geomagnetic field [19]. Once produced, the radionuclide is incorporated into the environment through different deposition and transport processes depending on its geochemical behavior. As a consequence, each cosmogenic radionuclide preferentially enters specific environmental compartments. This fact can be used to trace the behavior of the environment. For instance, cosmogenic noble gases can be used to trace the atmospheric dynamics, $^3$H or $^{36}$Cl the hydrosphere or the biosphere, etc. To make this possible, the radionuclide must be conservative regarding the spatial and temporal evolution of the traced system. This means that the radionuclide must behave exactly as the system it is tracing. In other words, the radionuclide and the system must have identical physical and dynamic behavior. For example, once it is formed, $^3$H rapidly incorporates into $H_2O$ molecules and therefore starts to participate in hydrological cycling. Therefore, it is expected that their spatial and time dynamics will be the same as those of the water mass we wish to trace.

Cosmogenic radionuclide half-lives range from days to millions of years (see Chap. 3). This offers a very wide group of time markers in the environment that cover the time evolution of many environmental processes. In fact, a very important application of cosmogenic radionuclide is the dating of several natural processes. Some of them follow.

$^{14}$C dating is a very well-known method for dating archaeological remains belonging to the Holocene. It was introduced by Libby [20], who discovered that $^{14}$C is produced in the atmosphere according to the process described in Table

3.4. $^{14}$C and rapidly oxidizes to form $CO_2$ or $CO$ and starts to participate in the global carbon cycle. This means that the biosphere is marked $^{14}$C. Living beings exchange C with the environment in such a way that its $^{14}$C-specific activity is constant. This exchange ceases once the living being dies, and the specific activity of $^{14}$C will decrease as it decays according to its half-life and is not replaced by external $^{14}$C inputs (see Fig. 4.5) [21]. Thus, if $A(0)$ is the $^{14}$C activity at the time of death, it turns out that the actual $^{14}$C activity, $A(t)$, in a sample taken from the system studied is

$$A(t) = A(0)e^{-\lambda_C t} \tag{4.7}$$

where $t$ is the time elapsed since death, and $\lambda_C$ is the decay constant of $^{14}$C. In this way, an estimation of the system age is given. Despite the apparent simplicity of (4.7), method has several difficulties. First, the determination of $A(t)$ is a typical low-level counting problem since the method involves the measurement of low or very low $^{14}$C activity concentrations. As we will see in Chaps. 14 and 15, a new family of counting techniques, Low-Level Counting Techniques, was developed to address the problem of measuring such low levels of activity. Today, such techniques are of application to many scientific disciplines but were originally developed to solve the $^{14}$C problem [22]. However, $A(0)$ is not constant. In fact, it is known that it varies due to changes in solar activity and the intensity and orientation of the geomagnetic field [21]. In addition, some anthropogenic effects have introduced variations in $A(0)$. Indeed, the Suess effect describes the decrease

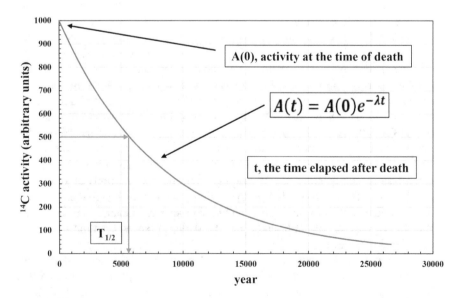

**Fig. 4.5**  $^{14}$C scheme of decay after death

in atmospheric [14]C-specific activity as a result of the introduction of fossil C into the atmosphere during the industrial revolution [23, 24]. In contrast, nuclear detonation tests from the 1960s of the XX century have increased [14]C-specific activity [25].

To overcome this problem, the [14]C dates are calculated by using the $A(0)$ corresponding to the atmospheric [14]C-specific activity in 1950. This is the consequence of a general agreement within the radiocarbon community. The agreement also includes the use of the so-called [14]C Libby half-life, 5568 y, originally determined by the Libby group [21], instead of the generally accepted [14]C half-life, 5730 y, which can be found in the nuclear data tables [28]. This means that the [14]C date is not real and has to be converted into actual calendar dates with the help of a calibration curve. This curve is constructed on the basis of dendrochronological data and well-known age marine corals and other systems [26, 27]. As shown in Fig. 4.6, the calibration curve relates the apparent [14]C data to the real data obtained for the calibration samples. Once the radiocarbon date is obtained, the real calendar date is deduced by interpolation. The application of the [14]C dating method is, in fact, limited by the time period covered by the calibration curve, which today goes back to 55,000 years [26]. This time interval increases periodically as the radiocarbon community investigates new older samples. From a technical point of view, with the help of AMS (see Chap. 18), it would be possible to date samples close to $10^5$ years old. Thus, the expansion of the calibration curve is currently an active research field that promises to extend the frontiers of [14]C dating.

[10]Be is used for the dating of marine sediments. Unlike [14]C, it does not form gases in the atmosphere once produced, as described in Table 3.4. Thus, its atmospheric residence time is very short, and it is rapidly removed by wet or dry fallout. Once it reaches the ocean surface, it readily associates with suspended matter and is transported into the sediment. In general terms, the concentration of activity of [10]Be in a sediment layer of age t, $A_{Be}(t)$, would be

$$A_{Be}(t) = A_{Be}(0)e^{-\lambda_{Be}t} \tag{4.8}$$

with an obvious meaning for the notation. If the average sedimentation rate is $a$ and $\Phi$ the flux of [10]Be in the sediment, it turns out.

$$A_{Be}(t) = \frac{\phi}{a}e^{-\lambda_{Be}\frac{d}{a}} \tag{4.9}$$

where $d$ is the depth of the sediment layer. Equation 4.9 corresponds to the very simple case in which $a$ is constant. However, it is very frequently not constant, and we can apply a similar approach to that of the [210]Pb CRS model. There, it was assumed that P, the atmospheric flux of [210]Pb, is constant within the time period during which the [210]Pb dating method can be applied, approximately 100–150 y. Of course, the [10]Be atmospheric production rate is not constant but is modulated by solar activity and the orientation and intensity of the geomagnetic field [19]. In fact, it is known that changes in the [10]Be flux are significant within the time period where the [10]Be is of application, some $10^6$ y [19]. Fortunately, the residence time

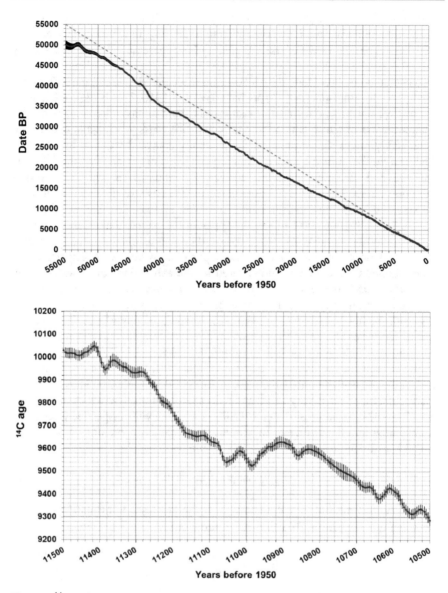

**Fig. 4.6** $^{14}$C calibration curve. The second picture shows a detailed view of the complete calibration curve to highlight the fluctuations that make it difficult in some age ranges to provide one
unique date. Figures are reprinted from [26, 27]

of $^{10}$Be in seawater is large enough to uniformly distribute $^{10}$Be in water. This helps to homogenize its content in seawater. In practice, $\Phi$ could be considered constant for a sufficiently long period of time. For that, a similar Equation to (4.6) can be applied for $^{10}$Be. Now, the integral activities per unit time and surface correspond to the $^{10}$Be integral activities.

A more accurate approach consists of the simultaneous application of $^{10}$Be and $^{26}$Al (see Table 3.4) for sediment dating. Now, the appropriate equation is

$$\frac{A_{Al}(t)}{A_{Be}(t)} = \frac{A_{Al}(0)}{A_{Be}(0)} e^{(\lambda_{Be} - \lambda_{Al})t} \tag{4.10}$$

The advantage of this approach is that the $\frac{A_{Al}(0)}{A_{Be}(0)}$ ratio could be considered independent of the cosmic-ray flux.

$^{10}$Be or $^{10}$Be–$^{26}$Al dating can provide information on sedimentary processes within a time period that extends up to approximately $10^6$ years. It is a very interesting time interval in which geological changes can be traced in the sediment. In addition, the methods provide a powerful chronological tool that can be used to interpret the information preserved in the sediment layers.

From a radiological viewpoint, the only cosmogenic radionuclides of interest are $^3$H and $^{14}$C, since they can easily be incorporated into the human body by inhalation or ingestion. Certainly, the amounts of $^3$H or $^{14}$C naturally produced are not radiologically significant. However, they are also released into the environment during the operation of nuclear reactors and during nuclear detonations [8]. The amounts produced this way can be locally very high compared to cosmogenic production and for that reason could have radiological consequences [29].

### 4.1.3 NORM Materials and Non-nuclear Industries

NORM is the acronym for Naturally, Occurring Radioactive Materials, although there are a vast number of rocks, minerals, etc., containing radioactivity, the term is applied only to materials that require regulatory control [30]. In Table 4.8, we have already presented the most relevant examples of these rocks and minerals. We can see the composition and radioactivity concentration ranges of the material. It is important to note that the majority of these rocks and minerals have industrial uses that contribute, as we will see, to the release of some amounts of natural radioactivity in the environment that can be radiologically relevant.

Rare earth phosphates, such as monazites or xenotimes, contain up to 9% $ThO_2$ or 45,000 Bq/kg $^{232}$Th and up to 0.8% $U_3O_8$ or 7300 Bq/kg $^{238}$U. Bastnaesite is a rare earth carbonate-fluoride compound with negligible amounts of U but with up to 2% $ThO_2$. Zircon silicates can contain up to 250 ppm U or 200 ppm Th, indicating activities of 5000 Bq/kg $^{238}$U or 500 Bq/kg $^{232}$Th. The group of Ti oxides as rutiles or ilmenites (FeO: $TiO_2$) also has a high radioactivity content, with concentrations reaching 1200 Bq/kg $^{238}$U or 1600 Bq/kg $^{232}$Th. Spodumene, used for Li extraction, contains up to 10 Bq/kg $^{238}$U or no more than 1 Bq/kg

$^{232}$Th. Phosphate minerals are known to contain important amounts of radioactivity, depending on their geological origin. Apatite, for example, contains mainly U with concentrations ranging from 5 to 300 ppm. Sedimentary phosphate minerals, phosphorites, are significantly more radioactive, with concentrations of up to 5000 Bq/kg $^{238}$U and up to 100 Bq/kg $^{232}$Th. Bauxites, AlO, are known to contain both U and Th with concentrations of ~100 Bq/kg $^{238}$U and $^{232}$Th. Finally, fossil fuels contain radioactivity that, as we shall see, makes the fuel production process and use radiologically relevant. The radioactivity concentrations in coals, for instance, vary depending on the origin. On average, $^{238}$U and $^{232}$Th concentrations can be said to be approximately 20 Bq/kg with a range of 5 to 350 Bq/kg. Interestingly, in the case of natural gas, the relevant radionuclide is $^{222}$Rn, with concentrations reaching 400 Bq/kg on average [15, 16].

Of course, the radioactivity concentrations in these rocks and minerals depend on their geological origin, as seen in [31], where a very complete table can be found with the radioactivity concentrations determined for many materials. The industrial use of these materials is of radiological interest. In fact, it can lead to the concentration process in products, byproducts, or the production of wastes that can be released into the environment [32–34], producing in many cases a nonnegligible radiological impact [35].

Table 4.9 shows a list of potentially contaminating industrial processes using NORM. In addition, the direct use of some NORM, such as building materials, has a significant impact on house inhabitants [1]. In fact, building materials are mostly compounded with those presented in Table 4.8. Thus, granites can contain radioactivity concentrations of $^{238}$U and $^{232}$Th up to 60 or 8Bq/kg, respectively, although the $^{40}$K content reaches $10^3$Bq/kg [1]. The $^{238}$U and $^{232}$Th concentrations in sandstones can reach up to 6 or 8Bq/kg, respectively; in limestone concrete up to 31 and 9Bq/kg, the sandstone concrete is much less contaminated in $^{238}$U, close to 10Bq/kg, although.

The concentration of $^{232}$Th is approximately the same. Clay bricks can contain 100 and 40Bq/kg $^{238}$U and $^{232}$Th, respectively. Wood is virtually free of U and Th but contains important amounts of $^{40}$K, close to 3 kBq/kg [31, 36]. We will return to this problem in Chap. 5, where the Rn problem is discussed. The production of $^{222}$Rn indoors is a consequence of the building materials used and the geological substrate on which the house is placed. Ventilation of houses helps greatly reduce the radiation dose received by inhabitants [1].

As described in Table 4.9, there are many industrial activities that, at least potentially, should be considered for regulatory purposes. The use of zircon rocks is very general in the ceramic industry, the production of refractory materials, the glazing of tiles, and others. The same can be said of the production of pigment based on $TiO_2$ or the extraction of rare earths from monazites and other similar rare earth phosphates.

The case of phosphoric acid production using phosphorite as a raw material is especially important from an environmental (an occupational) point of view [32]. In Fig. 4.7a, b, the distribution of natural radionuclides in estuarine waters affected by the operation of phosphoric acid production in southwestern Spain

**Table 4.9** Industrial processes using NORM. Mining activities can produce important occupational radiation exposure depending on the treated mineral. The waste products from mining activities can reach the environment

| Industrial activity | Raw material |
|---|---|
| *Mining and mineral processing* | |
| U | |
| Cu | |
| Zr | |
| Phosphates | |
| Fe | |
| Zn | |
| Granites | |
| Limestones | |
| Au | |
| Ni | |
| Ag | |
| Coal | |
| *Industrial processes* | |
| Gas mantles | |
| Lamps | |
| Ceramic industries | Zircon |
| Tiles | Zircon |
| $TiO_2$ pigments | Ilmenite, Rutile |
| Rare earth extraction | Monazite, Bastnaesite, Xenotime, … |
| Phosphoric acid production | Phosphorites and apatites |
| Metal scrap processing | Scraps containing radioactive traces. Melting of old radiation sources, pipes from natural gas |
| Steel production | Iron minerals and coal |
| Al extraction (red mud wastes) | Bauxite |
| Coal-fired energy production (ash from coal burning) | Coal |
| Steel and iron foundries | Iron minerals and coals |
| Oil and Gas | Petroleum and natural gas |
| *Building materials* | |
| Granite use | Granites with high radioactivity content |
| Limestone concrete | Limestones with high radioactivity content |
| Phosphogypsum use | Phosphogypsum: wastes from phosphoric acid production contaminated with U and Ra isotopes |

(continued)

**Table 4.9**  (continued)

| Industrial activity | Raw material |
| --- | --- |
| Cement with coal ash | Coal ash containing radioactive traces |
| *Agricultural practices* | |
| Soil amendment | Phosphogypsum |
| Soil fertilization | Phosphoric acid, phosphogypsum |

is presented. It is very obvious that the release of waste is redistributing natural radionuclides in the environment. Radionuclide concentrations measured over time show a clear decrease as a result of changes in waste management practices that reduce the environmental impact [37]. However, river sediments accumulate radioactivity over time, and it was shown that they can become sources of radioactivity as radionuclides can be redissolved and incorporated into the water phase.

The production of phosphoric acid through the wet acid method generates phosphogypsum, a relevant byproduct that is a NORM material. Phosphogypsum contains mainly $^{226}$Ra [38] and has been used in many countries as a complementary building material [39]. In addition, it has been used as a Ca soil amendment

**Fig. 4.7** $^{226}$Ra activity concentrations in **a** waters and **b** sediments, both recorded over time. Changes in waste practices have decreased the concentration in waters and sediments. However, the $^{226}$Ra accumulated in the sediment could be dissolved and incorporated into the water. Figures reprinted from [37] with permission from Elsevier

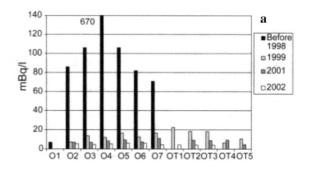

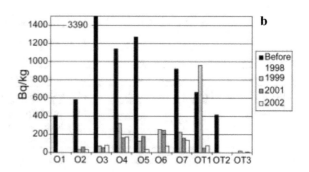

for sodic soils [40]. The agricultural use of phosphogypsum poses the need for radiological control of agricultural products [41].

Bauxite is the main mineral used for Al production. During the industrial process, so-called red mud is produced, a waste product that may contain up to twice the original radioactivity concentration of the mineral used. In some cases, red mud is used as a construction material. Environmental concerns about the fate of these wastes are important [42].

Although the radioactivity in the coals is not too high, it is known that the concentration of radioactivity in the ashes after coal burning is approximately ten times higher [43, 44]. This poses the need to control the gas and ash emissions from coal power plants.

The presence of $^{222}$Rn in natural gas is also important. $^{222}$Rn diffuses from the geological formation into the gas deposit and is transported through the pipeline system to the houses. The concentration of $^{222}$Rn is very dependent on the origin of the deposit, with values ranging from 10 to 50,000 Bqm$^{-3}$ of gas [16, 45, 46].

Drilling works for oil extraction remove radioactive rocks associated with the deposit and groundwater that normally coexist with the oil. This groundwater usually contains radioactivity because it can dissolve the natural radionuclides present in the rock that forms the oil deposit. Water is extracted and used for the transport of oil to storage tanks. During this process, some scales are formed on the inside of pipes that have been found to be highly radioactive [47]. The most abundant radionuclide on the scales is $^{226}$Ra, and typical concentrations have been found to range from 1 to $10^3$ kBq/kg. In some special cases, concentrations up to $1.5 \times 10^4$ kBq/kg have been reported in the literature [48]. It is shown [49] that the abundance of $^{226}$Ra in the scales is due to coprecipitation with Ba of the dissolved Ra in the water as sulfate. These scales have to be treated as NORM, and the radioactive waste protocols must be applied. In many cases, some iron scraps from these pipes are used for Fe melting and steel production. As a consequence, during steel production, some concentrations of radioactivity can occur in the obtained product, although the original Fe ores used are generally very low in radioactivity, <50 Bq/kg for $^{238}$U and $^{232}$Th [50]. For this, the use or recycling of iron scraps produces accidental releases of radioactivity in the environment.

## 4.2 Sources of Artificial Radionuclides

### 4.2.1 The Start of the Nuclear Era. The Bomb Pulse

The discovery of neutron-induced nuclear fission (see Chap. 3) opened the door to the reintroduction in Nature of extinct radionuclides, as it made it possible to generate the fission products. The massive introduction of these radionuclides into the environment started during the test that preceded the Hiroshima and Nagasaki bombings during August 6th and 9th 1945, respectively. Once this happened, a long run among the political powers started to achieve and improve the new bomb.

This run unfortunately has not stopped, although on September 1996, a Comprehensive Nuclear Test-Ban Treaty was open for countries signing. The CNTBT prohibits the execution of nuclear tests by and in the countries that signed the treaty. To date, 178 of the 195 existing countries have signed the CNTBT [29].

Soon after the first explosions, the important radiological impact that the nuclear test had on the environment, including the biosphere, was recognized. Thus, during 1950, some negotiations started between the USA and the former Soviet Union to reduce and control nuclear tests, which diminished the environmental impact. On 5 August 1963, 18 years after the Hiroshima bombing, the Nuclear Ban Treaty, the Moscow treaty, was signed by the USA, USSR, and UK, the main nuclear powers at the time, banning all nuclear tests with the exception of underground tests. The reduction in atmospheric detonations was almost complete, although underground tests continued. In addition, some of the nonsigning countries, relevantly France and China, continued testing their nuclear weapons in the atmosphere. The last recorded atmospheric test took place in 1980, and after this year, some underground tests were carried out. The number of underground tests exceeds that of the atmospheric tests, but the total explosive yield (see next paragraphs) is clearly lower. The last recorded underground tests took place in 1998 and were declared by Pakistan. To date, 530 nuclear tests have taken place, including the Hiroshima and Nagasaki bombings. However, since not all countries have signed the CBTNT, it cannot be assured that nuclear testing has ceased. A very detailed account of the number of atmospheric and underground nuclear tests, as well as the responsible country and the power yield, can be seen in [29]. Here, we summarize the number of explosions that occurred over time and the corresponding yields (Mt) in Fig. 4.8.

At first, nuclear tests were based on nuclear fission of $^{235}$U or $^{239}$Pu. The energy released in this process is high enough, as we have already commented in Chap. 3. Of the total energy produced during fission, approximately 50% is released as a blast, and approximately 35% is released in the form of thermal energy. The remaining 15% is released as ionizing radiation, 5% as prompt radiation emitted some seconds after the explosion, and the rest is released as delayed radiation of the fission products. Of course, the result of a fission explosion is the release of a large amount of energy, in addition to the production of a high number of radionuclides, as presented in Fig. 3.4. The majority of these nuclides are short-lived and rapidly disappear from the environment, but some other remains in nature because of its longer half-life. In Table 4.10, we summarize the nuclear properties of the most important fission products.

In addition, activation radionuclides are produced during detonation as a result of the interaction of the intense neutron flux produced in the explosion with the components of the bombs, the elements in the atmosphere, the soil, or seawater. As we shall see below, in the case of nuclear fusion devices, the so-called thermonuclear bombs, activation products can be more relevant as the neutron flux is larger. The production of activation radionuclides depends on the environment in which the detonation takes place. For example, when the detonation takes place

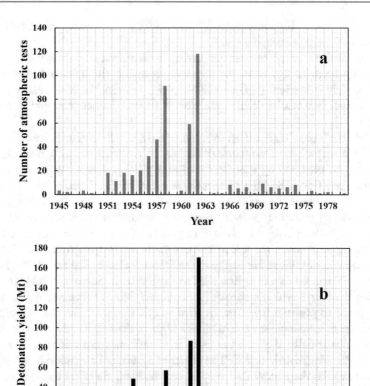

**Fig. 4.8** Number of atmospheric tests and the corresponding yield

**Table 4.10** Summary of relevant fission products, including $T_{1/2}$ and fission yield. Figure 3.4 gives a full account of the fission products that can be produced

| Fission products | $^{235}$U fission yield (%) | $^{239}$Pu fission yield (%) |
|---|---|---|
| $^{90}$Sr ($T_{1/2} = 28.90$ y) | 5.90 | 2.10 |
| $^{99}$Tc ($T_{1/2} = 2.111 \times 10^5$ y) | 6.11 | 6.14 |
| $^{129}$I ($T_{1/2} = 1.57 \times 10^7$ y) | 0.718 | 1.39 |
| $^{131}$I ($T_{1/2} = 8.052$ d) | 2.88 | 3.85 |
| $^{135}$Cs ($T_{1/2} = 2.3 \times 10^6$ y) | 6.53 | 7.62 |
| $^{137}$Cs ($T_{1/2} = 30.08$ y) | 6.27 | 6.73 |

close to the ground, high amounts of $^{45}$Ca and $^{55}$Fe are observed as a result of the activation of the soil components.

Bomb devices based on nuclear fusion were developed shortly after fission bombs appeared. In that case, the physical process is the nuclear fusion of light nuclei, mostly deuterium ($^2$H) and tritium ($^3$H) or even $^6$Li. The energy released per unit mass in a nuclear fusion reaction is slightly higher than that in a fission reaction [51].

To achieve fusion, nuclei must overcome the repulsive Coulomb barrier between them. In devices based on nuclear fusion, it is achieved with high temperatures, ~$10^6$ K, obtained by a nuclear fission explosion that serves as a trigger for the fusion reaction. The high temperatures provoke the dissociation of the electrons from the atoms, i.e., plasma production, and the fusion of the nuclei in the fuel by thermal agitation. Most of the energy produced is released in the explosion and as thermal radiation. However, part is released as the kinetic energy of the resulting neutrons in the fusion that can cause nuclear reactions and, therefore, the appearance of radioactive nuclei in the area affected by the explosion. In some devices, this neutron flux is in turn used to trigger the fission of a $^{238}$U blanket surrounding the bomb, which increases its explosive yield. Thermonuclear explosions are expected to produce fewer radioactive nuclei. However, the fission bomb that triggers the thermonuclear explosion, the high neutron flux, and the use of $^{238}$U blankets substantially increase the number of radionuclides released in the environment after the explosion [52]. The power of a nuclear detonation is measured in comparison with the energy delivered in conventional chemical explosions. Thus, a nuclear detonation delivering an energy equivalent to the explosion of $10^6$ Tm of TNT (trinitrotoluene) has an explosive yield of 1 megaton (Mt). Generally, such a detonation releases an energy close to $4.2 \times 10^{18}$ J. For comparison, it can be found that the explosive yield of complete fission of some 50 g of fissile material is 1 kiloton, kt [52]. Of course, as the yield increases, the production rate of radionuclides increases. In Table 4.11, an estimation of the activity produced per Mt in fission of the most relevant fission products is presented.

**Table 4.11** Radionuclide production (PBq) per Mt in fission devices

| Radionuclide | Activity produced per Mt (PBq) |
|---|---|
| $^{89}$Sr | 740 |
| $^{90}$Sr | 3.7 |
| $^{95}$Zr | 925 |
| $^{99}$Tc | $8.5 \times 10^{-4}$ |
| $^{103}$Ru | 684 |
| $^{106}$Ru | 10.7 |
| $^{131}$I | 4625 |
| $^{135}$Cs | $9 \times 10^{-5}$ |
| $^{137}$Cs | 5.92 |
| $^{144}$Ce | 137 |

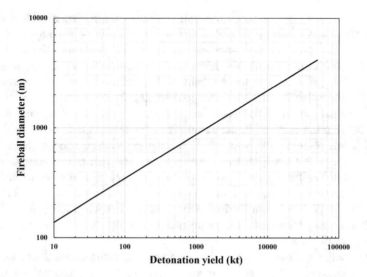

**Fig. 4.9**   Fireball diameter against yield. See (4.11)

Nuclear explosion produces a fireball that grows until it reaches a maximum height that depends on the explosive yield of the device, $W$. It can be found [52] that the final diameter of the fireball, $D$, in meters is

$$D = 55W^{0.4} \tag{4.11}$$

where $W$ is given in kilotons. A plot of this function is presented in Fig. 4.9. The diameter of the fireball increases, as does the explosive yield. In some cases, the size is large enough, so to reach the stratosphere. This is a key point regarding the dissemination of the fission or activation products in the environment, as we will see in the next section. In fact, should the fireball remain within the troposphere, the radioactive nuclei formed are deposited locally or regionally depending on the wind and meteorological conditions. However, once part of the fireball enters the stratosphere, the radioactive debris is globally dispersed within the corresponding hemisphere.

## 4.2.2   Radioactive Fallout

The distribution of radioactive debris produced in an atmospheric nuclear detonation depends not only on the explosive yield and the bomb design but also on the meteorological conditions and the direction of the wind at the time of the explosion. After detonation, radionuclides associate with solid particles in the atmosphere, previously present or removed from the soil. Large particles will rapidly fall out within the first hours, while intermediate or small particles will remain longer in the atmosphere, being distributed in the troposphere during the following few days after the detonation. While the first fraction is deposited locally, the second fraction will be distributed regionally depending on the local wind and meteorological conditions, especially rainfall. A strong latitudinal dependence of the distribution of this debris fraction can be found. In the event that the yield is large enough to make it part of the fireball entering the stratosphere, this radioactive debris fraction starts participating in the stratospheric dynamics and will spread globally throughout the hemisphere where the detonation occurred [29]. This fraction remains in the stratosphere for months and, as we will see in what follows, re-enters the troposphere following the exchange processes taking place between the troposphere and stratosphere. The radioactive composition of fractions is not uniform, as it depends on the refractory or volatile character of the radionuclides produced and their precursors [52]. According to (4.11) and Fig. 4.9, fireballs from W 100 kt remain in the troposphere, but when the yield is 500 kt, they almost completely enter the stratosphere (see Fig. 4.10 for a description of the atmospheric structure).

The so-called radioactive fallout is a compound of the fractions described above and is responsible for the massive reintroduction of extinct radionuclides into the environment.

Underground tests also introduce radionuclides into the atmosphere, but it depends on whether the explosion is vented or not. In this case, the major fraction of radionuclides produced remains, but venting can generate contamination mostly

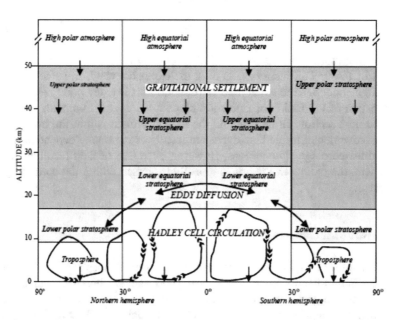

**Fig. 4.10** Structure of the atmosphere. Figure reprinted from [29] with permission from UNSCEAR

of the soils of the test site. The effects of soil resuspension become very important for the distribution of radionuclides in the environment. Some papers have documented that some contaminated soils from nuclear test sites in the Sahara Desert can arrive in Europe due to the combination of resuspension and specific meteorological conditions [53, 54].

Since the first tests, the presence of radionuclides in the atmosphere has been monitored in either rainwater or surface air samples. In Fig. 4.11, a long record of $^{137}$Cs concentrations is presented in air samples taken in the UK. The data cover the period from the very beginning of nuclear testing in 1953 until 1991 [55]. After 1963, the concentration of $^{137}$Cs globally decreased as the number and yield of atmospheric tests decreased. However, the rate of such decreases does not linearly correspond to the sudden decrease of the tests as a result of the Moscow treaty (see Fig. 4.8). As we shall see in the next paragraphs, once the radionuclides are injected into the stratosphere, their reincorporation into the troposphere is not immediate but follows an exponential law [56]. It is also observed in Fig. 4.11 that the general decrease in the $^{137}$Cs content is modulated by some activity intrusions related to the few atmospheric tests that took place after 1963 (see the behavior of the data series from 1970 to 1975). These specific contributions can easily be dated back to estimate the date of detonation. Indeed, since the fission yields of the radionuclides are known, it is possible to estimate the activity ratio, $R(0)$, of key radionuclides at the time of fission. Some examples are given in Table 4.12. Thus, the activity ratio at time $t$, $R(t) = \frac{A_A}{A_B}$, between two fission products $A$ and $B$ will be

$$R(t) = R(0)e^{(\lambda_B - \lambda_A)t} \qquad (4.12)$$

$\lambda$ is the decay constant of both radionuclides. In real samples, it is possible to measure such activity ratios, and specifically, the ratios $\frac{^{144}Ce}{^{137}Cs}$ and $\frac{^{95}Zr}{^{137}Cs}$ are very frequently used. By studying the behavior of the ratios over time, it is possible to find the time at which the theoretically expected ratio occurred in the atmosphere and to estimate the date of the explosion. These calculation exercises are presented in Fig. 4.11. Another interesting fact that can be found when analyzing atmospheric radioactivity data is the so-called spring effect [57]. Now, when studying the yearly evolution of the radionuclide concentration in air, a peak appears during the spring. It can be clearly observed in the 1978 data in Fig. 4.11 or in Fig. 4.12, taken from [58], where the spring effect for $^{239}$Pu is shown. The spring effect must be interpreted as a result of the stratosphere–troposphere exchange dynamics.

After 1963, the testing of nuclear devices in the atmosphere clearly decreased. Short-lived fission or activation radionuclides have rapidly disappeared from the environment. However, longer-lived radionuclides, such as $^{137}$Cs or $^{90}$Sr, remained, and it has been observed that their concentrations in the atmosphere decrease following an exponential decay. This can be seen in Fig. 4.11, where

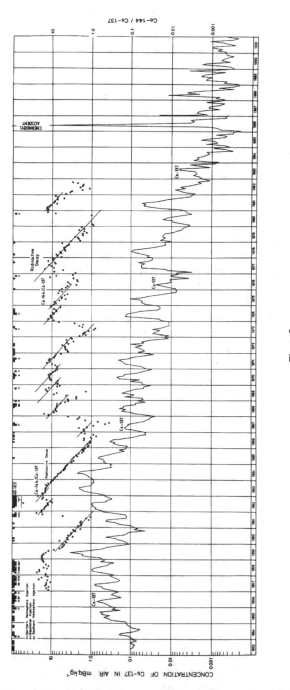

Ce-144 / Cs-137
CHERNOBYL ACCIDENT
Radioactive Decay
Ce-144/Cs-137
Cs-137
CONCENTRATION OF Cs-137 IN AIR mBq.kg-1
Figure 2
Concentrations of Cs-137 and Ce-144/Cs-137 Ratios in Air Near Ground Level at Chilton (1953-1991)

**Fig. 4.11** $^{137}$Cs in surface air up to 91. Together with the activity ratios of $^{144}$Ce/$^{137}$Cs or $^{95}$Zr/$^{137}$Cs. Figure reprinted from [55]

**Table 4.12**   Activity ratios at the time of fission of some fission products

| Radionuclides | Activity ratios |
| --- | --- |
| $\frac{^{90}Sr}{^{137}Cs}$ | 0.625 |
| $\frac{^{95}Zr}{^{137}Cs}$ | 145 |
| $\frac{^{99}Tc}{^{90}Sr}$ | $2.29 \times 10^{-4}$ |
| $\frac{^{99}Tc}{^{137}Cs}$ | $1.43 \times 10^{-4}$ |
| $\frac{^{140}Ba}{^{95}Zr}$ | 5 |
| $\frac{^{144}Ce}{^{137}Cs}$ | 30 |
| $\frac{^{141}Ce}{^{144}Ce}$ | 8.4 |
| $\frac{^{140}Ba}{^{141}Ce}$ | 2.9 |

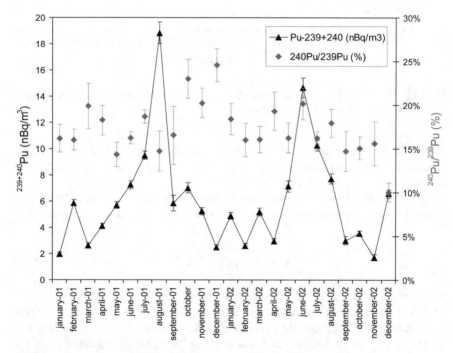

**Fig. 4.12**   Spring effect for $^{239}$Pu. Figure reprinted from [58] with permission from Elsevier

a general exponential decay dominates the yearly concentration of the radionuclides, despite the oscillations due to local inputs or spring effects. The exponential
decrease is evident when studying the concentrations of $^{14}$C content in tree rings
(see Fig. 4.13). Each ring represents a year of tree growth and integrates the
atmospheric $^{14}$C during this period of time [59]. The integration modulates the

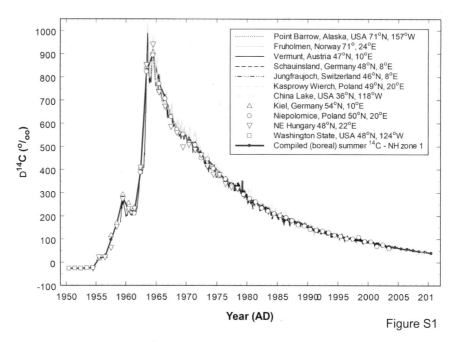

Figure S1

**Fig. 4.13** $^{14}$C in tree rings. The bomb pulse is clearly observed. Figure reprinted from [59] with permission from Cambridge University Press

variations that occur throughout the year, and exponential decay appears more clearly. This behavior can be found worldwide and corresponds to a pattern of radionuclide incorporation time from the stratosphere to the troposphere of exponential type [56]. This is the so-called residence time model in which the activity injected into the stratosphere at the beginning of $A(0)$ changes over time as

$$A(t) = A(0)e^{-(\lambda+k)t} \tag{4.13}$$

with $k = \frac{1}{\tau_s}$, where $\tau_s$ is the radionuclide stratospheric residence time. $A(t)$ is the activity at the time of observation, and $\lambda$ is the decay constant of the radionuclide. The stratospheric residence time, $\tau_s$, informs the removal time of the radioactive material from the stratosphere and is also related to the dynamic behavior of the atmospheric layers, as explained before. Residence times of 1–2 years [29, 60] have been found for different fission and activation products, and these estimations do not differ from the time constants measured for air masses or other materials injected into the stratosphere.

Either the spring effect or the removal time pattern from the stratosphere to the troposphere must be interpreted as a consequence of atmospheric dynamics. Figures 4.10 and 4.14 taken from [29] can help to understand these processes. The structure of the atmosphere and the main transport processes taking place are depicted in Fig. 4.10. In general terms, the atmosphere is divided into the strato-sphere and the troposphere. The borders between these two zones vary depending on the atmospheric region. Thus, in the polar region (from 90° to 30° latitude), the troposphere reaches some 9 km, while in the equatorial region (30° to 0° latitude), it reaches some 17 km. In fact, the tropopause, i.e., the troposphere limit, also varies with the season. Nevertheless, this is not necessary to be considered for the discussion presented here. The arrows indicate the prevailing transport processes among the different compartments that atmosphere can be divided. After detona-tion, the radioactive debris will be injected into the atmosphere. Equation 4.11 will give an idea of the different portions arriving at the troposphere and stratosphere. Radioactive debris injected into the troposphere can be distributed locally or regionally, while those arriving in the stratosphere can participate in global move-ments. Thus, they are distributed throughout the stratosphere, but essentially in the hemisphere where the detonation occurred. In Fig. 4.14, the exchange time pattern of material among the different atmospheric compartments is represented by the numbers in parentheses. They have been calculated by using different materials, including radionuclides, as tracers of air mass movements. According to [29], they are the removal half-time in moths of materials (radionuclides in our case) from one compartment to another. Four numbers are given in parentheses, each corre-sponding to a quarter of the year. The order of the quarters is March–April–May; June–July–August; September–October–November; December–January–February. For example, parentheses (8, 36, 24, 12) indicates a removal half-time for the first quarter of 8 months, 36 months for the second quarter, 24 months for the third, and 12 months for the fourth. In other words, if this parentheses represents the incorporation of radionuclides into the troposphere from the stratosphere in the Northern Hemisphere, the removal half-time is clearly shorter during the spring-time since half of the debris will fall into the troposphere in 8 months, while in the other seasons, the half-times are longer. A view of the picture confirms that this springtime effect also appears in the Southern Hemisphere, considering that in this case, the springtime covers the third quarter of the year. This is the explanation of the spring effect detected in Figs. 4.11 and 4.12. The removal times show that there is virtually no radionuclide exchange between the hemispheres where the removal half-times are very long. This explains how the release of radionuclides essentially affects the hemisphere where the detonation occurred. The radionuclide deposition in the Southern Hemisphere is an order of magnitude smaller than that in the Northern Hemisphere, as explosive yields have been remarkably lower [29].

As seen in Fig. 4.8, the number of atmospheric explosions and their yield dra-matically decreased after the 1963 ban treaty. It could be said that after such a year, stratospheric radioactivity injections were negligible. In Fig. 4.15, we present the deposition of $^{137}$Cs and $^{90}$Sr on the Earth's surface since the beginning of the nuclear tests. The peak corresponding to 1963 clearly dominates.

The cumulative deposition is also presented for both radionuclides in Fig. 4.16. Clearly, after 1963, the cumulative deposition also decreases as a consequence of the ban treaty but not suddenly as a result of (4.13).

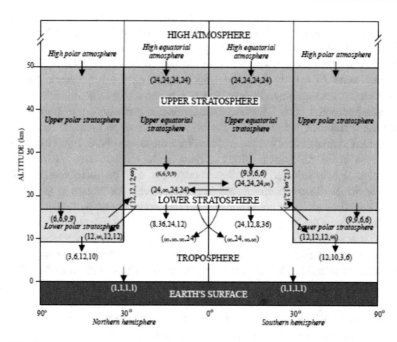

**Fig. 4.14**  Transport processes in the atmosphere and exchange mean times. Figure reprinted from [29] with permission from UNSCEAR

It is interesting to note the difference in deposition between the Northern and Southern Hemispheres. Because the number of tests in the Southern Hemisphere is very low compared to those that occurred in the Northern Hemisphere, the radionuclide deposition is smaller by one order of magnitude.

The large injection of 1963 can also be found in the human diet or even in human bones, as can be observed in Fig. 4.17 constructed with data from [61].

In some cases, this peak is named the bomb pulse. Somehow it could be said that Nature was marked with an outstanding radioactivity concentration in 1963. This can be used to date young sediments as a complementary method to that of $^{210}$Pb. In fact, once the radionuclide is injected into the troposphere, its residence time is approximately 1 or 2 months [59], and rapidly deposited on the surface of the Earth or on the surface of a lake. In this case, the radionuclide is removed toward the sediment and incorporated into each sediment layer. The sediment layer corresponding to 1963 is expected to contain a higher concentration than the rest and therefore could be clearly identified. This provides a time marker that, together with that of the sampling time, allows the calculation of an average sedimentation rate, $cmy^{-1}$, since the depth difference between both markers can be easily determined. Dating using this method is possible with several man-made radionuclides [62]. The most commonly used is $^{137}$Cs since it can be easily measured by γ-spectrometry. In Figs. 4.18a, b, some examples are given on the sediment profiles of $^{137}$Cs in a lacustrine environment. As in the case of $^{210}$Pb, the activity profile

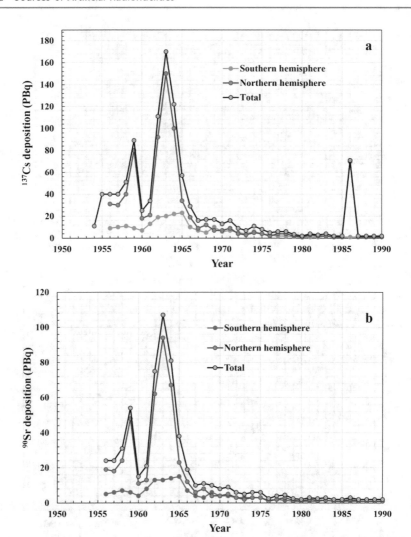

**Fig. 4.15** Global deposition of $^{137}$Cs (**a**) and $^{90}$Sr (**b**) over the Earth as a result of nuclear tests

can be modified by geological or biological processes, the chemical environment in the lake, or the geochemical behavior of the radionuclide that can make its diffusion likely in the sediment. Direct 'visual' dating of the sediment is generally difficult. Mathematical models are needed [62]. However, combining the information provided by $^{210}$Pb and these dating methods greatly helps to understand the time evolution of the studied system [9].

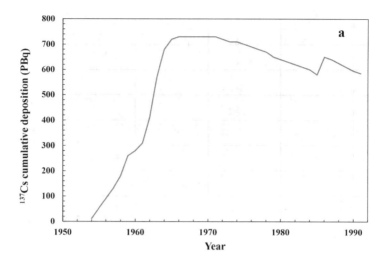

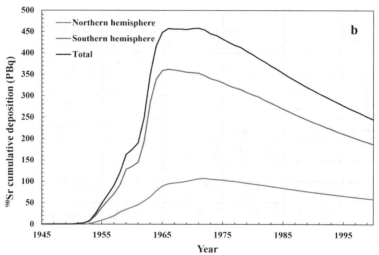

**Fig. 4.16**  Cumulative deposition of $^{137}$Cs (**a**) and $^{90}$Sr (**b**) over the Earth

## 4.2.3  Nuclear Fuel Reprocessing Plants

Nuclear fuel reprocessing contributes greatly to the presence of man-made radionuclides in the environment. It has been the main factor responsible for the reintroduction of artificial radionuclides in nature once the nuclear tests declined after 1963. To understand the role of reprocessing, it would be convenient to have a complete idea of the nuclear fuel cycle [63]. A simplified scheme is presented in Fig. 4.19. Nuclear fuel in a conventional thermal nuclear fission reactor is $UO_2$, where the isotopic abundance of the fissile isotope, $^{235}$U, is kept at the natural

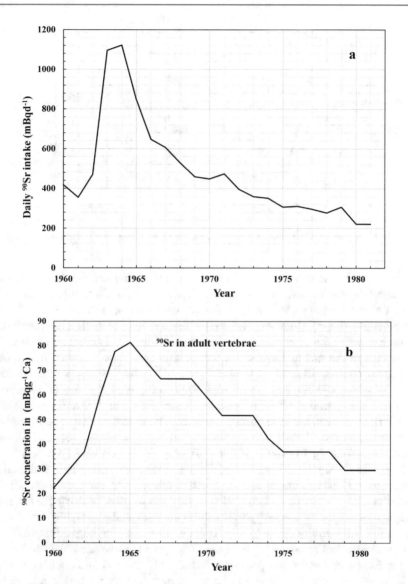

**Fig. 4.17**  $^{90}$Sr in the diet (**a**) and in human vertebrae (**b**) over time

level, 0.7%, or enriched to a few percent. This is the most common option, since enrichment makes it possible to reach useful energy production with a smaller fuel volume. After mining, U is converted to $UF_6$ prior to the enrichment process, in which the isotopic abundance of $^{235}U$ increases and the resulting U is produced as $UO_2$. Usually, the physical form of the nuclear fuel is a cylindrical ceramic pellet (1 cm in diameter and 1 cm in height) that is stacked into Zircaloy rods approximately 4 m long and 1 cm in diameter. Once the fuel is manufactured, it

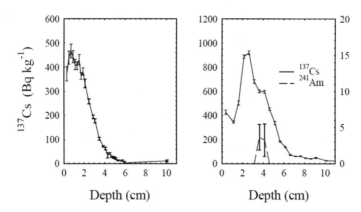

**Fig. 4.18** Examples of $^{137}$Cs activity profiles in lacustrine sediment. Figure reprinted from [9]

is transported to the nuclear power station, and the fission reaction is initiated by irradiating the nuclear fuel with neutrons. A fission chain reaction is produced as a result of the neutrons generated per fission, which are able to trigger the fission of the neighboring $^{235}$U nuclei. The consequence is the production of energy and fission and activation radionuclides, as we saw in Chap. 3. The fission chain reaction is kept in the fuel as the neutron flux is high enough to maintain it. As time passes, the concentration of $^{235}$U diminishes, but at the same time, the concentration of some nuclides with a very high neutron capture cross section increases. These are, for example, $^{135}$Xe ($\sigma = 2.665 \times 10^6$ barn) and $^{149}$Sm ($\sigma = 4.052 \times 10^4$ barn). Such nuclei tend to absorb a remarkable amount of fission neutrons. As the concentration of these neutron-poisoning nuclides increases, the neutron flux inside the nuclear fuel becomes too low to keep the fission chain reaction. This happens when a significant amount of $^{235}$U still remains in the fuel. This makes it economically interesting to recycle the fuel elements to extract the remaining $^{235}$U. For that, the spent fuel, which is highly radioactive, is carried to interim storage sites within the nuclear power station complex and allowed to cool down, i.e., let the radioactivity decay to an acceptable level for handling. After that, it is sent to nuclear fuel reprocessing facilities. The spent fuel contains $^{235}$U and $^{238}$U, but the fission and activation products, including $^{239}$Pu, a fissile nucleus of high strategic interest, can also be used to prepare new fuel elements. Of course, the extraction of $^{235}$U is a difficult process that can only be achieved in some countries and is carried out using a physical and chemical process in which the spent fuel elements are dissolved and purified from the fission and activation products. Some of the recovered radionuclides have applications in industrial processes, in research, or in medicine. However, some of the nuclides involved are unavoidable and are released into the environment. Furthermore, some gaseous or liquid residues are discharged into the environment according to radiation safety regulations [64]. Of course, most of the waste generated in the recycling process is temporarily or permanently deposited in radioactivity storage facilities.

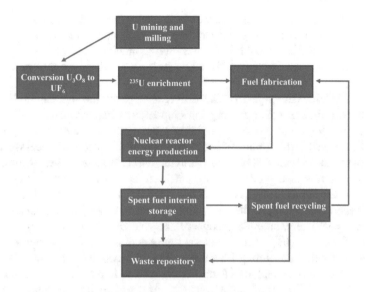

**Fig. 4.19** Schematic (and simplified) view of the nuclear fuel cycle

The composition of the spent fuels before recycling is varied. Of course, fission and activation products are present, but in a proportion that depends on the irradiation time of the fuel elements, the neutron flux, the neutron energy spectrum, the specific position of the element in the reactor, the original fuel $^{235}$U enrichment factor itself, the geometry of the fuel, the reactor burnup, the cooling time, and others. It can even vary for the same reactor and for the same fuel rod. However, some estimation can be done on the amount of radioactive products in spent fuel. Thus, for a burnup of 33 Gwd, approximately 34 kg of fission products [63] can be obtained per ton of U enriched with 3%.

A specific mention deserves the formation of actinides during the burning of nuclear fuel by neutron activation. $^{239}$Pu is a product of $^{238}$U neutron capture reactions and the subsequent $\beta^-$ decay of its daughters. Indeed,

$$^{238}\text{U} + n \rightarrow {}^{239}\text{U} \xrightarrow{\beta^-} {}^{239}\text{Np} \xrightarrow{\beta^-} {}^{239}\text{Pu}$$

Additionally, $^{239}$Pu can be produced by $^{238}$Pu neutron capture, which in turn is produced by $^{240, 241, 242, 243}$Pu can be formed by neutron capture of the preceding Pu isotope. $^{241}$Am is produced by $^{241}$Pu $\beta$-decay. $^{242,243,244}$Am are formed by the preceding neutron capture of the Am isotope. Cm isotopes are formed by the $\beta$-decay of Pu isotopes. For example, $^{242}$Cm is a product of the $\beta$-decay of $^{242}$Am. At the same time, Cm isotopes capture neutrons, and up to $^{246}$Cm is produced. Neutron activation is in fact the mechanism that leads to the production of actinides in nuclear fuel (see Fig. 3.5).

$$^{235}U + n \rightarrow {}^{236}U + n \rightarrow {}^{237}U \xrightarrow{\beta^-} {}^{237}Np + n \rightarrow {}^{238}Np \xrightarrow{\beta^-} {}^{238}Pu$$

Many of them are fissile and contribute to energy production despite their short half-life. However, the most significant are the Pu isotopes and specifically $^{239}$Pu ($T_{1/2} = 2.411 \times 10^4$ y), which is in fact used as nuclear fuel in a nuclear power station and for nuclear weapons. Thus, Pu isotopes produced in this way must be recovered to be used for energy production.

The environmental impact of the different steps of the nuclear fuel cycle is certainly very different. The mining, enrichment, and fuel fabrication of U mainly produce an operational radiological impact on workers, although care must be taken with regard to the storage of the byproducts of the different processes, as well as the tailing and waste sludge from U mining and milling activities. Reprocessing spent nuclear fuel also produces a significant radiological environmental impact, which is much higher than that of the other activities in the nuclear fuel cycle.

A detailed description of the spent fuel separation procedures of fission, activation, and, specifically, actinide products is beyond the objectives of this book. A good text for this is [63]. Throughout the reprocessing process, waste is generated. They are classified depending on their radioactivity content as Low-, Medium-, or High-Level Wastes and based on their physical forms as Solid, Liquid, and Gaseous Wastes. High-level solid waste is generated mainly during the dissolution of spent fuel elements. Dissolution residues contain activation products and nondissolved spent fuel pieces. These wastes are conditioned and stored at a special site. Of course, during the different operation steps, medium- and low-level solid wastes are produced that are classified according to the main emitted radiation, the physical shape and size, etc. After volume reduction and conditioning, they are stored at the disposal sites together with wastes originating from other activities.

High-level liquid wastes come essentially from the PUREX process. Plutonium uranium redox extraction (PUREX) is a conventional procedure for the separation of Pu and U isotopes from spent fuel (Ref) by using tributylphosphate (TBP) mixed with tetrapropylene (TPH) as the extracting agent. Basically, the previously dissolved spent fuel is mixed with TBP + TPH, and Pu and U are extracted into the organic phase. The remaining liquid phase contains fission products as well as Am, Cm, and, to a lesser extent, Np isotopes. Pu is backextracted with $HNO_3$, and the remaining organic phase still contains U, which is extracted by a subsequent $HNO_3$ backextraction. This process is repeated three times to recover as much Pu and U as possible. The remaining liquid phases are treated as high-level waste, especially those coming from the first separation, which is supposed to contain up to 99% of the original fission products in the spent fuel. These wastes are usually vitrified and stored in the same reprocessing plant.

Medium- and low-level liquid wastes are generated from different operations within the plant and are treated and conditioned before storage or, in some cases, when the activity is low enough, released into the environment according to national regulations.

Gaseous wastes appear in the pretreatment of the spent fuel prior to PUREX or other dissolution processes. In general, volatile radionuclides such as I and Ru isotopes are retained in the plant because the gas streams are purified before entering the stacks using charcoal, zeolite, or steel wool filters. Within noble gases, $^{85}$Kr ($T_{1/2} = 10.739$ y) is the most important, as the other Kr or Xe isotopes are very short-lived. $^{85}$Kr is retained in pressurized cylinders within the plant. Condensation of gases at $LN_2$ temperature and subsequent fractional distillation are carried out to separate Kr from the other gases.

$^3$H is also released during pretreatment of the spent fuel elements and their dissolution. The main form is HTO, although $^3$H$_2$ is also produced. $^3$H is difficult to retain, and part is released into the environment. There are procedures to trap $^3$H$_2$ or HTO. From an occupational point of view, the release of $^3$H inside the factories is an important issue. $^{14}$CO$_2$ is also released during the dissolution of the burned fuel. $^{14}$C is formed by neutron irradiation of the N present in the fuel elements. Some procedures also trap $CO_2$, but the release of $^{14}$C into the environment is significant, as we will see in what follows.

Spent fuel recycling plants are present in a number of countries, as seen in Table 4.13, where the main commercial facilities and some technical data on their activity are given. These plants cover not only the domestic needs of the country but also those of other countries [65]. As described above, the main part of the radioactive waste generated is stored at the site, although some releases are permitted according to national regulations. These limits have changed over time, as seen in [66]. A complete account of the liquid and gaseous releases over time from these reprocessing plants is provided in [29]. Some examples are given in Fig. 4.20. The release time pattern varies, although on a long-term basis, the released concentrations seem to have decreased over time.

**Table 4.13** Operating or planned spent nuclear fuel reprocessing facilities around the world. An estimation of the processing capacity of spent fuel mass (measured as tons of heavy metal, tHM) per year is also given

| Country | Facility | Activity | Capacity (tHMy$^{-1}$) |
|---|---|---|---|
| USA | Barnwell (Deferred)<br>Los Alamos Pu facility | Commercial<br>Laboratory | |
| Argentina | Ezeiza (Deferred) | Pilot plant | |
| Japan | JAEA Tokai<br>Rokkasho (Under construction) | Pilot plant<br>Commercial | |
| France | La Hague UP2-800<br>La Hague UP3 | Commercial<br>Commercial | 1000<br>1000 |
| China | Lanzhou (Under construction) | Pilot plant | |
| UK | NDA B205 Magnox reprocessing<br>NDA Thorp<br>NDA UKAEA Reprocessing Plant, Mox (Stand by) | Commercial<br>Commercial<br>Commercial | 1500<br>900 |
| North Korea | Radiochemical laboratory (Stand by) | Laboratory<br>Pilot Plant | |
| Russia | RIAR<br>RT1-Combined Mayak<br>RT2-Krasnoyarsk (Deferred) | Commercial<br>Commercial | 400 |

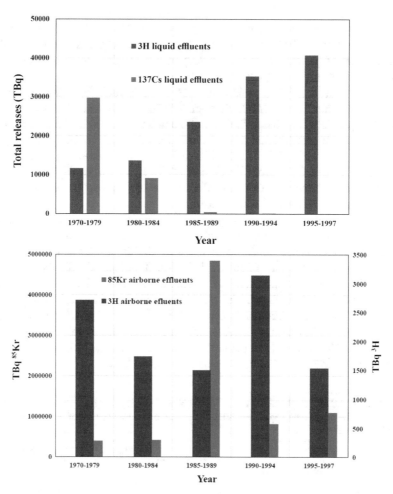

**Fig. 4.20** Examples of airborne effluents ($^3$H and $^{85}$Kr) and liquid effluents ($^3$H and $^{137}$Cs) effluents from nuclear recycling

The effluents of nuclear reprocessing plants produce local environmental impacts. Indeed, the $^{239+240}$Pu concentrations in various marine organisms taken on the coasts of the Irish Sea close to Sellafield can be seen in Fig. 4.21 [64], revealing a spatial distribution compatible with the wasting activities of the plant.

The impact can be global in nature, as has been found in many works on the determination of radioactivity in the environment. Examples are the study in [67], where $^{129}$I coming from Sellafield and La Hague is detected in atmospheric samples in southern Spain, and the study in [68], where $^{137}$Cs is detected in deep marine sediments collected in the Skagerrak area 20 km from the Swedish coast. In both cases, atmospheric and seawater circulations are responsible for the arrival of the activity to the sampling area from the release point.

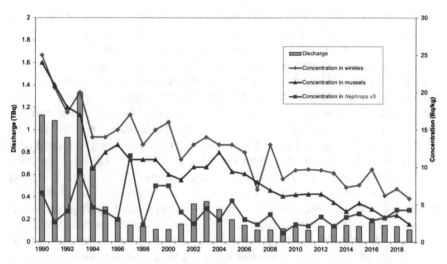

**Fig. 4.21**  Historic Sellafield $^{239}$Pu discharges and concentrations in winkles, mussels, and nephrops from the surroundings. Figure reprinted from [64]

### 4.2.4    Other Nuclear Facilities and Activities: Nuclear Power Plants

Nuclear power plants are also a source of artificial radionuclides to the environment. As of April 2020, there were 443 nuclear reactors operating in 30 countries worldwide with a total installed power of approximately 334 GWe [69]. Some 50 new power reactors are under construction, which will increase the installed power by an additional 53 GWe. The contribution of nuclear energy to electricity consumption varies and depends on the country. For example, on 31 December 2019, in France, nuclear energy contributed up to 71% of electricity consumption, while for example, in Iran, the contribution was approximately 2% [69].

Fission and activation products accumulate during the operation of nuclear power plants in amounts that depend on the reactor type and nuclear fuel composition (see Chap. 3). In general terms, the amount of fission products formed during nuclear fuel burning is much greater than that of activation products. Nevertheless, activation products are formed by neutron capture reactions of the impurities in the coolant and moderator and are therefore easily transported to the primary coolant and subsequently to liquid waste. Volatile fission products are released from the reactor core and transported by He used in the fuel elements to improve the heat transfer between the fuel and the fuel cladding. Diffusion of fission products through fuel cladding is also possible and therefore can contaminate the coolant, which can transport some radionuclides in dissolved or suspended form as well as in gaseous form. The purification of the coolant is carried out repeatedly to remove gases, suspended particles, and dissolved radionuclides. A gas stripper extracts the gases from the coolant, which are then filtered to recover the suspended particles.

For dissolved radionuclides, anion- or cation-exchange resins are used, which are able to remove dissolved fission or activation products from water to a high extent. This process is performed several times, and the waste is stored in the facility. In many cases, the coolant is reused. However, some of the liquid residues can be released into the environment following well-established regulations [29, 66]. However, coolant leakages are possible through valves, pumps, flanges, etc., that can reach the sumps and therefore enter the environment. Therefore, despite careful purification of the coolant, uncontrolled environmental releases are possible.

The gases produced in the reactor core or stripped from the coolant are passed through different filtration steps, stored in tanks, and allowed to decay. Most noble gases decay within 3 months after storage. The exception is $^{85}$Kr, which in some cases is released into the environment or condensed cryogenically and stored in the waste repository [63, 70]. However, some gas leakage occurs during the normal operation of nuclear reactors, and fission or activation gases are introduced into the reactor building. Periodically, the building is vented to allow workers to enter and carry out different operations. Consequently, these gases are released into the environment.

Therefore, despite the safety regulations within nuclear power plants, some radioactive releases occur in the environment. A full account of them can be found in [29]. In Table 4.14, some examples of airborne and liquid effluents from nuclear power plants are given.

As we will see in Chap. 6, there are radiological programs to control the environmental impact of nuclear power stations that include the measurement of fission and activation products in air, water, and biota in the surroundings of the facility. Typically, this is performed by γ-spectrometry, which provides a rapid and sensitive method for determining the radioactive impact. In special cases, when abnormal contamination of the environment is found, some more specific techniques, usually involving radiochemical separations, are used for non-γ-emitting radionuclides. In many cases, $^{14}$C is used as a marker of the historical impact of the installation in a close environment. Thus, $^{14}$C is measured in tree rings, which gives an idea of the time evolution of the radiological impact of the station in the

**Table 4.14** Some relevant radionuclides are released from NPP. The period spans from 1990 to 1997

| Radionuclide | Airborne releases (TBq) | Liquid releases (TBq) | Total (TBq) |
|---|---|---|---|
| $^3$H | 51.82 | 79.19 | 131.01 |
| $^{14}$C | 844 | | 844 |
| $^{131}$I | 1.034 | | 1.03 |
| Noble gases | 461.20 | | 461.20 |
| Other radionuclides ($^{137}$Cs, $^{90}$Sr, $^{99}$Tc, Pu isotopes, short-lived fission products...) | 103.50 (associated with particles) | 78.86 | 182.36 |

**Table 4.15**  Radionuclides used in medicine. Estimated activities from 1991 to 1996 worldwide

| Radionuclide | Therapeutic activity administered (MBq) | Diagnostic (MBq) |
| --- | --- | --- |
| $^{32}$P | 318 | 8662.5 |
| $^{99m}$Tc | | |
| $^{131}$I | 16,745 | |
| Other ($^{11}$C, $^{67}$Ga, $^{81m}$Kr, $^{123}$I, $^{127}$Xe, $^{128}$Au, $^{133}$Xe, ...) | | 14,373 |

environment. The radiological impact of nuclear power stations is essentially local, except in the case of accidents, as we will see in the next section.

The use of radionuclides in medicine practices is very common, as radio-pharmaceuticals are used mostly for patient diagnosis [71]. The commonly used radionuclides are short-lived, as the radiation dose on the patient must be minimized.

Many of them are fission products, such as $^{131}$I ($T_{1/2}$ = 8.05 d) or $^{133}$Xe ($T_{1/2}$ = 5.25 d), or derivatives, such as $^{99m}$Tc ($T_{1/2}$ = 6.01 h), which is a daughter of the fission product $^{99}$Mo and is by far the most used in diagnosis. More recently, some $\beta^+$-emitting radionuclides have been introduced in nuclear medicine for PET imaging [71]. For this, the pair of $\beta^+$ annihilation 511 keV photons emitted in opposite directions is used. In fact, differential attenuation of these $\gamma$-rays within the organs studied allows the formation of a tridimensional picture. The most commonly used currently are $^{11}$C ($T_{1/2}$ = 20.36 min), $^{13}$N ($T_{1/2}$ = 9.97 min), $^{15}$O ($T_{1/2}$ = 122.24 s), and $^{18}$F ($T_{1/2}$ = 109.77 min), which are produced in dedicated cyclotrons. Table 4.15 shows some estimations of the radionuclide amounts used worldwide in nuclear medicine. Of course, being all short-lived, the radiological environmental impact of their use is negligible. However, care must be taken with patients injected with radionuclides that could be considered a potential source of radionuclides to the environment and irradiation to family members. This is mainly applicable when $^{131}$I or $^{99m}$Tc has been used. For that, some protocols must be followed before their release from hospitals.

## 4.2.5  Nuclear Accidents

Since the beginning of the nuclear era, accidents have occurred in nuclear facilities. Some of them generated the dispersion of radioactivity in the environment. Their impact has been mostly local. Others, however, have reached a global dimension.

A list is included in Table 4.16 with the location of the accident, the affected area, and the main radionuclides released [72, 73]. The Chernobyl and Fukushima accidents have been the most important thus far and have released an important amount of radioactivity into the environment at a global scale. For that, two special paragraphs will be devoted to both accidents. Some comments follow on the most important accidents.

**Table 4.16** Nuclear accidents leading to relevant releases of radioactivity into the environment. See text for a more detailed description with location, affected area, and radionuclide released

| Accident | Date | Affected area | Radionuclides released |
|---|---|---|---|
| Windscale (UK) | October 1957 | Mostly local. Some findings in Germany and Norway | I isotopes, $^{89}$Sr, $^{90}$Sr, $^{210}$Po |
| SL-1 (USA) | January 1961 | Mostly local | Noble gases, $^{131}$I |
| Three Mile Island (USA) | March 1979 | Mostly local | $^{133}$Xe, $^{131}$I |
| SNAP 9A | April 1964 | Southern Hemisphere. Mostly Indian Ocean | $^{238}$Pu |
| Cosmos 954 | January 1978 | Northern Hemisphere. Mostly Canada | $^{90}$Sr, $^{131}$I, and $^{137}$Cs |
| Palomares (Spain) | January 1966 | Local. South-East Spain | Pu-isotopes, $^{235}$U, $^{3}$H |
| Thule (Greenland) | 1968 | Local. Greenland | Pu-isotopes, $^{241}$Am |
| Chernobyl (USSR) | April 1986 | Global. Northern Hemisphere | See Tables 4.17 and 4.18 |
| Fukushima (Japan) | March 2011 | Detected in many countries in Northern Hemisphere | $^{131}$I, $^{137}$Cs, see Tables 4.18 and 4.19 |

The accident of Windscale reactor number 1 occurred in October 1957. The Windscale facility, today Sellafield (see previous paragraph), was working by that time mostly on Pu production in two air-cooled nuclear reactors with graphite moderators. Neutron flux can cause the displacement of C atoms in graphite, breaking its crystalline structure. It is well known that during this process, some energy is stored in the C structure by the so-called Wigner effect [74], resulting in local temperature increases in the moderator that are dangerous since they could cause graphite combustion. The stored energy in this way must be liberated, and this is achieved by periodic annealing (heating) of the moderator. During annealing, the crystalline graphite structure is restored, and the Wigner energy is released, resulting in local temperature increases. Therefore, the annealing process must be carried out under very well-controlled conditions. In the case of the Windscale accident, part of the graphite moderator combusted during an annealing process since an anomalous temperature increase inadvertently took place. As a result, some 740 TBq of gaseous iodine was released into the environment, as well as some 40 TBq $^{137}$Cs, 4.8 TBq $^{89}$Sr, and 0.2 TBq $^{90}$Sr. Some studies indicate that approximately 9 TBq $^{210}$Po was also released into the environment as a consequence of the accident. Although the reactor was mostly dedicated to Pu production, it seems that it was also utilized for $^{210}$Po production by irradiating Bi with neutrons. $^{210}$Po is normally used as a source of $\alpha$ particles in $^{210}$Po-Be neutron sources (see Chap. 19), which serves as an n source in nuclear weapons. The accident also had some global dimension since radioactivity in the air was found in Germany and Norway.

**Table 4.17** Radionuclide activities released during the Chernobyl accident and the percentage released from the radionuclide inventory in the reactor core

| Radionuclide | Activity released (PBq) | Percentage (%) of the core inventory |
|---|---|---|
| *Gases* | | |
| $^{85}$Kr | 33 | 100 |
| $^{133}$Xe | 6500 | 100 |
| *Volatile elements* | | |
| $^{129m}$Te | 240 | |
| $^{132}$Te | 1150 | 10–60 |
| $^{131}$I | 1760 | 50–60 |
| $^{133}$I | 910 | |
| $^{134}$Cs | 47 | 33±10 |
| $^{136}$Cs | 36 | |
| $^{137}$Cs | 85 | 33±10 |
| *Intermediate volatility elements* | | |
| $^{89}$Sr | 115 | 3.5–4.5 |
| $^{90}$Sr | 10 | 3.5–4.5 |
| $^{103}$Ru | >168 | 3.5–6 |
| $^{106}$Ru | >73 | 3.5–6 |
| $^{140}$Ba | 240 | 3.5–6 |
| *Refractory elements* | | |
| $^{95}$Zr | 84 | 3.5 |
| $^{99}$Mo | >72 | 3.5–6 |
| $^{141}$Ce | 84 | 3.5 |
| $^{144}$Ce | 50 | 3.5 |
| $^{239}$Np | 400 | 3.5 |
| $^{238}$Pu | 0.015 | 3.5 |
| $^{239}$Pu | 0.013 | 3.5 |
| $^{240}$Pu | 0.018 | 3.5 |
| $^{241}$Pu | 2.6 | 3.5 |
| $^{242}$Pu | $4 \times 10^{-5}$ | |
| $^{242}$Cm | 0.4 | 3.5 |

Another important accident occurred in January 1961 at the USA Army Low-Power Reactor SL-1. The reactor was a 3 MWt boiling water reactor type that was shut down for maintenance in December 1960. After maintenance, the reactor resumed full power again in January 1961. The causes of the accident are not known today, but it appears that a control rod was improperly removed for unknown reasons. As a consequence, the reactor reached critical conditions, and

**Table 4.18** Comparison of global radionuclide releases from Chernobyl, nuclear tests, and the Fukushima accident

| Radionuclide | Chernobyl releases (PBq) | Global nuclear weapon releases (PBq) | Fukushima releases (PBq) |
|---|---|---|---|
| $^{3}$H | | 186,000 | |
| $^{14}$C | | 213 | |
| $^{54}$Mn | | 3980 | |
| $^{55}$Fe | | 1530 | |
| $^{89}$Sr | 115 | 117 000 | |
| $^{90}$Sr | 10 | 622 | 0.14 |
| $^{91}$Y | | 120,000 | |
| $^{95}$Zr | 84 | 148,000 | |
| $^{103}$Ru | >168 | 247 000 | |
| $^{106}$Ru | >73 | 12 200 | |
| $^{125}$Sn | | 741 | |
| $^{131}$I | 1760 | 675,000 | 158 |
| $^{140}$Ba | 240 | 759,000 | |
| $^{141}$Ce | 84 | 263 000 | |
| $^{144}$Ce | 50 | 30 700 | |
| $^{137}$Cs | 85 | 948 | 15.3 |
| $^{239}$Pu | 0.013 | 6.52 | $3.2 \times 10^{-5}$ |
| $^{240}$Pu | 0.018 | 4.35 | $3.2 \times 10^{-6}$ |
| $^{241}$Pu | 2.6 | 142 | |

**Table 4.19** Fukushima releases to the sea

| Radionuclide | Activity (PBq) | |
|---|---|---|
| | Direct release | Atmospheric deposition |
| $^{3}$H | 0.3–0.7 | |
| $^{90}$Sr | 0.04–1 | |
| $^{129}$I | 7–8 | |
| $^{131}$I | 9–13 | 57–100 |
| $^{137}$Cs + $^{134}$Cs | 3.5–5.6 | 5–11 |
| $^{239}$Pu + $^{240}$Pu | Negligible | |

an explosion occurred. Some 3 and 370 TBq $^{131}$I and noble gases, respectively, have been estimated to reach the atmosphere from the explosion.

The Three Mile Island Unit 2 accident occurred in March 1979. Unit 2 was an 850 MWt PWR-type reactor. A failure in the water pumps in the steam generating system causes an increase in the reactor core temperature and the opening of a relief valve in the reactor core. The safeguard system immediately ordered closing the valve nut to which it was stuck. Consequently, the coolant was draining, and the temperature of the reactor core increased. The emergency water pumps started working, but the valves were closed, and no water reached the core. As the temperature continued to increase, a redundant water pump system started to send water to the reactor core. Without apparent reason, a reactor operator manually shut down one of the pumps and reduced the water flux of the other. This meant that the core temperature continued to increase as the water level decreased, the fuel cladding failed, and the fuel started melting. Consequently, the temperature, the Zr fuel cladding started reacting with water coolant to produce $H_2$, which escaped to the reactor building, giving rise to an explosion. Six hours after the beginning of the accident, the core was covered by water. Some 370 PBq noble gases, mainly $^{133}Xe$, and 570 GBq $^{131}I$, are considered to have escaped into the atmosphere.

Some satellite and aircraft accidents have also been of environmental significance since radionuclides have been dispersed into the environment as a consequence. In April 1964, the SNAP 9A satellite failed during its reentry into the atmosphere. The satellite was powered by a 0.63 PBq $^{238}Pu$ ($T_{1/2} = 87.7$ y) generator. The accident occurred at a height of 46 km over the Indian Ocean, and $^{238}Pu$ was injected into the stratosphere, which was distributed globally, as described in the preceding sections. The majority of the activity was injected into the Southern Hemisphere, and it has been found that approximately 95% of the total $^{238}Pu$ was deposited onto the Earth's surface by 1970. By January 1978, the Cosmos 954 satellite re-entered the atmosphere over the Canadian territory. The satellite was powered by a nuclear reactor presumably containing approximately 20 kg of highly enriched U. The failure during reentry led to the dispersion of radioactive debris along the Canadian area, and it is estimated that approximately 95% of the reactor content remained in the stratosphere. Some estimations show [29] that the reactor contained 3.1 TBq $^{90}Sr$, 180 TBq $^{131}I$, and 3.2 TBq $^{137}Cs$ at the time of reentry. In January 1966, a B-52 bomber and a KC-135 tanker collided during a refueling operation. The bomber crashed into several pieces, and four hydrogen thermonuclear bombs were felt over Palomares in southeastern Spain [75]. Two bombs were recovered intact, while the other two suffered a conventional explosion when they hit the ground and provoked an extensive dispersion of Pu over a 2.5 km$^2$ area. According to the first estimations, the $^{239+240}Pu$ activity dispersed was within a range from 3.5 to 32 TBq (1.5 to 13.5 kg Pu). In addition, $^{235}U$ and $^3H$ were supposed to also be distributed correspondingly. In fact, some measurements showed the presence of abnormal $^{235}U$ activities in soil samples collected from the contaminated area several decades later. However, from a radiological point of view, neither $^{235}U$ nor $^3H$ was considered significant since the presumable dispersed $^{235}U$ was very low and the half-life of $^3H$ was very low to be considered a problem as time passed. Two years later, in 1968, a similar accident took place in Greenland, where a B-52 aircraft crashed 12 km from the Thule air base in Greenland. The aircraft carried 4 thermonuclear bombs and 1.4

TBq $^{239+240}$Pu (approximately 0.5 kg Pu) was considered scattered in the close environment [76].

*Chernobyl Accident*

The Chernobyl accident occurred on April 26, 1986, in Unit 4 of a nuclear complex situated in Chernobyl [72]. The reactor was a 1GWe water-cooled graphite moderated of RBMK type after the Russian acronym. The reactor consisted of a 12-m-diameter graphite cylinder and a 7 m height. Fuel was 115 kg of 2% enriched U distributed in 1661 fuel elements that were introduced vertically into the graphite cylinder. Twenty channels allow the control rods to also pass vertically, and finally, the coolant, light water, was introduced along the fuel channels from the top side of the reactor core. The hot water with some voids was then passed to the steam separator, and the resulting steam was passed to the turbine generators. The reactor core was not surrounded by a steel vessel, and the whole reactor was not inside a containment building. Both features were very much criticized by experts, as were some drawbacks in the design of this type of reactor. The reactor was inside a conventional building. The accident was the consequence of some failures resulting from the violation of some operating rules. In fact, a routine shutdown of Unit 4 was programmed, but before a test on the possibilities that the inertial working of the turbines would provide enough energy to feed the emergency control systems of the plant. For that, the power was reduced from the maximum 3.2 GWt to approximately 0.7 – 1 GWt of the power level programmed for the  test. Additionally, the emergency cooling system and the automatic reactivity control systems were disconnected. Everything was done according to the test protocol. However, electric power was reclaimed by the authorities, and Unit 4 was kept working for some 9 h in these conditions, something that did not follow the safety rules. The production of $^{135}$Xe, which, as already described, has a very high neutron capture cross section, made it difficult to manually control the power of the reactor in the absence of the automatic system. In fact, the power decreased to 30 MWt and then stabilized at approximately 0.2 GWt. In this situation, the reactor should have been shut down. Instead, further manipulation was performed to increase the void fraction in the coolant, which led to an increase in the thermal power until the generated steam exploded. The continuous increase in temperature caused the steam to react with the zircalloy fuel cladding, and a second explosion, this time due to the liberated H, occurred. The reactor cover and the containment building roof were seriously damaged by the explosion, and the incoming air started graphited combustion that lasted for two weeks before it was suffocated. A direct consequence of the explosion was the release to the environment of fission and activation products from the reactor core but also some pieces of the core that were ejected in the explosions [73]. The release of radioactivity occurred during the explosions, but it did not stop there since graphite combustion continued to inject radionuclides into the atmosphere. In fact, it is considered that radioactivity releases extended over some 10 days after the explosions with two intense episodes, the first corresponding to the initial explosions and the second one week later due to an exposition of the damaged reactor fuel to oxidation conditions. The releases consisted of gases, vapors, aerosols, and fuel fragments (radioactive, 'hot' particles, see Chap. 13). After the accident, several estimations have been made on the total activity released. In Table 4.17, the activities released in the accident that correspond to the estimated percentage of radionuclide inventory in the reactor core are presented [77, 78]. It is estimated that approximately 100% of the Noble gases were emitted into the environment, while 50–60% of $^{131}$I was released. Approximately 33% of $^{134}$Cs and $^{137}$Cs was also injected into the environment. Both $^{131}$I and $^{137}$Cs were the most relevant radionuclides from a radiological point of view, and some calculations were made on the collective radiation dose received by the population due to them. For the rest of the radionuclides, somewhere between 3 and 6% was supposed to be released, except for $^{132}$Te, where the estimation was 10 and 60%. However, its low half-life (3.204 d) makes it radiologically irrelevant.

Meteorological conditions varied greatly during the 10-day period of radioactivity emissions from Unit 4. The radioactive plume was consequently detected in different countries, depending on them. The first countries to report the arrival of radioactivity were Finland and Sweden just one day after the accident. Germany and Poland reported radioactivity during the following days. Subsequently, radioactivity was detected on the first days of May in Japan, China, Canada, India, and the USA.

It can be said that virtually all of the Northern Hemisphere was affected by the Chernobyl accident [79]. A map showing the dispersion of the radioactive plume over Europe is given in Fig. 4.22.

Of course, the main contaminated areas are located in Ukraine and the European zone of the former USSR. It has been estimated that approximately 1–2 EBq [29] was the total radioactivity released during the accident. The main radionuclides released are those presented in Table 4.17.

Some 56% of the $^{137}$Cs was deposited within the former USSR, 40% in Belarus, 35% in the Russian Federation, 24% in Ukraine, and the remaining 1% in other former USSR areas. Approximately 34% of $^{131}$I was dispersed around the same regions. The rest have been distributed around the world. In Table 4.18, a comparison is made on the radionuclide deposition from the Chernobyl accident and the global fallout for key radionuclides. Regarding $^{137}$Cs, the input from Chernobyl has meant some one-tenth of the global fallout. Some authors have proposed that some radioactive releases from Chernobyl reach the stratosphere, giving rise to a global dispersion of radioactive debris [80, 81]. An important feature of the accident was the release of radioactive particles [72, 73]. These particles were either fuel fragments or basically compounds of Ru. It is estimated that some 3.5% of the fuel was injected into the environment as a result of the first explosions. The particles were rich in refractory elements and depleted in volatile elements. They were found in the surrounding environment of the Chernobyl facility but also in close countries or regions such as Poland, Bulgaria, Hungary, Scandinavia, and Greece. The smallest fuel fragment particles were approximately 10 μm in size, which explains why most samples were found in a radius not more than a few tens of kilometers from the accident center. The Ru particles were supposed to have

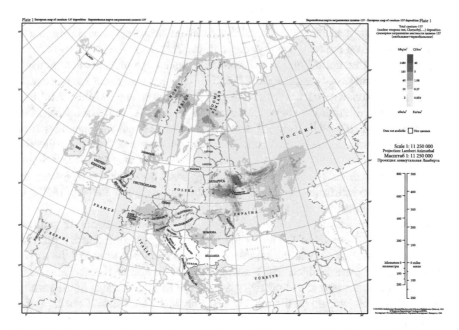

**Fig. 4.22** Chernobyl radioactive plume over Europe. With courtesy of De Cort et al. (1998): Atlas of Caesium Deposition on Europe after the Chernobyl Accident; EUR report nr. 16,733, EC, Office for Official Publications of the European Communities, Luxembourg

been formed by the condensation of metallic Ru in the atmosphere resulting from the reduction of Ru oxide vapors emitted from the reactor. This explains why they were found throughout Europe.

*The Fukushima Accident*

The Fukushima nuclear accident has been the second most important accident in the civilian use of nuclear energy. On March 11th, 2011, a great earthquake of 9 intensity took place with an epicenter 150 km offshore Sendai, east of Honshu, the main island of Japan. The earthquake was followed by a series of tsunamis with waves that reached 39 m in some areas. Some 20,000 lives were lost as a result of the earthquake and tsunamis.

Four nuclear power plant housing 11 nuclear reactors were in operation in the area. Soon after the earthquake, all reactors were automatically shut down, and the cooling systems started to work to remove the decay heat produced in the reactors by the radioactive decay of the fission products [73]. It can be said that the damages caused by the earthquake itself were not significant for the operating units and that the safety protocols worked. However, a large tsunami, 14 m high, impacted the Fukushima Daiichi Nuclear Power Plant 45 min after the quake, overpassing the tsunami barrier of the plant, which was approximately 6 m high. This is although the Fukushima Nuclear Power Plant was located 11 km from the coast and 10 m above sea level.

The Fukushima Daiichi Nuclear Power Plant hosted six BWR-type reactor units. Unit 1 provided 0.5 GWe, Units 2–5, 1.8 GWe, and 1.1 GWe Unit 6. At the time of the earthquake, Units 4–6 were shut down, and only Units 1–3 were in operation. The tsunami overflooded the Fukushima facility, submerging and damaging emergency cooling systems and emergency diesel generators, except one of them. Additionally, the tsunami caused an electrical shutdown in the area and damaged roads and ways, making it extremely difficult to reach the nuclear power plant. At the time of failure, approximately one hour before the shutdown of the reactors, the thermal power in the units in operation would have been reduced to approximately 1.5% of the nominal power. This means some 22 MWt in Unit 1 and 33 MWt in Units 2 and 3. Under these conditions, the heat is high enough to generate steam and to make the production of H very likely from the reaction of the hot steam with the Zr of the nuclear fuel cladding. In Unit 1, the water level of the coolant decreased to the top of the fuel, which increased the temperature, and part of the fuel melted. The pressure inside the reactor vessel increased rapidly, and venting was ordered. Thus, steam, radioactive gases, and H were released from the reactor vessel. However, due to the lack of electrical power, the liberated gases remained in the reactor building and did not reach the stack prepared for these events. As a consequence, an H explosion occurred on March 12th, blowing off part of the building roof. Several venting processes avoided the same fate as Units 2 and 3, although this boosted the injection of radioactivity into the environment. However, a H explosion also occurred in Unit 4, which in turn was shut down by that time on March 15th. The explosion was probably due to the intrusion of H in Unit 4 from the venting procedures in Unit 3. It is estimated that approximately 800–1000 kg H were produced in the units. Throughout 2011, it was possible to cool the reactors by the introduction of water, and the hydrogen was removed by injecting nitrogen into the building. In mid-January 2012, the temperature in the reactor cores ranged from 27 to 54 °C.

Fuel melting occurred in Unit 1 but also in Units 2 and 3. Thus, the fuel basically remained in the reactor cores. Thus, the radioactivity released consisted of volatile radionuclides that were injected into the atmosphere by venting processes and explosions. In addition, soluble species were introduced into the environment through leaks of cooling water, especially in Unit 2, which showed breaches in the containment building.

Releases to the atmosphere follow the sequence of venting procedures and explosions. It started on 11 March with two peaks during the H explosions and a rapid decline during the next two weeks. However, atmospheric releases persisted for several weeks. It is estimated [82] that between 100 and 500 PBq $^{131}$I and 6 to 20 PBq $^{137}$Cs have been injected into the atmosphere as a consequence of the accident. This means between 2 and 8% of the total $^{131}$I inventory and 1 to 3% for $^{137}$Cs in the fuel. This was 10% and 20% of the Chernobyl atmospheric input for $^{131}$I and $^{137}$Cs, respectively. The dispersion pattern depended on meteorological conditions, with wind directions changing during the same day and rainfall episodes provoking substantial deposition. The Fukushima accident impacted the marine environment, as the Fukushima complex is located close to the seashore. Indirectly, the atmospheric releases, driven by the West and Southwest Ward winds, were partially deposited over the Pacific Ocean. Additionally, direct effluents from the plant were released into the sea first by accidental leakage from the damaged Unit 2 and then under controlled conditions from all the Units. In Table 4.19, elaborated from [82], an account of direct and indirect releases from May to April 2011 to the ocean is presented for the more relevant radionuclides. Radionuclides incorporated into the ocean have participated in seawater vertical and horizontal movements. First, they moved eastward driven by the Kuroshio and the Kuroshio extension current, and radionuclide traces were found along the North American coasts in 2012, 2014, and 2015. On the other hand, the incorporation of $^{137}$Cs into coastal marine sediments is a relevant radiological problem, since it could be a long-term source of $^{137}$Cs for the close environment. The inventory of $^{137}$Cs in coastal sediments is estimated to be between 40 and 230 TBq.

Finally, in Table 4.18, the releases of relevant radionuclides from global fallout and the Chernobyl and Fukushima accidents are presented.

# References

1. G. Cinelli, M. De Cort, T. Tollefsen (Eds.), European Atlas of Natural Radiation, Publication Office of the European Union, Luxembourg (2019). https://remon.jrc.ec.europa.eu/About/Atlas-of-Natural-Radiation.
2. A. Martínez-Aguirre, M. García-León, Natural radioactivity in the Guadalquiver river at the South of Spain. J. Radioanal. Nucl. Chem. **178**, 337–350 (1994)
3. A. Martínez-Aguirre, M.C. Morón, M. García-León, Measurements of U- and Ra-isotopes in rainwater samples. J. Radional. Nucl. Chem. **152**, 37–46 (1991)
4. R.B.M. Sparovek, J. Fleckenstein, E. Schnug, Issues of uranium and radioactivity in natural mineral waters. Landbauforschung Völkenrode **4**, 149–157 (2001)
5. A. Martínez-Aguirre, M. García-León, Natural radioactivity enhancement by human activities in rivers of the southwest of Spain. J. Radional. Nucl. Chem. Lett. **155**, 97–106 (1991)
6. M. Ivanovich, R.S. Harmon, Uranium series disequilibrium: applications to environmental problems in Earth sciences (Clarendon Press, 1982)
7. D. Scholz, D. Hoffmann, 230Th/U-dating of fossil corals and speleothems. Q. Sci. J. **57**, 52–76 (2008)
8. M. Eisenbud, T.F. Gesell, *Environmental Radioactivity from Natural, Industrial and Military Sources*, Fourth Edition (Academic Press, 1997)
9. P.G. Appleby, Radiometric dating of sediment records in European mountain lakes. J. Limnol. **59**(Suppl. 1), 1–14 (2000)
10. P.G. Appleby, F. Oldfield, The calculation of $^{210}$Pb dates assuming a constant rate of supply of unsupported $^{210}$Pb to the sediment. Catena **5**, 1–8 (1978)
11. A. Arias-Ortiz, P. Masqué, J. García-Orellana, O. Serrano, I. Mazarrasa, N. Marbà, C.E. Lovelock, P.S. Lavery, C.M. Duarte, Biogeosciences **15**, 6791–6818 (2018)
12. J.A. Robbins, Geochemical and geophysical applications of radioactive lead, in *Biogeochemistry of Lead in the Environment*. ed. by J.O. Nriagu (Elsevier, 1978), pp.285–393
13. M. García-León, A. Martínez-Aguirre, R. Periánez, J.P. Bolívar, R. García-Tenorio, Levels and behaviour of natural radioactivity in the vicinity of phosphate fertilizer plants. J. Radioanal. Nucl. Chem. **197**, 173–184 (1995)
14. A. Martínez-Aguirre, M. García-León, Natural radioactivity in the Guadalquivir river at the South of Spain. J. Radioanal. Nucl. Chem. **178**, 337–350 (1994)

15. M. Eisenbud, T.F. Gesell, *Environmental Radioactivity from Natural, Industrial and Military Sources*, Fourth Edition, Chapter 6 (Academic Press, 1997)
16. L.C. Scholten, Approaches for regulating management of large volumes of waste containing natural radionuclides in enhanced concentrations, KEMA Nuclear report 40643-NUC 95-5354 (1995)
17. K.P. Severin, *Energy Dispersive Spectrometry of Common Rock Forming Minerals* (Springer, 2004)
18. M.S. Potgieter, Solar modulation of cosmic rays. Living Rev. Solar Phys. **10**, 3 (2013)
19. J.C. Gosse, F.M. Phillips, Terrestrial in situ cosmogenic nuclides: theory and application. Quatern. Sci. Rev. **20**, 1475–1560 (2001)
20. W.F. Libby, *Radiocarbon Dating*, 2nd edn. (University of Chicago Press, 1955)
21. W.F. Libby, History of radiocarbon dating, in *Radioactive Dating and Methods of Low-Level Counting. Proceedings of the Symposium Vienna*, STI/PUB/152, IAEA, pp. 3–26 (1967)
22. W.F. Libby, Dating by radiocarbon. Acc. Chem. Res. **5**, 289–295 (1972)
23. H.E. Suess, Radiocarbon concentration in modern wood. Science **122**, 415–417 (1955)
24. H.E. Suess, Secular variations of the cosmic-ray-produced carbon 14 in the atmosphere and their interpretations. J. Geophys. Res. **70**, 5937–5952 (1965)
25. W.S. Broecker, A. Walton, Radiocarbon from nuclear tests. Science **130**, 309–314 (1959)
26. P. Reimer, W. Austin, E. Bard, A. Bayliss, P. Blackwell, C. Bronk-Ramsey, S. Talamo, The IntCal20 Northern Hemisphere radiocarbon age calibration curve (0–55 cal kBP). Radiocarbon **62**, 725–757 (2020)
27. http://intcal.org/blurb.html
28. https://www.nndc.bnl.gov/nudat3/
29. , UNSCEAR 2000 Report: Sources and effects of Ionizing Radiation, Vol. 1, Annex C
30. *INTERNATIONAL ATOMIC ENERGY AGENCY, Naturally, Occurring Radioactive Material (NORM V), Proceedings Series* (IAEA, 2008), pp. 1–28
31. N. Tsurikov, 'Trade in radioactive materials' in *INTERNATIONAL ATOMIC ENERGY AGENCY, Naturally, Occurring Radioactive Material (NORM V), Proceedings Series* (IAEA, 2008), pp. 437–454
32. A. Martínez-Aguirre, M. García-León, M. Ivanovich, 'The distribution of U, Th and $^{226}$Ra derived from the phosphate fertilizer industries on an estuarine system in Southwest Spain. J. Environ. Radioact. **22**, 155–177 (1994)
33. E.O. Darko, D.O. Kpeglo, E.H.K. Akaho, C. Schandorf, P.A.S. Adu, A. Faanu, E. Abankwah, H. Lawluvi, A.R. Awudu, Radiation doses and hazards from processing of crude oil at the Tema oil refinery in Ghana. Radiat. Prot. Dosim. **148**, 318–328 (2012)
34. T.P. Ryan, A. Janssens, E. Henrich, J.-L. Daroussin, Z.K. Hillis, E.I.M. Meijne, Indistries giving rise to NORM discharges in the Europena Union-a review, in *INTERNATIONAL ATOMIC ENERGY AGENCY, Naturally, Occurring Radioactive Material (NORM IV), IAEA-TECDOC-1472* (IAEA, 2005), pp. 169–185
35. UNSCEAR 2000 Report: Sources and effects of Ionizing Radiation, Vol. 1, Annex B
36. K. Mamont-Ciesla, B. Gwiazdowski, M. Biernacka, A. Zak, Radioactivity of building materials in Poland, in *Natural Radiation Environment*, ed. by K.G. Vohra, U.C. Mishra, K.C. Pillai, S. Sadasivan (Wiley Eastern Ltd., 1982), pp. 551–556
37. A. Absi, M. Villa, H.P. Moreno, G. Manjón, R. Periañez, Self-cleaning in an estuarine area formerly affected by $^{226}$Ra anthropogenic enhancements. Sci. Total Environ. **329**, 183–195 (2004)
38. J.P. Bolívar, R. García-Tenorio, M. García-León, Enhancement of natural radioactivity in soils and salt-marshes surrounding a nonnuclear industrial complex. Sci. Total Environ. **173–174**, 125–136 (1995)
39. M.J. Gázquez González, J.P. Bolívar Raya, R. Garcia-Tenorio Garcia-Balmaseda, Rafael, F. Vaca Galán, A. Mena, Use of the red gypsum industrial waste as substitute of natural gypsum for commercial cements manufacturing. Mater. de Construc. **60**, 183–198 (2011)
40. J.M. Abril, R. García-Tenorio, S.M. Enamorado, M.D. Hurtado, L. Andreu, A. Delgado, The cumulative effect of three decades of phosphogypsum amendments in reclaimed marsh soils

from SW Spain: $^{226}$Ra, $^{238}$U and Cd contents in soils and tomato fruit. Sci. Total Environ. **403**, 80–88 (2008)

41. J.M. Abril, R. García-Tenorio, R. Periáñez, S.M. Enamorado, L. Andreu, A. Delgado, Occupational dosimetric assessment (inhalation pathway) from the application of phosphogypsum in agriculture in South West Spain. J. Environ. Radioact. **100**, 29–34 (2009)

42. Z. Sas, W. Schroeyers, G. Bator, M. Soutsos, W. Sha, R. Doherty, T. Kovacs, Radiological risks associated with building materials and industrial byproducts, in *Naturally Occurring Radioactive Material (NORM VIII)/International Atomic Energy Agency Proceeding Series*, STI/PUB/1832, pp. 110–115 (2016)

43. Z. Papp, Z. Dezső, S. Daróczy, Significant radioactive contamination of soil around a coal-fired thermal power plant. J. Environ. Radioact. **59**, 191–205 (2002)

44. Y.M. Amin, M.U. Khandaker, A.K.S. Shyen, R.H. Mahat, R.M. Nor, D.A. Bradley, Radionuclide emissions from a coal-fired power plant. Appl. Radiat. Isot. **80**, 109–116 (2013)

45. M.S. Hamlat, H. Kadi, S. Djeffal, H. Brahimi, Radon concentrations in Algerian oil and gas industry. Appl. Radiat. Isot. **58**, 125–130 (2003)

46. A. Abbasi, H. Zakaly, M.H. Hesham, M. Hessien, Radon concentration in compressed natural gas and liquefied petroleum gas and its release range in residential houses. Radiochim. Acta **109**, 793–798 (2021)

47. D.O. Kpeglo, J. Mantero, E.O. Darko, G. Emi-Reynolds, A. Faanu, G. Manjón, I. Vioque, E.H.K. Akaho, R. García-Tenorio, NORM waste from Ghanaian oil production, in *Naturally Occurring Radioactive Material (NORM VIII)/International Atomic Energy Agency Proceeding Series*, STI/PUB/1832, pp. 261–266 (2016)

48. F. Bou-Rabee, A.Z. Al-Zamel, R. Al-Fares, H. Bem, Technologically enhanced naturally occurring radioactive materials in the oil industry (TENORM). A review. Nucleonika **54**, 3–9 (2009)

49. L. Al Attar, B. Safia, B. Abdul Ghani, J. Al Abdulah, Recovery of NORM from scales generated by oil extraction. J. Environ. Radioact. **153**, 149–155 (2016)

50. F. Trotti, C. Zampieri, E. Caldognetto, R. Ocone, A. di Lullo, L. Magro, G. Jia, G. Torri, A study concerning NORM in integrated steelworks, in *International Atomic Energy Agency, Naturally, Occurring Radioactive Material (NORM V), Proceedings Series* (IAEA, 2008), pp. 351–354

51. H. A. Enge, *Introduction to Nuclear Physics*, Chapter 14 (Addison-Wesley Publishing Company, 1966)

52. M. Eisenbud, T.F. Gesell, *Environmental Radioactivity from Natural, Industrial and Military Sources*, Fourth Edition, Chapter 9 (Academic Press, 1997)

53. O. Masson, D. Piga, R. Gurriaran, D. D'Amico, Impact of an exceptional Saharan dust outbreak in France: PM10 and artificial radionuclides concentrations in air and in dust deposit. Atmos. Environ. **44**, 2478–3248 (2010)

54. M.K. Pham, E. Chamizo, M. Lopéz-Lora, J. Martín, I. Osvath, P.P. Povinec, Impact of Saharan dust events on radionuclides in the atmosphere, seawater, and sediments of the northwest Mediterranean Sea. J. Environ. Radioact. **214–215**, 106157 (2020)

55. K. Playford, J. Toole, I. ADsley, Radioactive fallout in air and rain: Results to the end of 1991, AEA-EE-0498, DOE/RAS/93.003, AEA Technology (1993)

56. M. Pourchet, F. Pinglot, Determination of the stratospheric residence time from the total β activity of Antarctic and Greenland snows. Geophys. Res. Lett. **6**, 365–367 (1979)

57. R.P. Parker, J.O. Crookall, Seasonal variation and age of radioactive fall-out. Nature **190**, 574–576 (1961)

58. E. Chamizo, M. García-León, S.M. Enamorado, M.C. Jiménez-Ramos, L. Wacker, Measurement of plutonium isotopes, $^{239}$Pu and $^{240}$Pu, in air-filter samples from Seville (2001–2002). Atmos. Environ. **44**, 1851–1858 (2010)

59. Q. Hua, M. Barbetti, A.Z. Rakowski, Atmospheric radiocarbon for the period 1950–2010. Radiocarbon **55**, 2059–2072 (2013)

60. G. Bennett, G. Burton, Worldwide dispersion and deposition of radionuclides produced in atmospheric tests. Health Phys. **82**, 644–655 (2002)

61. Environmental Measurement Laboratory report, EML-412, US Department of Energy (1982)

62. O. Evrard, P.-A. Chaboche, R. Ramon, A. Foucher, J.P. Laceby, A global review of sediment source fingerprinting research incorporating fallout radiocesium ($^{137}$Cs). Geomorphology **362**, 107103 (2020)

63. G. Choppin, J. Rydberg, J.-O. Liljenzin, C. Ekberg, *Radiochemistry and Nuclear Chemistry*, Fourth Edition, Chapter 21 (Elsevier, 2013)

64. Sellafield Ltd. Discharges and Environmental Monitoring Annual Report 2019, Published 2020

65. Nuclear fuel cycle information system: a directory of nuclear fuel cycle facilities, 2009 Edition, IAEA-TECDOC-1613 (IAEA, 2009)

66. S. Van der Stricht, A. Janssens, Radioactive effluents from nuclear power stations and nuclear fuel reprocessing sites in the European Union, 1999–2003, Directorate-General for Energy and Transport Directorate H-Nuclear Energy Unit H.4-Radiation Protection (2005)

67. J.M. López-Gutiérrez, F.J. Santos, M. García-León, Ch. Schnabel, H.-A. Synal, Th. Ernst, S. Szidat, Levels and temporal variability of $^{129}$I concentrations and $^{129}$I/$^{127}$I isotopic ratios in atmospheric samples from southern Spain. Nucl. Instrum. Methods Phys. Res. B 223–224, 495–500 (2004)

68. J.M. Abril, M. García-León, R. García-Tenorio, C.I. Sánchez, F. El-Daoushy, Dating of marine sediments by an incomplete mixing model. J. Environ. Radioact. **15**, 135–151 (1992)

69. Nuclear Power Reactors in the World, 2020 Edition, IAEA-RDS-2/40 (IAEA, 2020)

70. G. Choppin, J. Rydberg, J.-O. Liljenzin, C. Ekberg, *Radiochemistry and Nuclear Chemistry*, Fourth Edition, Chapter 20 (Elsevier, 2013)

71. UNSCEAR 2000 Report: Sources and effects of Ionizing Radiation, Vol. 1, Annex D

72. M. Eisenbud, T.F. Gesell, *Environmental Radioactivity from Natural, Industrial and Military Sources*, Fourth Edition, Chapter 12 (Academic Press, 1997)

73. V. Valkovic, *Radioactivity in the Environment*, Second Edition, Chapter 9 (Elsevier, 2019)

74. R.H. Telling, M.I. Heggie, Radiation defects in graphite. Philos. Mag. **87**(31), 4797–4846 (2007)

75. C. Sancho, R. García-Tenorio, Radiological evaluation of the transuranic remaining contamination in Palomares (Spain): a historical review. J. Environ. Radioact. **203**, 55–70 (2019)

76. H. Dahlgaard, M. Eriksson, E. Ilus, T. Ryan, C.A. McMahon, S.P. Nielsen, Plutonium in the marine environment at Thule, NW-Greenland after a nuclear weapons accident, in *Plutonium in the Environment*, ed. By A. Kudo (Elsevier, 2001), pp. 15–30

77. UNSCEAR 2008 Report: Sources and effects of Ionizing Radiation, Vol. 2, Appendix A

78. Chernobyl Ten years: Radiological and health impact. An Assessment by the NEA Committee on Radiation Protection and Public Health, OECD NUCLEAR ENERGY AGENCY, November 1995

79. UNSCEAR 2000 Report: Sources and effects of Ionizing Radiation, Vol. 2, Annex J

80. Z. Jaworowski, L. Kownacka, Nuclear weapon and Chernobyl debris in the troposphere and lower stratosphere. Sci. Total Environ. **144**, 201–215 (1994)

81. M. Aoyama, Evidence of stratospheric fallout of caesium isotopes from the Chernobyl accident. Geophys. Res. Lett. **15**, 327–330 (1988)

82. UNSCEAR 2020 report. SCIENTIFIC ANNEX B: Levels and effects of radiation exposure due to the accident at the Fukushima Daiichi Nuclear Power Station: implications of information published since the UNSCEAR 2013 Report

# Levels and Behavior of Environmental Radioactivity

**5**

**ABSTRACT**

In this chapter, the dynamics of radionuclides in the environment are studied. Thus, some general principles of radioecology and interesting parameters are introduced to study the transfer of radionuclides between the different environmental compartments. The levels and behavior of natural and man-made radionuclides in the environment are also reviewed in order to provide the reader with a general view on the presence of radioactivity in Nature.

## 5.1 Dynamics of Radioactivity in the Environment

### 5.1.1 General Concepts of Radioecology

Radioecology encompasses the analysis of the behavior and dynamics of radionuclides in the environment. Thus, the transfer of radionuclides among the different environmental compartments, as well as the transit times and distribution among them, is studied by radioecology. The bioavailability of radionuclides is also an important issue in radioecology, as it allows the assessment of radiological impacts on the biosphere and specifically on humans [1]. Radionuclide dynamics and fate are affected by the sources and transport scenarios, as well as the radionuclide chemical form of the molecules and their interactions with other environmental contaminants, which can lead to synergistic or antagonistic effects. Therefore, radioecology requires the information provided by different scientific disciplines. In turn, it is expected that radioecology also provides them with useful information [1]. In other words, the study of the environmental dynamics of radionuclides must integrate different scientific disciplines since many effects are involved. Thus Chemistry, Biology, Environmental and Nuclear Sciences take part in radioecology studies, but also Mathematics since mathematical modeling is also very powerful tool to cope several of the challenges posed by radioecology.

© The Author(s), under exclusive license to Springer Nature Switzerland AG 2022
M. García-León, *Detecting Environmental Radioactivity*, Graduate Texts in Physics, https://doi.org/10.1007/978-3-031-09970-0_5

## 5.1.2　Radionuclide Speciation in the Environment

Traditionally, radioecology studies have been based on the analysis of total radionuclide concentrations and their temporal and spatial evolution in the environment. However, it is known that the physical and chemical forms of radionuclides influence their mobility in the environment and hence their fate and bioavailability [1, 2]. So-called radionuclide speciation studies are a key point in radioecology as they provide the information needed to foresee the environmental evolution of radionuclides, including their interaction with the biosphere [2].

Radionuclides are expected to be present in different forms in the environment, as illustrated in Fig. 5.1 [1]. The physicochemical forms of radionuclides in water are described. Low-mass molecular forms are known to be more mobile and bioavailable in the environment. However, different processes can lead to changes in radionuclide form over time, reducing or increasing its mobility and bioavailability. This makes the study of radionuclide dissemination in the environment difficult; although in turn, it shows the necessity of carrying out speciation studies to thoroughly analyze the environmental behavior of radionuclides. In Table 5.1 [3], a more general classification of the different physicochemical forms that radionuclides can take in environmental compartments is presented.

As we will see in Chap. 13, there are several methods to separate radionuclides into their physicochemical forms, i.e., to conduct speciation studies of radionuclides. Methods start from sampling, which is of paramount importance to assess radionuclide species. Radionuclide fractionation studies are carried out on site

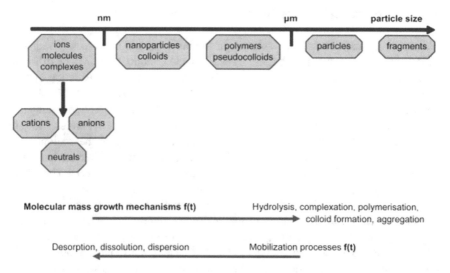

**Fig. 5.1** Physical and chemical forms of radionuclides in the environment. Figure reprinted from [1] with permission from Elsevier

**Table 5.1** Radionuclide speciation in different environmental compartments [3]

| Atmosphere | Water | Sediments, soils, rocks |
|---|---|---|
| Ions | Ions | Dissolved forms in interstitial water |
| Gaseous molecules | Ion pairs | Adsorbed forms |
| Aerosols | Molecules | Coprecipitated forms |
| | Colloids | Bound with organic matter |
| | Particulate | |

by filtration and ultrafiltration of waters or atmospheric aerosols or by sequential extraction procedures in the laboratory in the case of soils or sediments [4]. As stated above, the source term is crucial for radionuclide speciation. Chap. 4 provides an account of the different sources of natural and artificial radionuclides in the environment. Radionuclides from reactor effluents are in the form of colloids and low-mass molecular species. However, refractory radionuclides are in a good proportion released in the form of radioactive particles in nuclear accidents and weapon detonations. Radioactive particles, i.e., macroscopic particles containing radionuclides, will be studied in the next sections. The occurrence of radioactive particles has made it necessary to evolve traditional analytical procedures and include new techniques (see Chap. 20) to support speciation studies. An important aspect of radionuclide speciation studies lies in the study of future environmental scenarios. The long-term management of radioactive wastes is one of them. The eventual dissemination of radionuclides from radioactive waste deposits is determined by the physicochemical form of the radionuclide, which can change over time. Climate conditions indirectly influence the formation of environmental radionuclide species through changes in water reservoirs and the atmosphere that affect both soil weathering processes and radionuclide remobilization from sediments. How climate change affects these processes and, consequently, the diffusion of radionuclides in the environment is also a scenario that needs to be understood [1].

### 5.1.3 Exchange and Transport Processes. Transfer Parameters

The transport of radionuclides along the environment is driven by different processes depending on the compartment containing the radionuclides and their chemical forms. In waters, dissolved radionuclides are transported following the vertical and horizontal movements of water masses. In this case, the radionuclide is said to be conservative with respect to the water mass. Nonsoluble species do not follow the movements of the water mass but tend to interact with solid phases such as suspended matter and finally with sediment. Thus, some exchange takes place between the water and the suspended matter or sediment. In soils, radionuclides move vertically or horizontally according to their geochemical properties, weathering conditions, soil acid or basic nature, and the presence of organic matter. Biological displacements due to microorganisms or small living species, such as worms, are also very important in many cases. For sediments, whatever their kind, the mobility of radionuclides is boosted by the water–sediment interface and the penetrability of the water phase into the sediment. In marine sediments, the salinity of the water solubilizes some radionuclides [5]. In the case of lake sediments, the anoxic character of the waters influences the solubility of radionuclides from the sediments [6]. It is interesting to note that, within this context, sediments, which are considered sinks of radionuclides in water systems, can instead become sources. This can have radiological significance in some environments [7]. In any case, exchanges occur between the different environmental compartments [8]. The chemical form of radionuclides makes them available to biological systems in some cases. Thus, plants can absorb radionuclides from soils or from the atmosphere. Alternatively, terrestrial animals can take radionuclides from water by ingestion or from the atmosphere by breathing. Or fish can take radionuclides from surrounding waters, etc. [1].

Exchanges and transport processes drive radionuclides throughout the Earth ecosystem, and it is important to study the mechanisms governing such dissemination, including those addressing radionuclides to the biosphere and human beings.

Traditionally, several parameters have been defined to describe the exchange between environmental compartments. Thus, the exchange between water and suspended matter or between water and sediment is described on the basis of distribution coefficients, $K_d$, defined as the radionuclide concentration in the solid phase, $Bqkg^{-1}$, divided by the concentration in water, $Bql^{-1}$ or $Bqm^{-3}$. $K_d$ values can be found in the literature for different systems and radionuclides. However, the variability among the observed $K_d$ values is large [9]. This happens because $K_d$ defined in this way refers to equilibrium conditions in both phases, while the real situation in nature is far from being at equilibrium. In fact, the exchange between phases follows a kinetic process driven by time constants depending on several variables that are difficult to control in the laboratory and, of course, under natural conditions [10]. From this point of view, $K_d$ should be defined as time functions instead of constant values. Moreover, $K_d$ depends on the chemical and physical species of the radionuclide, i.e., it strongly depends on the chemical form and the molecular size. Even if the radionuclides are in radioactive particles, the measured $K_d$ will not be representative at all. In other words, $K_d$ should not be defined as the ratio between total radionuclide concentrations. Instead, a specific $K_d$ should be used as is done for modeling purposes. In Table 5.2, some $K_d$ coefficients can be found for the sake of comparison [9].

The transfer factor (TF) is used to describe exchanges between waters, sediments, or soils with vegetation and animal tissues [11]. It is defined as the total concentration of radionuclides in the vegetable, dry weight $Bqkg^{-1}$, divided by the concentration in the soil, $Bqkg^{-1}$. Some other parallel concepts are the transfer coefficient, $Bqkg^{-1}$ vegetation per $Bqm^{-2}$ soil or the concentration factor (CF), $Bqkg^{-1}$ animal per $Bqkg^{-1}$ fodder or $Bqkg^{-1}$ fish wet weight per $Bqm^{-3}$ water.

**Table 5.2** $K_d$ coefficients for different radionuclides [9]

| Radionuclide (element) | $K_d$ (lkg$^{-1}$) |
| --- | --- |
| U (soil) | 46–3,95,100 (sandy) |
| | 0.03–2200 (clayey) |
| U (sediment) | 0.8–332 (sandy) |
| | 8.6–100 (clayey) |
| Ra (soil) | 57–21,000 (sandy) |
| | 1262–5,30,000 (silty) |
| | 696–56,000 (clayey) |
| Ra (sediment) | 1.2–1700 (saline water) |
| | 5–38,000(fresh water) |
| Cs (soil) | 82–1,70,000 |
| Cs (sediment) | 540–2000 (marine) |
| | 17,000–51,000 (freshwater) |

These factors, as in the case of $K_d$, refer to total concentrations and not to bioavailable species, thus underestimating the real transfer. They also have seasonal variations [11]. Transfer factors and distribution coefficients although useful are not approximations of the real exchange process.

Today, efforts are being made to find transfer factors and distribution coefficients that more realistically describe the disequilibrium situation that occurs in nature, as well as the exchange of bioavailable species [1].

## 5.1.4 Mathematical Modeling

Mathematical modeling is a powerful tool to understand the dynamics of radionuclides in the environment. With models, it is possible to simulate the behavior of the radionuclide in the system. For that, some equations are used based on hypotheses that are tested with real experimental data. Nature is very complicated, and a mathematical model is just a simplification of reality. However, with a model, it is possible to select some of the involved processes and check how significant they are. However, models are of huge utility for decision-making purposes when evaluating the environmental impact of, for instance, an accidental release of radionuclides in a given environment.

The compartment or box model approach is the classical approach in which the studied system is divided into boxes that exchange radionuclides among them according to the transfer rates $k_{ij}$ (y$^{-1}$) involving boxes $i$ and $j$. An example is given in Fig. 5.2a, in which a marine environment in the North Sea is divided into boxes to study the dissemination of radionuclides from Sellafield. In the case of radionuclides, each box is supposed to contain an activity $A_i$ as a result of the exchange with other boxes and the presence of sources or sinks in the box itself [12]. Of course, $A_i$ also changes with time following its radioactive decay. These changes can be significant or not depending on the exchange times involved in the problem.

The time variation of $A_i$ is given by [13]

$$\frac{dA_i}{dt} = \sum_{j=1}^{n} k_{ji} A_j - \sum_{j=1}^{n} k_{ij} A_i - k_i A_i + Q_i \qquad (5.1)$$

This equation expresses the transfer of radionuclides between the $n$ boxes, which represent the whole system. Certainly $k_{ii} = 0$. On the other hand, $Q_i$ accounts for the presence of sources in the box, and $k_i$ accounts for that of sinks. Specifically, for radionuclides, $k_i$ can represent radioactive decay. Equation 5.1 is of application in the case where there is a uniform and instantaneous distribution of radionuclides in the box. Therefore, the validity of box models is very much

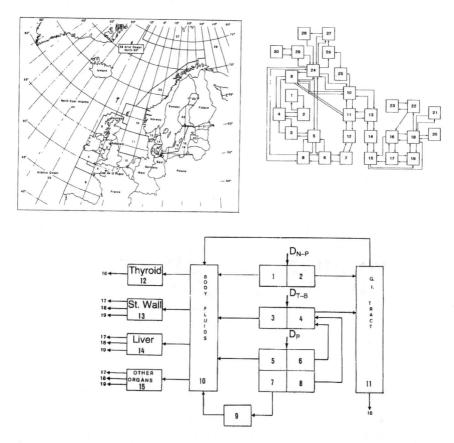

**Fig. 5.2** Two examples of box models for **a** the North Atlantic marine environment and **b** the human body. Figures reprinted from [12] with permission from Oxford University Press and from [14] with permission from Elsevier

dependent on the spatial and time scales involved in the system. In the case of marine simulations, it is frequent that spatial and time scales involve thousands of kilometers and years. The selection of box sizes is crucial in many cases. Equation 5.1 can include other effects such as radionuclide transfer between water and suspended matter or sediments, for example, in the case of a marine simulation. For that, the above defined $K_d$ values are used and included in the equation. Of course, it is expected that the times involved in radionuclide transport allow the use of equilibrium $K_d$. For deep waters, the box can be divided into layers to simulate vertical transport [13].

$k_{ij}$ are in some cases obtained by calibrating the model with experimental data [13]. Other approaches can be found from hydrological data [15], i.e., from known water streams in the area. On some occasions, the times involved are smaller than those needed to reach equilibrium conditions for exchange between the solid and water phases. This is the case for waters affected by tidal changes [16]. A dynamic approach is applied now in which $K_d$ is substituted by a kinetic coefficient, $k_1$, describing the exchange from the dissolved phase to the solid phase, and a kinetic coefficient, $k_2$, describing the opposite reaction [17]. A more realistic approach even takes into account the existence of different valence states of radionuclides, assigning a different coefficient for each of them [13].

Box models can be used for many purposes. A very common application lies in the calculation of the internal radiation dose received by humans after inhalation or ingestion of radionuclides (see Chap. 6). In Fig. 5.2b, a box model is depicted for the respiratory and gastrointestinal tracks [14]. Once incorporated through inhalation or ingestion, the radionuclide disseminates through the body according to some transfer rates determined by ICRP [18]. Radiation dose is also calculated by applying conversion factors that give the committed effective dose produced per activity unit of the incorporated radionuclide [19]. It is measured in $SvBq^{-1}$ (see Chap. 6).

Box models are easy to solve and very intuitive but have limitations. A more powerful approach is that of dispersion models in which radionuclide dissemination in the marine environment is solved as an advection/diffusion transport process. Advection represents radionuclide movement due to water circulation, while diffusion accounts for turbulent mixing phenomena producing local changes. In the case of conservative radionuclides, the time evolution of the radionuclide concentration C dissolved in the water is given by [13]

$$\frac{\partial C}{\partial t} + \frac{\partial uC}{\partial x} + \frac{\partial vC}{\partial y} + \frac{\partial wC}{\partial z} = K_h\left(\frac{\partial^2 C}{\partial x^2} + \frac{\partial^2 C}{\partial y^2}\right) + K_v\frac{\partial^2 C}{\partial z^2} \qquad (5.2)$$

$u$, $v$, and $w$ are the water speed components along the $x$, $y$, and $z$ directions, respectively, with $K_h$ and $K_v$ being the horizontal and vertical diffusion coefficients. In a first approximation, they can be assumed to be constant. In the case where the vertical transport is ignored or not relevant, (5.2) simplifies since both z terms disappear. Radionuclide sinks (such as decay rate $\lambda C$) and sources can be added to (5.2).

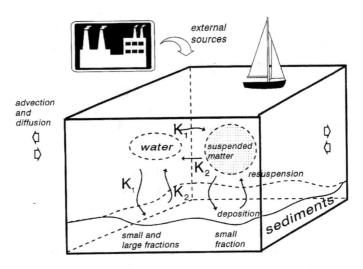

**Fig. 5.3** Exchange mechanisms for nonconservative radionuclides. Figure reprinted from [13] with permission from Springer

In the case of nonconservative radionuclides, i.e., when the exchanges between solid and water phases are important, the movement of suspended matter as well as the resuspension of sediments and the exchange between the different phases must be taken into account. Figure 5.3 illustrates the relevant mechanisms when nonconservative species are involved. In this case, additional equations describing the advection and diffusion transport of suspended matter must be included. The use of $K_d$ is possible if the time scales involved in the processes are compatible with the need to reach equilibrium between phases. In many cases, kinetic transfer coefficients, as defined before, are necessary [10].

The dissemination of radionuclides in sediments can also be studied by mathematical modeling. Sediments are expected to be a sink for radionuclides released in water systems. However, changes in environmental conditions could cause the remobilization of radionuclides from sediments, converting them into sources. On the other hand, resuspension effects are possible in dynamic systems, so the transport of sediment-borne radioactive material can be significant. Moreover, for sediment dating, as in the case of the $^{210}$Pb dating method (see Chap. 4), the transport of radionuclides along the sediment must be known to assign the correct date to each layer. The transport of radionuclides in sediments can also be treated as an advection–diffusion process, but the sediment is now considered a continuum medium. A certain activity input is supposed to have occurred in a given time, and its transport along the sediment column is modeled accordingly. In general terms, radionuclide transport depends on tree effects. First, the activity input diffuses along the sediment layers. Second, the sediment growths and a layer located at depth z will be displaced to deeper places as time passes. This transports activity along the sediment as well. Finally, the activity decays according to its decay constant, λ.

Following [20], the radionuclide activity in the z-layer, $A(z, t)$, changes with time as

$$\frac{\partial}{\partial t} A(z,t) = \frac{1}{\rho_m} \left( \frac{\partial}{\partial z} \left( k_b \rho_m \frac{\partial}{\partial z} A(z,t) \right) - \frac{\partial}{\partial z} (wA(z,t)) \right) - \lambda A(z,t) \qquad (5.3)$$

where $\rho_m$ is the dry matter density of the layer and $w$ is the mass sedimentation rate. $k_b$ is a diffusion coefficient that in general is $z$ dependent, although for the majority of specific applications, it can be considered a constant. Equation 5.3 has a general application to radionuclide transport problems in sediments. It can be simplified for particular cases. For example, for dating purposes very frequently, $\frac{\partial}{\partial t} A(z,t) = 0$ and $A$ change due to external inputs on the layer [21]. Equations 5.1, 5.2 and 5.3 have numerical solutions, and there are several methods to proceed with the calculations [13].

## 5.2    Levels and Behavior of Radioactivity in the Atmosphere

### 5.2.1    Radioactivity in the Air

The atmospheric structure was already described in Chap. 4. With regard to the dynamics and distribution of radionuclides, the important parts are the stratohsphere and the troposphere. The interactions between them were also described in Chap. 4, and it was found that both atmospheric layers exchange matter following a well-defined seasonal pattern.

Primordial radionuclides can be found in the troposphere as the result of direct inputs from industrial sources, the emanation of radioactive noble gases from the Earth or ocean surfaces, soil resuspension, or ocean spray transport. Cosmogenic radionuclides can be found in the stratosphere and the troposphere, since the nuclear reactions leading to their production can take place in both layers. Moreover, many of them are also produced during nuclear detonations or along the nuclear fuel cycle. Both sources contribute to its presence in the stratosphere and the troposphere. Artificial radionuclides can also be found both in the stratosphere and the troposphere as a consequence of nuclear detonations or tropospheric input from some steps of the nuclear fuel cycle.

Radionuclides in the atmosphere can be present in particulate or gaseous forms. In both cases, radionuclide exchanges between the stratosphere and the troposphere can take place following air mass movements. From a radiological viewpoint, it is very important to understand the radionuclide dynamics within the troposphere. Winds, of course, help transport radionuclides along the troposphere, while wet or dry fallout contributes to its deposition on the Earth's surface. Once deposited, radionuclides may return to the atmosphere by resuspension effects. They are relevant both at local and regional scales. In fact, there is wide documentation of the long-distance transport of dust from desert areas (see Chap. 4).

Rainwater helps transport radionuclides from the troposphere to the Earth's surface. Although it is neither a source nor a sink of radioactivity, rainwater is a good indicator of its presence in the troposphere. For that, the collection of rainwater and the measurement of some specific radionuclides are systematically included in the radiological protection programs (see Chap. 6). Specifically, rainwater is a direct indicator of the occurrence of accidental radionuclide releases. Rainwater transport can take place by mechanical displacement of solid particles or by dissolution of soluble chemical forms. Chapter 13 discusses the methods of atmospheric sample collection.

*Cosmogenic radionuclides*

The presence of cosmogenic radionuclides in the atmosphere is modulated by different long- or medium-term effects. Changes in solar activity, the Earth's magnetic field, seasonal changes, and, of course, local condition variability determine their concentration in the atmosphere. Table 5.3 gives some typical worldwide values for the most interesting cosmogenic nuclides in the troposphere. Many of the radionuclides in the Table are also produced and released to the environment during the nuclear fuel cycle or by nuclear detonations. This will be discussed in the following. In fact, the data presented in Table 5.3 [22] refer to cosmic-ray production and are derived from mathematical estimates. In any case, while the cosmogenic radionuclides naturally produced spread globally around the Earth, those released into the environment from nuclear power stations, or any other step of the nuclear fuel cycle, do so locally.

From a radiological point of view, only $^{14}$C and $^3$H are significant since they are part of the global carbon or water cycle.

As already discussed in Chap. 4, $^{14}$C undergoes oxidation to $^{14}CO_2$ or $^{14}CO$ once produced in the atmosphere. In this way, $^{14}$C enters the global carbon cycle. It is calculated that approximately 1.5 PBqy$^{-1}$ is produced in the atmosphere by the interaction of cosmic radiation with the atmosphere (see Chap. 4). Previously, for anthropogenic $^{14}$C input from the nuclear fuel cycle and nuclear detonations, it was estimated that maximum $^{14}$C atmospheric concentrations reached a value close to 50 mBqm$^{-3}$ [22].

Natural $^3$H appears in the atmosphere through the following reactions.

$$n +^{14} N \rightarrow^3 H +^{12} C$$

and

$$n +^{16} O \rightarrow^3 H +^{14} N$$

It is produced predominantly as $^3$HHO with an annual production rate of approximately $10^5$ TBqy$^{-1}$ [23]. Molecular $^3$HH can also be found in the stratosphere, but the production rate is at least one order of magnitude lower, $10^4$

**Table 5.3**   Table of representative cosmogenic radionuclide concentrations in the atmosphere [22]

| Radionuclide | Production rate (PBqy$^{-1}$) | Average concentration in the troposphere (mBqm$^{-3}$) |
|---|---|---|
| $^3$H | 72 | 1.4 |
| $^7$Be | 1 960 | 12.5 |
| $^{10}$Be | 0.000064 | 0.15 |
| $^{14}$C | 1.54 | 56.3 |
| $^{22}$Na | 0.12 | 0.0021 |
| $^{26}$Al | 0.000001 | 1.5 10–8 |
| $^{32}$Si | 0.00087 | 0.000025 |
| $^{32}$P | 73 | 0.27 |
| $^{33}$P | 35 | 0.15 |
| $^{35}$S | 21 | 0.16 |
| $^{36}$Cl | 0.000013 | 9.3 10–8 |
| $^{37}$Ar | 31 | 0.43 |
| $^{39}$Ar | 0.074 | 6.5 |
| $^{81}$Kr | 1.7 10 | 0.0012 |

TBqy$^{-1}$. The most abundant form, H$^3$HO, is present in the troposphere up to $4 \times 10^3$ TBq. The gaseous $^3$H concentration from the prenuclear era reached values not exceeding $10^{-1}$ mBqm$^{-3}$ [23] in the atmosphere, while in the rainwater, concentrations of 0.6–1 Bql$^{-1}$ are considered typical [23]. This has been globally modified after the start of atmospheric nuclear detonations and locally because of the liquid and gaseous effluent releases from nuclear installations. Frequently, the $^3$H concentrations are given in TU, tritium units, corresponding to 1 $^3$H atom per $10^{18}$ hydrogen atoms. It can be found that 1 TU is equivalent to 0.118 Bq l$^{-1}$.

*Primordial Radionuclides*
Primordial radionuclides are present in the lower parts of the troposphere as a result of resuspension effects and emanation of $^{222}$Rn from the Earth's crust. Furthermore, atmospheric effluents from industrial activities using NORM (see Chap. 4) are enriched in primordial radionuclides, which contributes to improving the concentration levels in the surroundings of the industry. From this point of view, the concentrations vary depending on local conditions. Regardless, it is possible to give global concentration figures or typical ranges as those presented in Table 5.4, taken from [22]. Concentrations of relevant radionuclides from the $^{238}$U and $^{232}$Th series together with $^{235}$U are given. $^{222}$Rn and $^{220}$Rn are not included in the table. They are studied in the next section due to their radiological significance. As expected, secular equilibrium does not hold among parents and daughters, as the response of the radionuclides to the geochemical processes involved in their transport is different. In general, inorganic particles are the main cause of primordial radionuclides in air.

**Table 5.4** Average
concentrations of primordial
radionuclides in the
atmosphere [22]

| Radionuclide | Concentration worldwide ($\mu$Bqm$^{-3}$) |
|---|---|
| $^{238}$U | 0.02–18 |
| $^{226}$Ra | 0.6–32 |
| $^{210}$Po | 10–80 |
| $^{232}$Th | 0.01–0.9 |
| $^{228}$Ra | <0.3–1.5 |
| $^{235}$U | 0.04 |

The radionuclide concentration values vary from region to region, and although some reference values are proposed in [22], the range is wide. For example, 1 $\mu$Bqm$^{-3}$ air is the reference value for $^{238}$U and 0.5 $\mu$Bqm$^{-3}$ for $^{232}$Th. However, the concentration values range from 0.02 to 18 $\mu$Bqm$^{-3}$ for $^{238}$U or 0.01 to 0.9 $\mu$Bqm$^{-3}$ for $^{232}$Th.

Specific events modify these concentrations. This was the case for the Mount Saint Helen eruption in 1980, which provoked an increase in the $^{238}$U atmospheric concentration detected in rainwater taken from the summer of 1980 to the fall of 1982. An increase was observed during the volcano activity period. Rainwater was estimated to deposit approximately $2.4 \times 10^2$ TBq on the surface as a consequence of this event [24, 25]. Nuclear detonations or accidents have changed the expected concentrations and the activity ratios between primordial radionuclides. Indeed, it has been found that the Cosmos 954 accident (see Chap. 4) caused the input of U isotopes into the atmosphere. In fact, anomalous $\frac{^{234}U}{^{238}U}$ and $\frac{^{235}U}{^{238}U}$ activity ratios close to 30 and 1, respectively, were observed in rainwater samples collected in Arkansas (USA) during 1980 [26]. These values, far from the expected natural ratios of 1 and 0.046, were attributed to satellite fallout and were also detected in Japan [27]. In 1970 and 1971, French nuclear tests were found to be responsible for the enhancement of U concentrations in rainwater collected in Japan during those years [28].

However, in the absence of specific events, the concentrations in the atmosphere are governed essentially by resuspension effects, and works can be found in the current literature demonstrating this fact. For example, rainwater samples collected in southwestern Spain were measured for U and Ra isotopes. The results show high variability depending on meteorological effects, i.e., wind direction and strength, precipitation conditions, and others, with $^{238}$U values ranging from 1 to 10 mBl$^{-1}$. $\frac{^{234}U}{^{238}U}$ The activity ratio is $1.09 \pm 0.30$, which is consistent with the existence of secular equilibrium. However, the $\frac{^{226}Ra}{^{238}U}$ average activity ratio is $2.27 \pm 1.22$, far from equilibrium, reflecting the different behavior of radionuclides in soils [29].

Inputs from industrial activities that use NORM are relevant mainly from a local point of view, since they can cause significant radiological impacts. In Table 5.5, taken from [22], the average annual atmospheric inputs of some industries are given. In global terms, atmospheric effluents are not relevant. However, the local environmental impact can be significant as the dissemination of effluents is reduced

**Table 5.5** Atmospheric releases of some selected primordial radionuclides from NORM industries [22]

| Industry | $^{238}$U (GBqy$^{-1}$) | $^{226}$Ra (GBqy$^{-1}$) | $^{222}$Rn (GBqy$^{-1}$) | $^{210}$Po (GBqy$^{-1}$) |
|---|---|---|---|---|
| Phosphoric acid | 0.07 | 0.09 | 820 | 0.14 |
| Iron steel production | 0.01 | 0.01 | 180 | 90 |
| Coal-fired power plant | 0.16 | 0.11 | 34 | 0.8 |
| Cement industry | 0.2 | 0.2 | 157 | 78 |
| Mineral sands handling | 0.97 | 0.73 | 0.73 | 0.73 |
| Titanium pigment | 0.001 | 0.001 | 6.2 | 0.001 |
| Gas-fired power plant (400 MWe) | | | 230 | |
| Oil extraction | | | 540 | |
| Gas extraction | | | 500 | |

to the surrounding environment. Of course, as we will see in Chap. 6, professional exposure to radiation in these industries poses an important radiological protection problem.

*Man-made Radionuclides*

Man-made radionuclides were reintroduced in nature as a consequence of nuclear weapon tests, which started in 1945. Table 4.9 shows the main radionuclides produced by fission or activation during detonation. Most radioactive species have a short half-life and have rapidly disappeared from the environment. However, some of them remain and still remain in the environment because of their long half-life. For that reason, $^{90}$Sr, $^{137}$Cs and Pu isotopes have traditionally been the most studied radionuclides in the environment, specifically in the atmosphere, after the first nuclear tests.

Table 5.6 gathers radionuclide production during atmospheric nuclear tests [30]. The activities produced in this way were dispersed in the atmosphere, and after the nuclear weapon ban treaty, their concentration progressively decreased in the atmosphere [31]. In the Northern Hemisphere, the $^{90}$Sr atmospheric concentration levels reached an average value close to 2 mBqm$^{-3}$ in 1963 but decreased up to approximately $10^{-2}$ mBqm$^{-3}$ during the middle 1980s prior to the Chernobyl accident. More than one order of magnitude lower concentrations have been observed in the Southern Hemisphere. Correspondingly, for the same period, the $^{137}$Cs atmospheric concentrations ranged from 2.6 to 0.009 mBqm$^{-3}$, also in the Northern Hemisphere, occurring the same in the Southern Hemisphere as in the case of $^{90}$Sr.

Pu isotope inputs from nuclear detonations have also been investigated. $^{239}$Pu is part of the nuclear fuel in weapons or is produced by neutron activation of $^{238}$U,

**Table 5.6** Radionuclide production from atmospheric nuclear weapon tests [30]

| Radionuclide | Global release (PBq) |
| --- | --- |
| $^3$H | 186,000 |
| $^{14}$C | 213 |
| $^{54}$Mn | 3980 |
| $^{55}$Fe | 1530 |
| $^{89}$Sr | 117,000 |
| $^{90}$Sr | 622 |
| $^{91}$Y | 120,000 |
| $^{95}$Zr | 148,000 |
| $^{103}$Ru | 247,000 |
| $^{106}$Ru | 12,200 |
| $^{125}$Sb | 741 |
| $^{131}$I | 675,000 |
| $^{140}$Ba | 759,000 |
| $^{141}$Ce | 263,000 |
| $^{144}$Ce | 30,700 |
| $^{137}$Cs | 948 |
| $^{239}$Pu | 6.52 |
| $^{240}$Pu | 4.35 |
| $^{241}$Pu | 142 |

as already described in Chap. 4. Other Pu isotopes are produced during detonation by subsequent neutron activation of the fission or other activation products. Additionally, in Chap. 4, an account of these processes is presented. $^{239}$Pu is an α-emitter and has historically been determined by α-spectrometry. It is interesting to note that $^{240}$Pu, another Pu isotope produced during nuclear fuel burning, emits σ-particles as well with energies very close to those of $^{239}$Pu. Even with modern Si α-spectrometers (see Chap. 11), these energies cannot be resolved. Hence, the $^{239}$Pu, or simply Pu, concentrations presented in many old papers truly correspond to $^{239+240}$Pu concentrations. Furthermore, in previous reports, the concentrations were inferred from those of $^{90}$Sr, which was easier to measure at that time [32]. However, at the beginning, Pu was not considered a hazard to humans and was not included in radiation monitoring programs. In general, it is assumed that the $\frac{^{239}\text{Pu}}{^{90}\text{Sr}}$ activity ratio in the fallout is 0.017 at the time of production. This value increases with time as $^{90}$Sr decays [32]. For that, it is estimated that the Pu concentration in the surface air peaked during 1963, reaching a value close to 30 μBqm$^{-3}$ [32] as a consequence of the atmospheric detonations of the USA and the USSR. Some data determined in New York during 1966 showed surface air concentrations close to 20 μBqm$^{-3}$, which is quite consistent with the previous estimate [33]. After the 1960s, the Pu concentrations progressively decreased over time, with some peaks attributed to the Chinese detonations that took place during the 1970s and first 1980s.

The French test in the Southern Hemisphere also contributed to some concentration peaks in this area [33]. Additionally, in Fig. 5.4 the time evolution of Pu concentration in the atmosphere is given, and it can be seen how the general levels have decreased. It can be said that today, the general level is approximately 1 nBqm$^{-3}$ in the absence of local sources. In the case of Pu isotopes, resuspension effects are important. In this way, Pu residing in soils is reincorporated into the lower atmosphere depending on local or regional weather conditions. It has been shown that resuspension effects can remobilize Pu in soils of the Sahara Desert and transport it to Europe [33, 34]. The $\frac{^{240}Pu}{^{239}Pu}$ isotopic ratio provides information on the origin of Pu. It reveals the nuclear fuel composition and the fuel burnup. In Table 5.7, the accepted typical ratios for different sources are presented. The fallout isotope ratio of Pu is estimated to be close to 0.18, which very clearly identifies the origin of Pu in the measured samples.

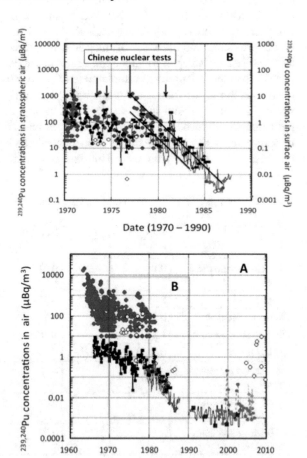

**Fig. 5.4**  Pu isotope concentrations in the stratosphere and troposphere over time. Figure reprinted from [33]. Temporal variations in $^{239,240}$Pu concentrations in stratospheric and surface air of the Northern Hemisphere. Closed red circles: the upper stratosphere (20–40 km height); open blue circles: the lower stratosphere (10.1–14.2 km); closed black, brown, green, purple, brown squares; and rhombic points: the surface air

Less studied has been $^{129}$I, $^{36}$Cl, or $^{99}$Tc. Although their half-life is longer than that of the traditionally studied radionuclides, the difficulties of their analysis have hindered the investigation of their presence in nature. The use of very low-level counting systems or mass spectrometry methods has facilitated their measurements, and some results are available in the more recent scientific literature. $^{129}$I ($T_{1/2} = 15.7 \times 10^6$ y) is a very volatile element that is produced in the atmosphere by spallation reactions with Xe and in the Earth's crust by spontaneous fission of $^{238}$U [35]. It is considered that some 250 kg [36] $^{129}$I originated from natural processes. However, $^{129}$I is also a fission product and is therefore produced during fission of $^{235}$U and $^{239}$Pu. It is part of the nuclear fuel cycle, and it is estimated that approximately 6000 kg has been released into the environment. Only 3% of this mass was produced during atmospheric weapon tests that took place between 1940 and 1980. Traditionally, $^{129}$I concentrations are reported as $\frac{^{129}I}{^{127}I}$, where $^{127}$I is the stable I isotope present in nature. The prenuclear isotopic ratios close to $10^{-12}$ have been slightly modified by radioactive fallout, and the present background is very variable, with values ranging from $10^{-11}$ to $10^{-8}$ in the Northern Hemisphere [37]. As we shall see later, this ratio is very much dependent on the proximity and influence of nuclear fuel reprocessing plants whose releases have masked the weapon modification of the prenuclear isotopic ratio [38].

$^{36}$Cl is another long-lived radionuclide ($T_{1/2} = 3.01 \times 10^5$ y) produced during the detonation of thermonuclear weapons by activation of $^{35}$Cl neutrons in marine waters [41]. It is present in the nuclear fuel cycle, but $^{36}$Cl is also a cosmogenic radionuclide produced by a spallation reaction with $^{40}$Ar [42]. It is very difficult to detect and only recently has it been possible to have a good set of data on its presence in the environment. This has been due to the possibility of using

**Table 5.7** Average $^{240}$Pu/$^{239}$Pu atom ratio and some Pu isotope activity ratios as a signature of the Pu source in the environment

| Source | $^{240}$Pu/$^{239}$Pu atom ratio | $^{238}$Pu/$^{239+240}$Pu activity ratio | $^{241}$Pu/$^{239+240}$Pu activity ratio |
|---|---|---|---|
| Integrated weapon test fallout | 0.18 | 0.026 | 15 |
| Weapon grade | 0.01–0.07 | 0.014 | 3 |
| Chernobyl accident | 0.40 | 0.5 | 85 |
| Fukushima accident | 1.2 | 108 | 0.3 |
| Nuclear reactors | 0.23–0.67 | | |
| Nuclear fuel reprocessing plants | | 0.25 | 25 |
| Fallout from the 21st Chinese test | | 0.03 | 11 |
| Fallout from the 26th Chinese test | | 0.02 | 5.5 |

Data compiled from [33, 38–40]

Accelerator Mass Spectrometry to determine $^{36}$Cl at environmental levels. It is estimated that approximately 20 atoms m$^{-2}$ s$^{-1}$ are produced by the interaction of cosmic rays with the atmosphere [43, 44]. The so-called bomb pulse multiplied this amount by 3000 from 1953 to 1964. Soon after this year, the $^{36}$Cl atom fallout recovers the prebomb values.

$^{99}$Tc is a fission product with a very long half-life of $2.11 \times 10^5$ y. It is produced in the spontaneous fission of $^{238}$U, and data exist on its presence in U ores [45]. $\frac{^{99}Tc}{^{238}U}$ isotopic ratios close to $10^{-12}$ have been documented in pitchblende samples [46]. However, it is mainly produced by the n-induced fission of $^{235}$U and $^{239}$Pu, and hence, it was reintroduced in nature during nuclear weapon atmospheric tests [47]. Thus, it is estimated that approximately 140–160 TBq $^{99}$Tc has been released during nuclear detonations. This amount is certainly negligible compared to the total production of $^{99}$Tc in the nuclear fuel cycle and at least one order of magnitude lower than the $^{99}$Tc released into the environment from the nuclear industry [47]. Consequently, the nuclear bomb pulse cannot be easily found today in nature since it is masked by the other releases. However, the measurement of $^{99}$Tc is difficult. Today, it is possible to use mass spectrometry techniques, but at the beginning, the only realistic method was the use of low-background β-counting [48]. Hence, data on its presence in the atmosphere during the time of nuclear detonations are very scarce. However, there are some. Rainwater samples from Texas collected from 1960 to 1970 were measured for $^{99}$Tc, and concentrations of approximately $10^{-2}$ mBql$^{-1}$ were obtained [49, 50]. Rainwater samples taken in southwestern Spain from 1984 to 1987 also showed concentrations of $10^{-3}$ to $10^{-1}$ mBql$^{-1}$ [51]. In the same area, surface air was sampled from 1965 to 1967. The time evolution pattern of the measured activity concentration showed three activity peaks corresponding to the summer periods of 1965, 1966, and 1967. This behavior was related to local weather conditions, such as rainfall events, and resuspension effects. The activity peaks were 2.63, 1.44, and 2.89 μBqm$^{-3}$ [52].

It is estimated that 183 EBq $^3$H has been introduced into the atmosphere as a result of nuclear weapon tests, mainly from fusion devices [23]. $^3$H, in fact, is part of the nuclear fuel and, in turn, is produced during the burning of fusion fuel. Despite its relatively short half-life ($T_{1/2} = 12.33$ y), this amount has masked the natural production of $^3$H for decades. In fact, $^3$H in rainwater reached levels close to $10^2$ Bql$^{-1}$ during 1963 [53], decreasing to 1 Bql$^{-1}$ before Chernobyl. In Fig. 5.5, the $^3$H concentration in the atmosphere is given for a long period starting from 1950 [53]. It can be seen that nuclear weapons largely contributed to the presence of $^3$H in the environment that masks natural production even today.

For $^{14}$C, some $2.2 \times 10^2$ PBq was introduced into the atmosphere as a result of the nuclear bomb tests. Mainly by the interaction of the neutrons produced in the detonation with the $^{14}$N present in the atmosphere (see Chap. 4). Although this total amount has been introduced into the atmosphere within a 30 y period, the resulting concentration is still very high compared to the naturally produced $^{14}$C per year. Figure 4.13 clearly illustrates the dimension of the input produced by nuclear detonations. In 1963, the $^{14}$C concentration in the atmosphere

**Fig. 5.5** $^3$H in atmospheric
samples collected in Japan
over decades. Figure
reprinted from [53]

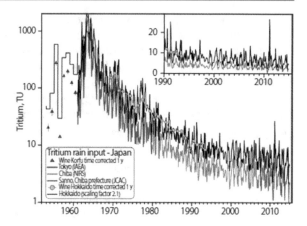

was 200 times higher than previously in the nuclear era. Today, levels are still
approximately 100 times higher [53].

The nuclear industry is today the most important source of man-made radionuclides to the atmosphere and the environment, in general. Nuclear reactors or
industrial activities carried out along the different steps of the nuclear fuel cycle
impact the environment and specifically the atmosphere. This impact is generally
local but, in many cases, has a global nature. Atmospheric or liquid effluents,
especially if released in the sea, are transported by wind or water currents and
distributed on local or regional scales [54].

With regard to atmospheric releases, there is ample scientific literature with
records of emissions from nuclear reactors or nuclear fuel reprocessing plants.
For atmospheric impact, target radionuclides are noble gases, mostly $^{133}$Xe and
$^{85}$Kr, $^3$H, $^{14}$C, $^{131}$I and some other fission products. Short-lived fission or activation products are of course emitted, but they are not relevant from a radiological
viewpoint since they rapidly disappear from the environment. Data records exist,
of course, and some examples can be found in [54, 55]. The exception is $^{131}$I
($T_{1/2} = 8.02$ d), which contributes greatly to the dose received by the population
during the first weeks of an emission [54]. $^{131}$I is systematically controlled due to
its radiological relevance.

UNSCEAR [30] reported the total activities released into the atmosphere of the
radionuclides mentioned above from nuclear reactors, even classifying the emission according to the type of reactor. It is interesting to review the data in the
provided reference. Some data are given in Table 5.8. For example, during the
period from 1990 to 1997, a total amount of 51.8 PBq $^3$H has been emitted to the
atmosphere by NPP worldwide, while approximately 1.03 TBq $^{131}$I was released
during the same period.

Very detailed information can also be found in a series of reports from the
European Union [56]. For instance, data from 2004 to 2008 reveal that the emission
of $^3$H to the atmosphere from the European NPP reached approximately 2.5 PBq.
Some data are collected in Table 5.9 with ranges of activity emitted from the

**Table 5.8**  Nuclear power plants atmospheric releases of $^3$H, $^{131}$I and noble gases in TBq from 1990 to 1997 [30]

| Radionuclide | 1990 | 1991 | 1992 | 1993 | 1994 | 1995 | 1996 | 1997 |
|---|---|---|---|---|---|---|---|---|
| $^3$H | 8624 | 6791 | 6448 | 10,400 | 6925 | 4196 | 4226 | 4212 |
| $^{131}$I | 121 | 127 | 205 | 158 | 131 | 94.2 | 102 | 95.4 |
| Noble gases | 85,500 | 73,810 | 62,760 | 75,570 | 77,440 | 40,040 | 24,820 | 21,260 |

**Table 5.9**  European nuclear power plants atmospheric range of releases for $^3$H, $^{14}$C, and $^{131}$I. Ranges are given of minimum and maximum from 2004 to 2008 [56]

| Year | $^3$H (GBqy$^{-1}$) | $^{14}$C (GBqy$^{-1}$) | $^{131}$I (GBqy$^{-1}$) |
|---|---|---|---|
| 2004 | $4.4 \times 10^2$–$7.50 \times 10^3$ | $1.3 \times 10^2$–$3.5 \times 10^3$ | $1.67 \times 10^{-3}$–$3.68 \times 10^0$ |
| 2005 | $6.58 \times 10^1$–$5.91 \times 10^3$ | $1.35 \times 10^1$–$4.03 \times 10^3$ | $4.90 \times 10^{-5}$–$2.60 \times 10^0$ |
| 2006 | $1.10 \times 10^2$–$2.99 \times 10^3$ | $5.00 \times 10^1$–$5.9 \times 10^2$ | $4.3 \times 10^{-4}$–$7.7 \times 10^0$ |
| 2007 | $1.03 \times 10^2$–$4.41 \times 10^3$ | $1.09 \times 10^2$–$8.6 \times 10^2$ | $5.46 \times 10^{-4}$–$1.81 \times 10^{-2}$ |
| 2008 | $1.8 \times 10^2$–$8.10 \times 10^3$ | $2.5 \times 10^1$–$2.1 \times 10^3$ | $3.42 \times 10^{-7}$–$4.19 \times 10^{-1}$ |

different European nuclear power plants for several radionuclides. For instance, in the case of $^{14}$C, a few $10^2$ GBqy$^{-1}$ are typically emitted from NPP to the atmosphere, with some specific peaks reaching approximately $3 \times 10^3$ GBqy$^{-1}$. In the case of noble gases, the most representative radionuclides are $^{41}$Ar ($T_{1/2} = 109$ min), $^{85}$Kr ($T_{1/2} = 10.74$ y) and $^{133}$Xe ($T_{1/2} = 5.25$ d). Emissions close to $10^3$ with occasional peaks of $10^6$ GBqy$^{-1}$ for $^{41}$Ar and $10^2$ with some peaks at $10^3$–$10^4$ GBqy$^{-1}$ for $^{85}$Kr and $^{133}$Xe have been observed for the period. $^{131}$I is a target radionuclide, as previously described. During the reported period, approximately $10^{-1}$ TBq was emitted to the atmosphere from NPP.

Data on transuranic atmospheric releases can also be found in official reports. This is the case of the previously mentioned European Union reports [56]. $^{238}$Pu, $^{239+240}$Pu, and $^{241}$Am activities between $10^{-6}$ and $10^{-5}$ GBq have been released annually according to a previous report.

Nuclear fuel reprocessing is a source of radionuclides to the environment. With respect to the atmosphere, once again, the relevant radionuclides are those that are longer-lived and potentially transported in gaseous forms or with volatile species. $^3$H, $^{14}$C, and $^{85}$Kr are the most important radionuclides. In addition, $^{129}$I is another important component of atmospheric releases. In Fig. 5.6, prepared with data from [30], the worldwide atmospheric releases for the above-mentioned radionuclides are given for a period covering from 1970 to 1997. According to the data, some 4.7 PBq $^3$H was released into the atmosphere throughout the world by nuclear reprocessing activities from 1990 to 1997. This is clearly lower than the amount found for NPP. In the case of $^{131}$I, some 0.13 TBq has been emitted, which is approximately 10% of the activity estimated for NPP. However, the $^{129}$I activity released into the atmosphere is twice that of $^{131}$I, reaching a value of approximately 0.29 TBq.

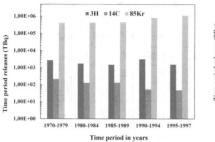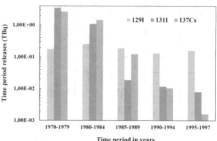

**Fig. 5.6** Nuclear fuel reprocessing plant atmospheric effluents worldwide from 1970 to 1997. Based on data taken from [30]

Recent data in Europe show that in the period from 2004 to 2008 [56], some 1.16 PBq $^3$H has been released into the atmosphere as a consequence of nuclear fuel reprocessing activities. Very close to half of the NPP emissions. The $^{85}$Kr releases are very relevant compared with the rest of the radionuclides. The total amount in the period reaches a value close to 1.43 EBq, certainly very high compared to that of $^3$H or $^{14}$C of 70 TBq. In the case of $^{131}$I, it is estimated that some $10^{-3}$ PBq has been released for the period, which is two orders of magnitude smaller than the NPP releases. For transuranics, some 0.16 and 0.12 GBq $^{239+240}$Pu and $^{241}$Am, respectively [56], were released into the atmosphere from 2004 to 2008, according to previous reports.

Nuclear fuel reprocessing plants are authorized by law to discharge radioactivity into the environment under some specific limits. As an example, in Fig. 5.7, the atmospheric discharges for some key radionuclides are given for the case of the Sellafield Nuclear Fuel Reprocessing Facility located close to the Irish Sea, UK [55]. Limits are also displayed. Reprocessing facilities are required by law to carry out monitoring programs in the vicinity of the installations [54]. It is known that the close environment is affected. In this specific case, environmental contamination is due to the direct pathway or to the combination of the different pathways. Indeed, according to the report in the reference, the radioactivity concentration in the air is due to atmospheric releases but also to the contribution of sea-to-land transport processes from the nearby marine environment where some releases take place. In general, radionuclide concentrations are below the limit of detection (see Chap. 14).

However, the long-range influence of the releases from nuclear fuel reprocessing plants is well documented in the literature. Either atmospheric or aquatic discharges can lead to global effects. In the case of the atmosphere, $^{129}$I measurements in the air have determined the influence of atmospheric releases as far as 150 km from the Tokai facility, Japan [57]. Additionally, for $^{129}$I, the Sellafield and La Hague (north of France) releases have been shown to influence the atmosphere of Central, Northern, and Southern Europe [58]. The authors can correlate the $^{129}$I time pattern atmospheric concentrations in the sampling area with stack emissions from the reprocessing facilities. However, in Northern Europe,

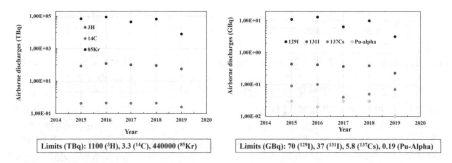

**Fig. 5.7**   Annual discharge for some key radionuclides with the limit permitted from 2015 to 2019. Based on data taken from [55]

atmospheric concentrations were related to liquid discharges to the marine environment. This shows the relevance of sea-to-land transport processes in the case of $^{129}$I, which is a very volatile species.

As described in the previous chapter, the Chernobyl accident produced an important input of radioactivity into the atmosphere. Table 4.16 gives an account of this. Table 5.10, based on [59], gives the ranges of the total deposition in European countries for several radionuclides of interest. A more detailed account is given in [59]. Of course, only long-lived radionuclides are currently of relevance with respect to the impact of the accident.

Just after the accident, the radioactivity concentration levels in the atmosphere suddenly increased around the accident area, as in other countries. In Monaco air samples, more than 28 gamma-emitting radionuclides were detected during the first days of May 1986. Approximately 1.6 Bqm$^{-3}$ $^{137}$Cs was detected in air filter samples collected on May 3, and this concentration was approximately 6 orders of magnitude higher than those found before the accident. The activity level dropped to 0.1% within the following two weeks and up to six orders of magnitude one year after the accident. In the case of $^{239+240}$Pu, the activity increased two orders of

**Table 5.10**   Radionuclide deposition over Europe from the Chernobyl accident

| Radionuclide | Range (kBqm$^{-2}$) |
|---|---|
| $^{137}$Cs | 0.02–154 |
| $^{95}$Zr | 0.1–600 |
| $^{103}$Ru | 0.04–570 |
| $^{106}$Ru | 0.01–154 |
| $^{131}$I | 0.07–3000 |
| $^{132}$Te | 0.004–1600 |
| $^{134}$Cs | 0.01–77 |
| $^{140}$Ba | 1.3–1000 |

Based on data from [59]

magnitude just after the accident, reaching values close to 5 $\mu$Bqm$^{-3}$. In rainwater, $^{239+240}$Pu activity concentrations of approximately 1 Bqm$^{-3}$ were detected to decrease rapidly in the following weeks. The short-lived $^{131}$I suddenly starts being detected, reaching activities of approximately 1 kBq l$^{-1}$. This activity decreased by approximately three orders of magnitude after two months [60]. Similarly, in western Spain, 70 mBqm$^{-3}$ $^{137}$Cs was detected in air filter samples collected on May 3rd [61]. A few days later, the activity of $^{137}$Cs was below the limit of detection. In rainwater taken in southern Spain from April to November 1986, the concentration of $^{137}$Cs activity increased by approximately two orders of magnitude, up to 125 mBql$^{-1}$, compared to previous levels [51]. In Japan, the atmospheric deposition of $^{137}$Cs increased by 4 orders of magnitude in May 1986, reaching values close to 100 mBqm$^{-2}$ [62]. By the end of 1986, the level of deposition recovered the value before the accident, which was close to 10$^{-2}$ mBqm$^{-2}$ [63].

It is important to note that the arrival of a Chernobyl radioactive cloud was easily identifiable, as the magnitude of the activity input was very high. Another important feature of the injected radioactivity was the characteristic activity ratios of some radionuclides. Specifically, the activity ratio for Cs isotopes was $\frac{^{134}Cs}{^{137}Cs} = 0.54$ at the time of emission [64, 65]. This reference ratio served to identify the origin of the activity increase observed in environmental samples.

In general terms, the Chernobyl accident affected the atmosphere of practically the whole Northern Hemisphere, suddenly increasing the concentration levels of many man-made radionuclides. The radionuclides were swept from the atmosphere by wet or dry deposition, and after some weeks, the activity levels in the lower atmosphere significantly decreased. However, the effects of resuspension and remobilization in general were relevant during the years after the accident, which posed an important radiological problem [59]. Even they are still important many years after the accident. Very interesting evidence of these effects is presented in Fig. 5.8, where $^{137}$Cs concentration levels in air samples taken at Thessaloniki are depicted. The peak observed during April 2020 was related to the forest fires that occurred during such dates in the Chernobyl exclusive zone [66].

The authors interpret this peak as the result of a dispersion of deposited activity in the forest due to some forest fires that took place during this year. These effects are crucially important for the protection of humans and the environment in general. In fact, although the majority of the Chernobyl-derived radionuclides dispersed in the environment have already decayed, longer-lived remains. They can be reincorporated into the different pathways that impact humans as the result of natural remobilization processes. Of course, they provide a temporal mark useful for dating natural compartments such as sediments or ice cores.

The Fukushima accident was also a very important event, although its environmental impact was clearly lower than that of Chernobyl. The injected radioactivity in the atmosphere followed a complicated time pattern already described in Chap. 4, which varies on a very short time scale of hours. The radioactivity deposition was determined by the weather conditions during the emission and was also very variable, with significant differences within distances of approximately 10 km. The radioactive emissions were mostly composed of Cs and I radioisotopes,

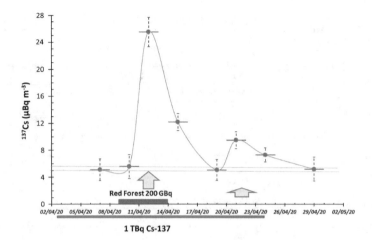

**Fig. 5.8**  $^{137}$Cs concentration in the atmosphere at Thessaloniki, Greece. The activity peak in the middle of April was interpreted as an input of resuspended $^{137}$Cs from the forest fires that took place during April 2020 the Chernobyl exclusive zone. Figure reprinted from [66] with permission from Elsevier

although other radionuclides were also present. Table 5.11 gives the estimated atmospheric emissions of Fukushima [67]. Chernobyl is also included for comparison. Approximately 20% of the emitted $^{137}$Cs was deposited in Honshu, the main island of Japan. The rest was transported to Northern Pacific Ocean areas. In Fig. 5.9, the $^{137}$Cs deposition over the main island is given [68]. The high variability of the observed deposition levels is evident. As in the case of the Chernobyl accident, activity ratios have been used as a signature of the accident. For instance, the $\frac{^{134}Cs}{^{137}Cs}$ activity ratio has been determined to depend on the reactor unit and ranges from 0.95 to 1.05 [69]. These ratios were used to identify the origin of the activity measured in the different samples. The Fukushima environmental impact extended almost globally, and there are data on the arrival of the radioactive plume to different countries. The arrival time and amount of transported radioactivity depended greatly on weather conditions and air circulation since the injection occurred at tropospheric levels. For example, high levels of $^{137}$Cs, $^{134}$Cs, and $^{131}$I were measured in the Canary Islands two weeks after the accident [70], with concentration peaks in the air of 0.408, 0.382, and 1.85 mBqm$^{-3}$, respectively. The levels represented an increase of one order of magnitude for $^{137}$Cs and two orders for $^{131}$I with respect to previous measurements. Soon after two weeks, activity concentrations decreased to previous levels. The $\frac{^{134}Cs}{^{137}Cs}$ ratio was 1.053, which is compatible with the expected value of radioactive debris from Fukushima.

In Lisbon [71], 1,4 mBqm$^{-3}$ was observed for $^{131}$I at the end of March 2011, while for $^{137}$Cs and $^{134}$Cs, levels of 0.139 and 0.153 mBqm$^{-3}$ were measured in air filters. Once again, the activity ratio of the Cs isotopes, 1.101, is compatible with the Fukushima signature. In less than one month, the concentration levels decreased to values obtained prior to the accident. There are many examples in

**Table 5.11** Some relevant radionuclides were released to the atmosphere from the Fukushima accident compared with the Chernobyl atmospheric inputs [67]

| Radionuclide (PBq) | Fukushima | Chernobyl |
|---|---|---|
| *Noble gases* | | |
| $^{85}$Kr | 44 | 33 |
| $^{133}$Xe | 14,000 | 6500 |
| *Volatile elements* | | |
| $^3$H | 15 | 1.4 |
| $^{129}$I | $5.5 \times 10^{-5}$ | $4 \times 10^{-5}$ |
| $^{131}$I | 65–380 | 1760 |
| $^{134}$Cs | 11.8–18 | 47 |
| $^{137}$Cs | 6–62 | 74–98 |
| *Intermediate volatility* | | |
| $^{89}$Sr | 0.2 | 115 |
| $^{90}$Sr | 0.02 | 10 |
| $^{106}$Ru | | >73 |
| *Refractory elements* | | |
| $^{95}$Zr | | 84 |
| $^{125}$Sb | | 0.23 |
| $^{238}$Pu | $2 \times 10^{-6}$–$5 \times 10^{-6}$ | 0.015 |
| $^{239}$Pu | | 0.013 |
| $^{240}$Pu | | 0.018 |
| $^{239+240}$Pu | $1.0 \times 10^{-6}$–$2.4 \times 10^{-6}$ | 0.031 |
| $^{241}$Am | | 0.0024 |
| $^{244}$Cm | | 0.0027 |

the current literature where data of the Fukushima environmental impact can be obtained. A good example is [72], where data from the USA, Central and Northern Europe, Central and Southern Asia, Iceland, etc. can be found. This shows the relevant impact of atmospheric releases.

Radioactivity dispersed by the Fukushima accident is still under control, as several monitoring works are carried out in Japan. Thus, it has been observed that after a rapid decrease in radioactivity in the air within the first two months after the event, $^{137}$Cs concentrations slowly decreased, but the actual levels in the air were still higher than those before the accident [73]. Figure 5.10 shows that some activity peaks appear over time in samples taken at several stations independent of the distance from the Fukushima NPP [73]. These peaks are related mainly to removal actions carried out by the responsible staff of the NPP but also to the resuspension of the soils contaminated by the activity deposited from the accident. This is additional evidence that soils can be sinks but also sources of radionuclides in the environment, which must be taken into account when estimating the radiation dose received by the population. A very interesting feature of the Fukushima accident was the emission of radioactive particles containing $^{137}$Cs [73]. They were found

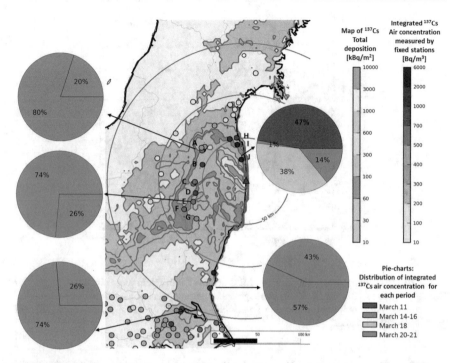

**Fig. 5.9** $^{137}$Cs deposition on Honshu Island in Japan. Figure reprinted from [68] with permission by Elsevier

in air and soil samples and identified by $\gamma$-spectrometry or autoradiography techniques (see Chap. 20). The size was approximately 2.6 $\mu$m and was made mainly of a $SiO_2$ glass matrix. Unlike the Chernobyl case, there is no evidence of large ($\approx$100 $\mu$m) radioactive particles consisting of fuel materials [73]. As in the case of the Chernobyl accident, it has been suggested that Fukushima will provide a new time marker for sediment or ice cores, which will be useful for dating work [74].

### 5.2.2   The Radon Problem

Rn is a noble gas with three natural isotopes: $^{222}$Rn, $^{220}$Rn, and $^{219}$Rn. $^{222}$Rn ($T_{1/2}$ = 3.82 d), called radon, is a member of the $^{238}$U radioactive series (see Chap. 3) and is produced by the $\alpha$ decay of $^{226}$Ra. $^{220}$Rn ($T_{1/2}$ = 55.6 s), called thoron, belongs to the $^{232}$Th radioactive series and is produced by the $\alpha$ of $^{224}$Ra. Finally, $^{219}$Rn ($T_{1/2}$ = 3.96 s) is called actinon and belongs to the $^{235}$U radioactive series, being an $\alpha$-decay product of $^{223}$Ra. From an environmental and, more specifically, radiological viewpoint, the only important isotope is $^{222}$Rn, since $^{220}$Rn and $^{219}$Rn are very short-lived. Nevertheless, in some cases, $^{220}$Rn must be taken into

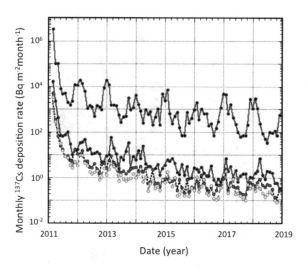

**Fig. 5.10**  $^{137}$Cs deposition after the Fukushima accident at different places in Japan. Deposition rapidly decreased after the accident, although some peaks appeared over time due to resuspension effects and removal actions by the Fukushima NPP staff. Figure reprinted from [73] with permission from Elsevier

account, especially in environments with high $^{232}$Th contents. In any case, the relevant isotope is $^{222}$Rn. This radionuclide is an $\alpha$-emitter and produces several radionuclides, as shown in Fig. 5.11. The descendants of $^{222}$Rn are all short-lived and rapidly decay to $^{210}$Pb ($T_{1/2} = 22.20$ y). As we studied in Chap. 4, $^{210}$Pb is an interesting radionuclide that gives rise to a very useful dating method.

Inhalation of $^{222}$Rn is considered the second leading cause of lung cancer after smoking [75]. In fact, the risk of lung cancer increases 16% per 100 Bqm$^{-3}$ $^{222}$Rn in inhaled air, although the cause of the cancer is the descendant and not directly $^{222}$Rn. Indeed, because they are chemically reactive, they can be trapped by aerosol particles and lung tissues. Therefore, $^{222}$Rn is a very important radionuclide to be considered within radiation protection programs, not only for exposed workers in mines and industries processing NORM materials but also for the general public. In

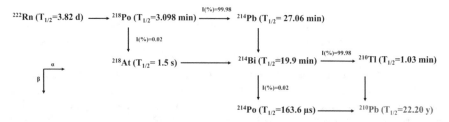

**Fig. 5.11**  $^{222}$Rn descendants

fact, $^{222}$Rn inhalation in dwellings is a very relevant matter within these protection programs [75].

The presence of $^{222}$Rn in air is derived from its gaseous character. Once it is produced by $^{226}$Ra α-decay in soils or rocks, $^{222}$Rn can emanate from grains due to the recoil energy gained in disintegration. Then, it can move across the pores toward the soil or rock surface and finally escape (see Fig. 5.12). The probability of emanation depends on several parameters [76]: $^{226}$Ra concentration and concentration gradient in the soil, pore size and shape, soil moisture, temperature, and composition. It is estimated that the emanation coefficient can range from 2.1 to 32% in rocks, 0.14 to 80% in soils, and 0.1 to 58% in building materials. The $^{222}$Rn concentration in the soil gas, $C_{Rn}$, is

$$C_{Rn} = C_{Ra} f \rho_s \left(\varepsilon^{-1} - 1\right)[m(K_t - 1) + 1]^{-1} \tag{5.4}$$

$C_{Ra}$ is the $^{226}$Ra concentration in the soil (Bqkg$^{-1}$), f is the emanation factor, $\rho$ is the soil grain density (kgm$^{-3}$), ε is the total porosity, which includes the air and water phases, m is the water-filled porosity fraction (the saturation fraction) and $K_t$ is the coefficient of partition Rn, that is, the Rn ratio between the air and water phases [22]. In the case of dry soils, $m = 0$ and (5.4) simplifies to the first term.

The $^{222}$Rn escape or exhalation rate from the soil is measured in Bqm$^{-2}$ s$^{-1}$. This $^{222}$Rn flux, $J_D$, is strongly dependent on the geology of the area and the physical and chemical properties of the soil. Thus, the porosity of the soil, its humidity, and temperature influence the exhalation rate. External factors also influence it. Atmospheric temperature, humidity, and pressure, as well as rainfall episodes, are important to estimate the $^{222}$Rn exhalation rate [75]. It is known that for dry soils [22].

$$J_D = C_{Ra}\lambda_{Rn} f \rho_s (1 - \varepsilon)L \tag{5.5}$$

**Fig. 5.12** Rn diffusion and emanation from soils

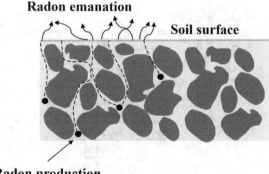

**Radon emanation**

**Soil surface**

**Radon production**

where $\lambda_{Rn}$ is the $^{222}$Rn decay constant and $L$ is the diffusion length, which is $\left(\frac{D_e}{\lambda_{Rn}}\right)^{\frac{1}{2}}$, with $D_e$ being the effective diffusion coefficient. Using representative values for the different parameters, it can be found that $J_D = 0.033$ Bqm$^{-2}$ s$^{-1}$, which is higher, although within the order of magnitude of the average $J_D$ world-wide, 0.016 Bqm$^{-2}$ s$^{-1}$ [22]. $^{222}$Rn in the open air and outdoors varies according to the geographical area, depending on the geological substrate, meteorological conditions, etc. Usually, concentration levels ranging from 2 to 20 Bqm$^{-3}$ can be found worldwide [22]. The lower values typically correspond to open areas such as islands or coastal areas. High values correspond to specific areas with high Rn exhalation rates or valleys that tend to accumulate $^{222}$Rn in the air. Concentration peaks of 100 Bqm$^{-3}$ have been found in some of these last cases [75]. However, from a radiological point of view, outdoor $^{222}$Rn is not a hazard. However, $^{222}$Rn indoors and especially in dwellings can reach values close to $10^5$ Bqm$^{-3}$ [75]. In Table 5.12 prepared with data from [22], $^{222}$Rn indoor average values are presented for selected regions around the world. Detailed information can be found in [22]. Additionally, in Fig. 5.13 from [75], a map of $^{222}$Rn indoors in Europe is depicted. The values are variable, but high concentration levels can be seen. In general terms, the concentration of $^{222}$Rn in buildings depends on the geological substrate and the construction materials, especially their permeability and mean temperature, and it also depends on the year season and ventilation habits [75].

A similar equation to (5.2) can be used to estimate the $^{222}$Rn flux from the building walls. According to [22], $J_D$ from the walls is

$$J_D = C_{Ra} \lambda_{Rn} f \rho L \tanh \frac{d}{L} \tag{5.6}$$

The meaning of the different parameters is the same as in (5.5) (the value for $L$ is that of concrete) except for $d$, which is the half-thickness of the wall or floor slab. Within the specialized literature, a reference house can be defined [22], and

**Table 5.12** Worldwide average $^{222}$Rn concentrations indoors.

| Region | Average values range (Bqm$^{-3}$) |
|---|---|
| Africa | 9–30 |
| North America | 34–46 |
| South America | 25–37 |
| East Asia | 10–57 |
| West Asia | 14–104 |
| North Europe | 53–120 |
| West Europe | 20–110 |
| East Europe | 41–140 |
| South Europe | 7–120 |
| Oceania | 11–20 |

Data taken from [22]

**European Indoor Radon Map, November 2021**

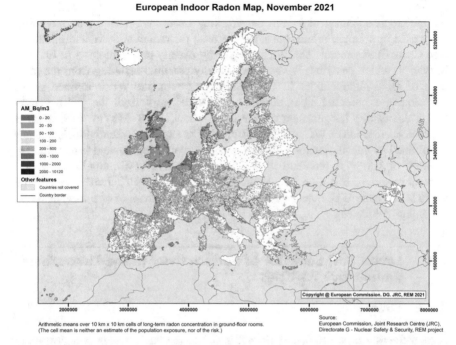

Arithmetic means over 10 km x 10 km cells of long-term radon concentration in ground-floor rooms.
(The cell mean is neither an estimate of the population exposure, nor of the risk.)

Source:
European Commission, Joint Research Centre (JRC),
Directorate G - Nuclear Safety & Security, REM project

**Fig. 5.13** $^{222}$Rn indoor map in Europe. Figure taken from [75]. European Atlas of Natural Radiation, Publication Office of the European Union, Luxembourg, 2019; https://remon.jrc.ec.europa.eu/About/Atlas-of-Natural-Radiation

the entry $^{222}$Rn in the house (Bqm$^{-3}$ h$^{-1}$) from the building elements can be calculated as

$$U = \frac{3.6 \times 10^3 S_B J_D}{V} \tag{5.7}$$

$S_B$ and $V$ are the house surface areas and volume (m$^3$), respectively. This formula can be used to calculate the entry of Rn in houses as a function of the physical characteristics of the building and the geological surroundings. Some results can be found in [22].

The connection between $^{222}$Rn concentration and lung cancer risk has been clearly demonstrated [75]. In fact, between 3 and 14% of lung cancer in countries is estimated to be directly related to Rn. This amount depends on the prevalence of smoking in the country. For that, countries must establish Rn reference levels for workplaces and buildings. Thus, the concentration of Rn indoors must be below such a level. The World Health Organization (WHO) recommends a reference level of 100 Bqm$^{-3}$ indoors and 300 Bqm$^{-3}$ [77] for workplaces and buildings when the previous limit is impossible to achieve. For example, European countries can freely determine their own reference levels as long as they are not higher than 300 Bqm$^{-3}$.

Finally, it is interesting to describe an old but still used unit when dealing with the radiological impact of $^{222}$Rn: the Working Level (WL). This unit was originally used in uranium mines, but it can be used for radiation dose determinations (see Chap. 6) in general. The WL is an energy density unit, measured in $Jm^{-3}$ or $MeVm^{-3}$, which describes the available energy potential resulting from the presence of some radionuclides in 1 L of air. For $^{222}$Rn, the WL is defined as the emission of a potential alpha energy of $1.3 \times 10^5$ meV from the $^{222}$Rn daughters. This quantity is equivalent to 20.8 $\mu Jm^{-3}$ or $1.3 \times 10^8$ $MeVm^{-3}$. In terms of $^{222}$Rn concentration, 1 WL = 37,000 $Bqm^{-3}$. The exposure of individuals to Rn is determined by calculating the potential energy or concentration and is measured in $Bqhm^{-3}$ or $Jhm^{-3}$. In old texts, the unit used is the WLM (Working Level Month) of clear meaning. It can be found that 1 WLM = $3.54 \times 10^{-3}$ $Jhm^{-3}$. According to the ICRP, each WLM means a dose equivalent (see Chap. 6) of 5 mSv or 6 nSv per $Bqhm^{-3}$ [22, 78].

## 5.3    Levels and Behavior of Radioactivity in the Lithosphere. Radioactive Particles

### 5.3.1  Soils

A high fraction of the radioactivity released into the environment is deposited in soils. In addition, natural radionuclides are present in soils at concentrations that depend on the geological substrate. On the other hand, cosmogenic nuclides can be produced on site by direct reactions of cosmic radiation with soil materials.

The soil is a dynamic system and cannot be considered a simple radionuclide sink. In fact, radionuclides deposited in the first cm can be resuspended and redistributed along the close environment or, in some cases, at a nearly global scale. However, radionuclides can migrate deep in soils and enter groundwater or can be incorporated into plants through the root system. In both cases, radionuclides follow a pathway that could impact humans [22]. Specifically, for agricultural soils, the uptake of radionuclides by edible vegetables poses a radiological problem of paramount importance.

Radionuclides present or deposited in the soil can migrate into deeper soil layers. There are several processes that favor or retard migration [79].

Water irrigation of soils or precipitation of rainfall facilitate the transport of the more soluble radionuclides downward. This is the so-called convection mechanism of radionuclide vertical transport. In the absence of water flux, diffusion is the main migration process to deeper layers. This is a very slow process and is only important in dry areas. Insoluble radionuclides can also be transported downward by particle migration. Thus, radionuclides present as precipitates in the soil solution or as colloids or adsorbed to soil particles can move down, provided the particle size is smaller than the soil pore size. Agricultural practices, such as plowing, help to transport radionuclides from the surface to deeper layers. However, repeated plowing helps the radionuclides return to the superficial layers. Biological transport must also be considered an important mobilization mechanism. In this case, some small animals, mostly worms, take radionuclides and excrete them downward or upward. Of course, since radionuclides are taken up by plant roots, their growth moves the radionuclides to deeper layers.

In contrast, migration can be retarded by the sorption process in the soil. In the case that sorption is not reversible, migration following water flux can be retarded. Radionuclides with higher distribution coefficients, $K_d$ (see previous paragraphs), are candidates to be retained longer at the soil surface. However, since $K_d$ depends on many factors, such as radionuclide type and chemical species, chemical composition of the soil, and external parameters, such as temperature, the migration rate can highly change with small variations in some of these parameters. Hence, the variability of the behavior of the radionuclides in soils and of the same radionuclide in the same soil type.

As in the case of the atmosphere, radionuclide concentrations in soils vary considerably from place to place. Nevertheless, some data are discussed below to provide a general view of the different sources and sinks of radionuclides in soils.

*Cosmogenic Radionuclides*

Cosmogenic radionuclides can also be produced by the interaction of cosmic radiation with elements present in the Earth's crust. This in situ production is possible since part of the cosmic-ray flux can reach the Earth's surface directly or, more likely, can produce secondary radiation along its path with reaching the Earth's surface. These processes are illustrated in Fig. 5.14. Radiation impinging on the Earth's surface, rocks, or soils generates nuclear reactions, so cosmogenic nuclides appear [80]. The nuclide production rate is small. In fact, the only realistic possibility of measuring radionuclide concentrations this way created is the use of very sensitive mass spectrometry methods, as we will see in Chaps. 16 and 18.

Spallation, (n, $\alpha$), (n, p), or (n, $\gamma$) reactions with rock components are the main mechanisms of radionuclide production in situ. In Table 3.4, some data are given on the mechanisms for relevant cosmogenic radionuclide production. In [80], a very complete account of production rates is given. Of course, the production rate depends on the rock type, altitude, and several other parameters. Hence, the value is variable, but in principle, it can be estimated that the figures in [80] are accepted by the specialized scientific community. Production rates between 4 and 6 atom $g^{-1}y^{-1}$ for $^{10}$Be and approximately 30 for $^{26}$Al or 50 for $^{36}$Cl reveal that the presence of cosmogenic nuclides in rocks or soils is not a radiological problem.

In addition, the production rate decays with the depth in the rock, as shown in Fig. 5.15. It is apparent that the penetration of cosmic rays can be modeled to be of exponential type, $e^{-\mu \rho z}$, where z is the depth, $\rho$ is the rock density, and $\mu$ is the mass absorption coefficient of the rock, which will depend on the rock material and the energy and type of cosmic radiation [80]. The production of cosmogenic radionuclides in situ is useful in geochemistry studies. A very popular example is the calculation of soil erosion rates $\varepsilon$ [81].

**Fig. 5.14** Cosmic radiation reaching the Earth's crust. Figure reprinted from [80] with permission by Elsevier

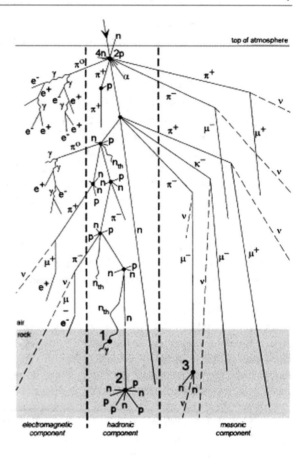

It is a matter of very simple mathematics to find that the concentration of radionuclide at a depth $z$ of a rock of age $t$, $C(z, t)$, is

$$C(z, t) = \frac{Pe^{-\mu\rho z}}{\lambda + \varepsilon\mu\rho}\left(1 - e^{-(\lambda+\varepsilon\mu\rho)t}\right) \tag{5.8}$$

where $P$ is the production rate of the radionuclide at the surface and $\lambda$ is the radionuclide decay constant. For old rocks, compared to the radionuclide half-life, i.e., for $\lambda t \gg 1$,

$$C(z, t) = \frac{Pe^{-\mu\rho z}}{\lambda + \varepsilon\mu\rho} \tag{5.6}$$

*Primordial Radionuclides*

The presence of primordial radioactivity in soils is strongly related to the geological substrate of the site. Furthermore, resuspension effects, which depend mainly

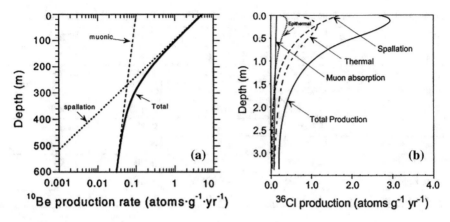

**Fig. 5.15** Decay of production rate with depth for **a** $^{10}$Be and **b** $^{36}$Cl. Figure reprinted from [80] with permission by Elsevier

on the local wind regime and meteorological conditions, contribute to modifying the concentrations of radionuclides in soils.

Table 5.13 collects the range and mean value of the concentrations of $^{40}$K, $^{238}$U, $^{226}$Ra, and $^{232}$Th worldwide [22]. The variability of the data is apparent, and any comparison is difficult. As said before, the geological substrates are mainly responsible for the concentrations found (see Table 4.1). A very informative map of the presence of $^{238}$U, $^{232}$Th, and $^{40}$K, in the form of $K_2O$, in European topsoil can be found in the European Atlas of Natural Radiation [75]. As far as it is known today, U and Th have no biological function; therefore, their toxicity is related to the radiation they and their descendants emit. U and Th incorporate the roots of the plant with very different transfer factors (see previous paragraphs). In fact, those for U are at least one order of magnitude higher than those for Th [75]. This is due to the higher solubility of U. In contrast, K is an essential nutrient for plants and animals, and its presence in soils reveals that of $^{40}$K has an isotopic abundance of 0.00117%. Thus, the incorporation of K into living beings must be taken into account to evaluate the radiological impact of natural radioactivity [75]. The case of Ra, specifically its isotopes $^{226}$Ra ($T_{1/2} = 1600$ y) and $^{228}$Ra ($T_{1/2} = 5.75$ y), both belonging to the $^{238}$U- and $^{232}$Th-series, respectively, deserves special attention. Indeed, its chemical behavior is similar to that of alkaline elements and therefore to that of Sr, Ba, and, interestingly, Ca, which is an essential nutrient for living beings. Ra can replace Ca in carbonate minerals and be easily taken up by plants [75, 82]. The concentration of $^{226}$Ra in the soil depends, as that of the other natural isotopes, on the host rock, that is, the geological substrate but also on other environmental circumstances. In Table 5.14, the typical $^{226}$Ra concentrations in various rocks are given. In Table 5.15, the usual concentrations are presented in different types of soil. The variability is large, as is the content in rocks. Furthermore, the different geochemical processes that affect the specific environment tend to change the concentrations and meteorological conditions.

**Table 5.13**  Ranges of natural radioactivity concentrations in soils $(Bqkg^{-1})$ worldwide [22]

| Region | $^{40}K$ | $^{238}U$ | $^{226}Ra$ | $^{232}Th$ |
|---|---|---|---|---|
| Africa | 29–1150 | 1–120 | 5–180 | 2–140 |
| North America | 6–700 | 4–140 | 8–160 | 4–130 |
| South America | 540–750 | | | |
| East Asia | 7–1800 | 2–690 | 2–440 | 1 – 360 |
| West Asia | 87–980 | 10–78 | 8–77 | 5–60 |
| North Europe | 140–1150 | 3–30 | 9–310 | 5–94 |
| West Europe | 40–3200 | 8 – 330 | 5–900 | 1–180 |
| East Europe | 40–1400 | 5–190 | 1–210 | 2–160 |
| South Europe | 15–1650 | 1–240 | 1–250 | 1–210 |

**Table 5.14**  Typical $^{226}Ra$ concentrations in rocks [82]

| Type of rock | $^{226}Ra$ $(Bqkg^{-1})$ |
|---|---|
| Volcanic rocks | 48–137 |
| Base lava | 33 |
| Granite | 0.037–185 |
| Basalt | 11–48 |
| Magmatic acid | 111 |
| Magmatic intermediate | 96 |
| Magmatic base | 11–48 |
| Shale (Alume slate) | 14.8–2220 |
| Bitumen slate | 629–1040 |
| Sedimentary rocks | 9.2–15 |
| Lime | 5–18 |
| Sandstone | 7–55 |
| Phosphate rocks | 148–1480 |
| Carbonate | 26–30 |
| Clay rocks | 55 |

They represent, in any case, the $^{226}Ra$ content in normal areas, since it is known that there are high natural radiation areas in the world (see Chap. 4) where elevated $^{226}Ra$ concentrations can be found. They are presented in Table 5.16. The effluents of the NORM industries (see Chap. 4) have a high impact on the specific environments and soils [83]. In Table 5.17 [22], the average effluents into water are presented for some industries. This information has to be added to Table 5.5 to provide a complete view of the environmental impact of NORM industries. In general, terms, the radionuclides released can be dispersed and enter different natural compartments. The radiological impact is mostly local, and global effects are not expected. With regard to soils, atmospheric effluents are the most important since they can be dispersed from the industrial site by winds and then deposited onto

**Table 5.15** Average $^{226}$Ra concentrations in different types of soils [82]

| Type of soil | $^{226}$Ra (Bqkg$^{-1}$) |
|---|---|
| Taiga Soddy-podzolic | 280 |
| Mixed forest Podzol | 30 |
| Mixed forest-Grey forest soil | 28 |
| Deciduous Forest-Grey forest | 17–33 |
| Wet subtropics-Red clay | 7.4 |
| Dry subtropics- Red earth | 40–54 |
| Desert | 70–126 |
| Atlantic maritime-Podzol soils | 30 |
| Mixed wood plains-Grey brown Luvisol | 20 |
| Boreal shield-Podzol soils | 20 |
| Mixed wood plains-Calcareous Brunisol soils | 30 |

**Table 5.16** $^{226}$Ra in high natural radiation areas [82]

| Area | $^{226}$Ra (Bqkg$^{-1}$) |
|---|---|
| *Asia* | |
| India, Kerala | 7.8–2500 |
| Indonesia, West Java | 2.4–422 |
| Islamic Republic of Iran Ramsar | 80–50 000 |
| Mahallat | 500–7300 |
| *Europe* | |
| Czech Republic Central Bohemia | 76–275 |
| France Ambazac | 950–8860 |
| Russian Federation Komi Republic | 259–71,000 |
| Spain South Galicia Arribes del Duero | 60–250 |
| *Brazil (South America)* | |
| Phosphate area P.E | 29–207 |
| Araxa | 700–42,400 |
| Tapira | 800–29,000 |

soils depending on different mechanisms, such as rainfall, gravitational fallout, wind directions, and strength. By far, coal-fired electricity generation and cement production are industrial activities that release more atmospheric effluents. Apart from $^{222}$Rn, which has already been studied, some 0.2 GBqy$^{-1}$ $^{238}$U are emitted by both industries or 78 GBqy$^{-1}$ $^{210}$Po during cement production. For releases to waters, either river or seawater, phosphoric acid production is the most relevant industry. Up to 336 or 737 GBqy$^{-1}$ $^{238}$U and $^{226}$Ra are estimated to be released into waters [22].

**Table 5.17**  Liquid effluents from NORM industries [22]. These data should be completed with Table 5.5

| Industry | $^{238}$U (GBqy$^{-1}$) | $^{226}$Ra (GBqy$^{-1}$) | $^{222}$Rn (GBqy$^{-1}$) | $^{210}$Po (GBqy$^{-1}$) |
| --- | --- | --- | --- | --- |
| Phosphoric acid | 336 | 737 | 0.18 | 997 |
| Iron steel production | | | | 8 |
| Mineral sands handling | 0.088 | 0.066 | 0.066 | 0.066 |
| Titanium-pigment | 0.002 | 0.002 | 0.002 | 0.002 |
| Oil extraction | | 174 | 174 | 174 |
| Gas extraction | | 32 | 32 | 32 |

A fraction of the emitted radioactivity is deposited in the surrounding soils, contributing to the enhancement of natural radiation levels. This has led to the development of radiation protection programs for these specific industries since the environmental impact can be significant. The enhancement of soil radioactivity due to NORM industrial activities is well documented in the literature. For example, $^{238}$U and $^{226}$Ra concentrations have been found [84] up to $10^3$ Bqkg$^{-1}$ in soil samples collected in the vicinity of a coal-fired power plant, well above the levels accepted as the average content in soils around the world (see Tables 5.14 and 5.15). The origin of this enhancement is the deposition of fly ash produced during combustion and released to the atmosphere. The deposition of the ashes is mainly governed by meteorological factors such as the intensity and direction of the wind. In any case, the deposition is clearly distance dependent from the source, as seen in [85]. Today, effective fly ash removal systems have greatly reduced the radioactivity emissions of coal-fired power plants. A very interesting study on soil contamination by NORM industries was published in [86]. In that paper, a deep investigation is carried out on the NORM signature in soil radioactivity profiles, which allows us to distinguish between natural variations and man-made inputs. The existence, or lack, of secular equilibrium is a clear hint of the origin of radioactivity. Indeed, the industrial manipulation of the raw material may include chemical processes that could break secular radioactive equilibrium. The wastes contain such a mark, and hence, the contaminated samples preserve such a signature. Even coal firing produces a radionuclide fractionation that can be observed in the surrounding soils where the released natural radionuclides will deposit. In Fig. 5.16, an example is given of the differences expected between an unaffected soil radioactivity profile and one that has received the input from a close NORM industry [86].

The filling of contaminated soils helps to decrease the concentration of radionuclide inputs from the NORM industries by diluting the excess into uncontaminated lands. In [87], it was found that the concentrations of $^{210}$Pb and $^{226}$Ra in uncultivated soils affected by coal mine emissions are systematically higher by 2 or 3 times compared to cultivated soils that have been plowed.

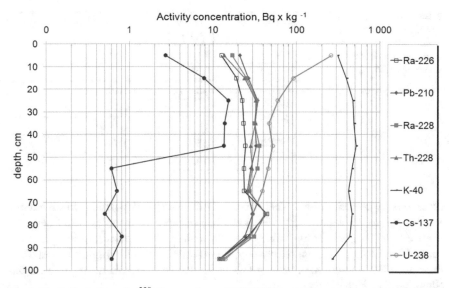

**Fig. 5.16**  Soils showing the $^{238}$U impact from a NORM industry. Figure reprinted from [86] with permission by Elsevier

*Man-made Radionuclides*

The detonation of nuclear weapons injected large amounts of artificial radioactivity into nature (see Chap. 4). A fraction of the released radionuclides was deposited onto soils at a global scale, although mainly in the Northern Hemisphere. In Table 5.18, an example is given on the concentration and deposition of key radionuclides in German soils [79]. In the case of $^{90}$Sr and $^{137}$Cs, concentrations of 7 and 10 Bqkg$^{-1}$, respectively, were found in samples collected during 1985, before the Chernobyl accident. $^{239+240}$Pu concentrations of 0.36 Bqkg$^{-1}$ were measured for the same samples. Since the main activity injection occurred in 1963, it can be considered that the Pu isotope concentrations have remained constant after deposition. However, the originally deposited $^{90}$Sr must have been higher due to its half-life ($T_{1/2} = 28.90$ y). By applying the exponential decay law, it can be found that at the time of deposition, the $^{90}$Sr activity was 11.9 Bqkg$^{-1}$. Similarly, for $^{137}$Cs ($T_{1/2} = 30.08$ y), the activity at the time of the original deposition was 18.1 Bqkg$^{-1}$. In both 1985 and 1963, the $^{137}$Cs activity ratio was close to 1.6, as expected from their fission yields (see Chap. 4). This shows a very similar behavior in that soil. The $\frac{^{239+240}\text{Pu}}{^{90}\text{Sr}}$ activity ratio obtained at the time of deposition is 0.019, close to the expected value of 0.017 [33, 88]. In 1985, the ratio was 0.030, close to the expected value of 0.029, once corrected by the $^{90}$Sr half-life.

In the Southern Hemisphere, $^{90}$Sr and $^{239+240}$Pu concentrations 3.5 and 3.9 lower than those in the Northern Hemisphere, respectively, are expected since they received approximately 20% of the total stratospheric fallout from the nuclear weapon test carried out until 1970. However, the so-called tropospheric fallout resulting from the detonations in French Polynesia and Australia noticeably

**Table 5.18** Data on the concentration and deposition of key radionuclides in soils from Germany before Chernobyl [78]

| Radionuclide | Concentration (Bqkg$^{-1}$) | Deposition (Bqm$^{-2}$) |
|---|---|---|
| $^{90}$Sr | 7 | 2000 |
| $^{106}$Ru | | |
| $^{129}$I | 0.3 | 0.09 |
| $^{131}$I | | |
| $^{137}$Cs | 11 | 3300 |
| $^{239+240}$Pu | 0.36 | 108 |
| $^{238}$Pu | 0.01 | 3.2 |

increased the concentrations. Therefore, in Chile, average depositions in soils ranging from 3 to 15 Bqm$^{-2}$ $^{239+240}$Pu have been found for samples taken between 2007 and 2009 [89]. These values are lower than the deposition found in Germany (see Table 5.18), approximately 100 Bqm$^{-2}$, but still slightly higher than expected. The authors have shown this to be due to tropospheric input from the Southern Hemisphere nuclear tests.

The Chernobyl accident contributed to increasing the activity in soils, as already shown in Chap. 4. This enhancement varied depending on the area [90]. For example, in certain areas of Sweden, some 188 kBqm$^{-2}$ was measured in soils, while some 4 kBqm$^{-2}$ was measured in Austrian areas or 0.44 kBqm$^{-2}$ in Italy. In the case of Germany, as presented in Table 5.18, the enhancement of the relevant radionuclide inventory was a factor of 6 for $^{137}$Cs but only a factor of $10^{-1}$ or $10^{-4}$ for $^{90}$Sr and $^{239+240}$Pu, respectively. These results are consistent with those found for soils collected in Monaco [60], where the authors claim that the concentration levels for $^{137}$Cs increased by 40% due to the Chernobyl accident.

The Palomares accident (see Chap. 4) had a direct impact on soil radioactivity. Early estimates [88] showed that Pu contamination was uneven in the area, with deposition depending on the site. It ranged from $1.18 \times 10^{-2}$ to 1.18 MBqm$^{-2}$ at the time of the accident, values clearly higher than those expected from nuclear weapon fallout (see Table 5.18, for example). After the accident, a remediation program started with an immediate intensive cleaning work of the contaminated areas, apparently approximately 2.3 km$^2$. Therefore, the radioactive contamination of Pu in the soils was significantly reduced. Today, the inventory of Pu isotopes has been reduced to approximately 0.5 kg. It is difficult to know the exact reduction percentage this means. However, considering a possible release of up to 13.5 kg of Pu [91] during the accident, it could be said that today only 4% of the originally dispersed remains in the area. According to [88], the released Pu was weapon grade. Thus, the atom ratio was $\frac{^{240}Pu}{^{239}Pu} = 0.05$. Recent determinations have found that this ratio ranges from 0.06 to 0.09 in soils collected in this area [38]. The ratios are far from being close to those expected for fallout in the Northern Hemisphere, approximately 0.17.

The SNAP 9A accident added a relevant amount of $^{238}$Pu to the inventory of such radionuclides from nuclear detonations. Of the 1.3 PBq $^{238}$Pu deposited on

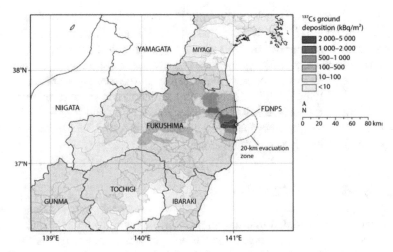

**Fig. 5.17** $^{137}$Cs deposition in the soil surrounding the Fukushima Daiichi NPP as measured months after the accident. Figure taken reprinted [93] with permission by UNSCEAR

the Earth's surface, approximately 0.5 PBq originated from the SNAP 9A accident already described in Chap. 4 [92]. Approximately 23% of the total activity was deposited in the Northern Hemisphere. This, together with $^{238}$Pu produced by weapon detonation, gives concentrations in soils close to 3 Bqm$^{-2}$ or $10^{-2}$ Bqkg$^{-2}$. The SNAP 9A accident slightly modified the $\frac{^{238}Pu}{^{239+240}Pu}$ activity ratio in the Northern Hemisphere, where a value of 0.03 was determined to be very close to the originally derived from weapon detonations, 0.024 [88]. However, in the case of the Southern Hemisphere, the variation was significant, reaching a value of approximately 0.18 [88].

The impact of the Fukushima accident on the soils was intense in the areas close to the accident site. Of course, it was detected in several countries as a result of atmospheric or seawater transport of effluents. However, the significant impact on soils was mainly limited to Japan. In Fig. 5.17, a $^{137}$Cs deposition map in soils is given based on measurements carried out months after the accident. $^{137}$Cs depositions close to or greater than $10^3$ kBqm$^{-2}$ are observed in the vicinity of the accident site, with values decreasing as the distance increases, as expected [93]. As discussed in the previous paragraphs, in some cases, $^{137}$Cs was found to be associated with radioactive particles [73] some $\mu$m in diameter with a high activity concentration of approximately $10^{11}$ Bqg$^{-1}$. From 2 to 80% of the $^{137}$Cs in soils was associated with these particles. $^{137}$Cs soil migration studies show how it depends on several factors, such as the presence of organic matter, soil temperature, bioturbation, and precipitation rates. Of course, ploughing activities helped mitigate the possible radiological impact, as they resulted in the dissolution of the deposited $^{137}$Cs.

Nuclear reprocessing plants include sampling and measurement of radioactivity in soils collected around facilities, as it is shown that radiological impact can exist, either from air or water effluents [54, 55]. For instance, in the case of Sellafield, $^3$H, $^{14}$C, $^{90}$Sr, $^{106}$Ru, $^{125}$Sb, $^{134}$Cs, $^{137}$Cs, $^{241}$Am, U and Pu isotopes are included in the surveillance program. In many cases, the radiological program is limited to the determination of γ-emitting nuclides because its measurements are relatively easy. However, this in turn reduces the information the measurements provide on the environmental impact. It is highly recommended to add more techniques, such as β-counting or α-spectrometry. The measurement of α- and β-emitters together with γ-emitters provides information on the source, age, and, of course, content of the measured radioactive contamination.

## 5.3.2  Radioactive Particles

In general, it is considered that the releases of radioactivity, either from weapon tests, accidental events, or current operational practices, are evenly distributed in the environment. However, a significant fraction of the releases occurs as radioactive particles [1], formerly called 'hot particles' after the high concentration of radioactivity found in them. These particles are aggregates of radionuclides of varied sizes, compositions, activity concentrations, and chemical forms. The presence of radioactive particles in a sample generates an inhomogeneous radioactivity distribution that can cause serious errors in determining the sample radioactivity concentration. Therefore, the possible existence of radioactive particles in environmental samples must be taken into account in radioactivity determination procedures. Moreover, the dynamics of the radionuclides contained in the particles, their solubility, mobility, and bioavailability are different from those of the radionuclides present in the environment as ions, molecules, or forming complexes. From a radiological viewpoint, this is certainly relevant, as the majority of dosimetry estimation models do not incorporate the specific behavior of radionuclides associated with radioactive particles. A radioactive particle picture taken by electron microscopy is given in Fig. 20.9.

According to [94], aggregates from submicrometer to 2000 μm diameters can be considered radioactive particles. Higher-size aggregates are called fragments. Specifically, in water, radioactive particles are defined as aggregates with diameters greater than 0.45 μm. They tend to settle down as a result of gravitational forces, provided its density is large enough. The aggregates with a size from 0.001 to 0.45 μm are radioactive colloids or pseudocolloids. In the case of aerosols, radioactive particles can be classified according to the average aerodynamic diameter (AMAD). However, it should be known that particles up to 10 μm can be relevant to the human respiratory system.

It is known that the radionuclide composition of radioactive particles, their radioactivity concentration, and isotopic ratios depend on the source. However, the size, structure, chemical composition, and oxidation states are also influenced by the release scenario and the environment where the particle is deposited. It has been shown that radioactive particles remain for a long time after release.

This method represents a local point of contamination that can be radiologically relevant on a long-term basis.

Although not easy, radioactive particles released to the environment can be classified into the following types [94].

1. Nuclear fuel or fission product particles formed by mechanical disruption of the original volume of radioactive material.
2. Particles produced by condensation or aggregation processes after emission, generally of volatile elements.
3. Particles produced within the nuclear fuel during normal operation or during the releases (an example is the Ru particles found in the Chernobyl accident or $^{137}$Cs particles from Fukushima, which will be described in what follows).
4. Colloid or pseudocolloid forms are directly emitted or formed after release.
5. NORM particles produced at the beginning of the nuclear fuel cycle during U or Th mining. NORM particles from nonnuclear industry operations or minerals with high U and Th contents.

Traditionally, the concept of radioactive particles has been associated with artificial radioactivity, as the first particles were found within scenarios affected by nuclear weapon or reactor accidents. However, the study of NORM radioactive particles and their radiological impact is currently completely accepted within this field [95].

*Nuclear Weapon Tests*
Following [94], there are four types of particles produced in the fireball during nuclear detonation.

Type 1 particles are those 'formed by condensation of vapors of fission and activation products and bomb materials, including fissile materials and other bomb construction materials'. These particles are spherical and usually have a diameter <1 μm.

Type 2 particles are 'formed by precipitation of the molten components of the bomb and the materials that are sucked into the fireball'. The particles are generally spherical and have diameters ranging from 0.2 to 2 mm.

Type 3 particles are 'formed from solid or partially molten components of the ground'. The shape of these particles is irregular. In addition, on the surface of the particles, condensed gaseous fission products or small type 1 particles can be found.

Finally, agglomerates of Type 1 to 3 particles can be found and are called Type 4 particles. The particle type depends on the detonation. Thus, particle Types 1 and 2 are typical of high-altitude detonations, while Type 3 is the result of ground detonations.

Radioactive particles have been found as a result of global fallout. They were injected or formed in the stratosphere and participated in the general atmospheric circulation pattern before ground deposition. These particles are clearly distinguishable from those found at the test sites as a result of local fallout. Although

**Table 5.19** Radioactive particles feature global fallout and local fallout [94]

| Location | Type | Diameter ($\mu$m) | Activity (Bq) |
|----------|------|-------------------|---------------|
| Sweden | Global | >0.2–10 | 0.1–3 |
| Germany | Global | <10 | <40 |
| Japan | Global | 8–14 | 200–3000 |
| USA | Global | 2–4 | 0.5–35 Bq/1000 m$^{-3}$ |
| Taranaki | Local | 100–1600 | 600–330,000, Pu |
| Taranaki | Local | 6–500 | 15, Am |
| Mururoa | Local | 150–650 | 200–500, Am |
| Mururoa | Local | 200–600 | 200–240,000, Am |

the former have diameters of approximately a few $\mu$m, with few exceptions having diameters from 30 to 60 $\mu$m, the local fallout particles can be fragment size, i.e., with diameters beyond 2 mm [94]. In fact, fragments of tenths of cm have been found in some test sites. In Table 5.19, taken from [94], some examples are given of the characteristics of radioactive particles, either from global or local fallout. It can be seen that in general, radioactive particles derived from global fallout are smaller than those originating from local fallout, which in some cases are fragments. In addition, the activity content is clearly higher for local fallout particles.

*Nuclear Reactor Accidents*
Radioactive particles are also produced during nuclear reactor accidents. The most studied case has been the Chernobyl accident, where three types of particles have been distinguished [94].

Monoelemental particles or Ru particles. They are compounded mainly by $^{103}$Ru and $^{106}$Ru but also by stable elements, mainly transition metals, as well as lower activities from $^{125}$Sb, $^{60}$Co and some volatile fission elements.

Fuel particles. U oxide particles are depleted in volatile isotopes ($^{131}$I, $^{132}$Te, $^{137}$Cs, and $^{134}$Cs) and rich in refractory elements ($^{239}$Np, $^{141}$Ce, $^{144}$Ce, $^{103}$Ru, $^{106}$Ru, $^{95}$Zr, $^{95}$Nb, and $^{99}$Mo).

Condensation particles. Particles formed by condensation of volatile elements in aerosols or soil particles.

Table 5.20 lists some examples of radioactive particles found in Europe as a consequence of the Chernobyl accident. An ample variety of sizes and activities can be found in both U-fuel-like and Ru-particles. Radioactive particles were also emitted during the Fukushima accident [73]. In general, they were found in dust and soil samples, mainly containing $^{137}$Cs and $^{134}$Cs. The particles associated with the dust samples were spherical in shape with a diameter of 2.6 $\mu$m, and the distribution of Cs isotopes was homogeneously embedded in silicates. In the case of soils, two kinds of particles were found. Both types had a nonspherical shape, and the Cs isotopes were inhomogeneously distributed. The matrix of the larger particles, some 2 $\mu$m in diameter, was glass $SiO_2$, while that of the smaller particles, approximately 10 nm in diameter, was a compound of Z and Fe. To date,

**Table 5.20**   Some examples of radioactive particles from the Chernobyl accident collected in Europe [94]

| Location | Type | Diameter ($\mu$m) | Activity (Bq) |
|---|---|---|---|
| Uusikaupunki | U fuel, Ru | 23, 5 | 120, 200 |
| Helsinki | U fuel | 24 | 140 |
| Nyköping | U fuel, Ru | 41, 27 | 750, 31,000 |
| Stockholm | U fuel, Ru | 33, 17 | 380, 7400 |
| Kjeller | Ru | 6.3 | 65 |
| Mikolajki and Krakow | U fuel, Ru | 35 58 | 2000, 308,000 |
| Kiev | U fuel | 45 | 950 |
| Brunswick | U fuel | 6.9 | 15 |
| Konstanz | U fuel, Ru | 16, 5.6 | 41, 270 |
| Vienna | U fuel | 18 | 66 |
| Budapest | U fuel, Ru | 33, 72 | 230, 570 |
| Athens, Thessaloniki | U fuel, Ru | 28, 30 | 240, 43,000 |

there is no evidence of the emission of fuel particles similar to those released from Chernobyl.

Other reactor accidents produced radioactive particles, and evidence has been found to demonstrate that they reach the environment. An interesting case was that of the Windscale reactor accident that resulted in the uncontrolled emission of radioactivity into the environment from 1952 to 1957. Once failure was detected, the surrounding contaminated tissue was stored. Years later, in 1992, the soils were analyzed, and flake-like radioactive particles were found in the samples. In 1957, a fire of graphite in one of the units produced the emission of $^{239}$Pu and U oxides. The soils located up to 4 km from the reactors were contaminated. Radioactive particles of 20–500 $\mu$m were found at such distances. Furthermore, according to [94, 95], particles up to 700 $\mu$m in length were emitted from the stacks at that time.

*Nuclear Weapon Accidents*

Radioactive particles from the Palomares and Thule accidents (see Chap. 4) have been collected in soils and sediments affected by them. The samples were taken almost thirty years after the accidents, which showed that the particles remained intact for a long time within the corresponding environment. As already described, the accidents were of similar nature, and it has been found that the particles have similar characteristics as well. Their size ranges from 1 to 50 $\mu$m, although fragments have been collected in the case of Palomares [91]. In either the Palomares or Thule particles, U and Pu are unevenly distributed. The $\frac{^{240}Pu}{^{239}Pu}$ isotopic ratios were found to be $0.061 \pm 0.006$ and $0.055 \pm 0.007$ for Palomares and Thule particles, respectively. These values are clearly far from those expected for global fallout, which is close to 0.18. The $\frac{^{235}U}{^{239}Pu}$ isotopes were also similar in both cases,

$0.78 \pm 0.14$ for Palomares and $0.62 \pm 0.13$ for Thule particles, with $^{235}$U being the main fissile material in the bomb. In fact, an atom ratio $\frac{^{238}U}{^{235}U} = 5.4 \pm 0.2$ was found for one Palomares particle, which revealed that the nuclear fuel was highly enriched in $^{235}$U. Note that such a ratio is 137.80 in natural samples.

*Depleted Uranium Ammunition*

Depleted uranium is known to have lower amounts of U isotopes different from $^{238}$U. This heavy isotope has been shown to be very useful in military applications since it is denser than Pb and is able to penetrate thick shielding. This ammunition has been used in several wars. Soil samples collected in war scenarios have been shown to contain depleted uranium particles with sizes close to 1 µm in the case of the Gulf War or of submicrometer to 30 µm in the Balkans War [96, 97]. The $\frac{^{235}U}{^{238}U}$ isotopic ratio was found to be close to 0.2%, which is 3 times lower than the natural isotopic ratio. $^{236}$U has been measured in the samples, showing that the procedure of the used U is certainly the reprocessing of nuclear fuel.

*Nuclear Reprocessing Plants*

Radioactive particles from nuclear fuel reprocessing facilities have been documented to exist in the environment surrounding installations. Particles from the Mayak complex in Russia have been found embedded in sediments and soils collected along the Techa River. Together with several elements, such as Fe, Ti, Ce, Pb, Mo, or Sr, $^{241}$Am and $^{137}$Cs have been detected as the only radioactive species in the sample. Radioactive particles from Sellafield have been collected in Irish Sea sediments. In this case, the matrix of the particles was U, and it is likely that they were U-fuel radioactive particles. They contain $^{241}$Am and $^{137}$Cs, and their sizes range from 5 to 20 µm. Radioactive particles in the form of colloids from Sellafield and La Hague have also been found.

*Nuclear Waste Dumped into the Sea*

Radioactive particles have been detected in the Kara Sea and Novaya Zembla fjords, which have been attributed to accidental or current dumping practices in the area. Particles $10 \times 20$ µm$^2$ sampled in sediments containing $^{60}$Co mixed with clay have been studied. In principle, the presence of U isotopes or any other α-emitting nuclides was not documented.

*NORM Sources*

Radioactive particles can be present in areas of high natural radioactivity or released into the environment from conventional industries that use NORM as raw material. In both cases, they potentially represent a high radiological risk. Sizes from 1 to 100 µm can be found, and special consideration must be paid to those relevant for the respiratory system ($\leq 10$ µm). The origin is varied. Indeed, products, byproducts, tailings, slag dumps, wastes, etc., from conventional industry using NORM as raw material (see Chap. 4) can populate the surrounding environment with radioactive particles. Thus, soils, atmospheric water currents, and, of course, the biosphere can be reached by the particles released during operation. Fine particles, 1–3 μm, can be easily transported into the air and inhaled by the population.

A very relevant update on radioactive particle investigation can be found in [98], where a very complete account of the different scenarios where radioactive particles can appear is presented.

There are several methods to characterize radioactive particles from a physical or chemical point of view. A full account of them is given in Chaps. 13 and 20.

Radioactive particles can have short- or long-term radiological significance. Respiratory significant particles have of course an immediate radiological impact, but those deposited in soils or sediments behave as a local source of activity that would be disseminated into the environment by different processes. First, the particles themselves can be resuspended and dispersed in the environment due to meteorological factors or water current movements. In addition, radionuclides can be mobilized from particles by weathering and dissolution processes. The mobility and consequently the bioavailability of the radionuclides will depend on the composition and structure of the particle matrix, the oxidation states of the radionuclides, and the chemical scenario of the environment in which the particle was deposited [99]. It is very important to know the behavior of radionuclides associated with particles to have a correct description of their environmental dispersion. As already commented on in previous paragraphs, the distribution and transfer coefficients are conventionally defined for homogeneously distributed radionuclide concentrations in water, sediments, soil, or vegetables and under equilibrium conditions. This is just an approximation to what is occurring in Nature. However, it is far from reality when radioactive particles are present.

The mobility and bioavailability of radionuclides in radioactive particles are studied by applying leaching or sequential extraction procedures (see Chap. 13) to determine the capacity of radionuclides to dissolve by some inorganic or organic agents [4]. Thus, the particles are exposed to abiotic agents such as rainwater, sea, and irrigation water, mimicking solutions simulating body fluids or to biotic agents such as rumen liquids from grazing animals. When these procedures are applied, it has been found that Palomares particles are very refractory and have remained unchanged for decades in soils. They are basically made up of $PuO_2$ and $UO_2$, which explains the low mobility found in soils with very low vertical migration. The Thule particles are similar but have been shown to be more sensitive to leaching experiments, and the Pu particles have been remobilized more easily than those of Palomares [98].

## 5.4 Levels and Behavior of Radioactivity in Fresh Waters

### 5.4.1 Rivers and Sediments

There are several processes that cause the presence of radioactivity in rivers. Atmospheric global fallout and liquid or air releases from local or nearby sources of radioactivity are the main suppliers of radionuclides for river systems. In addition, radionuclides from the river drainage area, including connected aquifer systems, also contribute to the radioactivity concentration in river waters. Rivers are dynamic systems, and radionuclide exchanges take place between their different components. In this way, there is an exchange between sediments and river water and between the suspended matter and soluble phase. Suspended matter deposits on the riverbed, introducing radionuclides into the sediments with radionuclides. Similarly, sediments can be resuspended, and the fine particle fraction can be transported following the vertical movement of the water. This was already illustrated in the previous sections and specifically in Fig. 5.3.

Rivers can also be considered a source of radioactivity for oceans or lakes. On the other hand, since rivers contain approximately 0.003% freshwater worldwide, they are also a potential source of radiation dose to the population. Indeed, river waters are used on many occasions for human consumption after purification. Generally, this cleaning process is not adequately designed for radionuclides, so a significant fraction can remain in purified water and enter the human food chain. In addition, when $^3$H is present in river waters, it is not possible to avoid its presence in purified water. Flooding episodes contribute to the transport of radionuclides from rivers to the nearby environment. They lead to the deposition of radionuclides in soils surrounding the river, which is very relevant for agricultural soils. In turn, the inverse process can reintroduce radionuclides in the water stream during floods. Thus, radionuclides in soils can be remobilized by flood waters.

In the case of estuarine waters, the interactions among the components of the river system are also affected by tidal effects. Seawater intrusion changes the chemical environment, and salinity may change the sediment–water exchange for certain radionuclides.

Since the nature of the river is a system where water flows the residence time of radionuclides in the waters is short compared to that in other freshwater systems, such as lakes. In addition, the river water flux changes during the season. Usually, the water flux increases during spring due to snow melting at the river head. Aquifer recharging can also increase water flow in connected rivers. In contrast, during summers, the flux decreases, and in some environments, the river completely dries. In these cases, the river bed is directly exposed to atmospheric global fallout or other atmospheric inputs. The river bed accumulates radionuclides that can be mobilized during the recharge of the river system in autumn and winter.

*Cosmogenic Radionuclides*

Cosmogenic radionuclides enter river systems directly from the atmosphere or through the river drainage area [100]. However, significant cosmogenic nuclides are also produced by nuclear fission or activation and, therefore, are released into the environment along the nuclear fuel cycle and during detonations. In fact, the bomb pulse masked the natural production of cosmogenic radionuclides. In the case of $^3$H ($T_{1/2} = 12.32$ y), the contribution from nuclear detonations has decreased, and today, natural levels are being recovered. However, since $^3$H is released during different steps of the nuclear fuel cycle, it is difficult to find pristine environments where the $^3$H cosmogenic concentrations can be measured without interference from artificial sources. Measurements carried out in several French rivers not affected by nuclear installations have shown a high variability in the obtained values, with concentrations ranging from 0.11 to 0.86 Bql$^{-1}$ [101]. In Indian rivers, an average value of 0.35 Bql$^{-1}$ was found, and the authors of this work do not exclude the possibility of some influence on these $^3$H values released during nuclear tests [102]. The main form of $^3$H in surface and subsurface waters is HTO, although organically bound $^3$H (OBT) has been discovered in river sediments [103]. However, this finding is mainly related to artificial inputs and, therefore, will be described in the next paragraphs.

Cosmogenic $^{10}$Be and $^{26}$Al are very interesting for river water systems. They can help to understand the time evolution of the river basin, as well as the sedimentation production and deposition process in rivers. This can be done by determining erosion or denudation rates using $^{10}$Be and $^{26}$Al measurements [104].

*Primordial Radionuclides*
In the absence of local sources of activity, the presence of primordial radionuclides in rivers is essentially governed by the geological substrate of the watershed. Hence, large variability was found for the concentrations of primordial radionuclides in river waters.

U is the most soluble of the primordial radionuclides. Th is less soluble and can be found at clearly lower concentrations in water than in the U. For example, $^{238}$U concentration levels of 10.7 to 32.1 mBql$^{-1}$ were found in the Guadalquivir River in southern Spain [5]. The $^{232}$Th concentrations for such waters were one order of magnitude smaller, ranging from 0.08 to 3.5 mBql$^{-1}$. An extensive review [105] shows that on average, the concentration of $^{238}$U in river waters worldwide is approximately 2 mBql$^{-1}$, but significant variations are closely related to the geological substrate and the water flux. It is important to note that the survey was carried out on a large number of river systems with water flows from 1 to 7000 km$^3$y$^{-1}$. Concentrations are very variable, however, and depend on the region. For example, $^{238}$U concentrations in German rivers have been reported to range from 0.5 to 310 mBql$^{-1}$. The connection to the geological substrate is very clear, and in granitic regions, levels can be measured up to 100 Bql$^{-1}$ [75].

Differences in solubility explain the disequilibrium found among the members of the same radioactive series. This is a very common effect, and one can find $\frac{^{230}Th}{^{234}U}$ activity ratios from 0.006 to 0.142 in the Guadalquivir River samples discussed above. In the same system, $\frac{^{234}U}{^{238}U}$ activity ratios from 1.23 to 1.38 were also obtained. These values are in agreement with the consensus average values found for rivers around the world from 1.2 to 1.3 [105]. The disequilibrium of U isotopes comes from the preferential dissolution of $^{234}$U from the bedrock that feeds the river water. In general, the $^{238}$U $\alpha$ decay recoil nucleus provokes a break of the crystalline structure that causes the daughter nuclide to appear under an unstable electronic configuration that makes it more soluble in the rock–water interface.

This effect explains the disequilibrium in the $\frac{^{234}U}{^{238}U}$ activity ratio but also in that of $\frac{^{228}Th}{^{232}Th}$.

The case of Ra is very interesting [82]. Ra is less soluble than U, although more soluble than Th. Ra has a noticeable tendency to associate with suspended matter and is finally found in the sediment phase. Thus, its concentration in water is low, although higher than that of Th. For instance, in the case we are describing, levels of approximately 3–6 mBql$^{-1}$ $^{226}$Ra can be found. However, in estuarine systems, several authors note an increase in $^{226}$Ra in dissolution. For example, in the commented case of the Guadalquivir River, the concentrations increase to 30 mBql$^{-1}$ [5]. Since there are no local sources, it cannot be assigned to a seawater input insofar as the concentration of Ra in seawater is also low. The increase is attributed to a redissolution effect from the sediment due to an increase in salinity. This shows the dynamic character of riverine systems with exchanges among the different phases. This makes the sediment a radionuclide source instead of a sink.

Rivers are, in some cases, affected by NORM industry waste releases. In this case, the concentrations in waters, sediments, and suspended matter clearly increase, and the environment is contaminated with radiologically significant activities.

In [106], the impact of phosphoric acid factories based on phosphate rock processing was found in river waters taken close to these industries. Concentrations of 780 mBql$^{-1}$ $^{238}$U were found, ten times higher than those observed in the same river at sampling stations far from the factories. Now, activity ratios $\frac{^{234}U}{^{238}U} = 1.00 \pm 0.02$ revealed that the U isotopes in water were in secular equilibrium. This is not typical in natural systems. In contrast, the typical equilibrium value found in the minerals shows that direct releases were made from the factory to the rivers. $^{226}$Ra concentrations close to 700 mBql$^{-1}$ were determined in such waters, far from the currently accepted values from nonaffected rivers, some few mBql$^{-1}$. $^{238}$U in sediments reach values of approximately 1 Bqkg$^{-1}$. Once again, the $\frac{^{234}U}{^{238}U}$ activity ratio was close to the equilibrium value. Very interesting was the fact that $^{235}$U could be measured as a result of the high activities involved. In the most contaminated samples, the $\frac{^{235}U}{^{238}U}$ The activity ratio was found to be 0.048 ± 0.007, clearly compatible with the accepted natural value of 0.046. In the sediments, the $^{226}$Ra activity reached values of approximately 1.4 Bqkg$^{-1}$, very far from those found in unaffected rivers, approximately 0.06 Bqkg$^{-1}$.

As an additional example, we can give that presented in [85], where a deep study is carried out about the influence of coal mines on the concentration levels of primordial radionuclides in some German rivers. In this study, the $^{226}$Ra average concentrations in German river waters were found to be close to 4 mBql$^{-1}$. In general, the levels were maintained over time, and the deviations were attributed to local geological conditions. However, it was clearly identified that the impact of coal mines on the river system was clearly identified as concentration levels reached values ten times higher or more, that is, 40 mBql$^{-1}$ downstream of the coal mines. The pit water from the mines was identified as the cause of the increase.

The yearly $^{226}$Ra releases from the mines were estimated to range from 40 to 40 $\times$ 10$^3$ GBqy$^{-1}$.

*Man-made Radionuclides*

Global fallout radionuclides can reach river waters directly from the atmosphere or from the drainage area. Nuclear detonations in the 1960s affected river streams around the world, and deposited radionuclides were transported to oceans or lakes and river sediments. Today, the global fallout signal has been lost in river waters, although the bomb pulse can be still observed in some sediment systems [107]. Indeed, long-lived radionuclides derived from nuclear tests can be identified in sediments. The case of Pu isotopes is a good example, as Pu forms very insoluble chemical species in an aqueous environment. In this case, Pu remains in the sediment at measurable levels, and it is possible to identify its origin by determining the $\frac{^{240}Pu}{^{239}Pu}$ atom ratio, which is accepted to be 0.18 for global fallout in the Northern Hemisphere (see Chap. 4). There is a huge amount of information in the scientific literature regarding this possibility. An example [108] is the case of the Liao River in China, where Pu isotopes were measured in sediments taken from 2012 to 2015 with values for $^{239+240}$Pu activities of 0.294 mBqg$^{-1}$ on average. The mean atom ratio was found to be 0.188, clearly consistent with the origin of Pu fallout in the sediments.

Today, global fallout is an interesting tool for understanding the dynamics of river catchment areas, especially sedimentation and erosion processes. This is possible because $^{137}$Cs and $^{90}$Sr still provide a time signal useful for determining the sediment input and output fluxes, suspended matter ages, sediment residence times, and many interesting parameters [109–111]. In general, the measurements of $^{137}$Cs are made together with those of $^7$Be and $^{210}$Pb in soils from the catchment area and river sediments. A very complete review of these kinds of work can be found in [112], which is highly recommended. For the same reason, the measurement of $^{137}$Cs in underground river sediments has allowed the determination of the sources of sediments to the rivers [113].

As said before, global radionuclide concentration levels in river waters are low, and in many cases, only long-lived species can be found in sediments. Recently, [114], it has been observed that the retreat of glaciers due to global warming releases significant amounts of radionuclides preserved in ice and sediments trapped by the glacier mass. These radionuclides reach downstream rivers, and unusually high activity concentrations are found in waters and sediments of $^{137}$Cs, $^{241}$Am, and others. In addition to radionuclides, other contaminants are also mobilized due to the melting process, which is becoming an important environmental issue.

Despite this specific case, in rivers not affected by other sources, the global fallout concentration levels are very low, generally close to a few μBql$^{-1}$ for $^{137}$Cs, $^{241}$Am and $^{239+240}$Pu [107], which makes their measurement a difficult experimental challenge that cannot be overcome with conventional methods (see Chaps. 14, 15, 16, 18, and 19). Only 2% of the highly contaminated area (approximately 180 kBqm$^{-2}$) after the Chernobyl accident was occupied by rivers and lakes [72]. Initially, the contamination of these aquatic systems was very high, although the deposited activity rapidly decreased due to sedimentation processes and water transport [115]. However, the catchment areas were also contaminated, and since the accident, they have acted as radionuclide sources to rivers and lakes. After the initial decrease takes place within the first years after the accident, the radionuclide concentration in dissolution and suspended matter is governed by the radionuclide soil dynamics in the catchment area. Based on $^{137}$Cs observations, it has been found that either in dissolution or in particulate matter, the concentration time pattern could be reproduced by a two-exponential type model. This behavior seems to be general and has also been observed in Fukushima river systems [116]. In Fig. 5.18, data are provided on the time evolution of $^{137}$Cs in rivers affected by the Chernobyl accident and the Fukushima accident. Although the first exponential term describes the processes taking place within the river itself soon after the accident, the second exponential term describes a kind of medium- or long-term $^{137}$Cs behavior in the river and is dependent on the catchment area features as well as radionuclide inventory. A very interesting study can be found in [117]. There, the temporal evolution of $^{137}$Cs in dissolution and suspended matter is analyzed for two rivers affected by the Fukushima accident from 2015 to 2017. The $^{137}$Cs concentrations are given in Table 5.21. Based on the time evolution of $^{137}$Cs in waters and applying an exponential model, the authors found an environmental half-life of 3.7 and 2.4 y for $^{137}$Cs in dissolution or in particulate form, respectively, in the Ukedo River. In the case of the Ohta River, the values were 2.3 and 1.6 y, which showed a different behavior depending on the investigated catchment area.

Following national regulations, nuclear fuel reprocessing and nuclear power plants can release wastes to the environment and to rivers specifically. Inadvertent releases can also occur. Radioactive effluents will be transported by water currents and, in many cases, poured into the sea. A fraction of the releases, depending on the radionuclide, can be deposited in river sediments. The Clinch River results, affected by the Chalk River nuclear facility, show that 20% of the released $^{137}$Cs and 0.2% $^{90}$Sr can be found in river sediments after several decades of operation [22]. This is interesting data about the differences in mobility between two key radionuclides in the environment. The Rhone River in France is a good example of a river system in which several nuclear installations are placed. Nuclear power plants and the Marcoule Nuclear Fuel Reprocessing Plant have historically released radioactivity into the ocean, which in turn is a recognized man-made radionuclide source to the Mediterranean Sea [118]. Although releases have decreased noticeably by 85% since the late 1990s, the Rhone River receives approximately 65% of radioactive effluents in freshwater in France. Approximately 85% of the waste comes from the Marcoule facility. In 1997, some dismantling activities of the oldest installations began, and this significantly decreased the impact on the river. In fact, the concentrations of $^{238}$Pu and $^{239\,+\,240}$Pu in the sediments before 1997 were almost two orders of magnitude higher than the levels in the rest of the French rivers, reaching values close to 10 Bqkg$^{-1}$ for both radionuclides. It was found that approximately 40% of the $^{239+240}$Pu and 80% of the $^{238}$Pu found in the sediments originated from the nuclear facility [107]. In water, concentrations soon

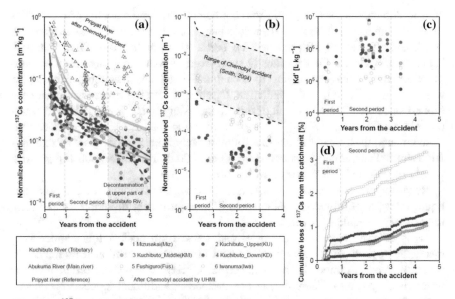

**Fig. 5.18**  $^{137}$Cs concentrations in dissolved and suspended matter in rivers affected by the Chernobyl and Fukushima accidents. The time evolution of the concentrations is very similar and follows an exponential decrease. Figure reprinted from [116]

**Table 5.21**  Temporal trends of $^{137}$Cs concentrations (Bqm$^{-3}$) in river water affected by the Fukushima accident in [117]

| River | 2015 | 2017 |
|---|---|---|
| Ukedo (catchment 151 km$^2$) | 175 (dissolved matter) | 125 (dissolved matter) |
| | 45.4 (particulate matter | 22.6 (particulate matter) |
| Ohta (catchment 57 km$^2$) | 90.4 (dissolved matter) | 52.8 (dissolved matter) |
| | 24 (particulate matter | 9.3 (particulate matter) |

after decommissioning began, and the levels decreased to 0.08 to 0.5 Bqkg$^{-1}$, respectively. Today, concentrations close to 0.1 Bqkg$^{-1}$ are currently found for $^{239+240}$Pu, levels very similar to the estimated contribution from the erosion of the Rhone watershed.

Regarding $^3$H, the Marcoule site clearly influences the environment of the Rhone River, since average concentrations of approximately 6 Bql$^{-1}$ can be found [54]. This value should be compared with the average levels found for other French rivers, ranging from $10^{-1}$ to 1 Bql$^{-1}$ [101]. It is estimated that approximately 300 TBqy$^{-1}$ is the annual $^3$H flux over the Mediterranean Sea from the Rhone River, which is approximately 100 times higher than the flux from other rivers.

A well-known case of a river system affected by releases from nuclear fuel processing plants is the Techa River in the Chelyabinsk region, where the Mayak nuclear complex is located [22]. From the beginning, large amounts of radioactivity have been released into the river. It is estimated that from 1949 to 1956, some 100 liquid effluents of PBq reached the Techa River, including 2 TBq α-emitters [119]. In 1957, some 74 PBq mostly short-lived radionuclides were released into the close river environment due to a chemical explosion that took place in the waste storage system of the installation. Discharges into the rivers have been very limited since then, but as a result of this waste history today, it is possible to determine levels from 7 to 23 $Bql^{-1}$ $^{90}Sr$ or from 0.6 to 0.23 $Bql^{-1}$ $^{137}Cs$. Of course, in sediments, the concentrations are comparatively higher, as they act as radionuclide sinks. Thus, it can be found from 40 to 2000 or 100 to 280,000 $Bqkg^{-1}$ dry weight of $^{90}Sr$ and $^{137}Cs$, respectively [22].

Regarding nuclear power plants, an extensive compendium of airborne and liquid releases worldwide is gathered in [22]. Intermittent discharges make their impact very variable over time.

$^{3}H$ is released from nuclear power plants and from nuclear fuel reprocessing plants, and estimated releases can be found in the above-mentioned reference. In water, the most common $^{3}H$ chemical species is $^{3}HHO$ and is therefore involved in water river dynamics. Therefore, in the absence of additional inputs, the concentration levels in the waters tend to rapidly decrease. However, it has been found to be $^{3}H$ organically bound to organic matter, OBT, present in river sediments, where it is preserved for longer times [103]. OBT from the watchmaking industry has been determined in sediments from the Rhone River. Indeed, the French and Swiss watchmaking industries used approximately 28 PBq $^{3}H$ from 1962 to 2008, taking advantage of the luminescent properties of some tritiated salts. Despite the stoppage of $^{3}H$ use and, at least, Swiss watch production being removed from the market and recovered from owners, concentrations $10^{3}$ to $10^{4}$ higher than natural $^{3}H$ levels are still found in Rhone River sediments under the OBT form [120].

In addition to these sources, there are other less intense anthropogenic radionuclides in river waters. The use of radionuclides in nuclear medicine, such as $^{131}I$ ($T_{1/2} = 8.025$ d), allows the discovery of traces in hospital sewage water and in sewage sludge from water treatment plants [121]. In river waters, levels of up to 20 $mBql^{-1}$ have been determined [122] and identified as originating in hospitals. Fortunately, nuclear medicine isotopes are short-lived, as $^{131}I$ is, or even shorter-lived. For that, they tend to disappear from the environment quickly.

## 5.4.2  Lakes and Sediments

Lakes contain approximately 0.3% freshwater worldwide. Radionuclides can be present in the dissolved phase, suspended matter, or sediments. The exchange among the different phases is a dynamic process that depends on the radionuclide but also on the chemical environment of the lake, which is very important because the water residence time in lakes is usually high and, of course, higher than that of rivers. Thus, redox conditions, salinity, and organic matter content are important parameters that should be taken into account when studying radionuclide exchange between phases. Such a high water residence time makes lakes interesting natural laboratories for determining the behavior of radionuclides in the environment, including their incorporation into the biota. Lakes are filled with water from incoming rivers, groundwater or aquifer systems, ice melting, soil runoff, and rainfall. In contrast, lakes lose water by evaporation or outflow mechanisms that provide water to subsequent rivers, aquifers, or the sea. Inflow and outflow rates determine the residence time of the water.

From a radiological viewpoint, lakes deserve investigation, provided that their waters can be used for human consumption. Furthermore, authorized (or accidental) discharges from nuclear installations are possible in lakes, which makes them a potential reservoir of radionuclides available to be distributed in the surrounding environment.

*Cosmogenic*

Cosmogenic radionuclides are present in lake water and sediments. The main interest of cosmogenic radionuclides within this topic is the utility some of them have to provide information on the time evolution of the lake or its surrounding environment. The measurements of $^{10}$Be and $^{26}$Al in sediments and lake shoreline soil or rock samples help to understand the time and geological evolution of the lake system [104]. $^{14}$C is also used to date paleo-lake environments [123]. $^{22}$Na has been used to study groundwater inflow in a lake [124]. Many other examples can illustrate how cosmogenic radionuclides are certainly useful for tracing lacustrine and, in general, natural processes. The large variety of half-lives and chemical behavior allow some radionuclides to trace specific natural compartments, and lakes are one of the potential examples.

$^{3}$H is a very important cosmogenic radionuclide that is incorporated into lake water as HTO and into lake sediment as OBT (see previous paragraphs). According to [121], the $^{3}$H concentration in surface waters of natural origin ranges from 0.20 to 0.90 Bql$^{-1}$, that is, some TU. This amount was surpassed by the contribution from nuclear detonations (see the following paragraphs) and in some specific systems by the releases from nuclear installations. Today, the natural $^{3}$H level is being recovered, and in the absence of specific input, such an amount can be considered representative of the present status.

*Primordial*

The presence of primordial radionuclides in a lake is basically due to their interaction with aquifers or groundwater reservoirs, incoming river water, the geological substrate, and soil runoff. $^{238}$U, $^{235}$U, $^{232}$Th and their descendants can be found in lakes at a high variety of concentrations due to the diversity of sources. From a radiological viewpoint, the relevant descendants are $^{210}$Po, $^{210}$Pb, $^{210}$Bi, $^{222}$Rn, $^{226}$Ra, and $^{228}$Ra [75]. In the absence of specific sources, the typical concentrations of these radionuclides in lake water are very similar to those in surface waters, which have already been described. Furthermore, the exchange of activity between the dissolved and sediment phases depends on several parameters, such as salinity, oxic, or anoxic conditions of the lake [125] and their combined action.

Thus, it is difficult to provide typical concentrations in lacustrine sediments. Nevertheless, the analysis of primordial radionuclides in water and sediments allows us to understand the lake dynamics and the behavior of radionuclides in them [126]. As an example, as already described in Chap. 4, the $^{210}$Pb dating method finds a very wide application field in the case of lacustrine systems.

Lake environments are affected by NORM industries. Oil and natural gas fields [127] or mining activities [128] contribute to the enhancement of the concentrations of primordial radionuclides. In these cases, the use of radionuclides as time markers helps to understand the time evolution and dynamics of the studied system. The fabrication of U fuels is also a potential source of additional U isotopes to the environment. This is the case for some French installations where the releases from such facilities are also well controlled with respect to natural radioactivity [54].

*Man-made*

Lakes received direct inputs from radioactive fallout from nuclear weapons. Nevertheless, the bomb pulse can hardly be observed today in waters, as different processes have contributed to the removal of man-made radionuclides from the water column. Radioactive decay and scavenging processes make it difficult to find significant radionuclide concentrations from fallout in lake waters. However, long-lived radionuclides persist in sediments that act as a repository of the radioactive bomb signature. Of course, the mobility of radionuclides in the sediment column modifies their concentrations, so it is difficult to give a representative figure directly connected to the large 1963 fallout in the corresponding sediment layer. However, it is possible to observe up to 100 mBq g$^{-1}$ $^{137}$Cs as a marker of nuclear weapon fallout in some sediment lakes [107]. Even in the case of $^3$H, lake sediments can act as a repository, since organically bound tritium can be found frequently [103]. In addition, fallout radionuclides provide a fingerprint that allows us to analyze the temporal evolution of sediments, lake dynamics, and the ecological history of the lake surroundings [129–131]. A very complete review on this possibility has been presented in [112, 132].

Radionuclide inputs from nuclear installations can be incorporated into lake waters and eventually reach the sediments. As can be understood, the effects of these discharges are mostly local and highly dependent on the specific environmental conditions. An interesting example is that presented in [107]. The influence of local inputs from nuclear power plants is compared to nuclear radionuclide fallout in different lake sediments. Concentrations ranging from 10 to 70 mBqg$^{-1}$ are found to come from nuclear installation releases, which are clearly lower than the fallout. In general terms, $^3$H and $^{14}$C are considered to be the most important contributors to the dose received by the population due to the operation of nuclear power plants. A very interesting review on this topic can be found in [53]. The impact of nuclear power plant operation in sediments is also studied.

The influence of the Marcoule facility in France on close lakes is regularly studied according to French regulations, and it is found that $^3$H concentrations in lake sediments have steadily decreased since 2009, reaching values close to

10 $Bql^{-1}$, one order of magnitude below previous levels at the beginning of the twentieth century [54]. Similarly, in [55], the effects of Sellafield were studied in its close lacustrine environments and insignificant activity concentrations of relevant radionuclides ($^3H$, $^{90}Sr$, $^{99}Tc$, $^{137}Cs$, Pu isotopes).

A very significant case study is that of the Chelyabinsk operation site in the former USSR [22] mentioned above. The lakes around this site have been found to contain 17 to 120 and 0.7 $mBql^{-1}$ of $^{90}Sr$ and $^{137}Cs$, respectively, in the waters, as well as 250–870 $mBqg^{-1}$ of $^{137}Cs$ in the sediments.

The Chernobyl accident had a global dimension, as already studied. Specifically, in Europe, Chernobyl radioactivity deposition has been well studied, and very complete results can be found in [133]. Lacustrine environments were affected as expected. In salt lake waters from the Crimea Peninsula, $^{90}Sr$ concentrations were found to reach 98.0–350.5 $Bqm^{-3}$ [134]. In Lake Sniardwy, Poland, the concentrations of $^{137}Cs$ in the waters reached values close to 3000 Bqm-3, while approximately 400 $mBqg^{-1}$ of sediment was found in the first cm as a consequence of the removal of $^{137}Cs$ from the water column [135]. Approximately 17,500 $Bqm^{-2}$ of $^{137}Cs$ was estimated to be introduced in Lake Constance due to the Chernobyl accident [136]. The high $^{137}Cs$ activity injected from Chernobyl made it possible to study its scavenging processes from the water column to the lake sediment. An exponential removal pattern was found by studying 16 European lakes affected by the accident. Interestingly, the authors also concluded that the presence of $K^+$ ions in the water prevented the deposition of $^{137}Cs$ [137]. In addition, some works appeared in the literature on the use of the Chernobyl $^{137}Cs$ fingerprint as a time marker for the date of lacustrine sediment. Some examples can be found in [138].

As already seen in Chap. 4, the Fukushima accident was important, although it does not reach the global dimension that the Chernobyl accident had. However, lake contamination was very clear in Japan, although it was variable. It was found that surface lake sediments were contaminated with $^{137}Cs$ activities ranging from 23 to 26,000 $mBqg^{-1}$ dry weight in Japanese lakes studied in [139, 140]. The measured activity due to the inputs from the Fukushima accident allowed for the conduct of very interesting radioecological studies on the dynamics of radionuclides, especially $^{134}Cs$ and $^{137}Cs$, in lacustrine environments [141, 142].

### 5.4.3  Groundwater

Groundwater covers almost 25% of freshwater in the world. The rest, up to 75%, is contained in ice and glaciers. Only 0.3% and 0.003%, as described before, are included in lakes and rivers, respectively. A relevant amount of underground water is used for human consumption [143]. In some European countries, for example, drinking water demand is supplied by underground water at a percentage close to 100%. In addition, its agricultural use is also very important in many countries. Thus, the study of radioactivity in underground waters is relevant from a radiological point of view. In most cases, underground waters are stored in aquifer systems that are recharged by infiltration from the surface and discharge the water through springs at the surface or under lakes or rivers. From this point of view, underground waters can incorporate radionuclides from external sources at the Earth's surface. These inputs are added to the natural radionuclides normally present in underground waters as a result of the interaction with the geological substrate containing the aquifer. However, the radionuclide residence times in these systems are longer than those in lakes and, of course, in rivers.

*Cosmogenic*

Cosmogenic radionuclides are present in underground waters mostly due to recharging and infiltration processes. They do not pose an important issue from a radiological viewpoint, although they provide a very interesting tool for tracing the dynamics of the aquifer system. In fact, underground water can be dated as $^3$H, $^{14}$C, $^{36}$Cl, or $^{81}$Kr [144–146].

*Primordial*

The $^{238}$U, $^{235}$U, and $^{232}$Th radionuclide series are present in underground waters in concentrations depending on the geological substrate. In addition to infiltration processes, aquifers contain primordial radionuclides as a result of the interaction of water with the bedrock. $^{226}$Ra activity concentrations in underground waters are generally higher than those in rivers or lakes, provided they are not affected by external sources of radioactivity. Thus, concentrations of up to 7 Bql$^{-1}$ [121] can be found in aquifers in the USA, although a more recent study concluded that activity concentrations ranging from 10 to $10^2$ Bql$^{-1}$ can be found in some saline aquifers in the USA [147]. An extensive study carried out in Finland [148] concluded that $^{226}$Ra in underground water ranged from 0.6 to 7.5 Bql$^{-1}$. In this work, some waters from the south of the country were found to contain high $^{238}$U activity concentrations, close to 7 Bql$^{-1}$.

$^{222}$Rn is present in aquifers, and its dynamics are controlled mostly by physical processes rather than by chemical processes since it is a noble gas. $^{222}$Rn originates from $^{238}$U and $^{226}$Ra in the aquifer environment. Consequently, its concentration of activity in waters is associated with the geological nature of the bedrock. From this point of view, U-rich rocks, such as granites (see Chap. 4), exhibit high $^{222}$Rn concentrations [149] with levels up to $10^4$ Bql$^{-1}$ in some aquifers in the USA [121]. In specific cases, it has been found to be up to 77,500 Bql$^{-1}$ $^{222}$Rn in granite-rich areas in southern Finland [148].

A good compilation of the presence of natural radioactivity in underground waters can be found in [150].

*Man-made*

The underground waters have incorporated artificial radioactivity from nuclear weapon detonations by infiltration or recharging processes. Most radionuclides, which are short-lived, are no longer present in the waters. Thus, it is difficult to find the nuclear fallout signature in waters today. Some works are available on the presence of $^3$H in underground waters, which reveal how the $^3$H fallout was incorporated into the aquifers essentially through vertical downward infiltration [151]. Water depth profiles in aquifers have shown the existence of activity peaks, which can be interpreted as corresponding to the 1963 large radioactive injection in the atmosphere. In [152–154], peak radioactivity concentrations ranging from 20 to 100 $Bql^{-1}$ were found for the Chinese and Japanese aquifers. Of course, the relatively short half-life of $^3H$ ($T_{1/2} = 12.32$ y) makes it difficult to interpret the absolute activity concentration values. However, it is true that the existence of activity peaks reveals the intrusion of nuclear weapon fallout in aquifers.

Underground nuclear detonations have been shown to produce radioactive contamination of the surrounding aquifers [155]. Soluble species are expected to be observed in the waters as a result of the detonation. From this viewpoint, examples are available in the literature of the measurement of $^{137}Cs$ and $^{90}Sr$ in underground water in the Fangataufa and Mururoa atolls, where 10 and 127 underground nuclear tests were carried out. Concentrations levels of up to 10 and 30 $Bql^{-1}$ were found for $^{137}Cs$ and $^{90}Sr$, respectively, in underground waters. For the same samples, $^3H$ levels reached values ranging from 0.1 to 10 $MBql^{-1}$. However, less soluble species, such as $^{238,239,240}Pu$, were almost undetectable [156]. Despite these results, it should be taken into account that colloids are present in many aquifer systems and can improve the transport of insoluble chemical species, as shown for Pu isotopes in [157].

Nuclear fuel cycle facilities are potential suppliers of radioactivity to surrounding aquifers. For that, extensive radiological protection programs are deployed around such installations, as seen in [54, 55, 121].

The Chernobyl accident affected the aquifers of the surrounding countries, as was shown soon after the accident [158]. It has been shown that in the years following the accident, $^{90}Sr$ increased from 50 $Bql^{-1}$ to 350 $Bql^{-1}$ in a five-year period starting in 1990, when a systematic investigation began. This posed a future problem for the use of these waters for either consumption or agriculture. Some $Bql^{-1}$ was also detected for Pu isotopes in dissolution. The relatively high levels of radioactivity, especially $^{90}Sr$, have allowed us to carry out an extensive research program on radionuclide dynamics in aquifers and their migration from soil to underground waters. To do so, a pilot site was constructed, the Chernobyl pilot site located 2.5 km southwest of Chernobyl [159]. Similar works have been performed far from the accident site [160], taking advantage of radioactivity determinations in Swiss rivers and adjacent aquifers shortly after the accident [161].

The groundwater was highly contaminated from the Fukushima accident, as stated in [67]. According to such a study, it is estimated that underground waters have become a $^{137}Cs$ reservoir from which annual discharges are produced to the ocean. In fact, preliminary calculations revealed that 19 TBq of $^{137}Cs$ was released from groundwater into the ocean from February 2012 to mid-2015. After this period, an impermeable wall was constructed [162], and the annual release was reduced to 0.5 TBq. There are many studies in the scientific literature on the presence of radionuclides from the Fukushima accident in groundwaters, including studies on the use of such waters for human consumption. Examples are [163,

164]. $^{137}$Cs concentrations of up to 292 mBql$^{-1}$ were detected in many of the studied wells and even in consumption waters. $^3$H concentrations close to 20 mBql$^{-1}$ on average were identified as a result of the Fukushima releases that occurred in contaminated tanks from 2013 to 2014. As in the case of Chernobyl, the high activities introduced in aquifers have allowed the study of their dynamics and interaction with surrounding water and soil systems [165].

## 5.5   Levels and Behavior of Radioactivity in Oceans

### 5.5.1   Global Circulation

Oceans contain approximately $1.37 \times 10^{21}$ L of water, which means four orders of magnitude more water than rivers and lakes. Most water has no direct interaction with human beings because it is deep water. In fact, only 01% of the water volume belongs to coastal seas where important economic activities, including food production, occur. However, oceans contain a diversity of environments, all of which are directly or indirectly relevant to humans. In turn, humans affect these environments mostly through pollution, including radioactivity, which has been especially relevant since the beginning of the industrial revolution at the end of the eighteenth century. Unlike the case of river and lake water, the mineral composition of marine water is fairly constant and homogeneous. However, some differences appear that contribute to the displacements of vertical and horizontal water masses and, consequently, to the transport of radionuclides.

Oceans contribute significantly to the thermal regulation of the Earth through the global water circulation pattern. Horizontal water movements refer to surface and subsurface water displacements and are mostly forced by atmospheric circulation patterns and sunlight absorption variations. Winds are mainly responsible for surface water displacements. Vertical displacements are determined by temperature and salinity gradients that affect water densities. Denser waters tend to sink, while lighter water tends to reach the surface. This vertical movement is called thermohaline circulation and occurs when warm waters are displaced by winds from the equator to the poles where they cool. Some of this water freezes, losing salt that passes to the remaining liquid water, making them denser. These cool and dense water sinks to the ocean bottom, coming back to the surface as they are heated and displaced by cooler waters. This happens once the water mass circulates across the so-called ocean conveyor belt (see Fig. 5.19) [166]. It is considered that this cycle takes approximately 1000 y.

There are different sources of radionuclides to the ocean. The most important global source of man-made radionuclides is the stratospheric fallout produced by nuclear weapon tests (see Chap. 4). In addition, there are some local sources that mostly affect specific locations, although it has been shown how ocean currents can distribute radioactive inputs, giving rise to long-range contamination of ocean waters, sediments, and biota. These sources are essentially nuclear fuel reprocessing plants, the most important being located in the Irish Sea (Sellafield, UK),

(a)

(b)

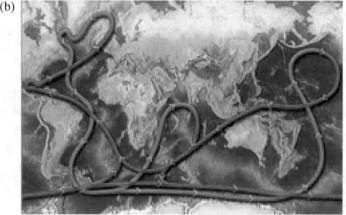

**Fig. 5.19** Ocean global circulation: (**a**) a general view and (**b**) a detailed view of the transport of cold (blue) and hot (red) water currents. Figure reprinted from [166]. Available at https://oceans ervice.noaa.gov/education/tutorial_currents/05conveyor2.html

the English Channel (La Hague, France) and the Mediterranean Sea (Marcoule, France). Their liquid effluents contain soluble radionuclides that are part of the local (or global) water circulation. However, some of the released radionuclides are less soluble and tend to associate with sediment and suspended matter, remaining in the vicinity of the marine ecosystem of the nuclear plant.

The Chernobyl accident affected the Baltic, Black, and Mediterranean Seas. The Fukushima accident essentially affected the Pacific Ocean as described in Chap. 4. Local accidents, such as the case of Palomares and Thule events (see Chap. 4), did not have a global impact. The satellite accidents SNAP 9A and Cosmos 954 also affected the marine environment, especially in the case of SNAP 9A, where the releases of radioactivity had a global nature. Although not well characterized, the dumping of radioactive wastes to the Artic, North Atlantic, and North Pacific Oceans must also be considered a source of radioactivity to marine ecosystems.

In general terms, river inputs are also a source of mostly natural radionuclides to the oceans. The interaction of seawater with the ocean crust helps remove natural radionuclides from the bottom sediments. The placement of industries near the sea, especially in river or estuarine areas, contributes in some cases to enhancing the radioactivity in the close ecosystem. The case of NORM industries is very relevant, and some data will be given below.

The presence of radionuclides in the oceans can help to understand their dynamics. In fact, the use of radioactive tracers is a great tool in following and documenting ocean currents, as well as matter exchange between water and sediments or biota. This provides essential information about the operation of the ocean ecosystem. Conservative tracers [167] for ocean waters are, for example, $^3$H ($T_{1/2} = 12.33$ y), $^{90}$Sr (28.79 y), and $^{137}$Cs (30.07 y), which can trace processes occurring on time scales within their relatively short half-lives. In contrast, $^{99}$Tc ($2.11 \times 10^5$ y) and $^{129}$I ($1.57 \times 10^7$ y) as well as $^{233}$U ($1.59 \times 105$ y), $^{236}$U ($2.34 \times 10^7$ y), or $^{237}$Np ($2.14 \times 10^6$ y), which are also conservative, are able to trace processes that take place over longer expansion time scales. Radionuclides such as $^{239}$Pu ($2.41 \times 10^4$ y), $^{240}$Pu ($6.35 \times 10^3$ y) and $^{241}$Am (432.2 y) are non-conservative and can be used to trace particle transport processes. Some natural radionuclides are also used as tracers. A remarkable example is the use of radionuclide pairs such as $^{238}$U–$^{234}$Th or $^{210}$Pb–$^{210}$Po to estimate the particle dynamics in oceans.

## 5.5.2  Seawater

*Cosmogenic*
Cosmogenic radionuclides are injected into ocean waters from the atmosphere but can also be produced in situ as described in [168]. For example, the atmospheric injection of $^3$H in the ocean surface is $1.39 \times 10^{-1}$ atomcm$^{-2}$ min$^{-1}$, while the in situ production is $1.2 \times 10^{-2}$ atomcm$^{-2}$ min$^{-1}$. In the case of $^{14}$C, the corresponding figures are $1.2 \times 10^2$ and $9.0 \times 10^{-3}$ atomcm$^{-2}$ min$^{-1}$ respectively, while for $^{10}$Be, the production rates are 2.7 and $1.8 \times 10^{-3}$ atomcm$^{-2}$ min$^{-1}$. In general terms, the concentrations of cosmogenic radionuclide activity in ocean waters range from $10^{-1}$ μBql$^{-1}$ to 1.7 mBql$^{-1}$. Specifically, typical concentrations of $^{10}$Be in seawater are close to 0.1 μBql$^{-1}$, and that of $^{14}$C is approximately 1 mBql. In the case of $^{36}$Cl, the values are approximately 10 μBql$^{-1}$ [75].

With the exception of $^3$H and $^{14}$C, cosmogenic radionuclides are radiologically irrelevant. However, they find a very significant application in oceanographic tracing. A very complete account of such applications is presented in [168–170]. Thus, $^7$Be is a useful tracer of mixing in water surfaces; $^{10}$Be and $^{26}$Al can trace sedimentation processes and interactions of solid particles in the water column; $^{32}$P and $^{33}$P can help trace nutrients in surface waters; $^{14}$C and $^{36}$Cl can trace water inventory, etc.

*Primordial*

Primordial radionuclides are present in seawater as a result of direct inputs from riverine areas, underground waters, the atmosphere, and exchange with the ocean bed.

$^{40}$K is very abundant in seawater, with radioactivity concentrations of approximately 10 Bql$^{-1}$ [121]. The total $^{40}$K inventory in marine waters is estimated to reach $1.6 \times 10^7$ PBq [171].

Unlike Th, uranium is very soluble in water, so $^{238}$U concentrations of approximately 40–45 mBql$^{-1}$ [172, 173] in seawater can be found around the world. The $^{234}$U/$^{238}$U activity ratios are found to be very uniform along the seawater at the horizontal and vertical scales. This reflects the solubility of U in water. In the absence of local sources, the ratio is close to 1.14 [174]. This slight deviation from equilibrium is generally explained on the basis of the unsupported $^{234}$U existing in the sources to the ocean. $^{235}$U is found to have natural abundance in seawater. The residence time of U in seawater is $4 \times 10^5$ y, which is longer than the time it takes for the water to circulate around the conveyor belt. This makes them very interesting radiotracers for water mass exchanges. $^{232}$Th concentrations are very small compared to U due to the tendency of tits to associate with solid particles and sink into the sediments. According to [175], levels of 0.02 to 1.7 µBql$^{-1}$ are frequently measured in water, which is consistent with a very short residence time of 0.7 y [176]. $^{226}$Ra and Ra isotopes, in general, are present in low concentrations in waters since their immediate parents, Th isotopes, are prone to associate with particles. Thus, activity concentrations close to 1 mBql$^{-1}$ are typically found in marine waters [177]. Diffusion from marine sediments is the main source of Ra into water. In coastal waters, concentration increases are very common due to input from rivers and dissolution of sediments as water salinity increases [5]. As of 1988, the inventories of $^{238}$U and $^{226}$Ra in seawater are estimated to be $5.5 \times 10^4$ and $4.8 \times 10^3$ PBq, respectively [171].

NORM industries incorporate natural radionuclides into the marine environment. In [178], an interesting review was performed on the natural radioactivity releases of several NORM industries, mostly phosphoric acid and oil production, in the European marine environment. Since 1981, 65 TBq of $^{210}$Po + $^{226}$Ra and 32 TBq of $^{210}$Pb have been released because of these industrial activities. In the same paper, it can be found that oil and natural gas industries can introduce in the ocean, liquid effluents with a $^{226}$Ra concentration close to 10 Bql$^{-1}$. Because these concentrations are certainly higher, the dilution in the seawater makes them decrease to levels that cannot be distinguished from the natural content of open ocean waters. In [22], a good compilation is given on the releases from NORM industries to waters around the world. General releases do not present a radiological hazard because the inputs are dilute in ocean waters. However, on a local scale, activity concentrations in waters can be highly enhanced. This is the case for the estuarine system studied in [179], where a phosphoric acid industry was releasing liquid wastes from the industrial process. It was found that the concentrations of radionuclides in the $^{238}$U series reached levels in estuarine waters up to an order of magnitude higher than those in an uncontaminated environment. Tidal movements

**Table 5.22** Inventory (PBq) of some relevant radionuclides in the marine environment as of 2000. The total inventory is given as well as the specific contribution from the different sources [180]

| Radionuclide | Global fallout | Local fallout | Fuel reprocessing | Chernobyl | Total (PBq) |
|---|---|---|---|---|---|
| $^{90}$Sr | 150.7 | 45.0 | 3.9 | 0.0 | 199.6 |
| $^{137}$Cs | 241.2 | 72.0 | 23.5 | 11.5 | 348.2 |
| $^{239+240}$Pu | 6.6 | 2.0 | 0.0 | 0.0 | 8.6 |

and the end of waste activities make the environment recover from this relatively high contamination of waters [7]. Nevertheless, as already mentioned [16], the sediments accumulated radioactivity during such years. After a time, they became a source of radioactivity, causing the enhancement of radioactive contamination of waters.

*Man-made*

In [180], an excellent compendium on man-made radioactivity present in oceans worldwide before the Fukushima accident was found. Table 5.22, inspired by the data presented in that study, gives the total inventory as of 2000 together with the sources of the radioactivity. In addition to such data, the contribution of the Fukushima accident to the inventory of radioactivity in the Pacific Ocean can be seen in Table 4.19. It is not possible to establish a characteristic radioactivity concentration of seawater, as it is influenced by ocean currents and the effluent release pattern of local sources of radioactivity. Thus, $^{137}$Cs concentrations in ocean waters can be found to change from $10^{-1}$ mBql$^{-1}$ in the Antarctic area to 60 mBql$^{-1}$ in the Irish Sea [181]. This is due to the direct influence of Sellafield releases. The impact of such a nuclear fuel reprocessing facility is apparent and well studied. From this point of view, a very clear latitudinal dependence of the radioactivity concentration of $^{137}$Cs can be found in the North Atlantic area with values of 10 mBql$^{-1}$, close, although clearly above the levels expected from the fallout, in the Arctic Ocean and approximately $10^2$ mBql$^{-1}$ in the Northern Irish Sea [182]. In the South Atlantic area, average $^{137}$Cs concentrations of approximately $10^{-1}$ mBql$^{-1}$ are commonly found. The waters of the South Pacific Ocean show concentrations of 0.4 and 0.6 mBql$^{-1}$ for $^{90}$Sr and $^{137}$Cs, respectively. These values increase northward to 1.8 mBql$^{-1}$ for $^{90}$Sr and 2.8 mBql$^{-1}$ for $^{137}$Cs.

The case of $^{239,240}$Pu is very interesting. Being a very reactive element, i.e., with a high tendency to associate with the solid phase, its concentration in water is low. For example, some 10–20 $\mu$Bql$^{-1}$ can be found in surface waters in North Atlantic areas, while 1–10 $\mu$Bql$^{-1}$ can be found in South Atlantic areas. However, the North Atlantic environment is strongly affected by the releases from the nuclear fuel reprocessing plants there located, i.e., Sellafield and La Hague, and a very variable concentration can be observed depending on local conditions. For example, concentrations of up to 50 $\mu$Bql$^{-1}$ can be found in the Irish Sea or up to 3 $\mu$Bql$^{-1}$ in the Baltic Sea. In the Pacific Ocean, concentrations from 0.8

to 8 $\mu Bql^{-1}$ can be found, while in the Mediterranean Sea, where the Marcoule reprocessing plant releases part of its waste, up to 12 $\mu Bql^{-1}$.

Before the Fukushima accident, the radioactivity concentration of $^{137}$Cs in coastal seawater typically ranged from 56 to 2.6 $mBql^{-1}$ in a series of measurements carried out from 1971 to 1983. In the case of $^{90}$Sr, the concentrations ranged from 33 to 2.2 $mBql^{-1}$ [183] for the same time period. Lower levels are considered background concentrations before the Fukushima accident. They should be compared to those appearing soon after the Fukushima accident close to $1 Bql^{-1}$, which rapidly decreased to $10^2$ $mBql^{-1}$ within the first year after the accident [181], although it remained higher than preaccident levels at least until 2015 [67].

A comprehensive repository of marine radioactivity concentrations is the Marine Radioactivity Information System (MARIS) of the IAEA [184]. There, a very complete dataset is gathered that should be consulted to obtain a clear idea of the immense variability of concentration values present in the oceans.

### 5.5.3 Marine Sediments

*Cosmogenic*
A fraction of the cosmogenic radionuclides entering seawater, either from the atmosphere or produced in situ, migrate toward marine sediments. Their average residence time in waters is very much dependent on the particle reactivity of radionuclides and, of course, on the water depth. Radionuclides introduced in this way in sediment are mainly used for dating purposes, as seen in [168]. The use of $^{10}$Be as a time marker to study the history of the geomagnetic field is well known and is one of the most popular applications [185]. However, the chronological tool provided by cosmogenic radionuclides is very wide, as the variety of available half-lives is also very ample. This, together with the improvement of the determination techniques (see Chaps. 16 and 18), has opened the door to many possible applications. Climate, paleoclimate, or paleoenvironment reconstruction are some of the possible studies that benefit from the use of cosmogenic dating methods [186]. Figure 5.20, taken from [186], illustrates the time range covered by the cosmogenic radionuclide dating of marine sediments.

*Primordial*
Primordial radionuclides are present in marine sediments. These are formed by the accumulation of suspended matter, which in a high fraction comes from terrestrial inputs from rivers or runoff processes in coastal waters, atmospheric injection of dust from resuspension effects, and others. The exchange of radioactivity between the soluble and solid phases in the ocean, which depends on the particle reactivity of radionuclides, contributes to enriching the suspended matter radioactivity content and consequently that of the sediment.

In general terms, the $^{238}$U content in deep sea sediments is known to range from 10 to 40 $mBqg^{-1}$ [187], very likely due to diffusion processes in the interphase water sediment [188]. An average value of $^{232}$Th close to 200 $mBqg^{-1}$ is found

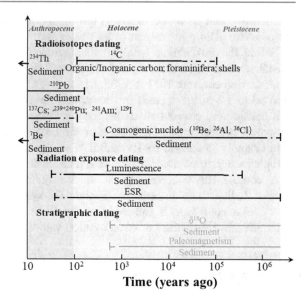

**Fig. 5.20** Cosmogenic and primordial radionuclides applications for marine sediment dating. The time ranges covered by each radionuclide are also shown. Figure reprinted from [186] with permission by Elsevier

for the same scenarios [187]. This is fairly consistent with the fact that Th is more reactive with solid particles and tends to be less soluble than U, as previously mentioned. Typical $^{226}$Ra activity concentrations ranging from 6 to 60 mBqg$^{-1}$ are generally found in marine sediments not affected by external sources of activity [189]. For $^{40}$K, an average worldwide value of 500 mBqg$^{-1}$ is present in marine sediments [190].

The use of primordial radionuclides for sediment dating is well known and is already mentioned in Chap. 4. It is recommended to read [186] to see how the determination of $^{234}$Th, $^{210}$Pb, $^{226}$Ra, and $^{230}$Th is very useful to obtain information about the time evolution of sediments and the surrounding marine environment. In general, it is possible to take advantage of the disequilibria between the natural series member, and from this viewpoint, there are several possibilities; therefore, an ample age range can be covered (see Fig. 5.20). Coastal sediments are affected by local inputs of radioactivity. With regard to NORM industrial activities, we have given some data previously [178]. The case study of [179] can also help illustrate the radiological impact of some NORM industries in their close marine environment.

*Man-made*

The sources of man-made radioactivity to the marine environment have already been described. Conservative radionuclides remain in the water column longer than nonconservative radionuclides. In any case, sooner or later, a fraction of the injected radionuclides reaches the sediment. Furthermore, once the radionuclide is incorporated into the sediment, the variation in the surrounding conditions may favor its redissolution.

The sediment inventory strongly depends on the sources of radionuclides to the ocean. In the case of the Pacific Ocean, the main sources are global or local fallout from nuclear weapons. Additionally, the SNAP-9A accident introduced some significant $^{238}$Pu activity. More recently, the Fukushima accident has caused radionuclide dissemination to the northeast Pacific Ocean, as already described. An interesting review of the sources and distribution of some artificial radionuclides in the Pacific Ocean before the Fukushima accident can be found in [191, 192]. As of 2000, sediment inventories for $^{137}$Cs and $^{90}$Sr in the northwest Pacific Ocean range from 7 to 300 and from 11 to 120 Bqm$^{-2}$, respectively, with concentrations in surface sediments between 3 and 7 for $^{90}$Sr or 2 and 30 mBqg$^{-1}$ for $^{137}$Cs [180]. In the case of $^{239+240}$Pu, concentrations in surface sediments can be found from 0.1 to 10 mBqg$^{-1}$, corresponding to inventories from 3 to 130 Bqm$^{-2}$.

The Fukushima accident produced a relevant incorporation of man-made radioactivity into the Northwest Pacific Ocean. According to [67], $^{137}$Cs in coastal Japanese sediments ranged from 10 to 1000 mBqg$^{-1}$ as an immediate consequence of the accident that should be compared with concentrations of approximately 1 mBqg$^{-1}$ prior to the accident. It is estimated that up to 230 TBq of $^{137}$Cs has been injected into sediments from the Pacific Ocean 30 km off Japan. Data suggest that approximately 1% or 2% of the total $^{137}$Cs released from the accident ended up in sediments off Japan. Moreover, sediment trap measurements suggest that $^{137}$Cs was rapidly deposited in the deep sea of the Northwestern Pacific after the accident [193]. Regarding Pu isotopes, measurements carried out during the 2011 summer in sediments 30 km off Japan revealed concentrations and activity ratios similar to preaccident levels [194].

In addition to global fallout, the main input of radioactivity to the Atlantic Ocean originates from the Sellafield and La Hague Nuclear Fuel Reprocessing Plants, present in the North Atlantic area and the English Chanel, as well as from the Chernobyl accident. This situation gives rise to a relatively large variety of radioactivity levels strongly dependent on the site and local sources, with significant variations even within a few kilometers in the same area.

Early estimations [195] suggested that sediment inventories in the North Atlantic areas were close to 800, 11 and 9 MBqm$^{-2}$ for $^{137}$Cs, $^{239,240}$Pu, and $^{241}$Am, respectively. More recent data [196] for the North Atlantic area also provide data of surface sediments of 0.3 to 3 mBqg$^{-1}$ for $^{239,240}$Pu and 2 to 3 mBqg$^{-1}$ for $^{137}$Cs corresponding to inventories in the first cm of sediments of 45 and 80 Bqm$^{-2}$ for $^{239,240}$Pu and $^{137}$Cs, respectively. These data should be compared with some results found for the Mediterranean area of 5 and 75 mBqg$^{-1}$ for $^{239,240}$Pu and $^{137}$Cs, respectively. The influence of Sellafield on the environment of the Irish Sea is very well described in [197]. In this work, we present a very complete time series of radionuclide concentrations in sediments from the Irish Sea, showing how documented plant releases are perfectly correlated with the radioactive content of the sediments. The authors demonstrate that the activity concentration in sediment decreases with the distance to the plant and along the time series from 1968 to 1988. Thus, they document how $^{137}$Cs concentration decreases by a factor close to 10 in 10 years. Indeed, concentrations by 1970 and close to the English coasts were 10 $Bqg^{-1}$ for $^{137}Cs$ and 1 $Bqg^{-1}$ during 1988. In the case of $^{239,240}Pu$, the concentrations decreased from 2 $Bqg^{-1}$ to 0.2 $Bqg^{-1}$ over the same period.

In [198], radioactivity concentrations in several places in the North Atlantic seas are compared. During an expedition from 1992 to 1994, $^{137}Cs$ was found to reach approximately 480 $mBqg^{-1}$ in the Irish Sea and 10.5 $mBqg^{-1}$ in the Skagerak zone. This value is very similar to that measured in the North Sea, 13.5 $mBqg^{-1}$, although lower than the concentration in the Norwegian Sea, close to 25 $mBqg^{-1}$. In the north of Finmark, as well as in Svalbard and the Barent Sea, concentrations of 3 to 8 $mBqg^{-1}$ were typically obtained. In the case of $^{239+240}Pu$, apart from the value found in the Irish Sea, 43 $mBqg^{-1}$ in the rest of the samples, all collected in the same areas, and the values were approximately 1 $mBqg^{-1}$, with a maximum in Skagerak, where 2.8 $mBqg^{-1}$ was measured.

The contribution of the Thule accident to the Pu isotopes present in Greenland sediments was measured [182]. Samples taken during 1974 and 1979 revealed $^{239+240}Pu$ concentrations close to $10^4$ $Bqm^{-2}$ around the accident area, values that decreased rapidly with the distance from the zone. Approximately 40 km from the accident area, the levels of Pu isotopes reached the expected fallout concentrations.

Data in the Southern Atlantic Ocean are very scarce. An example can be studied in [199]. Coastal sediments taken in southwestern Brazil ranged from 0.30 to 1.79 $mBqg^{-1}$ for $^{137}Cs$, from 15 to 150 $\mu Bq\,g^{-1}$ for $^{238}Pu$, and from 18 to 117 $\mu Bqg^{-1}$ for $^{239+240}Pu$. These levels revealed the influence of the SNAP-9A accident in this area.

## 5.6    Levels and Behavior of Radioactivity in the Biosphere

### 5.6.1    Plants, Animals

The incorporation of radioactivity into plants follows several pathways. Radionuclides can be directly incorporated from the atmosphere by deposition on leaves. This pathway is relevant for fallout radionuclides or those introduced in the lower layer of the atmosphere by resuspension from soils. In an indirect way, radionuclides can be incorporated through the plant roots that uptake nutrients from the surrounding soils. Radionuclides must first be deposited into the soil or infiltrated from nearby aqueous deposits. This is an important pathway when the radionuclide has been solubilized in the soil. Therefore, it will depend on the chemical behavior of radionuclides and the type of soil. Root uptake is a slow process. Thus, short-lived radionuclides are not considered when studying this pathway. Correspondingly, plants can in turn transfer radionuclides to soils or the atmosphere by transpiration, weathering, translocation, or some other plant–soil transfer. A very good review on radioactivity transfer to plants can be found in [200].

The presence of primordial radionuclides in plants depends on the geological substrate, and for that, attention should be given to specific sites for vegetables used for human consumption. The influence of the soil substrate is very clear in areas contaminated by NORM industries in which apparent enhancement in radioactivity can be seen in plants. This is the case of [201], where Spartina densi-flora plants are studied, as they grow in a salt marsh area affected by the operation of a phosphoric acid production factory. $^{226}$Ra levels of 2–15 mBqg$^{-1}$ in dif-ferent samples revealed a strong correlation with the concentrations of the soil substrate where the plant is growing. The authors were able to model the transfer of $^{226}$Ra from the soil to the plants with good precision, as the activity obtained was high and well determined. Regarding transfer to fruit plants, it is very inter-esting to study the results of [202]. A comprehensive review of the transfer of key radionuclides such as $^{137}$Cs and $^{90}$Sr to plants of human consumption is carried out.

Vegetables are good bioindicators for radionuclide fallout, as has been shown in several papers. The carpets of lichens have been measured for Pu isotopes and $^{99}$Tc [203, 204] within a more general study on the subarctic lichen-reindeer food chain. The authors found a clear time correlation between the radionuclide content in carpets and the nuclear weapon fallout deposition pattern. The max-imum activity for $^{239+240}$Pu, 100 mBqg$^{-1}$, was found for samples collected in 1963. In the case of $^{99}$Tc, a maximum activity of 60 mBqkg$^{-1}$ was found for the sample collected in 1967. A similar study was carried out on the Antarctic Peninsula with inventories of lichen and moss carpets of 130 and 3.7 Bqm$^{-2}$ for $^{137}$Cs and $^{239+240}$Pu, respectively [182]. Interestingly, $^{238}$Pu was easily detected, with an inventory of 0.9 Bqm$^{-2}$ as a consequence of the SNAP-9A input that is known to have mostly affected the Southern Hemisphere. More recently, lichens have been found to preserve Chernobyl inputs in Northern Europe [205]. Average concentrations of approximately 2500 mBqg$^{-1}$ have been detected for samples collected during 2000 in the same area as a signature of the Chernobyl accident according to the authors. Samples of heath, grasses, and trees around the same area show the same behavior. In Central Europe, mushrooms or moss were also used as indicators of the Chernobyl accident [206]. Concentrations of up to 300 or 1200 mBqg$^{-1}$ were measured. The radioactive impact of the Chernobyl and Fukushima accidents on vegetation has been widely studied. A very informative comparison is found in [72], where concentrations in plant material are presented for both events organized by distance from the accidental area. For the closest sam-pling location, activity concentrations of 227,900, 1,015,000, and 1457 mBqg$^{-1}$ are presented for $^{137}$Cs, $^{90}$Sr, and $^{239+240}$Pu as a result of the Chernobyl accident. In the case of Fukushima, the corresponding figures are 1,290,000, 1140, and 0.49 mBqg$^{-1}$. Furthermore, approximately 15,000 mBqg$^{-1}$ was detected for $^{131}$I as a consequence of the Fukushima accident.

Fukushima and Chernobyl affected forest ecosystems, which caused the need to model the evolution of radionuclides in forests as a way to estimate the radio-logical impact on the ecosystem and on humans [207] due to the incorporation of huge radioactivity in them. In fact, the use of some forest areas for cellulose pro-duction, for example, might suppose a massive introduction of radioactivity into the industrial process in the case of an accident [208]. A very good review of the impact of the Fukushima accident on plants and its agricultural implications can be found in [209]. Relevant radioactivity inputs in different areas, including those of agricultural interest, made it possible to conduct field studies on the transfer of radionuclides from soil to plants.

Animals are affected by radioactivity and incorporate radionuclides from different sources. However, to date, the problem of the presence of radioactivity in animals has been studied only within the framework of its possible impact on humans through the food chain. Thus, the development of control tools has mainly been oriented to protect humans against ionizing radiation with the belief that this is the only condition to protect the environment. A modern approach integrates this point of view within a general concept of global protection of the environment that includes that of the biota [210]. This integrated approach requires the development of dose assessment tools for wildlife [211]. An interesting paper on this concept is that presented in [209]. In fact, this approach is taken into account when modeling the radioactive contamination of forest ecosystems [207], where the uptake of radionuclides by animals is a submodel for the transfer of radionuclides between forest compartments.

A very complete account of radioactivity in wildlife from polar regions is gathered in [182]. Some results are presented for polar bear and seal bird flesh collected during 1980 in Arctic areas. The determined concentrations of $^{137}$Cs concentrations determined were approximately 6, 1, and 2 mBqg$^{-1}$ for these samples, while for $^{239+240}$Pu, it was only possible to obtain a measurable activity of 4 $\mu$Bqg$^{-1}$ for the flesh of polar bears. In Antarctica, the author presented a more complete set of results with data on leopard seal, crabeater seal, Weddell seal, Adelie penguin, chinstrap penguin, donkey penguin, krill, etc. The $^{137}$Cs concentrations in seal kidney ranged from 0.2 to 0.8 mBqg$^{-1}$. In penguin flesh, the measured concentrations were close to 0.1 mBqg$^{-1}$. The highest activity was found in the flesh of blackfish icefish, with a value of 4 mBqg$^{-1}$. These measurements basically account for the radionuclide concentrations in wildlife derived from nuclear weapon fallout. In other previously commented papers, the radioactivity uptake by reindeer in a subarctic environment was determined [203, 204]. As an example, $^{99}$Tc concentrations in reindeer bones or liver were found to be 194 and 49 mBqkg$^{-1}$, respectively.

The Chernobyl accident affected wild animals, as seen in several publications. Wild boars in some Serbian mountains studied during 2007 were found to contain up to 30 mBqg$^{-1}$ [206]. Reindeer in Norway were studied, and $^{137}$Cs concentrations of 4680 and 17,700 mBqg$^{-1}$ were found in the liver and kidney for samples taken during 2002 [205]. In both cases, the authors concluded that the Chernobyl effects were still present in the investigated environments. In fact, areas affected by the Chernobyl accident in Ukraine have been systematically studied, and a very interesting report is that presented in [212], where wild boars have been investigated since 1991. The authors conclude that higher $^{137}$Cs activities are found in muscle tissue and the kidney with concentrations up to 600 Bqg$^{-1}$. $^{90}$Sr was found to concentrate in bone tissues with values as high as 18 Bqg$^{-1}$. The effect of the Chernobyl accident was detected in migratory birds arriving in Spain shortly after the accidents [213]. Song thrushes (Turdus philomelos) were measured for $^{134}$Cs, $^{137}$Cs, and $^{90}$Sr. $^{137}$Cs concentrations up to 100 times higher than those of domestic bird species were found. In the case of $^{90}$Sr, the concentrations were

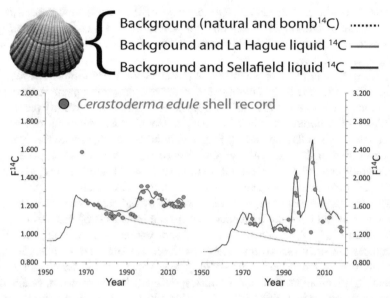

**Fig. 5.21**  $^{14}$C concentrations in shells collected in the North European marine environment. The concentrations show the influence of the bomb pulse and of the Sellafield and La Hague discharge on the marine environment. Figure reprinted from [214] with permission by Elsevier

approximately 10 times higher than those for the local bird species. In [209], a detailed study on the radioactivity concentrations in Japanese birds and boars was conducted in areas highly contaminated by the Fukushima accident. For bird samples collected 1.5 years after the accident, activities of $^{134}$Cs, $^{137}$Cs, and $^{110m}$Ag close to 223, 309, and 25 Bqg$^{-1}$ were determined, respectively.

## 5.6.2   Seaweed and Other Marine Bioindicators

Algae and other biological specimens are good indicators of radioactivity in the environment. Figure 5.21 taken from [214] shows the $^{14}$C radioactivity concentration in bivalve shells collected in the environment of the North Atlantic Ocean, showing the impact of releases from La Hague in the English Channel and Sellafield in the Irish Sea.

With regard to seaweed, there is ample record in the current literature on its radioactive content as an indicator of radioactive inputs in its close environment. Some examples are given here to demonstrate how seaweed can inform about the radioactive impact on the marine environment.

$^{137}$Cs pre-Chernobyl radioactivity concentrations seem to be quite uniform in seaweed collected along the Northern Hemisphere. In [182], the levels for Arctic regions were found to be approximately 1.5 mBqg$^{-1}$ for brown algae. These results are very similar to those obtained in the Mediterranean area [60, 215].

For that, it could be considered representative of the background caused by the nuclear weapon fallout. In Antarctica, the levels of $^{137}$Cs drop to 0.4 mBqg$^{-1}$ [179] as a consequence of the lower injections of radioactivity produced in the Southern Hemisphere.

In [216], a beautiful demonstration is presented on the utility of seaweed as a bioindicator of radioactive contamination of the marine environment. A time series of radioactivity concentrations in the Danish Strait extended over 40 years shows the impact of different sources of activity on such an environment. Specifically, the releases from Sellafield and La Hague and the Chernobyl fallout are very precisely recorded in the analyzed samples. Figure 5.22 taken from that paper shows the temporal evolution of $^{137}$Cs concentrations in water and seaweed in addition to the Sellafield and La Hague discharge patterns. The results are apparent and point to the use of seaweed as a very precise bioindicator. The $^{99}$Tc results in seaweed collected from the Irish coast are also an indication of Sellafield releases, as shown in [217]. Peak concentrations of approximately 8–10 Bqg$^{-1}$ for Fucus vesiculosus are recorded, corresponding to the maximum releases from Sellafield occurring during the period 1995–1996. These values should be compared with concentrations of approximately 0.3 mBqg$^{-1}$ found in southwestern Spain [215] attributed to radioactive fallout from nuclear weapons to evaluate the impact of nuclear fuel reprocessing plants in their close environment. In [218], $^{129}$I concentrations are provided in Fucus vesiculosus sampled in the English Channel close to 50 mBqg$^{-1}$ as a direct result of the La Hague discharges. As previously mentioned, the influence of nuclear fuel reprocessing plants on the marine environment extends over a long spatial and temporal range, since the releases contain conservative radionuclides. Paper [219] gives results on the concentrations of $^{129}$I and $^{137}$Cs in Fucus vesiculosus collected in the Skagerrak and Kattegat areas, demonstrating the influence of the reprocessing plants mentioned above in the Northern European marine environment.

Fucus vesiculosus was also used in [215] to analyze the influence of the Palomares accident (see Chap. 4) on the coastal marine environment of southwestern Spain. $^{239+240}$Pu concentrations of approximately 2 mBqg$^{-1}$ were found in Fucus vesiculosus collected in Palomares during 1989 and must be compared with levels close to 3 μBqg$^{-1}$ obtained for the same seaweed species on the Atlantic southwestern coast of Spain during the same year.

$^{137}$Cs was measured in brown algae taken during June 2011 on the eastern Japanese coast, showing concentrations up to 200 mBqg$^{-1}$ as a consequence of the Fukushima accident [220]. A longer record is presented in [221]. The authors measured $^{137}$Cs concentrations up to 3 Bqg$^{-1}$ in brown algae during May 2011. These levels decreased rapidly during the following summer of 2011, although concentrations close to 100 mBqg$^{-1}$ were still measurable during 2013.

A notable fact is the generalized use of seaweed for human consumption [220]. This poses an interesting challenge in radioecology as algae are very good bioindicators and, therefore, tend to concentrate radionuclides from their close environment. This is relevant not only for man-made radioactivity but also for

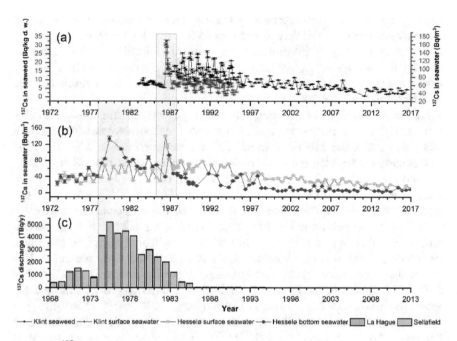

**Fig. 5.22**   $^{137}$Cs concentration time evolution in seaweed and seawater at Klint (**a**) and in seawater at Hesselö (**b**) both in Denmark. In addition to the discharge records of Sellafield and La Hague, the influence of the discharges on the $^{137}$Cs concentrations in both seawater and seaweed is shown. Figure reprinted from [216] with permission by Elsevier

natural radionuclides. Specifically, $^{210}$Po is of great significance, as it is considered the most important contributor to the dose received by humans from the consumption of marine food [222].

## 5.7   Levels and Behavior of Radioactivity in Foods

### 5.7.1   Drinking Water

As already described in Chap. 4, natural radioactivity can be present in drinking waters at concentrations that depend on the origin of the waters and the purification processes they can undergo. Some data are provided in Table 4.3. In Table 5.23, a compilation of natural radionuclide concentrations is listed in drinking water worldwide. They are based on data presented in [22, 75]. The variability of the data is high, even at the regional level. This is related to the origin of the water, as explained in Chap. 4.

Man-made radionuclides can enter aquifers and water reservoirs through different pathways. In this manner, man-made radionuclides from radioactive fallout, nuclear fuel cycle releases, or nuclear accidents are introduced into drinking waters. They are purification processes that can remove most of the bulk and, in fact, have been shown to be very efficient, as seen in [223], where the authors claim that conventional purification processes can be up to 95% efficient for Pu isotopes, although they are not very efficient for $^{90}$Sr and $^{137}$Cs. Under accidental situations in which significant radioactivity inputs could take place in the environment, this must be taken into account.

Nuclear weapon fallout has been detected very clearly in drinking waters. Pioneering work on this matter was carried out at the US Environmental Measurement Laboratory, and some data taken from [224] are presented in Fig. 5.23. The 1963 peak is apparent for $^{90}$Sr and $^{137}$Cs radioactivity concentrations in US tap water.

The Chernobyl accident caused the appearance of higher levels of activity in drinking water. Before the accident, the concentrations of $^{137}$Cs were generally approximately 10 mBql$^{-1}$. However, in northern Europe, concentrations close to 10 Bql$^{-1}$ were measured shortly after the accident [122]. $^3$H is a relevant radionuclide that must be investigated within all radiation protection programs. Specifically, in tap water or mineral waters, it is supposed to be present as it is part of the global water cycle. Concentrations typically close to 1 Bql$^{-1}$ before 1990 [122]. However, the data in [36] show concentrations in mineral and tap water close to 0.3–0.4 Bql$^{-1}$. This is a consequence of the short $^3$H half-life.

The Fukushima accident affected both tap water. In Fig. 5.24, we present some data taken from [225] for $^{131}$I and $^{134+137}$Cs taken after the accident for some months. $^{131}$I rapidly disappears from tap water, mainly due to its short half-life ($T_{1/2} = 12.32$ y) and the absence of new inputs. Regarding $^{134+137}$Cs, only a few samples were above the regulatory limit of 10 Bqkg$^{-1}$ a month after the accident.

## 5.7.2  Foodstuffs and Food Raw Materials

Radionuclides can enter the human food chain through different pathways. Atmospheric inputs can contaminate agricultural soils and then the crops growing in the area. Artificial irrigation with contaminated water can also introduce natural or artificial radionuclides into agricultural soils. Grazing on contaminated grasslands leads to the intake of radionuclides by cattle that can then be ingested by humans. The consumption of seafood also contributes to the ingestion of radionuclides. There are also many other direct or indirect pathways of radionuclide ingestion by

**Table 5.23** Ranges or average values in mBql$^{-1}$ for some primordial radionuclides in drinking water worldwide [22, 75]

| Region | $^{238}$U | $^{230}$Th | $^{226}$Ra | $^{210}$Pb | $^{210}$Po | $^{232}$Th | $^{228}$Ra | $^{235}$U |
|--------|-----------|------------|------------|------------|------------|------------|------------|-----------|
| North America | 0.3–77 | 0.1 | 0.4–1.8 | 0.1–1.5 | | 0.05 | 0–0.5 | 0.04 |
| Asia | 0.1–700 | | 0.2–120 | | | 0.04–12 | | |
| Europe | 0.04–150,000 | 1.4 | 1–49,000 | 0.2–20,000 | 0.2–7600 | 0.04–9.3 | 18–570 | 0–50 |

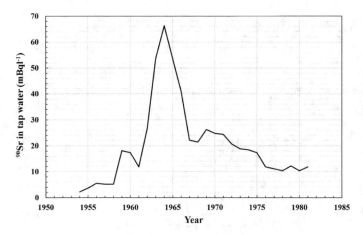

**Fig. 5.23** $^{90}$Sr concentrations in tap water collected in New York from the beginning of the 1950s. The bomb pulse is very clear. Detailed data can be found in [224]

humans. The need for food control for radioactivity is apparent, and the methodology will be discussed in Chap. 13. Here, we provide some typical radionuclide levels found in the average diet worldwide.

Extensive data collection can be found in [22, 75, 122]. Based on that, some data are given in Table 5.24. Natural radioactivity is currently measured in milk and milk-derived products, as well as in meat, fish, leafy, grain, root vegetables, and fruits. Average concentrations are very variable. For example, the concentrations of $^{238}$U and $^{226}$Ra in milk are certainly variable throughout the world. $^{238}$U levels from 0.7 to 17 $\mu$Bqg$^{-1}$ can be found in North America and India, respectively. In the case of $^{226}$Ra, it is possible to find concentrations of 2 to 130 $\mu$Bqg$^{-1}$ within the same country. In meat, $^{238}$U concentrations are within the same order of magnitude, but for $^{226}$Ra, the differences are remarkable, with values ranging from 2 to 220 mBqg$^{-1}$.

For man-made radionuclides, we already described the effects of nuclear weapon radionuclide fallout in Chap. 4, and in fact, Fig. 4.17 shows how the 1963 $^{90}$Sr radioactive peak appears in the diet of the USA population [224]. On the other hand, as seen in [54, 55], the discharges of nuclear fuel reprocessing plants are also monitored in agricultural sites, in milk, meat, seafood, etc., since they can reach the human food chain. The Chernobyl accident introduced a huge amount of radioactivity in the atmosphere that spread worldwide. The Chernobyl signature also appears in food samples. Examples are the measurements of $^{131}$I, $^{137}$Cs, and $^{90}$Sr in milk samples collected over Europe, the USA, and Japan that can be seen in [72]. $^{131}$I levels up to 40,000 Bql$^{-1}$ were measured in cow milk collected in Germany. In the case of $^{137}$Cs, concentrations of up to 400 Bql$^{-1}$ were also found in the UK for cow milk. In the USA, the Chernobyl signature was observed in milk shortly after the accident [90]. In fact, $^{131}$I reached a maximum of 1.47 Bql$^{-1}$ on May 17, 1986, in New York. In the first days of May,

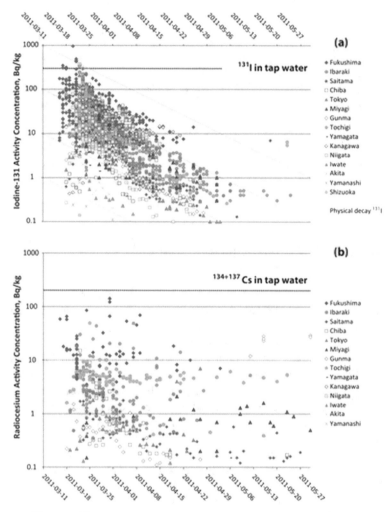

**Fig. 5.24**  $^{137}$Cs and $^{131}$I in Japanese tap water collected after the Fukushima accident. Figure reprinted from [225]

the $^{131}$I concentration was below the MDA (see Chap. 14) of approximately 0.2 Bql$^{-1}$. In the case of $^{137}$Cs, the maximum concentration was detected on May 19, approximately 0.7 Bql$^{-1}$. For this sample, 0.3 Bql$^{-1}$ of $^{134}$Cs was measured. Prior to the Chernobyl accident, the levels of $^{137}$Cs were also below the MDA, 0.3 Bql$^{-1}$.

In the case of the Fukushima accident, $^{131}$I levels of up to 100,000 Bql$^{-1}$ in cow milk have been documented in the USA [72]. In the case of $^{137}$Cs, the concentrations were higher within Japan, with values close to 5000 Bql$^{-1}$ in Fukushima prefecture. $^{90}$Sr was detected as a release from the accident. As in the case of $^{137}$Cs, the highest concentrations were detected in Japan, with levels of 60 Bql$^{-1}$

**Table 5.24**  Primordial radionuclide concentrations ($mBqkg^{-1}$) in food materials worldwide [22]

| Region | $^{238}U$ | $^{230}Th$ | $^{226}Ra$ | $^{210}Pb$ | $^{210}Po$ | $^{232}Th$ | $^{228}Ra$ | $^{228}Th$ | $^{235}U$ |
|---|---|---|---|---|---|---|---|---|---|
| *Milk* | | | | | | | | | |
| North America | 0.7 | 0.4 | 5.7 | 11 | | 0.27 | | | 0.05 |
| Asia | 0.55–13 | | 6–12 | 16 | 13–15 | 0.3–1.2 | 21 | | 0.06 |
| Europe | 0.1–4.9 | 1.2 | 0.4–200 | 5–280 | 2–220 | 1.2 | | 56 | |
| *Meat* | | | | | | | | | |
| North America | 0.8–2.3 | 0.5–3 | 20 | 18 | | 0.3–2 | | | 0.02 |
| Asia | 10–13 | | 36–41 | 140 | 120–440 | 2.3–4.3 | 120 | | 0.5 |
| Europe | 1–20 | 0.7–3 | 2–220 | 15–3700 | 37–67,000 | 0.5–3.6 | | 22–93 | |
| *Fish* | | | | | | | | | |
| North America | 13–1900 | 1.2–29 | 30–59 | 14–1800 | 150–55,000 | 1.2–3 | | | 0.4–90 |
| Asia | 12 | | 39 | 3500 | 4900 | 1.3 | 320 | | 0.5 |
| Europe | 2.5 | | 8.5–7400 | 20–4800 | 50–120,000 | | | 56–700 | |
| *Fruit* | | | | | | | | | |
| North America | 0.9–7.7 | 0.2–1.1 | 7–47 | 8–150 | | 0.08–1.4 | | | 0.1 |
| Asia | 0.4–26 | | 11–63 | 27 | 16–140 | 2.3–4.7 | 110 | | 0.6 |
| Europe | 0.9–2900 | 0.7–7.5 | 5–9400 | 18–4900 | 12–5200 | 0.4–7.1 | | 22 | |

in Ibaraki. However, $^{90}Sr$ was also detected in the USA at concentrations of 0.7 $Bql^{-1}$. An interesting review of the impact of Fukushima on food products in Japan is presented in [225]. Before the accident, radionuclide concentrations were generally well below the regulatory levels—above the regulatory levels, the authorities should start intervention practices to avoid the population receiving higher radiation doses than permitted (see Chap. 6). After the accident, all the analyzed foods showed activities above the regulatory levels up to the middle of 2012.

## Exercises for Chapters 4 and 5

1. In Table 4.2, the $^{238}U$ concentrations are presented in $mBql^{-1}$ or $mBqg^{-1}$. Find the conversion factor to give the concentrations in $\mu gl^{-1}$ and $\mu gg^{-1}$ and calculate them in the new units.
2. Demonstrate (4.1).
3. The paper "Radiometric dating of sediment records in European mountain lakes" by Peter Appleby [9] gathers several applications of $^{210}Pb$ (CIC and CRS models) and $^{137}Cs$ dating methods to lake sediments. Use data in Table 2 of such paper to reconstruct the unsupported $^{210}Pb$ profile in Fig. 2.b. Assume that the total $^{210}Pb$ at depth 0 is 1000 Bq/kg. Try a CIC model and explain the differences found with the experimental data in Fig. 2b, in the case there are.

Hint: Take an average sedimentation rate.

4. Demonstrate (4.5) and (4.6).

5. $^{10}$Be is a very useful cosmogenic radionuclide for dating marine sediments. Unlike $^{14}$C, it does not form gaseous compounds in the atmosphere. Therefore, its residence time in the atmosphere is very short. Once it reaches the ocean surface, it readily associates with suspended matter and is transported into the sediment. Suppose that $\phi$ is the $^{10}$Be flux over the sediment and r is the sedimentation rate. Suppose that we can measure the $^{10}$Be concentration profile in the sediment. Find the equations of the dating method for
   (a) $\phi$ and r are constant.
   (b) $\phi$ is constant and r variable.

6. Based on (4.11) find the minimum detonation yield to make the fireball reach the stratosphere.

7. The $^{144}$Ce/$^{137}$Cs activity ratio in the atmospheric sample is found to be 10. Find the time elapsed between the nuclear explosion that presumably injected these radionuclides into the atmosphere and the measurement instant.

8. Plot the time evolution of the $^{144}$Ce/$^{137}$Cs and $^{95}$Zr/$^{137}$Cs activity ratios in the atmosphere after a nuclear explosion and discuss the range of application of both ratios to estimate the time of explosion.

9. Use the data in Fig. 4.13 to estimate the bomb-produced 14C average residence time in the stratosphere.

10. Use the data in Fig. 4.15 to explain the spring effect in the Southern Hemisphere.

11. According to the data in Fig. 4.15, the Chernobyl accident did not affect the Southern Hemisphere. Explain that. However, a $^{137}$Cs increase after Chernobyl is clearly observed in the Northern Hemisphere. Compare that input with the total input from the nuclear weapon tests. Consider only the decade of the 1960s for this last case. Why the Chernobyl peak is not observed for $^{90}$Sr.

12. Suppose the body fluids in Fig. 5.2b contain a certain activity of a radionuclide as a consequence of the exchanges between the compartments connected to it. Following (5.1) write (without solving) the equation for the time variation of the activity in body fluids describing the meaning of each of the used coefficients.

13. In the figure, we have an activity concentration profile (activity as a function of depth) of $^{99}$Tc in a lake sediment core collected Sweden during November 1986. If $^{99}$Tc ($T_{1/2} = 2.11 \times 10^5$ y) is a $^{235}$U or $^{239}$Pu fission product:
   (a) Discuss the form of the activity profile, explaining the different zones and activity peaks. Could the sediment be dated with the $^{99}$Tc data?
   (b) Find the mean sedimentation rate of the sediment.

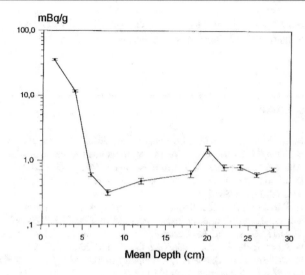

14. Using the data in Table 4.17 calculate, the total inventory of the different radionu-clides in the Chernobyl nuclear fuel. Compare the activity ratios in the inventory with the ratios in Table 4.12.

15. In Table 5.7, the $^{240}$Pu/$^{239}$Pu atom ratios are given for different cases. Find the corresponding activity ratios and compare them with those for the other Pu isotopes in the table.

16. The $^{99}$Tc/$^{238}$U atom ratio has been found to be $\sim 10^{-12}$ in pitchblende samples. Compare this ratio with that expected from $^{238}$U spontaneous fission.

17. In Fig. 5.5, the $^3$H concentration in the atmosphere is given for a long period starting from 1950. Compare the inputs from the nuclear tests of the sixties with the natural $^3$H produced by cosmic radiation.

18. With the data provided in Chap. 5 compare the inputs of $^3$H in the atmosphere from 1990 to 1997 from nuclear power plants worldwide with the $^3$H remaining from the nuclear tests for the same period. Use data in Fig. 5.5.

19. According to the data in the text, find the actual signatures for the Chernobyl and Fukushima $^{134}$Cs/$^{137}$Cs activity ratios.

20. Calculate the equivalency between the $^{222}$Rn Working Level (WL) and its concentration in Bqm$^{-3}$.

21. The $^{235}$U/$^{238}$U atom ratio was found to be 5.4 in soil samples taken after the Palomares accident. Find the corresponding activity ratio and compare it with the expected natural value.

22. In the so-called ocean belt, the water takes approximately 1000 y to completely circulate it. According to their half-lives, propose some good candidates to trace their movement between the natural or man-made radionuclides present in the marine environment. Consider their chemical behavior.

23. Explain the time range of application for the different radionuclides appearing in Fig. 5.20 to date marine sediment. The important data is the half-life.

24. Select a sea environment and use the MARIS database to construct a map of transuranic concentrations in waters and sediments.

# References

1. B. Salbu, Challenges in radioecology. J. Environ. Radioact. **100**, 1086–1091 (2009)
2. B. Salbu, O.C. Lind, L. Skipperud, Radionuclide speciation and its relevance in environmental impact assessments. J. Environ. Radioact. **74**, 233–242 (2004)
3. H. R. von Gunten1, P. Benes, Speciation of radionuclides in the environment. Radiochim. Acta **69**, 1–29 (1995)
4. L. Skipperud, B. Salbu, Sequential extraction as a tool for mobility studies of radionuclides and metals in soils and sediments. Radiochim. Acta **103**, 187–197 (2015)
5. A. Martínez-Aguirre, M. García-León, Natural radioactivity in the Guadalquivir river at the South of Spain. J. Radioanal. Nucl. Chem. **178**, 337–350 (1994)
6. E. Sholkovitz, A. Carey, J. Cochran, Aquatic chemistry of plutonium in seasonally anoxic lake Waters. Nature **300**, 159–161 (1982)
7. A. Absi, M. Villa, H.P. Moreno, G. Manjón, R. Periañez, Self-cleaning in an estuarine area formerly affected by [226]Ra anthropogenic enhancements. Sci. Total Environ. **329**, 183–195 (2004)
8. J. Lan, T. Wang, S. Chawchai, P. Cheng, K. Zhou, K. Yu, D. Yan, Y. Wang, J. Zang, Y. Liu, L. Tan, L. Ai, H. Xu, Time marker of [137]Cs fallout maximum in lake sediments of Northwest China. Quatern. Sci. Rev. **241**, 106413 (2020)
9. A. Kumar, S. Rout, V. Pulhani et al., 'A review on distribution coefficient ($K_d$) of some selected radionuclides in soil/sediment over the last three decades'. J. Radioanal. Nucl. Chem. **323**, 13–26 (2020)
10. J.M. Abril, Basic microscopic theory of the distribution, transfer and uptake kinetics of dissolved radionuclides by suspended particulate matter — Part I: Theory development. J. Environ. Radioactivity **41**, 307–324 (1998)
11. 'Handbook of parameter values for the prediction of radionuclide transfer in terrestrial and freshwater environments', STI/PUB/472, International Atomic Energy Agency, 2010.
12. L. Hallstadius, E. García-Montaño, U. Nilsson, An improved a validated dispersion model for the North Sea and adjacent Waters. J. Environ. Radioact. **5**, 261–274 (1987)
13. R. Periáñez, *Modelling the Dispersion of Radionuclides in the Marine Environment: An Introduction* (Springer, 2005)
14. E. Garcia-Montaño, M. García-León, L. Hallstadius, A new approach to the dosimetry of [99]Tc in the human body after inhalation or ingestión. Radiat. Prot. Dosim. **32**, 253–257 (1990)
15. J.M. Abril, M. García León, A marine dispersion model for radionuclides and its calibration from non-radiological information. J. Environ. Radioact. **16**, 127–146 (1992)
16. R. Periáñez, J.M. Abril, M. García-León, Modelling the dispersion of non-conservative radionuclides in tidal waters—Part 1: conceptual and mathematical model. J. Environ. Radioact. **31**, 127–141 (1996)
17. A. Laissaoui, J.M. Abril, R. Periáñez, M. García-León, E. García-Montaño, Kinetic transfer coefficients for radionuclides in estuarine waters: reference values from[133]Ba and effects of salinity and suspended load concentration. J. Radioanal. Nucl. Chem. **237**, 55–62 (1988)
18. ICRP Publication 30, *Limits for the Intake of Radionuclides by Workers—Part 1, Annals of the ICRP*, 2nd Edition (Pergamon Press, 1982)

19. UNSCEAR 2000 report, United Nations Scientific Committee on the Effects of Atomic Radiation, Annex A
20. J.M. Abril, M. García-León, The integrated atmospheric flux effect in a radiogeochronological model. J. Environ. Radioact. **24**, 65–79 (1994)
21. J.M. Abril, M. García-León, R. García-Tenorio, C.I. Sánchez, F. El-Daoushy, Dating of marine sediments by an incomplete mixing model. J. Environ. Radioact. **15**, 135–151 (1992)
22. UNSCEAR 2000 report, United Nations Scientific Committee on the Effects of Atomic Radiation, Annex B
23. K. Kozak, Enviromental tritium contamination on global and local scale, in *Low-Level Measurements and Their Applications to Environmental Radioactivity*, ed. M. García-León, G. Madurga (World Sci. Pub. Co., 1988), pp. 392–406
24. P.K. Kuroda, I.O. Essien, D. Sandoval, Fallout of uranium isotopes from the 1980 eruption of Mount St. Helens. J. Radioanal. Nucl. Chem. **84**, 23–32 (1984)
25. P.K. Kuroda, T. Barbod, S.N. Bakhtiar, Effect of the eruptions of Mount St. Helens and El Chichon on the ratios of thorium and uranium isotopes in rain. J. Radioanal. Nucl. Chem. **111**, 137–146 (1987)
26. Y. Sakuragi, L. Meason, P.K. Kuroda, Uranium and plutonium isotopes in the atmosphere. J. Geophys. Res. **88**, 3718–3724 (1983)
27. S.N. Bakhtiar, S. Salaymeh, P.K. Kuroda, Uranium fallout from the nuclear-powered satellites and volcanic eruptions. J. Radioanal. Nucl. Chem. **100**, 57–63 (1986)
28. T. Matsunami, A. Mizohata, T. Mamuro, Uranium in rain wáter from the French nuclear explosions of 1970 and 1971, in *Annual Report Radiation Center of Osaka Prefecturev*, vol. 18 (Sakai, Osaka, Japan, 1977)
29. A. Martínez-Aguirre, M.C. Morón, M. García-León, Measurements of U- and Ra-isotopes in rainwater samples. J. Radional. Nucl. Chem. **152**, 37–46 (1991)
30. UNSCEAR 2000 report, United Nations Scientific Committee on the Effects of Atomic Radiation, Annex C
31. K. Playford, J. Toole, I. Adsley, Radioactive fallout in air and rain: results to the end of 1991, AEA-EE-0498, DOE/RAS/93.003 (AEA Technology, 1993)
32. E.P. Hardy Jr., J. Rivera, Health and Safety Laboratory, Fallout Programme, Quarterly Summary Report, HASL-184 (USAEC, 1968)
33. K. Hirose, P. Povinec, Sources of plutonium in the atmosphere and stratosphere troposphere mixing. Sci. Rep. **5**, 15707 (2015)
34. E. Chamizo, M. García-León, S.M. Enamorado, M.C. Jiménez-Ramos, L. Wacker, Measurement of plutonium isotopes, $^{239}$Pu and $^{240}$Pu, in air-filter samples from Seville (2001–2002). Atmos. Environ. **44**, 1851–1858 (2010)
35. F.J. Santos, J.M. López-Gutiérrez, E. Chamizo, M. García-León, H. A. Synal, Advances on the determination of atmospheric $^{129}$I by Accelerator mass spectrometry (AMS). Nucl. Instrum. Methods Phys. Res. **B249**, 772–775 (2006)
36. Y. Fan, L. Zhang, J. Lan, X. Hou, W. Zhou, Pre-nuclear values for $^{129}$I/$^{127}$I in Chinese sediments and their geochronological implications. Palaeogeogr. Palaeoclimatol. Palaeoecol. **568**, 110312 (2021)
37. T. Jabbar, G. Wallner, P. Steier, A review on $^{129}$I analysis in air. J. Environ. Radioact. **126**, 45–54 (2013)
38. E. Chamizo, M. García-León, H.-A. Synal, M. Suter, L. Wacker, Determination of the $^{240}$Pu/$^{239}$Pu atomic ratio in soils from Palomares (Spain) by low-energy accelerator mass spectrometry. Nucl. Instrum. Methods Phys. Res. **B249**, 768–771 (2006)
39. T. Warneke, I.W. Croudace, P.E. Warwick, R.N. Taylor, A new ground-level fallout record of uranium and plutonium isotopes for northern temperate latitudes. Earth Planet. Sci. Lett. **203**, 1047–1057 (2002)

40. K. Hirose, Y. Igarashi, M. Aoyama, T. Miyao, Long-term trends of plutonium fallout observed in Japan, in *Plutonium in the Environment*, ed. by A. Kudo (Elsevier, 2001), pp. 251–266
41. H.W. Bentley, F.M. Phillipst, S.N. Davis, S. Gifford, D. Elmore, L.E. Tubbs, H.E. Gove, Thermonuclear $^{36}$CI pulse in natural wáter. Nature **300**, 737–739 (1982)
42. F.J. Santos, J.M. López-Gutiérrez, M. García-León, Ch. Schnabel, H.-A. Synal, M. Suter, Analysis of 36Cl in atmospheric samples from Seville (Spain) by AMS. Nucl. Instrum. Methods Phys. Res. **B223–224**, 501–506 (2004)
43. O.A. Chaeffers, O. Thompson, L. Lark, Chlorine-36 radioactivity in rain. J. Geophys. Res. **55**, 4013–4016 (1950)
44. M.D. Keywood, L.K. Fifield, A.R. Chivas, R.G. Cressw, Fallout of chlorine 36 to the Earth's surface in the southern Hemisphere. J. Geophys. Res. **103**, 8281–8286 (1998)
45. D. Curtis, J. Fabryka-Martin, P. Dixon, J. Cramer, Nature's uncommon elements: plutonium and technetium. Geochim. Cosmochim. Acta **63**, 275–285 (1999)
46. B.T. Kenna, P.K. Kuroda, Isolation of naturally occurring technetium. J. Inorg. Nucl. Chem. **23**, 142–144 (1961)
47. M. García-León, $^{99}$Tc in the environment: sources, distribution and methods. J. Nucl. Radiochem. Sci. **6**(3), 253–259 (2005)
48. E. Holm, J. Rioseco, M. García-León, Determination of $^{99}$Tc in environmental simples. Nucl. Instrum. Methods Phys. Res. **223**, 204–207 (1984)
49. M. Attrep, J.A. Enochs, L.D. Broz, Atmospheric Technetium-99. Environ. Sci. Technol. **5**, 344–345 (1971)
50. K.C. Ehrhardt, M. Attrep, Technetium-99 in the atmosphere. Environ. Sci. Technol. **12**, 55–57 (1978)
51. M. García-León, G. Manjón, C.I. Sánchez-Angulo, $^{99}$Tc/$^{137}$Cs activity ratios in rainwater samples collected in the South of Spain. J. Environ. Radioact. **20**, 49–61 (1993)
52. M. Garcia-León, C. Piazza, G. Madurga, $^{99}$Tc in surface air samples during the years 1965–1967. Int. J. Appl. Radiat. lsot. **35**, 961–963 (1984)
53. X. Hou, Tritium and $^{14}$C in the environment and nuclear facilities: sources and analytical methods. J. Nucl. Fuel Cycle Waste Technol. **16**, 11–39 (2018)
54. Radiological report of the French environment from 2015 to 2017, IRSN, Mission report (2018)
55. Sellafield Ltd. Discharges and Environmental Monitoring Annual Report 2019, Published 2020
56. S. Van der Stricht, A. Janssens, Radioactive effluents from nuclear power stations and nuclear fuel reprocessing sites in the European Union, 2004–2008, Directorate-General for Energy Directorate D—Nuclear Energy Unit D.4—Radiation Protection (2010)
57. C. Toyama, Y. Muramatsu, Y. Uchida, Y. Igarashi, M. Aoyama, H. Matsuzaki, Variations of $^{129}$I in the atmospheric fallout of Tokyo, Japan: 1963–2003. J. Environ. Radioact. **113**, 116–122 (2012)
58. J.M. Gómez-Guzmán, S.M. Enamorado-Báez, A.R. Pinto-Gómez, J.M. Abril-Hernández, J.M. López-Gutiérrez, M. García-León, Anthropogenic $^{129}$I concentration and $^{129}$I/$^{127}$I ratio in rainwater from Seville (Spain) in the period 2005–2008 as affected by airborne releases from Sellafield and La Hague facilities. Atmos. Environ. **56**, 26–32 (2012)
59. UNSCEAR 2008 report, United Nations Scientific Committee on the Effects of Atomic Radiation, Vol. II
60. S. Ballestra, J. Gastaud, J.J. López, Radiochemical procedures used at IAEA-ILMR Monaco for measuring artificial radionuclides resulting from the Chernobyl accident, in *Low-Level Measurements of Man-made Radionuclides in the Environment,* ed. M. García-León, G. Madurga (World Sci. Pub. Co., 1991), pp. 395–415
61. A. Baeza, L.M. del Río, J.M. Paniagua, E. Navarro, Radiological impact of the Chernobyl nuclear planta accident on two regions of Spain: Extremadura and Valencia, in *Low-Level*

*Measurements of Man-made Radionuclides in the Environment*, ed. M. García-León, G. Madurga (World Sci. Pub. Co., 1991), pp. 416–424

62. M. Aoyama, K. Hirose, Y. Suzuki, High level radioactive nuclides in Japan in May. Nature **321**, 819–820 (1986)

63. T. Kinase, K. Adachi, T.T. Sekiyama et al., Temporal variations of $^{90}$Sr and $^{137}$Cs in atmospheric depositions after the Fukushima Daiichi Nuclear Power Plant accident with long-term observations. Sci. Rep. **10**, 21627 (2020)

64. A. Aarkrog, The radiological impact of the Chernobyl debris compared with that from nuclear weapons fallout. J. Environ. Radioact. **6**, 151–162 (1988)

65. USSR State Committee on the Utilization of Atomic Energy (1986). The accident at the Chernobyl nuclear power plant and its consequences. Information compiled for the IAEA Experts Meeting 25–29 August 1986, Vienna

66. S. Stoulos, A. Besis, A. Ioannidou, Determination of low $^{137}$Cs concentration in the atmosphere due to Chernobyl contaminated forest-wood burning. J. Environ. Radioact. **222**, 106383 (2020)

67. UNSCEAR 2020 Report, Annex B, Levels and effects of radiation exposure due to the accident at the Fukushima Daiichi Nuclear Power Station: implications of information published since the UNSCEAR 2013 Report

68. A. Mathieu, M. Kajino, I. Korsakissok et al., Fukushima Daiichi–derived radionuclides in the atmosphere, transport and deposition in Japan: a review. Appl. Geochem. **91**, 122–139 (2018)

69. Y. Nishizawa, M. Yoshida, Y. Sanada, T. Torii, Distribution of the $^{134}$Cs/$^{137}$Cs ratio around the Fukushima Daiichi nuclear power plant using an unmanned helicopter radiation monitoring system. J. Nucl. Sci. Technol. **53**, 468–474 (2016)

70. M. López-Pérez, R. Ramos-López, N.R. Perestelo, X. Duarte-Rodriguez, J.J. Bustos, S. Alonso-Pérez, E. Cuevas, J. Hernández-Armas, Arrival of radionuclides released by the Fukushima accident to Tenerife (Canary Islands). J. Environ. Radioact. **116**, 180–186 (2013)

71. F.P. Carvalho, M.C. Reis, J.M. Oliveira, M. Malta, L. Silva, Radioactivity from Fukushima nuclear accident detected in Lisbon, Portugal. J. Environ. Radioact. **114**, 152–156 (2012)

72. G. Steinhauser, A. Brandl, T.E. Johnson, Comparison of the Chernobyl and Fukushima nuclear accidents: a review of the environmental impacts. Sci. Total Environ. **470–471**, 800–817 (2014)

73. K. Hirose, Atmospheric effects of Fukushima nuclear accident: a review from a sight of atmospheric monitoring. J. Environ. Radioact. **218**, 106240 (2020)

74. F. Wang, J. Ming, R. Edwards, Z. Li, N. Wang, The fallout from Fukushima Daiichi nuclear accident profiles a new dating reference in ice and comparison with the Chernobyl accident. Environ. Res. Lett. **15**, 084016 (2020)

75. G. Cinelli, M. De Cort, T. Tollefsen, (Eds.), European Atlas of Natural Radiation, Publication Office of the European Union, Luxembourg (2019). https://remon.jrc.ec.europa.eu/About/Atlas-of-Natural-Radiation

76. Y. Ishimori, K. Lange, P. Martin, Y.S. Mayya, M. Phaneuf, Measurement and Calculation of Radon Releases from NORM Residues. IAEA, Technical Reports Series 474 (2013)

77. https://www.who.int/news-room/fact-sheets/detail/radon-and-health

78. International Commission on Radiological Protection. 'Protection Against Radon-222 at Home and at Work', ICRP Publication 65. Annals of the ICRP 23 (Pergamon Press, Oxford, 1993)

79. K. Bunzl, The migration of radionuclides in the soil, in *Low-Level Measurements of Man-made Radionuclides in the Environment*, ed. M. García-León, G. Madurga (World Sci. Pub. Co., 1991), pp. 328–353

80. J.C. Gosse, F.M. Phillips, Terrestrial in situ cosmogenic nuclides: theory and application. Quatern. Sci. Rev. **20**, 1475–1560 (2001)

81. W.M. White, *Isotope Geochemistry*, Chapter 4 (Wiley and Sons, 2015)

82. The environmental behaviour of radium, IAEA Technical reports series STI/DOC/010/476 (2014)
83. NORM V: Proceedings of the Fifth International Symposium on Naturally, Occurring Radioactive Material, University of Seville. IAEA STI/PUB/1326 (2008)
84. Z. Papp, Z. Dezső, S. Daróczy, Significant radioactive contamination of soil around a coal-fired thermal power plant. J. Environ. Radioact. **59**, 191–205 (2002)
85. Y.M. Amin, M.U. Khandaker, A.K.S. Shyen, R.H. Mahat, R.M. Nor, D.A. Bradley, Radionuclide emissions from a coal-fired power plant. Appl. Radiat. Isot. **80**, 109–116 (2013)
86. B. Michalik, NORM contaminated area identification using radionuclides activity concentration pattern in a soil profile. J. Environ. Radioact. **173**, 102–111 (2017)
87. A. Komosa, S. Chibowski, M. Reszka, Natural radioisotope level differentiation in arable and non-cultivated soils at Łęczna-Włodawa Lake District', in IAEA, Naturally, Occurring Radioactive Material (NORM IV), IAEA-TECDOC-1472 (2005), pp. 117–126
88. J.H. Harley, Plutonium in the environment—a review. J. Radiat. Res. **21**, 83–104 (1986)
89. E. Chamizo, M. García-León, J.I. Peruchena, F. Cereceda, V. Vidal, E. Pinilla, C. Miró, Presence of plutonium isotopes, $^{239}$Pu and $^{240}$Pu, in soils from Chile. Nucl. Instrum. Methods Phys. Res. **B269**, 3163–3166 (2011)
90. A compendium of the Environmental Measurements Laboratory´s research projects related to the Chernobyl nuclear accident, EML-460 report (1986)
91. C. Sancho, R. García-Tenorio, Radiological evaluation of the transuranic remaining contamination in Palomares (Spain): a historical review. J. Environ. Radioact. **203**, 55–70 (2019)
92. E.P. Hardy, P.W. Krey, H.L. Volchok, Global inventory and distribution of Pu-238 from SNAP-9A, HASL-250 Health and Safety (TID-4500), US Atomic Energy Commission (1972)
93. United Nations Scientific Committee on the Effects of Atomic Radiation UNSCEAR 2013 Report to the General Assembly, with scientific annexes, Volume I: Report to the General Assembly, Scientific Annex A
94. Radioactive particles in the Environment: Sources, Particle Characterization and Analytical Techniques, IAEA-TECDOC-1663 (IAEA, 2011)
95. B. Salbu, V. Kashparov, O.C. Lind, R. García-Tenorio, M.P. Johansen, D.P. Child, P. Roos, C. Sancho, Challenges associated with the behaviour of radioactive particles in the environment. J. Environ. Radioact. **186**, 101–115 (2018)
96. B. Salbu, K. Janssens, O.C. Lind, K. Proost, L. Gijsels, P.R. Danesi, Oxidation states of uranium in depleted uranium particles from Kuwait. J. Environ. Radioact. **78**, 125–135 (2004)
97. P.R. Danesi, A. Markowicz, E. Chinea-Cano, W. Burkart, B. Salbu, D. Donohue, F. Ruedenauer, M. Hedberg, S. Vogt, P. Zaharadnik, A. Ciurapinski, Depleted uranium particles in selected Kosovo simples. J. Environ. Radioact. **64**, 143–154 (2003)
98. B. Salbu, A. Ulanowski, S. Fesenko (Eds.), IAEA: coordinated research on radioactive particles. J. Environ. Radioact. **216** (2021)
99. M. Cook, B. Etschmann, R. Ram et al., The nature of Pu-bearing particles from the Maralinga nuclear testing site, Australia. Sci. Rep. **11**, 10698 (2021)
100. H. Wittmann, F. von Blanckenburg, The geological significance of cosmogenic nuclides in large lowland river basins. Earth Sci. Rev. **159**, 118–141 (2016)
101. L. Ducros, F. Eyrolle, C. Della Vedova, S. Charmasson, M. Leblanc, A. Mayer, M. Babic, C. Antonelli, D. Mourier, F. Giner, Tritium in river waters from French Mediterranean catchments: background levels and variability. Sci. Total Environ. **612**, 672–682 (2018)
102. M.A. Ansari, H.V. Mohokar, A. Deodhar, N. Jacob, U.K. Sinha, Distribution of environmental tritium in rivers, groundwater, mine water and precipitation in Goa, India. J. Environ. Radioact. **189**, 120–126 (2018)
103. F. Eyrolle, L. Ducros, S. Le Dizès, K. Beaugelin-Seiller, S. Charmasson, P. Boyer, C. Cossonnet, An updated review on tritium in the environment. J. Environ. Radioact. **181**, 128–137 (2018)

104. G. Cazes, D. Fink, A.T. Codilean, R.-H. Fülöp, T. Fujioka, K.M. Wilcken, '$^{26}$Al/$^{10}$Be ratios reveal the source of river sediments in the Kimberley, NW, Australia. Earth Surf. Process. Land. **45**, 424–439 (2020)

105. M.R. Palmer, J.M. Edmond, Uranium in river water. Geochim. Cosmochim. Acta **57**, 4947–4955 (1993)

106. A. Martínez-Aguirre, M. García-León, Natural radioactivity enhancement by human activities in rivers of the Southwest of Spain. J. Radional. Nucl. Chem. Lett. **155**, 97–106 (1991)

107. E. Klemt, V. Putyrskaya, S. Röllin, J.A. Corcho-Alvarado, H. Sahli, Radionuclides in sediments of the Aare and Rhine river system: fallouts, discharges, depth-age relations, mass accumulation rates and transport along the river. J. Environ. Radioact. **232**, 106584 (2021)

108. K. Zhang, S. Pan, Z. Liu, G. Li, Y. Xu, Y. Hao, Vertical distributions and source identification of the radionuclides $^{239}$Pu and $^{240}$Pu in the sediments of the Liao River estuary, China. J. Environ. Radioact. **181**, 78–84 (2018)

109. J.T. Smith, S.M. Wright, M.A. Cross, L. Monte, A.V. Kudelsky, R. Saxén, S.M. Vakulovsky, D.N. Timms, Global analysis of the riverine transport of $^{90}$Sr and $^{137}$Cs. Environ. Sci. Technol. **38**, 850–857 (2004)

110. O. Evrard, J.P. Laceby, S. Huon, I. Lefèvre, O. Sengtaheuanghoung, O. Ribolzi, Combining multiple fallout radionuclides ($^{137}$Cs, $^{7}$Be, $^{210}$Pb) to investigate temporal sediment source dynamics in tropical, ephemeral riverine systems. J. Soils Sed. **16**, 1130–1144 (2016)

111. C. Froger, S. Ayrault, O. Evrard et al., Tracing the sources of suspended sediment and particle-bound trace metal elements in an urban catchment coupling elemental and isotopic geochemistry, and fallout radionuclides. Environ. Sci. Pollut. Res. **25**, 28667–28681 (2018)

112. O. Evrard, P.-A. Chaboche, R. Ramon, A. Foucher, J.P. Laceby, A global review of sediment source fingerprinting research incorporating fallout radiocesium ($^{137}$Cs). Geomorphology **362**, 107103 (2020)

113. A.S. Murray, R. Stanton, J.M. Olley, R. Morton, Determining the origins and history of sedimentation in an underground river system using natural and fallout radionuclides. J. Hydrol. **146**, 341–359 (1993)

114. P.N. Owens, W.H. Blake, G.E. Millward, Extreme levels of fallout radionuclides and other contaminants in glacial sediment (cryoconite) and implications for downstream aquatic ecosystems. Sci. Rep. **9**, 12531 (2019)

115. Radiological conditions in the Dnieper river basin: assessment by an international expert team and recommendations for an action plan, STI/DOC/1230 (IAEA, 2006)

116. K. Taniguchi, Y. Onda, H.G. Smith, W. Blake, K. Yoshimura, Y. Yamashiki, T. Kuramoto, K. Saito, Transport and redistribution of radiocesium in Fukushima fallout through rivers. Environ. Sci. Technol. **53**, 12339–12347 (2019)

117. T. Nakanishi, K. Sakuma, Trend of $^{137}$Cs concentration in river water in the medium term and future following the Fukushima nuclear accident. Chemosphere **215**, 272–279 (2019)

118. F. Eyrolle, D. Claval, G. Gontiera, C. Antonellia, Radioactivity levels in major French rivers: summary of monitoring chronicles acquired over the past thirty years and current status. J. Environ. Monit. **10**, 800–811 (2008)

119. O.C. Lind, D.H. Oughton, B. Salbu, L. Skipperud, M.A. Sickel, J.E. Brown, L.K. Fifield, S.G. Tims, Transport of low $^{240}$Pu/$^{239}$Pu atom ratio plutonium-species in the Obi and Yenisey Rivers to the Kara Sea. Earth Planet. Sci. Lett. **251**, 33–43 (2006)

120. A. Morereau, H. Lepage, D. Claval, C. Cossonnet, J.P. Ambrosi, B. Mourier, T. Winiarski, Y. Copard, F. Eyrolle, Trajectories of technogenic tritium in the Rhône River (France). J. Environ. Radioact. **223–224**, 106370 (2020)

121. M. Eisenbud, T.F. Gesell, *Environmental Radioactivity from Natural, Industrial and Military Sources*, Fourth Edition (Academic Press, 1997)

122. M. Pöschl, L.M.L. Nollet (Eds.), *Radionuclide Concentrations in Food and the Environment* (CRC Press, 2007)

123. H.-P. Hu, J.-L. Feng, F. Chen, Sedimentary records of a palaeo-lake in the middle Yarlung Tsangpo: implications for terrace genesis and outburst flooding. Quatern. Sci. Rev. **192**, 135–148 (2018)

124. A. Sakaguchi, Y. Ohtsuka, K. Yokota, K. Sasaki, K. Komura, M. Yamamoto, Cosmogenic radionuclide $^{22}$Na in the Lake Biwa system (Japan): residence time, transport and application to the hydrology. Earth Planet. Sci. Lett. **231**, 307–316 (2005)
125. N. Shadrin, N. Mirzoeva, I. Sidorov, A. Korotkov, E. Anufriieva, Natural radionuclides in bottom sediments of the saline lakes. What factors determine their concentration? Environ. Earth Sci. **79**, 168 (2020)
126. R. Nomura, K. Nakamura, K. Seto, M. Inoue, H. Kofuji, Opening of the closed water area and consequent changes of $^{228}$Ra/$^{226}$Ra activity ratios in coastal lagoon Nakaumi, southwest Japan. Appl. Radiat. Isot. **81**, 310–314 (2013)
127. F. Humbatov, B. Suleymanov, M. Ahmedov, V. Balayev, Radium isotopes in an oil-field produced lake near Baku, Azerbaijan. J. Environ. Prot. **7**, 1149–1156 (2016)
128. R. Thomas, J. Mantero, S.M. Pérez-Moreno, C. Ruiz-Canovas, I. Vioque, M. Isaksson, E. Forssell-Aronsson, E. Holm, R. García-Tenorio, $^{226}$Ra, $^{210}$Po and lead isotopes in a pit lake water profile in Sweden. J. Environ. Radioact. **223–224**, 10638 (2020)
129. C. Schuler, E. Wieland, P.H. Santschi, M. Sturm, A. Lueck, S. Bollhalder, J. Beer, G. Bonani, H.J. Hofmann, M. Suter, W. Wolfli, A multitracer study of radionuclides in Lake Zurich, Switzerland: 1. Comparison of atmospheric and sedimentary fluxes of $^{7}$Be, $^{10}$Be, $^{210}$Pb, $^{210}$Po, and $^{137}$Cs. J. Geophys. Res. Oceans **96**, 17051–17065 (1991)
130. E. Wieland, P.H. Santschi, J. Beer, A multitracer study of radionuclides in Lake Zurich, Switzerland: 2. residence times, removal processes, and sediment focusing. J. Geophys. Res. Oceans **96**, 17067–17080 (1991)
131. M. Eriksson, E. Holm, P. Roos, H. Dahlgaard, Distribution and flux of $^{238}$Pu, $^{239,240}$Pu, $^{241}$Am, $^{137}$Cs and $^{210}$Pb to high arctic lakes in the Thule district (Greenland). J. Environ. Radioact. **75**, 285–299 (2004)
132. P.G. Appleby, Three decades of dating recent sediments by fallout radionuclides: a review. The Holocene **18**, 83–93 (2008)
133. M. De Cort, G. Dubois, Sh.D. Fridman, M.G. Germenchuk, Y.A. Izrael, A. Janssens, A.R. Jones, G.N. Kelly, E.V. Kvasnikova, I. Matveenko, I.M. Nazarov, Y.M. Pokumeiko, V.A. Sitak, E.D. Stukin, L.Y. Tabachny, Y.S. Tsaturov, S.I. Avdyushin, Atlas of caesium deposition on Europe after the Chernobyl accident, Luxembourg, Office for Official Publications of the European Communities, CG-NA-16-733-29-C (1998)
134. N. Mirzoyeva, L. Gulina, S. Gulin et al., Radionuclides and mercury in the salt lakes of the Crimea. Chin. J. Ocean. Limnol. **33**, 1413–1425 (2015)
135. J.A. Robbins, A.W. Jasinski, Chernobyl fallout radionuclides in Lake Sniardwy, Poland. J. Environ. Radioact. **26**, 157–184 (1995)
136. J.A. Robbins, G. Lindner, W. Pfeiffer, J. Kleiner, H.H. Stabel, P. Frenzel, Epilimnetic scavenging of Chernobyl radionuclides in Lake Constance. Geochim. Cosmochim. **56**, 2339–2361 (1992)
137. J.T. Smith, R.N.J. Comans, D.G. Elder, Radiocaesium removal from European lakes and reservoirs: key processes determined from 16 Chernobyl-contaminated lakes. Wat. Res. **33**, 3762–3774 (1999)
138. N.V. Kuzmenkova, M.M. Ivanov, M.Y. Alexandrin, A.M. Grachev, A.K. Rozhkova, K.D. Zhizhin, E.A. Grabenko, V.N. Golosov, Use of natural and artificial radionuclides to determine the sedimentation rates in two North Caucasus lakes. Environ. Pollut. **262**, 114269 (2020)
139. T. Fukushima, H. Arai, Radiocesium contamination of lake sediments and fish following the Fukushima nuclear accident and their partition coefficient. Inland Waters **4**, 204–214 (2014)
140. K. Matsuda, K. Takagi, A. Tomiya et al., Comparison of radioactive cesium contamination of lake water, bottom sediment, plankton, and freshwater fish among lakes of Fukushima Prefecture, Japan after the Fukushima fallout. Fish. Sci. **81**, 737–747 (2015)
141. T. Basuki, S. Miyashita, M. Tsujimoto et al., Deposition density of $^{134}$Cs and $^{137}$Cs and particle size distribution of soil and sediment profile in Hibara Lake area, Fukushima: an investigation of $^{134}$Cs and $^{137}$Cs indirect deposition into lake from surrounding area. J. Radioanal. Nucl. Chem. **316**, 1039–1046 (2018)

142. S. Ochiai, S. Nagao, M. Yamamoto, T. Itono, K. Kashiwaya, K. Fukui, H. Iida, Deposition records in lake sediments in western Japan of radioactive Cs from the Fukushima Dai-ichi nuclear power plant accident. Appl. Radiat. Isot. **81**, 366–370 (2013)

143. I. Zektser, L. Everett, G. Lorne (Eds.), *Groundwater Resources of the World and Their Use* (UNESCO, Paris, 2004)

144. L.N. Plummer, E. Busenberg, J.K. Böhlke, D.L. Nelms, R.L. Michel, P. Schlosser, Groundwater residence times in Shenandoah National Park, Blue Ridge Mountains, Virginia, USA: a multi-tracer approach. Chem. Geol. **179**, 93–111 (2001)

145. P. Shand, W.M. Edmunds, *The Baseline Inorganic Chemistry of European Groundwaters. Natural Groundwater Quality* (Blackwell Publishing, 2009)

146. N.C. Sturchio, X. Du, R. Purtschert, B.E. Lehmann, M. Sultan, L.J. Patterson, Z.T. Lu, P. Müller, T. Bigler, K. Bailey, T.P. O'Connor, L. Young, R. Lorenzo, R. Becker, Z.E. Alfy, B.E. Kaliouby, Y. Dawood, A.M.A. Abdallah, One million year old groundwater in the Sahara revealed by krypton-81 and chlorine-36. Geophys. Res. Lett. **31**, 19234 (2004)

147. P.W. Spaite, G.R.J. Smithson, *Technical and Regulatory Issues Associated with Naturally, Occurring Radioactive Materials (NORM) in the gas and oil Industry* (Gas Research Institute, Chicago, USA, 1992)

148. L. Salonen, $^{238}$U series radionuclides as a source of increased radioactivity in groundwater originating from Finnish bedrock, in *Future Groundwater Resources at Risk*, ed. by J. Soukko, IAHS Publication No. 222 (IAHS Press, 1994), pp. 71–84

149. D.S. Vinson, A. Vengosh, D. Hirschfield, G.S. Dwyer, Relationships between radium and radon occurrence and hydrochemistry in fresh groundwater from fractured crystalline rocks, North Carolina (USA). Chem. Geol. **260**, 159–171 (2009)

150. N.D. Chau, M. Dulinski, P. Jodlowski, J. Nowak, K. Rozanski, M. Sleziak, P. Wachniew, Natural radioactivity in groundwater-a review. Isot. Environ. Health Stud. **47**, 415–437 (2011)

151. V. Cidzikien, V. Jakimaviĺi, M. Rasel, G. Ddien, J. Deika, R. Petrošius, Assessment of tritium activity in groundwater at the nuclear objects sites in Lithuania. Int. J. Nucl. Ener. **2014**, 201623 (2011)

152. R.D. Mutch Jr., J.D. Mahony, A study of tritium in municipal solid waste leachate and gas. Fusion Sci. Technol. **54**, 305–310 (2008)

153. K. Miyamoto, K.-I. Kimura, S. Hongo, Transfer model of tritium in a local hydrosphere. Fusion Technol. **28**, 910–917 (1995)

154. R. Lin, K. Wei, Tritium profiles of pore water in the Chinese loess unsaturated zone: implications for estimation of groundwater recharge. J. Hydrol. **328**, 192–199 (2006)

155. M.S. Garber, L.E. Wollitz, Measuring underground-explosion effects on water levels in surrounding aquifers. Groundwater **7**, 3–7 (1969)

156. S. Mulsow, M. Coquery, C. Dovlete, J. Gastaud, Y. Ikeuchi, M.K. Pham, P. Povinec, Radionuclide concentrations in underground waters of Mururoa and Fangataufa Atolls. Sci. Tot. Environ. **237–238**, 287–300 (1999)

157. A.B. Kersting, D.W. Efurd, D.L. Finnegan, D.J. Rokop, D.K. Smith, J.L. Thompson, Migration of plutonium in groundwater at the Nevada Test Site. Nature **397**, 56–59 (1999)

158. V.M. Shestopalov (Ed.), *Chernobyl Disaster and Groundwater* (A. A Balkema Publishers, 2002)

159. L. Dewiere, D. Bugai, C. Grenier, V. Kashparov, N. Ahamdach, $^{90}$Sr migration to the geosphere from a waste burial in the Chernobyl exclusion zone. J. Environ. Radioact. **74**, 139–150 (2004)

160. H.R. Von Gunten, U.E. Waber, U. Krihenbiihl, The reactor accident at Chernobyl: A possibility to test colloid-controlledtransport of radionuclides in a shallow aquifer. J. Contam. Hydrol. **2**, 237–240 (1988)

161. U. Waber, H.R. Von Gunten, U. Krähenbühl, The impact of the chernobyl accident on a river/groundwater aquifer. Radiochim. Acta **41**, 191–198 (1987)

162. A.H. Gallardo, A. Marui, The aftermath of the Fukushima nuclear accident: measures to contain groundwater contamination. Sci. Total Environ. **547**, 261–268 (2016)

163. K. Kashiwaya, Y. Muto, T. Kubo et al., Spatial variations of tritium concentrations in groundwater collected in the southern coastal region of Fukushima, Japan, after the nuclear accident. Sci. Rep. **7**, 12578 (2017)
164. K. Shozugawa, M. Hori, T.E. Johnson et al., Landside tritium leakage over through years from Fukushima Daiichi nuclear plant and relationship between countermeasures and contaminated wáter. Sci. Rep. **10**, 19925 (2020)
165. K. Sakakibara, S. Iwagami, M. Tsujimura, Y. Abe, M. Hada, I. Pun, Y. Onda, Groundwater age and mixing process for evaluation of radionuclide impact on water resources following the Fukushima Daiichi nuclear power plant accident. J. Contam. Hydrol. **223**, 103474 (2019)
166. https://oceanservice.noaa.gov/education/tutorial_currents/05conveyor2.html
167. H.D. Livingston, P.P. Povinec, A millennium perspective on the contribution of global fallout radionuclides to ocean science. Health Phys. **82**, 656–668 (2002)
168. D. Lal, An overview of five decades of studies of cosmic ray produced nuclides in oceans. Sci. Tot. Environ. **237**(238), 3–13 (1999)
169. J.M. Schaefer, A.T. Codilean, J.K. Willenbring et al., Cosmogenic nuclide techniques. Nat. Rev. Methods Primers **2**, 18 (2022)
170. T. Nakanishi, M. Kusakabe, T. Aono, M. Yamada, Simultaneous measurements of cosmogenic radionuclides $^{32}$P, $^{33}$P and $^{7}$Be in dissolved and particulate forms in the upper ocean. J. Radioanal. Nucl. Chem. **279**(3), 769–776 (2009)
171. R.J. Pentreath, Radionuclides in the aquatic environment, in *Radionuclides in the Food Chain. ILSI Monographs*, ed. By J.H. Harley, G.D. Schmidt, G. Silini (Springer, London, 1988)
172. J.H. Chen, R. Lawrence Edwards, G.J. Wasserburg, $^{238}$U, $^{234}$U and $^{232}$Th in seawater. Earth Planet. Sci. Lett. **80**, 241–251 (1986)
173. D. Delanghe, E. Bard, B. Hamelin, New TIMS constraints on the uranium-238 and uranium-234 in seawaters from the main ocean basins and the Mediterranean Sea. Mar. Chem. **80**, 79–93 (2002)
174. R.M. Dunk, R.A. Mills, W.J. Jenkins, A reevaluation of the oceanic uranium budget for the Holocene. Chem. Geol. **190**, 45–67 (2002)
175. C. Huh, W.S. Moore, C. Kadko David, Oceanic $^{232}$Th: A reconnaissance and implications of global distribution from manganese nodules. Geochim. Cosmochim. Acta **53**, 1357–1366 (1989)
176. P. McDonald, M.S. Baxter, E.M. Scott, Technological enhancement of natural radionuclides in the marine environment. J. Environ. Radioact. 67–90 (1996)
177. Y. Nozaki, J. Zhang, A. Takeda, $^{210}$Pb and $^{210}$Po in the equatorial Pacific and the Bering Sea: the effects of biological productivity and boundary scavenging. Deep Sea Res. Part II **44**, 2203–2220 (1997)
178. M. Betti, L. Aldave de las Heras, A. Janssens, E. Henrich, G. Hunter, M. Gerchikov, M. Dutton, A.W. van Weers, S. Nielsen, J. Simmonds, A. Bexon, T. Sazykina, Results of the European Commission Marina II Study Part II—effects of discharges of naturally occurring radioactive material. J. Environ. Radioact. **74**, 255–277 (2004)
179. J.P. Bolívar, R. García-Tenorio, F. Vaca, Radioecological study of an estuarine system located in the south of Spain. Water Res. **34**, 2941–2950 (2000)
180. Worldwide marine radioactivity studies (WOMARS): Radionuclide levels in oceans and seas, IAEA-TECDOC-1429 (2005)
181. C.R. Benitez-Nelson, K. Buesseler, M. Dai, M. Aoyama, N. Casacuberta, S. Charmasson, J.M. Godoy, A. Johnson, V. Maderich, P. Masqué, W. Moore, P.J. Morris, J.N. Smith, Radioactivity in the marine environment: cosmogenic and anthropogenic radionuclides. ASLO e-lect. Limnol. Oceanogr. **8**, 114–169 (2018)
182. E. Holm, Anthropogenic radionuclides in marine polar regions, in *Low-Level Measurements of Man-made Radionuclides in the Environment*, ed. M. García-León, G. Madurga (World Sci. Pub. Co., 1991), pp. 425–445

183. A. Yamato Distribution and behaviour of artificial radionculides in coastal environment in Japan, in *Low-Level Measurements of Man-made Radionuclides in the Environment*, ed. M. García-León, G. Madurga (World Sci. Pub. Co., 1991), pp. 314–327

184. https://maris.iaea.org/home

185. M. Frank, B. Schwarz, S. Baumann, P.W. Kubik, M. Suter, A. Mangini, A 200 kyr record of cosmogenic radionuclide production rate and geomagnetic field intensity from $^{10}$Be in globally stacked deep-sea sediments. Earth Planet. Sci. Lett. **149**, 121–129 (1997)

186. W. Li, X. Li, X. Mei, F. Zhang, J. Xu, C. Liu, C. Wei, Q. Liu, A review of current and emerging approaches for Quaternary marine sediment dating. Sci. Total Environ. **780**, 146522 (2021)

187. D. Heye, Uranium, Thorium, and Radium in ocean water and deep-sea sediments. Earth Planet. Sci. Lett. **6**, 112–116 (1969)

188. G.P. Klinkhammer, M.R. Palmer, Uranium in the oceans: where it goes and why. Geochim. Cosmochim. Acta **55**, 1799–1806 (1991)

189. A. Hosseini, N.A. Beresford, J.E. Brown, D.G. Jones, M. Phaneuf, H. Thørring, T. Yankovich, Background dose-rates to reference animals and plants arising from exposure to naturally occurring radionuclides in aquatic environments. J. Radiol. Prot. **30**, 235–264 (2010)

190. J. Li, B. Hu, J. Zhao, F. Bai, Y. Dou, L. Wang, L. Zou, X. Ding, Evaluation of Natural Radioactivity in Marine Sand Deposits from Offshore China. Open Journal of Marine Science **7**, 357–378 (2017)

191. T.F. Hamilton, J.C. Millies-Lacroix, G.H. Hong, $^{137}$Cs($^{90}$Sr) and Pu isotopes in the Pacific Ocean: sources and trends, in *Radionuclides in the Oceans: Inputs and Inventories*, ed. By P. Guéguéniat et al. (IPSN, Les Editions de Physique, Les Ulis, France, 1996), pp. 29–58

192. K.O. Buesseler, Fukushima and ocean radioactivity. Oceanography **27**, 92–105 (2014)

193. M.C. Honda, H. Kawakami, S. Watanabe, T. Saino, Fukushima-derived radiocesium in western North Pacific sediment traps. Biogeosci. Discuss. **10**, 2455–2477 (2013)

194. J. Zheng, T. Aono, S. Uchida, J. Zhang, M.C. Honda, Distribution of Pu isotopes in marine sediments in the Pacific 30 km off Fukushima after the Fukushima Daiichi nuclear power plant accident. Geochem. J. **46**, 361–369 (2012)

195. J.K. Cochran, H.D. Livingston, D.J. Hirschberg, L.D. Surprenant, Natural and anthropogenic radionuclide distributions in the northwest Atlantic Ocean. Earth Planet. Sci. Lett. **84**, 135–152 (1987)

196. M. Villa-Alfageme, E. Chamizo, F.J. Santos-Arévalo, J.M. López-Gutierrezb, I. Gómez-Martínez, S. Hurtado-Bermúdez, Natural and artificial radionuclides in a marine core. First results of $^{236}$U in North Atlantic Ocean sediments. J. Environm. Radioact. **186**, 152–160 (2018)

197. M. McCartney, P.J. Kershaw, D.S. Woodhead, D.C. Denoon, Artificial radionuclides in the surface sediments of the Irish Sea, 1968–1988. Sci. Tot. Environ. **141**, 103–138 (1994)

198. A.L. Rudjord, D. Oughton, T.D. Bergan, G. Christensen, Radionuclides in marine sediments—distribution and processes, in *Marine Radioecology Final Reports from Subprojects within the Nordic Nuclear Safety Research Project EKO-1*, ed. by S.E. Palsson, NKS-8: EKO-1 (98)TR1 (2001), pp. 81–106

199. R.C.L. Figueira, M.G. Tessler, M.M. de Mahiques, I.L. Cunha, Distribution of $^{137}$Cs, $^{238}$Pu and $^{239+240}$Pu in sediments of the southeastern Brazilian shelf–SW Atlantic margin. Sci. Total Environ. **357**, 146–159 (2006)

200. O. J. Nielsen, *A Literature Review on Radioactivity Transfer to Plant and Soil* (Risø National Laboratory, RISØ-R-450, 1981)

201. A. Martínez-Aguirre, R. Periáñez, Soil to plant transfer of $^{226}$Ra in a marsh area: modelling application. J. Environ. Radioact. **39**, 199–213 (1998)

202. F. Carini, Radionuclides in plants bearing fruit: an overview. J. Environ. Radioact. **46**, 77–97 (1999)

203. E. Holm, R.B.R. Persson, Fall-out plutonium in Swedish reindeer lichens. Health Phys. **29**, 43 (1975)
204. E. Holm, J. Rioseco, $^{99}$Tc in the subarctic food chain lichen-reindeer-Man. J. Environ. Radioact. **5**, 343–357 (1987)
205. L. Skuterud, E. Gaare, I.M. Eikelmann, K. Hove, E. Steinnes, Chernobyl radioactivity persists in reindeer. J. Environ. Radioact. **83**, 231–252 (2005)
206. B. Mitrovic, G. Vitorovic, D. Vitorovic, G. Pantelic, I. Adamovic, Natural and anthropogenic radioactivity in the environment of mountain region of Serbia. J. Environ. Monit. **11**, 383–388 (2009)
207. A. Diener, P. Hartmann, L. Urso, J. Vives i Batllé, M.A. Gonze, P. Calmon, M. Steiner, Approaches to modelling radioactive contaminations in forests: overview and guidance. J. Environ. Radioact. 178–179, 203–211 (2017)
208. F Vaca, G. Manjón, M. García-León, The presence of some artificial and natural radionuclides in a Eucalyptus forest in the south of Spain. J. Environ. Radioact. **56**, 309–325 (2001)
209. T.M. Nakanishi, K. Tanoi (Eds.), *Agricultural Implications of the Fukushima Nuclear Accident* (Springer, 2013)
210. ICRP. The 2007 recommendations of the international commission on radiological protection. ICRP Publication 103. Ann. ICRP 37 (2/3) (2007)
211. C.-M. Larsson, An overview of the ERICA integrated approach to the assessment and management of environmental risks from ionising contaminants. J. Environ. Radioact. **99**, 1364–1370 (2008)
212. A.V. Gulakov, Accumulation and distribution of $^{137}$Cs and $^{90}$Sr in the body of the wild boar (Sus scrofa) found on the territory with radioactive contamination. J. Environ. Radioact. **127**, 171–175 (2014)
213. A. Baeza, M. del Rio, C. Miró, A. Moreno, E. Navarro, J.M. Paniagua, M.A. Peris, Radiocesium and Radiostrontium levels in Song-Thrushes (Turdus philomelos) captured in two regions of Spain. J. Environ. Radioact. **13**, 13–23 (1991)
214. M. Castrillejo, R. Witbaard, C.A. Richardson, R. Dekker, C. Welte, L. Wacker, M. Christl, Impact of nuclear fuel reprocessing on the temporal evolution of marine radiocarbon. Sci. Total Environ. **738**, 139700 (2020)
215. G. Manjón, M. García-León, S. Ballestra, J.J. López, The presence of man-made radionuclides in the marine environment in the south of Spain. J. Environ. Radioact. **28**, 17–l89 (1995)
216. J. Qiao, K. Andersson, S. Nielsen, A 40-year marine record of $^{137}$Cs and $^{99}$Tc transported into the Danish Straits: significance for oceanic tracer studies. Chemosphere **244**, 125595 (2020)
217. V. Smith, M. Fegan, D. Pollard, S. Long, E. Hayden, T.P. Ryan, Technetium-99 in the Irish marine environment. J. Environ. Radioact. **56**, 269–284 (2001)
218. C. Frechou, D. Calmet, $^{129}$I in the environment of the La Hague nuclear fuel reprocessing plant—from sea to land. J. Environ. Radioact. **70**, 43–59 (2003)
219. J.M. Gómez-Guzmán, E. Holm, N. Niagolova, J.M. López-Gutiérrez, A.R. Pinto-Gómez, J.A. Abril, M. García-León, Influence of releases of $^{129}$I and $^{137}$Cs from European reprocessing facilities in Fucus vesiculosus and seawater from the Kattegat and Skagerrak áreas. Chemosphere **108**, 76–84 (2014)
220. Z. Baumann, N. Casacuberta, H. Baumann, P. Masqué, N.S. Fisher, Natural and Fukushima-derived radioactivity in macroalgae and mussels along the Japanese shoreline. Biogeosci. Discuss. **10**, 2617–2633 (2013)
221. H. Kawai, A. Kitamura, M. Mimura, T. Mimura, T. Tahara, D. Aida, K. Sato, H. Sasaki, Radioactive cesium accumulation in seaweeds by the Fukushima 1Nuclear Power Plant accident—two years' monitoring at Iwaki and its vicinity. J. Plant Res. **127**, 23–42 (2014)

222. A. Tejera, L. Pérez-Sánchez, G. Guerra, A.C. Arriola-Velásquez, H. Alonso, M.A. Arnedo, G. Rubiano, P. Martel, Natural radioactivity in algae arrivals on the Canary coast and dosimetry assessment. Sci. Total Environ. **658**, 122–131 (2019)

223. T. Gäfvert, C. Ellmark, E. Holm, Removal of radionuclides at a waterworks. J. Environ. Radioact. **63**, 105–115 (2002); T. Warneke, I.W. Croudace, P.E. Warwick, R. Taylor

224. Environmental Measurements Laboratory, EML-412-UC-11 (1982)

225. S. Merz, K. Shozugawa, G. Steinhauser, Analysis of Japanese radionuclide monitoring data of food before and after the Fukushima nuclear accident. Environ. Sci. Technol. **49**, 2875–2885 (2015)

# Radiological Impact. Radiation Dosimetry

**6**

**ABSTRACT**

The main objective of the chapter is the description of radiation effects in material systems, especially in biological systems and specifically in the human body. They are of paramount importance to determine the injury caused by environmental radioactivity in the population. The radiation dose is defined in terms of the deposited energy in a system per unit mass, and related to that concept, some dosimetry magnitudes are defined to quantify it. The main guidelines of radiological protection programs are described to prevent the impact of radiation on the population. The consequent radiation protection regulations are also described.

## 6.1 Radiation Dosimetry

The determination of radionuclides in the environment is of interest for many reasons. It helps to understand the environment and its dynamics but also allows one to better protect the public against the harmful effects of radiation. Somehow, the main objective of this book serves as a tool for radiological protection studies and for that it is convenient to introduce the main concepts associated with it. Of course, the reader should study other texts as in [1] to have a deeper understanding of the matter, which is today an independent discipline.

### 6.1.1 Radiation Exposure, Absorbed Dose and Dose Equivalent: Magnitudes and Units

As we will see in Chap. 7, nuclear radiation deposits energy in its pathway through matter causing ionization and excitation: Both effects are, in turn, very useful for radiation detection. However, ionization or excitation of matter can generate

© The Author(s), under exclusive license to Springer Nature Switzerland AG 2022
M. García-León, *Detecting Environmental Radioactivity*, Graduate Texts
in Physics, https://doi.org/10.1007/978-3-031-09970-0_6

changes in its structure and in the whole physical system in which it participates. These changes can be chemical, physical, or both and occur immediately or on a medium- or long-term basis. This applies to biological tissues as well, although in this case some repairing mechanisms appear that can solve the modification produced by the energy deposit. We will return to this problem in the following paragraphs.

In any case, according to the experience gained over the years, the effects on the systems depend on the amount of deposited energy, i.e., on the radiation dose received. To evaluate these effects, some magnitudes and their corresponding units have been defined.

Pioneering work in radioactivity led the scientific community to introduce the concept of radiation exposure to evaluate the amount of radiation received by people exposed to radiological tests using X-ray tubes [2]. Since the radiation used at that time was electromagnetic radiation, the concept was defined specifically [3]. Radiation exposure, $X$, is the absolute amount of charge, $dq$, of the same sign released by photons in air per unit mass, $dm$.

$$X = \frac{dq}{dm} \tag{6.1}$$

See Fig. 6.1 for an easier understanding of the concept.

The historical unit for $X$ is the Roentgen, R, defined as the exposure corresponding to a released electric charge of 1 e.s.u. in 0.001293 g of air. This mass corresponds to 1 cm$^3$ of air under normal temperature and pressure conditions. The unit in the international system is the Coulomb per kilogram of air (Ckg$^{-1}$). It is very easy to find that 1 R corresponds to the release of $2.08 \times 10^9$ ion pairs per cm$^3$. Thus, 1 R = 258 $\mu$C kg$^{-1}$.

The exposure rate can be defined as

$$\dot{X} = \frac{dX}{dt} \tag{6.2}$$

**Fig. 6.1** The charge created within a $dm$ will produce a radiation exposure to electromagnetic radiation of $X = \frac{dq}{dm}$

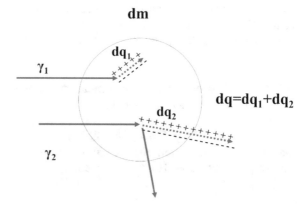

which is a relevant magnitude since, as we will see, for radiological protection, the rate at which the energy is delivered in the system is also important.

Radiation exposure is defined specifically for photons and a specific medium, the air. A more general concept is needed, and this is the radiation absorbed dose, $D$. The definition of $D$ applies to any type of radiation and to any medium [3]. It is defined as

$$D = \frac{dE}{dm} \tag{6.3}$$

where $dE$ is the amount of energy deposited by the radiation in a given mass element, $dm$, of a determined medium, i.e., the amount of energy deposited per unit mass. The historical unit of the absorbed dose is rad. It is defined as 1 rad $= 100$ ergg$^{-1}$. The International System unit is Gray, Gy, which is 1 Jkg$^{-1}$. According to that, 1 Gy $= 100$ rad. Similar to $\dot{X}$, the absorbed dose rate can be defined as

$$\dot{D} = \frac{dD}{dt} \tag{6.4}$$

$X$ and $D$ are both energy deposits, and therefore, there is a relationship between them. It is found that in air, an exposure $X$ in Roentgen corresponds to an absorbed dose in Gy of

$$D_{air}(Gy) = 87.7 \times 10^{-14} X(R) \tag{6.5}$$

Being the absorbed dose a general concept, including the different types of radiation and media, it is not yet enough to inform about the effects the radiation can induce in biological systems or tissues [4]. It is well known that identical $D$ produces different effects in biological tissues depending on the type of radiation and its energy. To take these differences into account, some weighting factors, called quality factors in the past, are introduced, and a new magnitude, the dose equivalent, $H$, is defined as

$$H = D \cdot w \tag{6.6}$$

$w$ is a weighting factor that accounts for the differences mentioned earlier [5]. As in the case of $\dot{D}$, a dose equivalent rate, $\dot{H}$, can be defined. The unit for $H$ is the Sievert (Sv) when $D$ is measured in Gy. Dimensionally, $H$ and $D$ are identical, but $H$ contains additional factors that take into account the differences each radiation has in the induced biological effects. The historical unit for $H$ was rem and 1 Sv $= 100$ rem.

As defined thus far, $D$ and, consequently, $H$ account for a dose received at a given point of an organ or system. However, from a radiological point of view, it is more interesting to refer to the average values that are more significant for the evaluation of a given effect. Equation 6.6 gives $H$ for an individual radiation type.

**Table 6.1** Weighting factors, $w_R$, for different radiation and energy levels

| Radiation and energies | Radiation weighting factor, $w_R$ |
|---|---|
| Photons, all energies | 1 |
| Electrons and muons, all energies neutrons, energy $E_n$ | 1 |
| $E_n < 10$ keV | 5 |
| 10 keV $< E_n < 100$ keV | 10 |
| 100 keV $< E_n < 2$ meV | 20 |
| 2 meV $< E_n < 20$ meV | 10 |
| $E_n > 20$ meV | 5 |
| Protons, $E_p > 2$ meV | 5 |
| Alpha particles, fission fragments, heavy nuclei of all energies | 20 |

To generalize the definition, it can be said that the total dose equivalent received by a tissue $T$, $H_T$, irradiated by different radiation types, will be

$$H_T = \sum_R w_R \cdot D_{T,R} \tag{6.7}$$

where $R$ refers to the type of radiation and $T$ refers to the tissue. In Table 6.1, we present the values of $w_R$ for different radiation and energy levels [6]. They have been proposed by the International Commission for Radiological Protection (ICRP) to represent the values of the relative biological efficiency (RBE) of the radiation to induce stochastic effects (see the next sections). RBE serves to compare the absorbed doses produced by different radiation types necessary to induce the same biological effect. Thus, the RBE of two types of radiation is defined as the quotient between the absorbed doses of each radiation that produces the same biological effect.

### 6.1.2   Effective and Committed Doses and Other Magnitudes

Furthermore, the probability of occurrence of a biological effect (of stochastic nature, see next sections) also depends on the irradiated organ or tissue. It is interesting to define another magnitude related to $H_T$, which considers the different irradiated tissues. In this way, the evaluation of a possible effect in the whole human body from the irradiation of one or several organs or tissues can be readily done. Thus, an effective dose, $E$, is defined as

$$E = \sum_T w_T \cdot H_T \tag{6.8}$$

**Table 6.2** Tissue or organ weighting factors, $w_T$, values

| Tissue | Tissue weighting factor, $w_T$ |
|---|---|
| Gonads | 0.20 |
| Breast | 0.05 |
| Colon | 0.12 |
| Red bone marrow | 0.12 |
| Lungs | 0.12 |
| Stomach | 0.12 |
| Urinary bladder | 0.05 |
| Liver | 0.05 |
| Esophagus | 0.05 |
| Thyroid | 0.05 |
| Bone surface | 0.01 |
| Skin | 0.01 |
| Remaining | 0.05 |

where the summation extends over the irradiated tissues and $T$ and $w_T$ correspond to a tissue weighting factor representing the contribution of the tissue to the total effect. We present some values for $w_T$ in Table 6.2 [7].

In other words, the occurrence of a given effect in the human body, because of an absorbed dose, depends on the radiation types and their energies, through $w_R$, and on the distribution of the dose around the body, through $w_T$. In addition, it is well known that the effect also depends on the dose rate, the chemical environment, and other parameters. The old ICRP recommendations included all these possible contributions as a factor $N$ in (6.6) or (6.7) [8]. Currently, the ICRP prefers to include them during the evaluation of the probability of appearance of a given effect after receiving a given radiation dose [9].

Organs and tissues can be externally or internally irradiated. If the radiation source is incorporated into the human body by the intake of material containing radionuclides, a dose equivalent will be induced in the affected organs or tissues. The irradiation will extend during the time the radionuclides remain there (see Exercise 11 for the concept of biological half-life). The dose equivalent, in this way induced, will change over time as the radionuclides decay or are biologically eliminated from organs or tissues. In other words, the energy deposit will change over time. To account for this, some new magnitudes are defined, the committed dose equivalent, $H_T(\tau)$, and the committed effective dose, $E(\tau)$, as

$$H_T(\tau) = \int_0^\tau \dot{H}_T(t)dt \tag{6.9}$$

and

$$E(\tau) = \sum_T w_T H_T(\tau) \tag{6.10}$$

where the integration extends over a period $\tau$. The meaning of (6.9) is immediate, being the equivalent dose accumulated in the organ or tissue during a period $\tau$. By adding the calculated $H_T(\tau)$ over all affected organs, we find $E(\tau)$ as the effective dose over such a period of time. Following the recommendations of the ICRP, $\tau$ is systematically taken as 50 y unless otherwise indicated [10].

The magnitudes described above serve to calculate the radiation dose received by the individuals. In many cases, it is interesting to evaluate the radiation dose received by the public in general or by a specific group of people who are occupationally exposed to the same radiation sources. In other words, a collective magnitude is needed to evaluate the average dose received by the group. The collective effective dose, $S$, is defined as

$$S = \sum_{i=1}^{n} E_i N_i \qquad (6.11)$$

where $E_i$ is the average effective dose received during a period of time period $\Delta t$ by population subgroup $i$, and $N_i$ is the number of individuals who belong to this subgroup. The collective effective dose is measured in person Sievert [11]. Interpreting the values obtained for $S$ is not easy within a complete radiological protection system. The ICRP recommends discussing the obtained results, taking into account several variables, such as the number of exposed individuals, their age and sex, and their geographical distribution, in the case the group is not confined in a given space.

Another interesting magnitude is the annual limit of intake (ALI) of a given radionuclide. This magnitude is especially relevant in environmental studies, as it gives the maximum allowed intake through ingestion or inhalation of a radionuclide present in the air, water, or foodstuff that could be consumed by the human being. ALI is defined as the incorporated activity in Bq of a given radionuclide that produces the annual effective dose permitted to exposed workers (see next sections), $E_{\mathrm{limit},w}$. ALI for a given radionuclide is expressed as

$$\mathrm{ALI(Bq)} = \frac{E_{\mathrm{limit},w}}{e(50)} \qquad (6.12)$$

$e(50)$ is the effective dose coefficient for internal exposure that gives, in a reference person [12], the committed effective dose produced per activity unit of the incorporated radionuclide. It is measured in $\mathrm{SvBq}^{-1}$.

ALI is the maximum permissible concentration of a given radionuclide in air or water. The MPC values give the maximum allowable amount of radionuclide in an environmental compartment compatible with producing ALI. For example, the MPC in air would be the concentration of air activity in $\mathrm{Bqm}^{-3}$ of a radionuclide that produces an intake equal to ALI. Taking a breathing rate of 1.1 $\mathrm{m}^3\mathrm{h}^{-1}$ and 2000 working hours per year for a reference person, this MPC turns out to be [13].

$$\mathrm{MPC}_{\mathrm{air}} = \frac{\mathrm{ALI}}{2200} \qquad (6.13)$$

Similarly, an MPC in drinking water can be defined by assuming a standard water consumption for a reference person of 2 L per day. In the same way, an MPC related to food consumption can be defined. Some data can be found in [14] of MPC for different radionuclides. We will return to this concept in the section devoted to regulations since in some cases MPC is taken as reference data to construct a complete system of radiation protection regulations. Today, the ICRP does not recommend ALI values, as it considers that to comply with dose limits it is the total dose from external and internal exposure that must be considered. However, ALI values can be useful for specific situations when the administration bodies need a tool to control the radiation hazard, for example, from an accidental release.

## 6.2    Biological Effects of Radioactivity

The harmful effects of radiation on human beings derive from the transformations that biological tissues undergo under irradiation, all of them induced by the energy deposited by the radiation, or, in other words, by the absorbed dose. Humans must be protected from external and internal irradiation [15]. External irradiation is produced by radiation sources located externally to the human body, that is, medical X- or γ-ray medical sources, very active radioactive sources, mainly γ-radiation sources, etc. Internal irradiation, or contamination, is produced by the intake of radionuclides into the human body by inhalation, ingestion, contact with wounds or scares in the skin. The presence of radionuclides in the environment contributes to both irradiation pathways, which is why its measurement is included in all radiological protection programs. Some examples of such programs are given in the following paragraphs. Before, we provide some details of the biological effects of radiation.

### 6.2.1    Stochastic and Deterministic Effects

Biological systems, like any other, undergo modifications under irradiation. The importance of such alterations depends on the absorbed dose. In many cases, the absorbed dose is not sufficient to cause changes in the system. In fact, the biological system tends to respond to modification and, in some cases, can restore the original situation by repairing the damage [16]. In the case of the human body, irradiation (either internal or external) can affect specific organs or the whole organism as a consequence. This can lead to the immediate appearance of diseases or to genetic modifications that in the medium or long term could induce other diseases in the irradiated body or in the descendants. Everything depends on the absorbed dose and on the absorbed dose rate, as we will see later. These macroscopic effects are the consequences of some basic phenomena occurring at a microscopic level.

**Table 6.3** Different steps of the radiation effects in biological systems. Cell water ionization triggers the process

| Direct damage | Indirect damage |
|---|---|
| *Molecular breaking* (carbon hydrates, fats, proteins,…) | Cell water is ionized, producing $H_2O^+ + e$ <br> $H_2O^+ \rightarrow OH + H^+$ <br> $e + H_2O \rightarrow H_2O^- \rightarrow H + OH^-$ <br> mainly, $H +$ and $OH^-$ induce physicochemical reactions that lead to chemical changes in the cell and, finally, biological effects in the individual or damage the genetic load |
| *DNA damage* (single and double strand breaks; base damage; protein–DNA and protein–protein cross-links) | |

Let us give some details. In Table 6.3, we give the radiation effects produced at the microscopic level in biological systems that will lead to macroscopic modifications of the organism [17].

Let us start from the beginning. Nuclear radiation has enough energy to ionize the medium that is passing through. It can also excite, but from a biological viewpoint, the relevant effect is mostly ionization. In Fig. 6.2, the effect of low and high linear energy transfer (LET, see Chap. 7) radiation passing through DNA is schematically presented. Ionization affects the main component of the human cell, water. Ionization of water molecules induces a series of physicochemical reactions that can lead to the breaking of essential molecules. The reason lies in the chemical reactivity of $H_2O^+$ that rapidly dissociates into $H^+$ and OH (see Table 6.3). On the other hand, the electron liberated in the ionization can be captured by a $H_2O$ molecule that is ionized to $H_2O^-$. This ion dissociates, producing H and $OH^-$. Both $OH^-$ and $H^+$ free radicals are very reactive from a chemical point of view and are produced within the first $10^{-12}$ s in these series of reactions. They can move rapidly along the medium, causing chemical changes in molecules during the first $10^{-6}$ s. These changes lead to microscopic and macroscopic effects that can be immediate or delayed for days, years, or even passed to descendants.

In other cases, some molecules can be broken directly due to a large local energy deposit. Molecules such as carbon hydrates, fats, and proteins can be affected in this way. In general, if the number of affected molecules is small, there are repairing mechanisms that can restore them, and no relevant effect appears at the microscopic or macroscopic level. The case of DNA is different. Of course, when the number of molecules affected by DNA is small, the damage can be repaired, but even in this case, if the damage is not perfectly repaired, microscopic or macroscopic effects can appear [18].

As mentioned above, the effects depend on the dose received by the biological systems and can be classified in different manners. The effects can be immediate or delayed, somatic or hereditary, and stochastic or deterministic. The latter being the official classification used currently, since, in general, the information

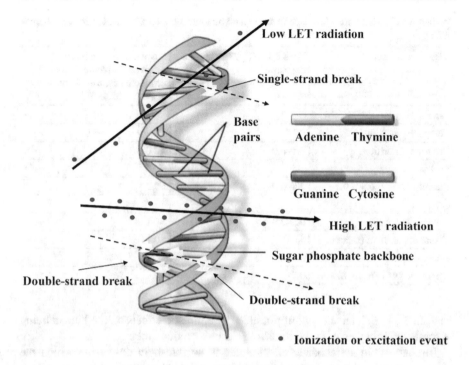

**Fig. 6.2** Schematic representation of the effect of radiation passing through DNA. High LET radiation affects in a different manner than low LET radiation. In addition to ionization, direct breaking of strands and other molecules can take place. Figure based on the US National Library of Medicine picture

we have on a given effect is the probability of occurrence. For that, the more general classification refers to this concept [19].

A deterministic effect appears once a threshold dose has been passed and the intensity of the effect is proportional to the received dose. Information on deterministic effects has been derived from the secondary effects observed in radiotherapy practices, from those observed in the population affected by Nagasaki and Hiroshima nuclear detonations or major nuclear accidents [20]. Thus, although the effects of radiation on the human body are very much affected by the health of individuals, it has been observed that in healthy people, the probability of an effect occurring is practically 0% for dose equivalents of hundreds or thousands of millisieverts. This probability increases rapidly to 100% for dose equivalents above a threshold. Since $w_R$ have been defined on the basis of stochastic effects (see next paragraphs), the dose equivalent concept is not always the adequate magnitude to analyze deterministic effects. For this, the thresholds for the effect are given in some cases as the absorbed dose [21]. Thus, the results show that for most tissues, deterministic effects are not observed below an annual absorbed dose of 0.5 Gy. However, gonads, crystallin, and bone marrow have lower thresholds for sterility, opacity, or blood production degradation, respectively, 0.15, 0.15, and 0.4 Gy per

**Table 6.4**  Threshold dose for some deterministic biological effects of radiation in some human organs

| Tissue and effect | Total dose (Gy) single brief exposure | Annual dose rate $(Gy y^{-1})$ received in fractionated or protracted exposures for many years |
|---|---|---|
| *Tests* | | |
| Temporary sterility | 0.15 | 0.4 |
| Permanent sterility | 3.5–6.0 | 2.0 |
| *Ovaries* | | |
| Sterility 2.5–6.0 6.0 > 0.2 | 2.5–6.0 | >0.2 |
| *Lens* | | |
| Detectable opacities | 0.5–2.0 | >0.1 |
| Visual impairment (cataract) | 5.05 | >0.15 |
| *Bone marrow* | | |
| Depression of hematopoiesis | 0.5 | >0.4 |

year. In Table 6.4, an account of possible deterministic effects on the human being is presented together with the threshold dose to produce them.

In contrast, in a stochastic effect, there is no threshold dose from which the effect appears, and it is assumed that the probability of occurrence is proportional to the received dose. Stochastic effects are very difficult to study since they are generally associated with low doses and do not necessarily appear in the irradiated individual but in its progeny. On the other hand, it is difficult to distinguish whether the effect has been produced for reasons other than irradiation. According to the present information, there do not appear to be other stochastic effects different from cancer in the irradiated individual as a product of the irradiation of somatic cells [22]. It is also very characteristic that the appearance of the effect occurs after a period of latency [23]. If irradiation affects reproductive cells or genetic material, it is possible to produce a genetic effect on the descendants. Many of these genetic changes are not harmful to the progeny but, in some cases, genetic diseases, malformations, etc., can result from exposure to radiation. More detailed information is given in the next section.

As commented before, stochastic effects are difficult to study. In addition to data from humans, investigations with animal models have provided information about them [24]. It is known that the probability of the appearance of a stochastic effect at low doses depends not only on the received dose but also on the type of radiation, especially if the radiation is of high or low LET (see Chap. 7). However, the form of the risk–dose curve at low doses is not clear. In Fig. 6.3, we give some of the possible shapes of the risk–dose curves accepted in the literature [25]. Since the dose equivalent limits we will see in the next section are selected according to the occurrence of stochastic effects, knowledge of the curve form is crucial. A

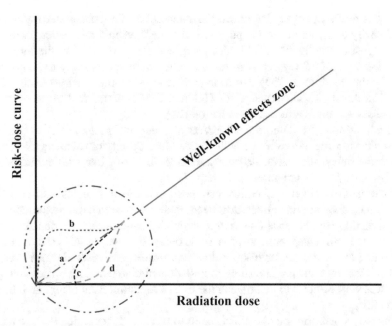

**Fig. 6.3** Probability of risk versus radiation dose for stochastic effects. Curves a, b, and c give the different hypothetical response curves to low doses. The known area corresponds to experimental data obtained from accidental releases, medical treatments with radiation, or nuclear bomb survivors

conservative hypothesis leads us to accept that the relationship is also proportional for low doses, although this assumption is recognized to have some uncertainty.

## 6.2.2 Radiation Effects on Human Health

There are several specialized texts on the effects of radiation on human health. We refer the reader to them for a deeper understanding of the topic [26], as a deep study of this matter is far beyond the objectives of this text. For that, only some basic information on the problem is included in what follows.

The irradiation of humans can lead to somatic or hereditary effects, i.e., effects directly appearing in the irradiated individual or effects appearing in its offspring. According to current knowledge, deterministic effects lead to somatic consequences, while stochastic effects can produce somatic or hereditary consequences [27].

The response of human organs or tissues to irradiation varies greatly. As presented in Table 6.4, the threshold for the appearance of a given deterministic effect depends on each organ or tissue. Additionally, it depends on the irradiation conditions. For example, as seen in [28], the threshold for temporary sterility in men is 0.15 Gy for a short exposure time. However, under prolonged exposure conditions,

0.4 Gy is needed per year to induce the same effect. Permanent sterility appears at 3.5–6 Gy under short-term exposure, while the threshold rate under prolonged-term exposure conditions is 2 Gy per year. In the case of women, the thresholds are 2.5–6 Gy and 0.2 Gy per year, respectively. These values apply to people with good health conditions. They can change depending on the previous health situation. The thresholds also change for children. It is assumed that growing organs and tissues are more affected by radiation [28].

The thresholds for deterministic effects also depend on the radiation that produces the exposure. An example is the human lens opacity or cataracts that appear with some delay after an acute exposure to 2–10 Gy of low LET radiation and 1–2 Gy of high LET radiation (see Chap. 7).

There is also evidence of human carcinogenesis due to ionizing radiation [29]. In fact, data showing this effect have been obtained from radiation-exposed workers, Hiroshima and Nagasaki survivors, radiology patients, etc. Of course, cancer in irradiated individuals appears after some delay. For that, in this case, it is more reasonable to use the cancer risk per dose unit instead of threshold doses to evaluate this effect on humans. The study of radiation carcinogenesis is epidemiological since it is very difficult to distinguish the role of radiation from other carcinogenic agents [30].

The only stochastic somatic effect seems to be cancer. According to the data, the risk of cancer development increases with absorbed dose, although an important incertitude still exists for low doses, as deduced from Fig. 6.3. In this region, it is difficult to distinguish the origin of carcinogenesis. Risk factors per unit dose for high doses and dose rates can be found in [31] for different tissues. These values simply help to evaluate the probability of radiation carcinogenesis, and of course, they are not approximations of risk estimates. Taking this into account, an estimate of the risk of developing a fatal cancer during the lifetime is $5 \times 10^{-2}$ $Sv^{-1}$ for a whole-body exposure [32]. High and acute total body exposure produces acute radiation syndrome. Within this context, the concept of lethal dose (LDx,y) is defined. Thus, a dose of LD is lethal for x% of the population within the next y days. For example, an acute exposure of 3–5 Sv is lethal for 50% of the exposed population within the next 2 months after exposure in the absence of medical treatment. For that $LD_{50,60} = 3$–5 Gy. The concept of LD100 is also used to define the lethal dose for 100% of the population. Radiation syndrome has several stages, as shown in Fig. 6.4. Within the first two days after exposure, individuals suffer from nausea, vomiting, fatigue, and fever. During the following 2 or 3 weeks, a general health improvement appears to appear. However, after this latent period and until the following 6–8 weeks, exposed individuals suffer from diarrhea, vomiting, fatigue, bleeding, bloody stools, and death. After that, the survivors recover. An acute dose above 8 Sv is lethal for almost 100% of the exposed group [33].

Hereditary effects are caused by the damage to the genetic charge transmitted to the individual offspring. They have been observed in animal models but have not been confirmed in humans with a sufficient statistical degree of confidence [34]. In fact, the analysis of the descendants of Japanese bomb survivors shows

| SYNDROME | DOSE | PRODROMAL STAGE | LATENT STAGE | MANIFEST ILLNESS STAGE | RECOVERY |
|---|---|---|---|---|---|
| Hematopoietic (Bone Marrow) | > 0.7 Gy (> 70 rads) (mild symptoms may occur as low as 0.3 Gy) | • Symptoms are anorexia, nausea and vomiting. • Onset occurs 1 hour to 2 days after exposure. • Stage lasts for minutes to days. | • Stem cells in bone marrow are dying, although patient may appear and feel well. • Stage lasts 1 to 6 weeks. | • Symptoms are anorexia, fever, and malaise. • Drop in all blood cell counts occurs for several weeks. • Primary cause of death is infection and hemorrhage. • Survival decreases with increasing dose. • Most deaths occur within a few months after exposure. | • in most cases, bone marrow cells will begin to repopulate the marrow. • There should be full recovery for a large percentage of individuals from a few weeks up to two years after exposure. • death may occur in some individuals at 1.2 Gy • the LD50/60 is about 2.5 to 5 Gy |
| Gastrointestinal (GI) | > 10 Gy (some symptoms may occur as low as 6 Gy) | • Symptoms are anorexia, severe nausea, vomiting, cramps, and diarrhea. • Onset occurs within a few hours after exposure. • Stage lasts about 2 days. | • Stem cells in bone marrow and cells lining GI tract are dying, although patient may appear and feel well. • Stage lasts less than 1 week. | • Symptoms are malaise, anorexia, severe diarrhea, fever, dehydration, and electrolyte imbalance. • Death is due to infection, dehydration, and electrolyte imbalance. • Death occurs within 2 weeks of exposure. | • the LD100 is about 10 Gy |
| Cardiovascular (CV)/ Central Nervous System (CNS) | > 50 Gy (some symptoms may occur as low as 20 Gy) | • Symptoms are extreme nervousness and confusion; severe nausea, vomiting, and watery diarrhea; loss of consciousness; and burning sensations of the skin. • Onset occurs within minutes of exposure. • Stage lasts for minutes to hours. | • Patient may return to partial functionality. • Stage may last for hours but often is less. | • Symptoms are return of watery diarrhea, convulsions, and coma. • Onset occurs 5 to 6 hours after exposure. • Death occurs within 3 days of exposure. | • No recovery is expected. |

**Fig. 6.4**  Stages of acute radiation syndrome

that an important group does not present genetic disorders. Hence, experts have no doubt on the existence of such effects, although more research is needed. However, risk factors are provided by assuming several models. Thus, within the indirect method that excludes so-called multifactorial disorders, it can be said that the risk of genetic affectation is $0.3 \times 10^{-2}$ in the following two generations per Sv absorbed by the reproductive segment of the population. In the other approach, the direct method, which considers clinically important diseases in the first generation, the risk estimate ranges from 0.2 to $0.4 \times 10^{-2}$ per Sv absorbed by the reproductive part of the population [35]. Although they are different approaches, the similarity of the results is apparent.

## 6.3  Radiological Impact

### 6.3.1  Radiation Protection Programs

Radiation protection programs are designed to protect the public from the harmful effects of radiation. They are implemented in all countries through the regulations we will see in the next section. In general, although with some differences, they are very similar since they are based on international recommendations provided by the International Commission on Radiological Protection (ICRP). This is an independent and international body formed by well-recognized and competent scientists. The ICRP recommendations are not compulsory but, in practice, are followed by all the countries, inspiring all the regulations on the matter.

The radiological protection system is based on three basic principles [36].

1. Principle of justification: No practice involving the use of radiation sources will be accepted unless there is no benefit to the human being.
2. Principle of optimization: The exposure of individuals to radiation will be as low as reasonably achievable taking into account social and economic factors (ALARA principle).
3. Principle of application of dose limits: The equivalent dose received by the individuals should not exceed the maximum limits recommended by the ICRP for each situation.

These principles are applicable to any source of radiation, regardless of its origin and size. Natural sources, such as primordial or cosmogenic radionuclides and their descendants, with an emphasis on Rn isotopes and cosmic radiation are included as well. From this point of view, the use of NORM materials must be taken into account properly during the operation of conventional industries and the management of waste from them. According to the ICRP, these principles must be applied, distinguishing three categories of exposure to radiation [36].

1. Occupational exposure is the radiation exposure of workers as a result of their work. In this case, the responsibility for worker protection lies with the employer.
2. Public exposure encompasses all radiation exposure of the public in general, different from the occupational or medical exposure each individual could particularly receive. Natural radiation sources are the main culprit for this exposure, although the contribution of artificial sources could be more relevant in accidental situations (see Chap. 4). Environmental surveillance programs are addressed to keep this method of exposure under control. Regulations for occupational and public exposures are different. However, the exposure of the embryo and fetus of pregnant workers is regulated as public.
3. Medical exposure of patients refers to the exposure of patients to radiation sources used in treatments, diagnostic and therapeutic practices, surgical interventions, etc. In this case, the use of radiation sources results in a benefit for the individual. For this, there exist specific regulations in this area [37].

Of course, this leads to the definition of three types of exposed individuals: workers, members of the public, and patients. Thus, a given individual can be exposed as a worker, as a general public, as a patient, or any other possible combination among the exposure categories. This is an important consideration when applying the regulations, especially the dose limits that we will see in the next paragraphs.

Each category has its own regulations and dose equivalent limits that derive from the joint application of Principles 1 and 2 to a given practice or activity, that is, once the benefit of a given practice or activity compatible with a reasonable minimization of the radiation exposure has been analyzed.

With respect to the objectives of this book, the most important category is public exposure. The ICRP defines the public as individuals who receive exposure that is neither occupational nor medical [38]. This means that a wide set of natural and artificial sources of radiation must be taken into account for the evaluation of the total exposure. On the other hand, the variability in age, size, diet, environment, and habits of members of the public advised the ICRP to introduce the concept of a 'critical group' as a way to take into account the exposure differences that arise from all these parameters. Currently, this concept has been substituted by that of 'Representative Person' [39]. The 'Representative Person' is the individual who receives a dose representative of members of the population who are subject to higher exposures.

To achieve the third principle and avoid inacceptable risks due to radiation exposure, some effective dose limits must be applied. Based on the experience gained over time, the ICRP [40] recommends the limits in Table 6.5. For occupational exposure, the limit is 20 mSvy$^{-1}$ averaged over a 5-year period, that is, 100 mSv over five years, with the additional condition that the effective dose must not exceed 50 mSv in any of the years in the period. Regarding public exposure, the limit is 1 mSvy$^{-1}$. In some special circumstances, as defined by the national authorities, this limit can be exceeded. However, in a 5-year period, the average effective dose must not exceed 1 mSvy$^{-1}$. As seen in Fig. 6.3, below these values, stochastic effects are very unlikely. However, it is only very unlikely. In fact, limits are not thresholds below which there is no risk. From this point of view, we must recall the second principle, which recommends reducing the dose as much as possible. In fact, the previous limits do not guarantee the absence of deterministic effects on skin, crystallin, hands, and feet. For that, some general limits are also imposed to protect such organs. They are also given in Table 6.5.

It is very interesting to check how the limits have been evolving over time, as they derive from the knowledge the ICRP has acquired along its existence. In Table 6.6, we can see the evolution of the limits at the beginning of the twentieth century and more recently [41].

**Table 6.5** Accepted dose limits for workers and the general population

| Annual limit | Occupational | Public |
|---|---|---|
| *Effective dose* | 20 mSv averaged over a 5-year period | 1 mSv |
| *Dose equivalent* | | |
| Lens of the eye | 150 mSv | 15 mSv |
| Skin | 500 mSv | 50 mSv |
| Hands and feet | 500 mSv | |

**Table 6.6**  Historical evolution of the dose limits proposed by ICRP or other bodies

| Year | Occupational limit (mSv) | Public limit (mSv) |
|------|--------------------------|--------------------|
| 1925 | 520 | |
| 1934 | 460 | |
| 1949 | 150 | |
| 1950 | 150 | |
| 1956 | 50 | |
| 1957 | 50 | |
| 1990 | 20 averaged over a 5-year period | 1 |

**Table 6.7**  Radon limits or reference levels in air, together with limits for radiation exposure during the manipulation of NORM materials

| Exposure to | Limits or reference levels |
|-------------|----------------------------|
| *Radon* | |
| Houses | $<10$ mSvy$^{-1}$ ($<600$ Bqm$^{-3}$) |
| Occupational | $<10$ mSvy$^{-1}$ ($<1500$Bqm$^{-3}$) |
| *NORM materials* | Between 1 and 20 mSv |

The case of Rn is very interesting, and the ICRP has developed specific recommendations for it [42]. Of course, Rn is present in houses depending on the construction materials used, the geological substrate, and the ventilation habits. Epidemiological work in mine workers and domestic studies has shown that there is a risk of lung cancer due to exposure to $^{222}$Rn even at low concentrations [43]. This has allowed us to define radon reference levels in houses (600 Bqm$^{-3}$ air) and workplaces (1500 Bqm$^{-3}$ air) from which actions have to be carried out to decrease the effective dose. Attention is also paid to NORM materials (see Chap. 4) in the ICRP recommendations. Some values are included in Table 6.7.

Furthermore, it is convenient to recall the existence of threshold doses for deterministic effects that must be considered when planning practices or activities involving radiation sources. Some of them are listed in Table 6.5.

The average level of exposure to natural radiation is an important contributor to the total effective dose received by the population. In Table 6.8, taken from [44], average levels around the world are presented.

**Table 6.8**  Average world effective dose due to natural sources

| Natural source | Worldwide average annual effective dose (mSv) | Range (mSv) |
|----------------|-----------------------------------------------|-------------|
| Cosmic rays | 0.4 | 0.3–1.0 |
| Terrestrial γ-radiation | 0.5 | 0.3–0.6 |
| Inhalation (mainly Rn) | 1.2 | 0.2–10 |
| Ingestion | 0.3 | 0.2–0.8 |

However, the inhomogeneity of these values is considerable, since it depends on many regional parameters, such as the geological substrate and height above sea level. In fact, natural radionuclide concentrations depend on the geological substrate and the irradiation levels of cosmic rays at height [45]. In any case, it is very clear that in a normal situation, the main contributor to the average dose received by the population is natural radiation sources and the exposure produced by medical practices. This can be seen in the pictorial representation of the comparative average radiation exposure from natural, occupational, medical, and man-made sources presented in Fig. 6.5 [46].

Reducing exposure from natural sources is difficult, although necessary, and the dose received from medical practices, although relatively high on average, produces a clear benefit for the population. Therefore, it is important to control and reduce radiation exposure as much as possible originating from other man-made sources, the most important being nuclear energy generation. In fact, as described in Chap. 4, radionuclide releases into the environment can potentially occur at the different steps of nuclear energy production. This also includes the nuclear fuel cycle during which nuclear wastes are produced and radionuclide releases to the environment occur and are even permitted according to certain regulations [47]. As shown in Chap. 4, the environmental impact of nuclear energy production can

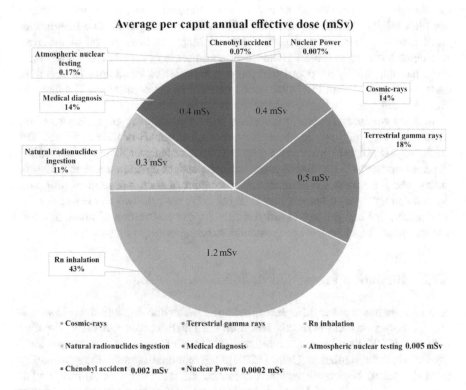

**Fig. 6.5**  Average annual effective dose from natural and anthropogenic sources

affect different environmental compartments. In this way, man-made radionuclides can appear in the atmosphere, hydrosphere, lithosphere, etc., and, of course, be incorporated into the human food chain.

Accidental intrusions of radionuclides into the environment are also an important source of radiation exposure, although they are reduced in many cases to a limited period of time (see Chap. 4).

In any case, it is important to establish radiological protection systems in the countries to monitor the average radiation exposure of the population. Within this context, the continuous measurement of radionuclides in the environment is an important source of information for the deployment of such systems [48]. In addition, periodic control of environmental radionuclide concentrations helps to detect accidental releases early and thus start possible interventions of the authorities.

From this viewpoint, different countries have implemented environmental surveillance programs that include the measurement of key radionuclides in different environmental compartments and foodstuffs.

In Table 6.9, we present the content of a typical surveillance program in which some radionuclides are measured in atmospheric, soil or drinking water samples, as well as in milk from local diet samples, all collected weekly, monthly, or even yearly [48].

As we will see in Chap. 14, the minimum detectable activity (MDA) is a measure of the sensitivity of the experimental method used for the determination of radionuclides. MDAs are also part of surveillance programs. The laboratories participating in them must meet the required MDA. In fact, several of the measurements included in the Surveillance Program serve to calculate the intake of a given radionuclide by the population through ingestion or inhalation. These values must be compared with the MPC levels included in the regulations. For that, the precision and sensitivity of the measurement must be ensured.

In many countries, nuclear installations also have to implement an Environmental Impact Program Study to assess the radiological impact of their activities. The results of this program are evaluated by the National Nuclear Safety Organization. In many cases, these organisms carry out independent environmental impact programs, and the results are compared. The content of such independent programs depends on the type of installation. In Table 6.10, we give two examples: one for a uranium mine and the associated facilities for the preparation of uranium oxide and the other for a low-level radionuclide waste repository [49].

## 6.3.2  Radiation Protection Regulations

Although there are some differences, radiation protection regulation worldwide is very homogeneous and is mostly based on the ICRP recommendations. Together with the ICRP, there are other international bodies, such as the International Commission for Radiation Units (ICRU), International Atomic Energy Agency (IAEA), United Nations Scientific Committee on the Effects of Atomic Radiation (UNSCEAR), World Health Organization (WHO), Nuclear Energy Agency

**Table 6.9** Conventional radiation surveillance program for the environment

| Sample | Analysis | Frequency | MDA |
|---|---|---|---|
| Aerosol | Total $\alpha$ | Weekly | $5 \times 10^{-5}$ Bqm$^{-3}$ |
| | Total $\beta$ | Weekly | |
| | $^{90}$Sr | Quarterly | $2 \times 10^{-5}$ Bqm$^{-3}$ |
| | $^{131}$I | Weekly | $1 \times 10^{-3}$ Bqm$^{-3}$ |
| | $\gamma$-spectrometry* | Weekly | |
| Tap water | Total $\alpha$ | Monthly | |
| | Total $\beta$ | Monthly | |
| | $^3$H | Monthly | |
| | $^{90}$Sr | Monthly | |
| | $^{137}$Cs | Quarterly | |
| | $\gamma$-spectrometry* | Monthly | |
| Soil | Total $\beta$ | Yearly | $5 \times 10$ Bqkg$^{-1}$ dry weight |
| | $^{90}$Sr | Yearly | |
| | $\gamma$-spectrometry* | Yearly | 1 Bqkg$^{-1}$ dry weight |
| Milk | $^{90}$Sr | Monthly | |
| | $^{137}$Cs | Quarterly | |
| | $\gamma$-spectrometry* | Monthly | |
| Diet | $^{14}$C | Quarterly | |
| | $^{90}$Sr | Quarterly | |
| | $^{137}$Cs | Quarterly | |
| | $\gamma$-spectrometry* | Quarterly | |
| | *$\gamma$-spectrometry includes natural $^7$Be, $^{40}$K, $^{208}$Tl, $^{212}$Pb, $^{214}$Bi and $^{214}$Pb and artificial nuclides $^{51}$Cr, $^{54}$Mn, $^{58}$Co, $^{60}$Co, $^{59}$Fe, $^{65}$Zn, $^{95}$Nb, $^{95}$Zr, $^{103}$Ru, $^{106}$Ru, $^{131}$I, $^{134}$Cs, $^{137}$Cs, $^{140}$Ba, $^{140}$La, and $^{144}$Ce | | From 1 to 8 Bqkg$^{-1}$ dry weight in soils for the different nuclides |

of the Organization for Economic Cooperation and Development (OCDE), Food and Agriculture Organization of the UN (FAO), United Nations Environmental Programme (UNEP), and European Atomic Energy Community (EURATOM) interested in the protection of the population and the environment against ionizing radiation from different viewpoints. They also periodically report, individually or in collaboration, new findings that serve as a reference for the elaboration of national regulation or norms. In the majority of the cases, they are independent reputed experts who advise these organizations on the produced reports and, in fact, write and analyze them. Examples of these reports are the [50] on international basic safety norms on radiological protection and the ICRP Publication number 103, which gathers the ICRP 2007 recommendations [9]. The IAEA report

**Table 6.10**  Independent control program of two nuclear installations

| U mine and U oxide manufacturing | | | Low-level waste repository | | |
|---|---|---|---|---|---|
| Sample | Frequency | Analysis | Sample | Frequency | Analysis |
| Surface water | Yearly (1 sample) | Total $\alpha$<br>Total $\beta$<br>U-isotopes<br>$^{230}$Th<br>$^{226}$Ra<br>$^{228}$Ra<br>$^{210}$Pb | Soil | Yearly | $\gamma$-spectrometry<br>$^{89,90}$Sr |
| Underground water | Yearly (3 samples) | Total $\alpha$<br>Total $\beta$<br>U-isotopes<br>$^{230}$Th<br>$^{226}$Ra<br>$^{228}$Ra<br>$^{210}$Pb | Surface water | Yearly (2 samples) | $\gamma$-spectrometry<br>$^{89,90}$Sr<br>Total $\beta$<br>$^{3}$H<br>$^{14}$C<br>$^{63}$Ni<br>$^{129}$I<br>$^{99}$Tc |
| | | | Sediments | Yearly | $\gamma$-spectrometry<br>Total $\beta$<br>$^{63}$Ni |
| | | | Underground water | Yearly | $\gamma$-spectrometry<br>$^{89,90}$Sr<br>Total $\beta$<br>$^{3}$H<br>$^{14}$C<br>$^{63}$Ni<br>$^{129}$I<br>$^{99}$Tc |
| | | | Fish | Yearly | $\gamma$-spectrometry<br>$^{89,90}$Sr |
| | | | Meat | Yearly | $\gamma$-spectrometry |
| | | | Honey | Yearly | $\gamma$-spectrometry |
| | | | Bioindicator | Yearly | $\gamma$-spectrometry<br>$^{89,90}$Sr<br>Total $\beta$<br>$^{3}$H<br>$^{14}$C |

includes the recommendations of ICRP Publication 103, and it can be said that most regulations worldwide are based on them. Furthermore, since EURATOM collaborated in the elaboration of the IAEA document, the European Union is compelled to apply the recommendations appearing in the report.

According to the documents described above, all regulations should be based on several principles [51].

1. Principle of responsibility for safety, by which the responsibility of safety is endorsed to the person or the organization in charge of the radioactive installation.
2. Principle of the role of government by which a legal and efficient system for safety against ionizing radiation must be implemented. An important part of the system is the creation of a national independent regulatory body.
3. The principle of leadership and management for safety by which the leadership of the organization in charge of radiological safety must be established and sustained.
4. Principle of justification of installations and activities by which the installations or activities involving the use of radiation sources must yield an overall benefit.
5. The principle of optimization by which the protection must be optimized to provide a safety level as high as reasonably achievable.
6. Principle of the limitation of risks to the population by which control practices must guarantee avoiding inacceptable risks due to radiation exposure.
7. Principle of radiological protection for present and future generations by which the environment, people, and future generations must be protected against radiological risks.
8. The principle of accident prevention by which all possible efforts must be deployed to prevent nuclear or radiological accidents and mitigate their consequences.
9. Principle of preparedness and response to emergency situations by which they must be adopted dispositions for preparedness and response to nuclear or radiation emergencies.
10. The principle of protection measures to reduce existing or unregulated radiological risks by which protection measures against existing or unregulated radiological risks have to be justified and optimized.

These principles contain those on which the Radiological Protection Principles described above are based. Indeed, principles 4, 5, 6, 7, and 10 are included in the Principles of Radiological Protection. The others are related mostly to the management of the safety system, as well as nuclear or radiological installations and accidental situations.

International regulations distinguish between three types of radiation exposure: planned exposure, emergency exposure, and so-called existing exposure. A planned exposure derives from the planned use of a radiation source for a given practice. From this point of view, the exposure limits can be calculated and planned based on the time of source utilization, the proper design of the installation, the formation of the operators, etc. When there exists only the possibility of radiation exposure during the operation of a source, it is preferable to talk of a potential exposure that can also be estimated. Emergency exposure is the result of an accident or delinquent activity. These situations must be foreseen, as well as the appropriate measures to be taken to mitigate their consequences.

Exposure to natural radiation or to residual radioactive materials utilized in the previously described planned situations or from accidental releases is called existing exposure.

For the three types of exposure, the radiation dose has to be limited to a level as low as reasonably achievable, taking into account economic and societal factors. These restrictions or dose limits are also included in the legislation as proposed in the ICRP document described above [9]. The limits are different for the public and for those exposed occupationally. Furthermore, there exist some specific limits for the case of medical exposure that are included within the planned exposure to radiation. It is also interesting to note that Rn is treated individually in the regulations. In fact, there are some specific levels of Rn within the existing exposure to the public and workers [42]. Some of the limits are given in Table 6.7.

The different national regulations generally include the definitions of concepts and units of interest for radioactivity and radiation protection: activity, absorbed dose, etc. The formation of radiological protection professionals is also regulated, and in many countries, the normative in this regard is included in the same text or in other norms developing the main regulation [52]. Finally, the regulations include all the necessary tables containing the dose limits of application to the different scenarios, as well as other magnitudes such as, for instance, $e(50)$ of (6.12). In Table 6.11, some values of $e(50)$ and MPC-derived MPC in air by ingestion or inhalation of some radionuclides are given. As seen in [53], $e(50)$ is different for infants (1–2 years), children (up to 17 years), and adults (>17 years). This corresponds to specific differences in radionuclide dynamics in organisms. In general, $e(50)$ is higher for infants than for children, and this, at the same time, is higher than for adults.

Some texts for specific national regulations can be found in [54]. The case of EU is very interesting since the text provided describes the minimum rules that the countries belonging to the European Union must accomplish. However, more strict rules are possible but not less exact.

**Exercises**

1. Roentgen is defined as $1\,R = 2.58 \times 10^{-2}\,\mu C kg^{-1}$. Show that this definition is equivalent to 1 esu cm$^{-3}$
2. Find the number of ion pairs produced in 1cm$^3$ of air exposed to 1 R.
3. Demonstrate (6.5).
4. In the human body, there are approximately 250 g of natural K, of which 0.0012% is $^{40}$K ($T_{1/2} = 1.3 \times 10^9$ y). This radionuclide emits $\beta$ radiation ($I = 0.89$) with $E_{max} = 1.3$ meV. Find the $^{40}$K activity in the human body in Bq.
5. Find the equivalent dose rate and the annual equivalent dose absorbed by the human body due to 40 K. Suppose that $\frac{2}{3}$ of $\beta$ decay energy is carried by neutrinos and that the mass of the human body is 70 kg on average. Discuss the result based on the dose limits regulations.

**Table 6.11** e(50) for some radionuclides of environmental interest. The values for the cases of inhalation and ingestion are included. For $^{222}$Rn MPC, see Table 6.7

| Radionuclide | Inhalation (nSvBq$^{-1}$) | Ingestion (nSvBq$^{-1}$) | MPC in air (Bqm$^{-3}$) |
|---|---|---|---|
| $^3$H | 0.018 | 0.018 | $5 \times 10^5$ |
| $^{14}$C | | 0.58 | |
| $^{40}$K | | 6.2 | |
| $^{55}$Fe | 0.77 | 0.33 | $9 \times 10^3$ |
| $^{60}$Co | 9.6 | 3.4 | $1 \times 10^3$ |
| $^{89}$Sr | 1 | 6 | $6 \times 10^3$ |
| $^{90}$Sr | 24 | 28 | $3 \times 10^2$ |
| $^{95}$Zr | 2.5 | 0.88 | $3 \times 10^3$ |
| $^{99}$Tc | 0.29 | 0.78 | $2 \times 10^4$ |
| $^{106}$Ru | 8.0 | 7.0 | $9 \times 10^2$ |
| $^{125}$Sb | 1.4 | 1.1 | $5 \times 10^3$ |
| $^{131}$I | 7.6 | 22 | $8 \times 10^2$ |
| $^{134}$Cs | 6.8 | 19 | $9 \times 10^2$ |
| $^{137}$Cs | 4.8 | 13 | $1 \times 10^3$ |
| $^{144}$Ce | 34 | 5.2 | $4 \times 10^2$ |
| $^{238}$U *series* | | | |
| $^{238}$U | 2900 | 45 | 10 |
| $^{234}$U | 3500 | 49 | 10 |
| $^{230}$Th | 14,000 | 210 | |
| $^{226}$Ra | 3500 | 280 | 4 |
| $^{222}$Rn | | 3.5 | |
| $^{210}$Pb | 1100 | 690 | 8 |
| $^{210}$Po | 3300 | 1200 | 10 |
| $^{232}$Th *series* | | | |
| $^{232}$Th | 25,000 | 230 | $3 \times 10^{-1}$ |
| $^{228}$Ra | 2600 | 690 | 5 |
| $^{228}$Th | 40,000 | 72 | $4 \times 10^{-1}$ |
| $^{235}$U *series* | | | |
| $^{235}$U | 3100 | 47 | 10 |
| $^{231}$Pa | 34,000 | 710 | |
| $^{227}$Ac | 220,000 | 1100 | |
| $^{237}$Np | 2100 | 110 | $6 \times 10^{-1}$ |
| $^{239}$Np | 0.9 | 0.8 | $8 \times 10^3$ |
| $^{238}$Pu | 46,000 | 230 | $3 \times 10^{-1}$ |
| $^{239}$Pu | 50,000 | 250 | $3 \times 10^{-1}$ |
| $^{240}$Pu | 50,000 | 250 | $3 \times 10^{-1}$ |
| $^{241}$Pu | 900 | 4.7 | 10 |
| $^{241}$Am | 42,000 | 200 | $3 \times 10^{-1}$ |
| $^{242}$Cm | 4800 | 12 | 1 |
| $^{244}$Cm | 25,000 | 120 | $5 \times 10^{-1}$ |

6. Suppose that a standard person accumulates $10 \, mBq \, ^{226}Ra$ in the body. Calculate the equivalent dose accumulated during one year due to this activity as well as the equivalent dose rate.

7. Calculate the temperature increase experimented by a person absorbing a whole-body radiation dose of approximately 4 Gy (this is supposed to be lethal in most cases). Take a mass of 70 kg and suppose that the specific heat of the human body is that of water. The energy equivalent of heat is $4.18 \, Jcal^{-1}$.

8. A total of $1 \, \mu Ci \, ^{14}C$ was injected into the human brain (m = 1.5 kg). Calculate the absorbed radiation dose assuming that the radioisotope is uniformly distributed in the organ.

9. Make the same calculation if $1 \, \mu Ci \, ^{238}U$ was distributed homogeneously in a human kidney (m = 150 g). For calculations, consider only the $\alpha$ particle emitted by $^{238}U$.

10. Assuming 70% of the human body is water, calculate the total absorbed by a standard individual (mass 70 kg) if 60% of the $H_2O$ is $^3H_2O$.

11. A patient is injected with a radiopharmaceutical labeled with $3 \, MBq \, ^{197}Hg$. This is homogeneously distributed in the kidney, which is supposed to have a mass of 260 g. If the biological half-life of the radiopharmaceutical in the kidney is 15 h, calculate the total dose equivalent absorbed by the kidney.

    (The biological half-life, $T_b$, of a substance injected into an organ corresponds to the time required to remove half of the substance from the organ. In the case that the substance is radioactive, the physical half-life, $T_{1/2}$, must be combined with the biological half-life to obtain the effective elimination half-life, $T_{eff}$. Since $\lambda_{eff} = \lambda_{phys} + \lambda_b$, $T_{eff} = \frac{T_{1/2}T_b}{T_{1/2}+T_b}$)

# References

1. M.G. Stabin, *Radiation Protection and Dosimetry: An Introduction to Health Physics* (Springer, 2010)

2. W. Röntgen, Ueber eine neue Art von Strahlen. Vorläufige Mitteilung. Aus den Sitzungs-berichten der Würzburger Physik.-medic. Gesellschaft Würzburg, 137–147 (1895); R. Behling, X-ray tube development—IOMP history of medical physics. Med. Phys. Int **6**, 8–55 (2018)

3. H.E. Johns, J.B. Cuningham, *The Physics of Radiology*, Chapter 7 (Charles C. Thomas Pb., 1983)

4. H.E. Johns, J.B. Cuningham, *The Physics of Radiology*, Chapter 15 (Charles C. Thomas Pb., 1983)

5. ICRP, Publication 26, *Recommendations of the International Commission on Radiological Protection* (1977)

6. ICRP, Publication 103, The 2007 Recommendations of the International Commission on Radiological Research Protection (2007), p. 272

7. ICRP, Publication 103, The 2007 Recommendations of the International Commission on Radiological Research Protection (2007), p. 65

8. ICRP, Publication 9, *Recommendations of the International Commission on Radiological Protection* (1966)

9. ICRP, Publication 103, *The 2007 Recommendations of the International Commission on Radiological Protection* (2007)
10. ICRP, Publication 103, *The 2007 Recommendations of the International Commission on Radiological Protection* (2007), p. 248.
11. ICRP, Publication 103, *The 2007 Recommendations of the International Commission on Radiological Protection* (2007), p. 75
12. ICRP, Publication 103, *The 2007 Recommendations of the International Commission on Radiological Protection* (2007), p. 77
13. ICRP, Publication 103, *The 2007 Recommendations of the International Commission on Radiological Protection* (2007), p. 20
14. See https://www.nrc.gov/reading-rm/doc-collections/cfr/part020/part020-appb.html
15. S. Christofides, D.R. Dance A.D.A. Maidment ID. McLean K.-H. Ng (Eds.), *Diagnostic Radiology Physics: A Handbook for Teachers and Students*, Chapter 20 (IAEA, 2014)
16. S. Christofides, D.R. Dance A.D.A. Maidment ID. McLean K.-H. Ng (Eds.), *Diagnostic Radiology Physics: A Handbook for Teachers and Students*, Chapter 20 (IAEA, 2014), p. 503
17. S. Christofides, D.R. Dance A.D.A. Maidment ID. McLean K.-H. Ng (Eds.), *Diagnostic Radiology Physics: A Handbook for Teachers and Students*, Chapter 20 (IAEA, 2014), p. 501
18. S. Christofides, D.R. Dance A.D.A. Maidment ID. McLean K.-H. Ng (Eds.), *Diagnostic Radiology Physics: A Handbook for Teachers and Students*, Chapter 20 (IAEA, 2014), p. 502
19. S. Christofides, D.R. Dance A.D.A. Maidment ID. McLean K.-H. Ng (Eds.), *Diagnostic Radiology Physics: A Handbook for Teachers and Students*, Chapter 20 (IAEA, 2014), p. 499
20. ICRP, Publication 103, The 2007 Recommendations of the International Commission on Radiological Research Protection (2007), p. 36
21. UNSCEAR (1993) Report, United Nations Scientific Committee on the Effects of Atomic Radiation, 'Sources and effects of ionizing radiation', p. 14
22. ICRP, Publication 103, The 2007 Recommendations of the International Commission on Radiological Research. Protection, Chapter 3 (2007)
23. ICRP, Publication 103, The 2007 Recommendations of the International Commission on Radiological Research. Protection, Annex A (2007)
24. UNSCEAR (2001) Report, United Nations Scientific Committee on the Effects of Atomic Radiation, "Hereditary effects of radiation". Chapter V
25. R. Takam, E. Bezak, E. Yeoh, L. Marcu, Assessment of normal tissue complications and second cancer risks following prostate cancer irradiation—a review. Austral. Asian J. Cancer **7**(3), 171–184 (2008)
26. S. Christofides D.R. Dance A.D.A. Maidment ID. McLean, K.-H. Ng (Eds.) *Diagnostic Radiology Physics: A Handbook for Teachers and Students* (IAEA, 2014)
27. ICRP, Publication 103, The 2007 Recommendations of the International Commission on Radiological Sciences Protection (2007)
28. UNSCEAR, Report, United Nations Scientific Committee on the Effects of Atomic Radiation, "Sources and effects of ionizing radiation" (1993), p. 15
29. S. Christofides D.R. Dance A.D.A. Maidment ID. McLean, K.-H. Ng (Eds.) *Diagnostic Radiology Physics: A Handbook for Teachers and Students* (IAEA, 2014), p. 509
30. UNSCEAR Report, United Nations Scientific Committee on the Effects of Atomic Radiation, "Hereditary effects of radiation" (2001), p. 52
31. UNSCEAR, Report, United Nations Scientific Committee on the Effects of Atomic Radiation, "Sources and effects of ionizing radiation" (1988) p. 17
32. UNSCEAR, Report, United Nations Scientific Committee on the Effects of Atomic Radiation, "Sources and effects of ionizing radiation" (1988) p. 21
33. UNSCEAR, Report, Scientific Committee of the United Nations on Atomic Radiation, "Sources and effects of ionizing radiation", Annex G (1988)
34. UNSCEAR, Report, United Nations Scientific Committee on the Effects of Atomic Radiation, 'Hereditary effects of radiation', Chapter III (2001)
35. UNSCEAR, Report, United Nations Scientific Committee on the Effects of Atomic Radiation, "Sources and effects of ionizing radiation" (1993), p. 17

36. ICRP, Publication 103, The 2007 Recommendations of the International Commission on Radiological Protection (2007), p. 88
37. ICRP, Publication 103, The 2007 Recommendations of the International Commission on Radiological Protection, Chap. 5 (2007)
38. ICRP, Publication 103, The 2007 Recommendations of the International Commission on Radiological Protection (2007), p. 86
39. ICRP 2006. Assessing the Dose of the Representative Person for the Purpose of Radiation Protection of the Public. ICRP Publication 101a. Ann. ICRP 36 (3).
40. ICRP, Publication 103, The 2007 Recommendations of the International Commission on Radiological Protection (2007), p. 98.
41. ICRP, Recommendations of the International Commission on Radiological Protection (1966, 1977, 1991, 2003, ...)
42. ICRP, Protection against radon-222 at home and at work. ICRP Publication 65. Ann. ICRP **23**(2) (1993b)
43. S. Darby, D. Hill, H. Deo et al., Residential radon and lung cancer—detailed results of a collaborative analysis of individual data on 7148 persons with lung cancer and 14,208 persons without lung cancer from 13 epidemiologic studies in Europe. Scand. J. Work Environ. Health **32**(1), 1–84 (2006)
44. UNSCEAR, Report, United Nations Scientific Committee on the Effects of Atomic Radiation, 'Sources and effects of ionizing radiation', Annex B (2000)
45. G. Cinelli, M. De Cort, T. Tollefsen (Eds.), *European Atlas of Natural Radiation* (Publication Office of the European Union, Luxembourg, 2019). https://remon.jrc.ec.europa.eu/About/Atlas-of-NaturalRadiation
46. UNSCEAR, Report, United Nations Scientific Committee on the Effects of Atomic Radiation, "Sources and effects of ionizing radiation", Annex B, C and D (2000)
47. See, for instance, *Monitoring Our Environment: Discharges and Environmental Monitoring* (Sellafield Ltd., Annual report 2020) (https://www.gov.uk/government/organisations///sellafield-ltd)
48. Environmental radiological surveillance programmes in Spain. https://www.csn.es/valores-radiologicos-ambientales-pvra-rem
49. Environmental surveillance programs around nuclear facilities in Spain. https://www.csn.es/red-de-vigilancia-en-el-entorno-de-instalaciones-pvra
50. IAEA, Radiation Protection and Safety of Radiation Sources: International Basic Safety Standards, STI/PUB/1578 (2014)
51. IAEA, Radiation Protection and Safety of Radiation Sources: International Basic Safety Standards, STI/PUB/1578 (2014), p. 3
52. See, for instance, Regulation on Sanitary Protection against Ionizing Radiations Published in the Spanish Official State Gazette number 178, of the 26th of July 2001
53. UNSCEAR, Report, United Nations Scientific Committee on the Effects of Atomic Radiation, 'Sources and effects of ionizing radiation', Annex B and C (2000)
54. Regulation on Prevention of Ionizing Radiation Hazards (Ministry of Labour Order No. 41 of September 30, 1972) Japan; Radiation Protection Regulations, Minister of Justice at the following address: http://laws-lois.justice.gc.ca, Canada; European Council Directive 2013/59/Euratom on basic safety standards for protection against the dangers arising from exposure to ionizing radiation and repealing Directives 89/618/Euratom, 90/641/Euratom, 96/29/Euratom, 97/43/Euratom and 2003/122/Euratom. OJ of the EU. L13; 57: 1–73 (2014)

# Principles of Radiation Detection: Interaction of Radiation with Matter

7

**ABSTRACT**

The main objective of the chapter is the description of the mechanisms governing the radiation interaction with matter. Emphasis is placed on those mechanisms useful for the design of α-, β-, and γ-radiation detectors. The case of neutrons is also studied since they can be used as an analytical tool for the determination of environmental radioactivity (see Chap. 19).

## 7.1  Interaction of Gamma Radiation with Matter

Gamma radiation can interact with matter in many ways. The relative probability of each to occur depends on the energy of the photon and the Z of the material through which the radiation is passing. In general terms, photons can undergo the following [1].

*Absorption Processes*
Photon absorption by atomic electrons (photoelectric effect) or by nucleons (photonuclear reaction) is the most relevant effects within this group. Pair production, in the presence of either the nuclear electronic field, in which an electron–positron pair ($e^--e^+$) is created, or in the presence of the electronic field, the so-called triplet production, is also process to be considered. Production of nucleon–antinucleon pairs is also included in this group.

*Dispersion Processes*
Elastic dispersion with atomic electrons (Thomson and Rayleigh effects) or inelastic dispersion with atomic electrons (Compton effect) together with elastic and inelastic collisions with nucleons ($\gamma$, $\gamma$) or ($\gamma$, $\gamma'$), respectively, forms part of this group. In addition, the elastic dispersion with the electric field around the nucleus (Delbruck dispersion) can also be considered.

© The Author(s), under exclusive license to Springer Nature Switzerland AG 2022
M. García-León, *Detecting Environmental Radioactivity*, Graduate Texts
in Physics, https://doi.org/10.1007/978-3-031-09970-0_7

However, for the usual energy range and typical materials used in detectors, the relevant interaction mechanisms for γ radiation are the photoelectric and Compton effects as well as pair production.

## 7.1.1  Photoelectric Effect

In this case, the photon transfers its energy, $E_\gamma$, to an atomic electron. The energy gained by the electron is used to overcome the atomic binding energy $E_B$. The rest is taken by the electron as kinetic energy, $E_e$, given by

$$E_e = E_\gamma - E_B \tag{7.1}$$

For this to occur, the electron must be bound to the atom since a free electron cannot absorb the energy of the photon and at the same time meet the principle of momentum conservation. A third body, the atom, is needed to absorb the recoil momentum. On the other hand, as a result of the produced atomic vacancy, X-rays or Auger electrons are emitted because of the photoelectric effect. The calculation of the photoelectric effect cross section is complicated given the complexity of the Dirac wave function for atomic electrons. It can be found that the photoelectric effect is more probable for strongly bound electrons. Thus, it essentially involves K-shell electrons. Assuming low energies, $E\gamma \ll m_0c^2$, the electron rest mass, a Born approximation can be used to calculate the effective cross section for $K$ electrons, and it turns out [1]:

$$\sigma_{ph}^{K)} = 4\sqrt{2}\sigma_T \alpha_{ef}^4 Z^5 \alpha^{-\frac{7}{2}} \tag{7.2}$$

where $\sigma_T = \frac{8\pi}{3}r_0^2 = 6.651 \times 10^{-25}$cm$^2$ is the Thomson cross section (see next section), $r_0 = \frac{e^2}{m_0c^2} = 2.82 \times 10^{-13}$cm is the classical radius of the electron, $\alpha_{ef} = \frac{e^2}{\hbar c} \approx \frac{1}{137}$ is the fine structure constant, and $\alpha = \frac{E_\gamma}{m_0c^2}$. $Z$ is the atomic number. The photoelectric cross sections for less bound electrons, L, M shells, etc., are relevantly lower, so the most important contribution to the total photoelectric cross section $\sigma_{ph}$ comes from $\sigma_K$ [1]. In fact, it can be found that:

$$\frac{\sigma_{ph}}{\sigma_K} = 1 + 0.01481(\ln Z)^2 - 0.00079(\ln Z)^3 \tag{7.3}$$

for $E_\gamma > E_B^{K)}$ . In other words, $\sigma_K$ contributes up to 90% to the total photoelectric effect cross section. These calculations do not consider relativistic effects. Therefore, the dependence of $\sigma$ on $Z$ and $E_\gamma$ is not exact. Furthermore, due to the shielding of the nucleus by inner electrons, the exponent of $Z$ depends on $E_\gamma$ [2]. As seen in Fig. 7.1, a good value for it lies between 4 and 5 in the case of human tissue [3]. This applies to other materials. Thus, for $Z = 5$–10, the exponent ranges from 4.4 to 4.8 and from 4 to 4.5 for $Z = 40$–80 [1].

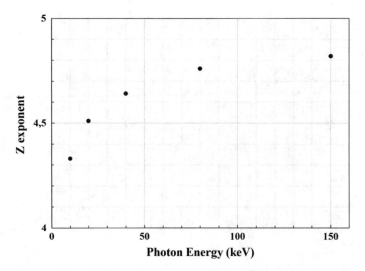

**Fig. 7.1** $Z$ exponent for the photoelectric cross section in the case of human tissue. This calculation is of interest for radiation dosimetry

Finally, in Fig. 7.2, we observe the $\sigma_{ph}$ dependence on $E_\gamma$ for $Z = 82$ and $Z = 26$. As predicted by (7.2), $\sigma_{ph}$ decreases drastically as energy increases. The energy exponent varies depending on the energy region, as remarked before. Not surprisingly, typical discontinuities in the excitation function appear, corresponding to the absorption edges of the electronic shells in the atom.

## 7.1.2 Compton Effect

The Compton effect consists of the inelastic scattering of photons by weakly bound atomic electrons [1]. According to the classical interpretation, electromagnetic radiation can be elastically dispersed by atomic electrons. Thus, when electromagnetic radiation hits weakly bound electrons, they vibrate and emit electromagnetic radiation with the same wavelength (energy). This is known as the Thomson effect, which is adequately described within the framework of the classical theory of electromagnetism. However, the Thomson interpretation of this phenomenon is valid only for $E_\gamma < 100$ keV. When the radiation energy approaches that of the electron rest mass ($m_0c^2 = 511$ keV/c$^2$), the wavelength (energy) of the dispersed radiation changes. This is the Compton effect, which is the quantum analog of the Thomson effect. The Compton effect can be explained by assigning a corpuscular nature to electromagnetic radiation and introducing the concept of photons. Thus, a photon collides with an electron, as presented in Fig. 7.3. In this collision, the electron is supposed to be free and at rest regarding the energy received by the incident photon. In real terms, the electron involved is weakly bound to the atom with an orbital kinetic energy very low compared to the $\gamma$-radiation.

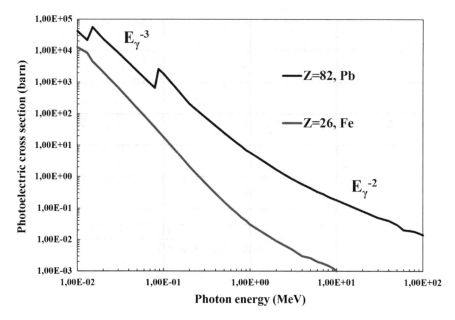

**Fig. 7.2** Photoelectric cross sections as a function of the photon energy for two $Z$ values

**Fig. 7.3** Schematic representation of the Compton effect

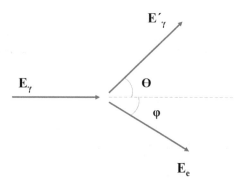

In the Compton effect, the photon changes its direction, as in the Thomson effect, and loses energy that is absorbed by the electron. Thus, by applying the laws of energy and momentum conservation, it can be found that the energy of the scattered photon and electron, $E'_\gamma$ and $E_e$, are, respectively, as follows: [1]

$$E'_\gamma = \frac{E_\gamma}{1 + \alpha(1 - \cos\theta)} \tag{7.4}$$

and

$$E_e = \frac{E_\gamma^2(1 - \cos\theta)}{m_0 c^2 + E_\gamma(1 - \cos\theta)} \tag{7.5}$$

$\theta$ is the photon scattering angle.

$E'_\gamma$ and $E_e$ can take a continuous energy range limited by the scattering angle, which can change from $\theta = 0$ to $\theta = \pi$. The maximum possible value for $E'_\gamma$ corresponds to the minimum value for $E_e$ at $\theta = 0$, while the minimum at $\theta = \pi$ corresponds to the maximum value of the electron energy. It can be readily found that

$$\frac{E_\gamma}{1 + 4E_\gamma} \le E'_\gamma \le E_\gamma \tag{7.6}$$

and

$$0 \le E_e \le \frac{4E_\gamma^2}{1 + 4E_\gamma} \tag{7.7}$$

In both Equations, we have taken $m_0c^2 \approx \frac{1}{2}$ MeV.

In Fig. 7.4, we depict the variation in $E_\gamma'$ with $E_\gamma$ for different $\theta$ values. For $E_\gamma < 10^{-1}$ meV, the Thomson effect applies, and $E_\gamma' = E_\gamma$. Insofar as $E_\gamma$ increases, the Compton effect starts being apparent, and $E_\gamma'$ becomes different from $E_\gamma$. It is also interesting to note that for high $E_\gamma$, $E_\gamma'$ only depends on $\theta$. Indeed, for large $E_\gamma$.

$$E'_\gamma \to \frac{m_0c^2}{1 - \cos\theta} \ne f(E_\gamma) \tag{7.8}$$

Furthermore, for

$$\frac{\pi}{2} \le \theta \le \pi \tag{7.9}$$

we have the following

$$\frac{m_0c^2}{2} \le E'_\gamma \le m_0c^2 \tag{7.10}$$

This result is especially interesting and very useful when interpreting the response of $\gamma$ radiation spectrometers, as we will see in Chaps. 10 and 11. Regarding this problem, it is important to know the spectrum shape of the Compton photons and electrons.

For that, it is convenient to analyze the Compton cross section as calculated by Klein–Nishina in 1929 [4] when solving the problem of photon scattering by unbound electrons. The angular distribution was found to be the following:

$$\frac{d\sigma(E_\gamma, \theta)}{d\Omega} = \frac{r_0^2}{2} \left\{ \frac{1 + \cos^2\theta}{[1 + \alpha(1 - \cos\theta)]^2} + \frac{\alpha^2(1 - \cos\theta)^2}{[1 + \alpha(1 - \cos\theta)]^3} \right\} \tag{7.11}$$

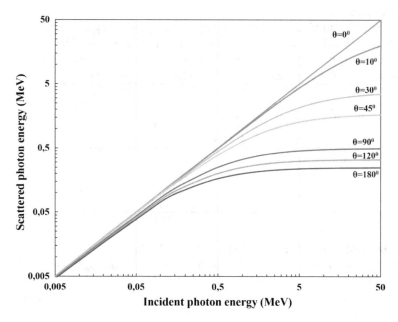

**Fig. 7.4** Energy of the scattered $\gamma$ radiation as a function of the incident $\gamma$ energy for different scattering angles in Compton scattering

which is the probability that a photon of energy $E_\gamma$ is scattered through an angle between $\theta$ and $\theta + d\theta$. At low $\gamma$ energy, $\alpha \rightarrow 0$, and the angular distribution corresponds to that expected for the Thomson effect, which has the form:

$$\frac{d\sigma}{d\Omega} = r_0^2 \frac{(1 + cos^2\theta)}{2} \qquad (7.12)$$

A view of the Klein–Nishina angular distribution is given in Fig. 7.5. For low energies, the Thomson effect dominates, and the probabilities of forward (0° to 90°) and backwards (90° to 180°) scattering are similar. As $E_\gamma$ increases, the probability of backscattering is lower than that of forward dispersion. Interestingly, when $E_\gamma$ is close to $m_0c^2$, the backscatter cross section becomes constant.

The integration of (7.11) for all possible angles gives the Compton scattering excitation function, and it turns out [1] that

$$\sigma_c = 2\pi r_0^2 \left\{ \frac{1+\alpha}{\alpha^2} \left[ \frac{2(1+\alpha)}{1+2\alpha} - \frac{1}{\alpha}\ln(1+2\alpha) \right] + \frac{1}{2\alpha}\ln(1+2\alpha) - \frac{(1+3\alpha)}{(1+2\alpha)^2} \right\} \qquad (7.13)$$

or, in other words, the scattering cross section decreases as the energy increases. All these results lead to the qualitative form of the Compton-dispersed photon and electron spectra presented in Fig. 7.6. As expected, the spectra are continuous. In the case of photons, an increase in the number of photons is observed in

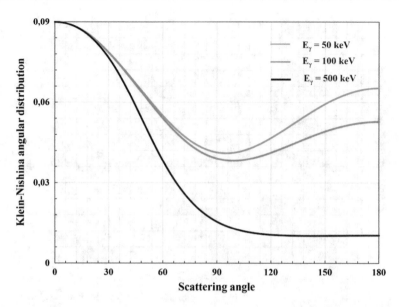

**Fig. 7.5** The Klein–Nishina angular distribution for the scattering of photons by unbound electrons

the high-energy area, corresponding to small-angle scattering, and at low energy, which corresponds to large dispersion angles. As the energy of the incident photon increases, the grouping of photons in the high-energy zone disappears but remains at low energy (backscattering processes). It is the so-called backscattering edge. This effect is a direct consequence of the shape of the angular distribution and the fact that the energy of the backscattered photons is very similar [see (7.6)].

However, it is even more interesting to study the spectrum of scattered or Compton electrons. In fact, the response of $\gamma$ radiation detectors is based on the electrons the photons release as they pass through matter. In Fig. 7.6b, we see how electrons accumulate in the high-energy zone, which corresponds to the backscattering edge. This is the so-called Compton edge, and it can be said that Fig. 7.6b is a picture of what we truly observe in a $\gamma$ spectrometer.

Finally, it should be noted that according to quantum electrodynamics, the Compton effect cross section increases proportionally to Z [1] when $\alpha \gg 1$.

### 7.1.3  Pair Production

In this case, the photon disappears when passing through the matter, and an $e^{-}$-$e^{+}$ pair is created. As in the photoelectric effect, the simultaneous conservation of energy and momentum requires the presence of a third body for this process to occur, a role that can be played by either the nucleus or an atomic electron. In any case, an electric field is necessary [1]. When a pair is created in the presence of

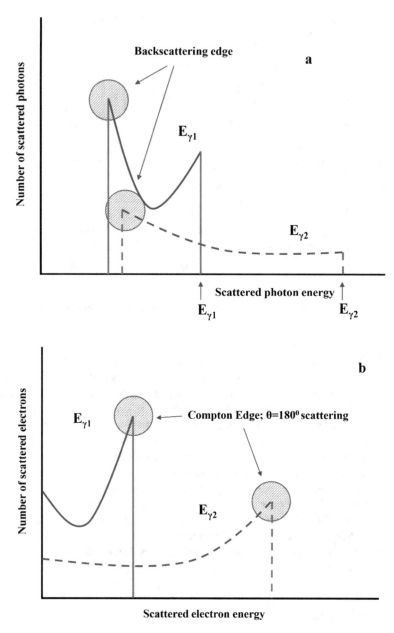

**Fig. 7.6** Qualitative shape of **a** spectrum of photons dispersed by the Compton effect and **b** spectrum of electrons dispersed by the Compton effect

an atomic electron, it is called triplet production. Pair production dominates the interaction of high-energy photons with matter, and by using the laws of energy and momentum conservation, the threshold energy for the creation of pairs is

$$E_\gamma = 1.022 \, \text{MeV}$$

in the presence of a nucleus, and

$$E_\gamma = 2.044 \, \text{MeV}$$

in the presence of an atomic electron.

The first process is by far the most likely, and the cross section of pair production, $\sigma_{pp}$, in the presence of a nucleus was calculated by Bethe and Heitler in 1934 [5]. Generally speaking,

$$\sigma_{pp} = 4Z^2 \alpha_{ef} r_0^2 \left\{ \frac{7}{9} \left[ \ln \frac{2E_\gamma}{m_0 c^2} - f(Z) \right] - \frac{109}{4} \right\} \tag{7.14}$$

$Z$ is the effective atomic number of the material, and $f(Z)$ is the Coulomb correction that considers the interaction of the electron and the positron with the nuclear electric field. The function within parentheses is often called $h(E_\gamma, Z)$, which monotonously increases when $E_\gamma$ increases and smoothly decreases as $Z$ increases. For high energies, i.e., $E_\gamma \gg 137 \, m_0 c^2 Z^{-1/3}$, we have

$$\sigma_{pp} = 4Z^2 \alpha_{ef} r_0^2 \left\{ \frac{7}{9} \left[ \ln 183 Z^{-\frac{1}{3}} - f(Z) \right] - \frac{1}{54} \right\} \tag{7.15}$$

that is, $h(E_\gamma, Z)$ becomes constant for a given $Z$. This makes $\sigma_{pp}$ energy independent and only dependent on the material. In this energy region, the pair-production effect dominates with respect to the Compton and photoelectric effects. In Fig. 7.7, $\sigma_{pp}$ is presented as a function of the photon energy for different significant $Z$ (I, Ge, and Si).

## 7.1.4 Attenuation and Absorption Coefficients

A photon beam passing through a material is attenuated because of the effects described above. Indeed, either by photoelectric, Compton, or pair production, the photons are removed from the beam, and clearly, such an attenuation effect will depend on the photon energy and the material properties. An attenuation coefficient could be defined to describe the capacity of a material to attenuate a photon beam, and such a coefficient must be related to the different photon interaction cross sections. The attenuation coefficient is a very interesting parameter for evaluating a given material as a good $\gamma$ detector or a good shielding against $\gamma$ radiation for radiological protection [6].

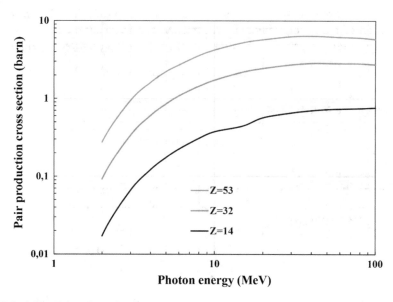

**Fig. 7.7** $h(E_\gamma, Z)$ in pair production [see (7.14)]

Let us suppose a monochromatic $\gamma$ beam $I(0)$ passing through a material layer with thickness x (see Fig. 7.8). The number of photons, $I(x)$, passing through the material layer will decrease since some photons can be removed from the beam by photoelectric, Compton, or pair-production effects. In any case, the surviving photons have the same energy as the beginning and travel along the same direction. In other words, $I(0) > I(x)$ as the beam can be attenuated by the photon interactions in the material.

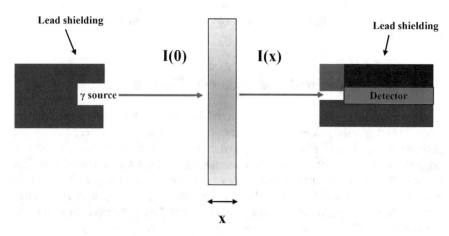

**Fig. 7.8** Schematic view of a $\gamma$ attenuation experiment

Taking a layer of thickness $dx$, it seems obvious that the relative change in the number of photons after passing the layer will depend on dx in the following manner.

$$\frac{-dI}{I} \propto dx \tag{7.16}$$

or

$$\frac{-dI}{I} = \mu dx \tag{7.17}$$

where $\mu$ is a proportionality constant that must be interpreted as the probability of photon interaction per unit length of photon interaction. Of course, the sign '−' must be used as the number of photons decreases. From this point of view, it is also apparent that

$$\mu = N_V \sigma = N_V \left( \sigma_{ph} + Z\sigma_C + \sigma_{pp} \right) \tag{7.18}$$

where $N_V$ is the atomic density of the material, and we have multiplied the Compton cross section by $Z$ to consider the number of electrons per atom. $\mu$ is called the linear attenuation coefficient and has dimensions of $[L^{-1}]$. Equation 7.17 is easily solved since $\mu$ is constant for a given $Z$ and $E_\gamma$. Thus,

$$I(x) = I(0)e^{-\mu x} \tag{7.19}$$

which is the exponential law of photon attenuation [6]. The attenuation coefficient contains all the information regarding the interaction of the photons with the material. The result is interesting because the experimental determination of $\mu$ is easy, although not trivial. The factor $e^{-\mu x}$ gives the probability that the photon passes through the layer of thickness x without undergoing any of the interaction effects. Therefore, it makes sense to define a photon mean free path, $\lambda$, which may well characterize the attenuation problem we are solving. It is obvious that:

$$\lambda = \frac{\int_0^\infty x e^{-\mu x} dx}{\int_0^\infty e^{-\mu x} dx} = \frac{1}{\mu} \tag{7.20}$$

$\mu$ depends on the density of the material and varies according to its aggregation and chemical state. To avoid this uncertainty, a mass attenuation coefficient is defined, $\frac{\mu}{\rho}$, where $\rho$ is the mass density. This coefficient has dimensions of $[L^2 M^{-1}]$, and (7.19) now becomes

$$I(x) = I(0)e^{-\frac{\mu}{\rho}(\rho x)} \tag{7.21}$$

where $\rho x$ is called the absorber mass thickness with $[ML^{-2}]$ dimensions. Figure 7.9 shows the dependence of the mass attenuation coefficient on the photon

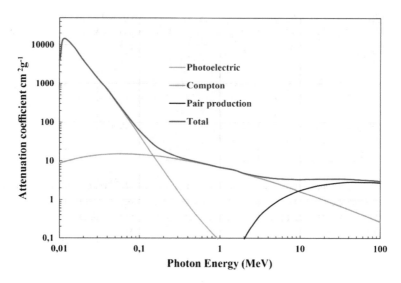

**Fig. 7.9** Attenuation coefficient as a function of the photon energy. The contribution of the three effects is included, and the calculation is performed for $Z = 32$ (Ge)

energy [2]. The partial coefficients associated with each of the interaction photon effects with matter are also depicted. In the case of a complex medium, the mass attenuation coefficient can be calculated using a Bragg addition rule [7]. In addition, so:

$$\frac{\mu}{\rho} = \sum_{i=1}^{n} \varepsilon_i \left(\frac{\mu}{\rho}\right)_i \qquad (7.22)$$

where $\varepsilon_i$ is the fraction by weight of the $i$-th component of the material that has an individual mass attenuation coefficient $\left(\frac{\mu}{\rho}\right)_i$.

The linear or mass attenuation coefficients must be distinguished from the corresponding absorption coefficients. In the latter, only energy-loss processes are considered. It is obvious that the partial attenuation and absorption coefficients for the photoelectric and pair-production effects coincide with each other. This is not the case for the Compton effect, since it includes elastic dispersions that contribute to the removal of the photon from the beam without depositing energy in the material. The absorption coefficient is clearly more relevant than the attenuation coefficient for various problems, among others, for dosimetry calculations or energy deposition in a calorimeter.

### 7.1.5 Designing Gamma Radiation Detectors

Generally, (see Fig. 7.9) at low $\gamma$ energies, the photoelectric effect dominates, whereas for high energies, the relevant process is pair production. Furthermore, when Z is large, the range of energies during which the photoelectric effect predominates is larger. Similarly, pair production begins to be relevant for energies closer to its threshold, 1.022 meV. As we will see in the following chapters, a radiation detector takes advantage of the mechanisms of radiation interaction with matter to provide information about radiation itself. The relevant mechanisms are those that produce a deposition of energy in the detector. For that, a given material is a good candidate for being a radiation detector if the mechanisms leading to an energy deposit are highly probable. In the case of $\gamma$ radiation, the interesting mechanisms are the photoelectric effect and pair production, where the photon deposits all its energy in the material. Therefore, a good $\gamma$ detector material must have a high Z. This condition becomes essential if we truly want to build a $\gamma$ spectrometer, an instrument whose main objective is not the simple detection of radiation but the measurement of the energy of the photons. As we will see in the following chapters, this is not the only condition for good $\gamma$ detectors. Additionally, the fingerprint left by $\gamma$ radiation in the detector should be readable.

## 7.2 Interaction of Charged Particles with Matter

The interaction of charged particles with matter is a complex problem governed by electromagnetic or strong interactions, or both at the same time. They consist basically of elastic or inelastic collisions with the electrons or nuclei of the material through which the particle is passing. The probability of each interaction channel depends on the charge of the particle, its energy, and the characteristics of the material, specifically, its effective charge and mass density. To simplify the discussion, we classify the mechanisms as follows [8]:

*Mechanisms Governed by the Electromagnetic Iteraction*
Elastic or inelastic collisions with electrons or groups of electrons or the nuclei.

*Mechanisms Governed by Electromagnetic and Strong Interactions*
Elastic or inelastic collisions with nuclei or groups of nuclei.

From the point of view of charged-particle detection, the most important mechanism, despite its low cross sections, $\sigma \sim 10^{-17}$ to $10^{-16}$ cm$^2$, is inelastic collision with atomic electrons. Here, the particle transfers energy to the medium, causing the excitation or ionization of its atoms. In the case of electrons and positrons, the so-called Cherenkov and Bremsstrahlung effects are also important. The first belongs to the group of inelastic collisions with electrons, and the second belongs to those with nuclei, both governed by the electromagnetic interaction. As we shall see later, both effects are formally possible for all charged particles but are only relevant for low-mass particles, i.e., electrons and positrons.

### 7.2.1  Ionization and Excitation

In this case, the charged particle deposits energy in the medium through inelastic collisions with electrons. Depending on the amount of deposited energy, the collision causes ionization, i.e., separation of the electron from the atom, or just excitation, in which the electron is promoted to a less bound orbit in the atom. In the first case, the electron moves in the material with an energy that is the difference between the energy transferred by the charged particle and the binding energy to the orbit. In the second case, the atom reaches an excited state, and the deexcitation usually gives rise to electromagnetic radiation. Both electron and electromagnetic radiation can be used to detect the pass of the charged particle through the material. This is a useful signal that allows for the design of charged-particle detectors.

From a quantitative viewpoint, the important data are the amount of energy transferred to the medium by the particle. For that, we must know the probability that the particle deposits an energy $E$ in the medium, that is, the cross section, $d\sigma(E)$, that a particle of mass $M$ and kinetic energy $T$ deposits an energy between $E$ and $E + dE$ when it collides with an electron. The calculation of the cross section is due to Bohr and Bethe [9], the classic and relativistic approaches that give the same information. Since the interaction is of Coulomb type, it is found that

$$d\sigma(E) = 2\pi z^2 r_0^2 \frac{m_0 c^2}{\beta^2} \frac{dE}{E^2} \tag{7.23}$$

where $\beta = \frac{v}{c}$, with $v$ being the speed of the particle of charge $z$. Equation 7.23 is valid for heavy charged particles, i.e., those with masses higher than that of the electron. It is also assumed that the particles are fast; i.e., the energy transferred is higher than the kinetic energy of the orbital electrons. This means, from a mathematical viewpoint, that the electron is free and at rest. In addition, since the particle is heavy and fast, it loses a very small amount of energy in the collision, and its trajectory will not be practically affected. Therefore, the trajectory of the particle practically turns out to be a straight line. As expected, the cross section decreases with the speed of the particle and increases with its charge. However, a specific comment should be made on the dependency on $z$. Indeed, there exists a nonnegligible probability, especially at low speed, that the particle captures electrons from the medium along its path. Thus, its effective charge could decrease in the trajectory. This also reduces the cross section. This effect is relevant for speeds close to that of orbital electrons [10]. Certainly, under these conditions, the hypothesis supporting (7.23) are no longer valid.

In the case of electrons and positrons, inelastic collisions with electrons are also described on the basis of electromagnetic interactions. However, the calculations are now much more complicated, as some of the assumptions cannot be applied. For example, the mass of the electron or positron is identical to that of the particle with which they collide, and it is easy to see that, from a classical point of view, the maximum energy transmitted in the collision could be

$$E_{\max} = \frac{2m_0 M^2 v^2}{(m_0 + M)^2} = \frac{1}{2}m_0 v^2 \tag{7.24}$$

In other words, the electron or the positron can lose all of its energy in a collision with the atomic electron. Consequently, in most collisions, the electron path is clearly affected, and the deviations from the incident direction are large. Furthermore, both the projectile and the target are identical and must be taken into account in the calculation. In this sense, it is accepted that, after the collision, the most energetic particle is the incident. By convention, it is assumed that the maximum loss of energy in a collision must be $\frac{T}{2}$, where $T$ is the kinetic energy of the incident electron or positron. With these considerations, Mott and Möller found [11]

$$d\sigma(T, E) = 2\pi r_0^2 m_0 c^2 \frac{(T + m_0 c^2)^2}{T(T + 2m_0 c^2)}$$

$$\left[ \frac{1}{E^2} + \frac{1}{(T - E)^2} - \frac{(2T + m_0 c^2)m_0 c^2}{E(T + m_0 c^2)^2 (T - E)} + \frac{1}{(T + m_0 c^2)} \right] dE \tag{7.25}$$

which is similar to (7.23), except that the parentheses are different. However, when $E \ll \frac{T}{2} \ll m_o c^2$, it is easy to see that the relevant term in the parentheses is $\frac{1}{E^2}$, and we again find the expression for heavy charged particles.

## 7.2.2  Stopping Power. The Bethe-Bloch Equation

Equations 7.23 and 7.25 must be understood in terms of the probability of losing a given amount of energy. However, the number of collisions per unit length along the particle trajectory is large, so the fluctuations in the total energy loss are small. Therefore, it makes sense to find the mean value of the particle energy loss per unit length to characterize the path of the charged particle through matter. This is what we call the stopping power $S_c$, specifically, the collisional stopping power. It is

$$S_c = \left( \frac{-dT}{dx} \right)_c \tag{7.26}$$

where $dT$ is the average loss of energy when crossing a distance $dx$. For heavy charged particles, it is easy to find [12]

$$S_c = 4\pi z^2 r_0^2 \frac{m_0 c^2}{\beta^2} \rho \frac{N_A}{A} Z ln \left( \frac{2m_0 c^2 \beta^2}{I} \right) \tag{7.27}$$

where $\rho$, $Z$, and $A$ are the mass density, effective atomic number, and mass number, respectively, of the medium. $N_A$ is the Avogadro number, and $I$ is the mean excitation energy, representing the minimum necessary energy to produce a transition in the atom or molecule forming the medium. This expression corresponds to the classic Bohr calculation and contains the necessary ingredients to comfortably describe the interaction of heavy particles. In the relativistic case, (7.27) changes

$$S_c = 4\pi z^2 r_0^2 \frac{m_0 c^2}{\beta^2} \rho \frac{N_A}{A} Z \left[ \ell n \left( \frac{2 m_0 c^2 \beta^2}{I} \right) + \ell n \left( \frac{1}{1-\beta^2} \right) - \beta^2 \right] \qquad (7.28)$$

which is known as the Bethe-Bloch equation, from which the Bohr equation, (7.27), can be deduced for $v \ll c$.

The mean excitation energy, $I$, can theoretically be calculated with quantum-mechanical arguments [8]. It is a logarithmic mean value of the excitation energies weighted by the oscillator intensities of the atomic levels. It is difficult to evaluate in real cases, as the oscillator intensities are not known for most of the elements. However, the following semiempirical formula deduced from experimental measurements of $S_c$ is very often used.

$$\frac{I}{Z} = 12 + \frac{7}{Z} eV \quad Z < 13 \qquad (7.29)$$

As in the case of the cross section, the Bethe-Bloch equation also takes a more complicated form for electrons. Indeed, now

$$S_c = 2\pi z^2 r_0^2 \frac{m_0 c^2}{\beta^2} \rho \frac{N_A}{A} Z \left[ \ell n \frac{m_0 c^2 \beta^2}{I^2} T + 2\ell n \gamma \frac{1}{\gamma^2} \left( \frac{9}{8} - \ell n 2 \right) + \frac{1}{\gamma} \left( 2\ell n 2 - \frac{1}{4} \right) + \frac{1}{8} - \ell n 2 \right]$$
$$(7.30)$$

still similar to that found for heavy particles.

In Fig. 7.10a, we can see the variation of $S_c$ with the energy for different heavy particles. In the nonrelativistic region, the term $\frac{1}{\beta^2}$ dominates. Insofar as the energy increases, this term becomes constant, and we pass through the minimum ionization to later observe the relativistic increase. For very low energies, the Bethe-Bloch equation does not account for different processes that take place, especially the effective charge effect, which was previously discussed and is very important for heavy nuclei [10]. In Fig. 7.10b, we show the stopping power of electrons in Al at different energies. As the energy of the electron increases, the stopping power increases until reaching a maximum at approximately 100 eV, which is on the Al conduction band. From this energy, $S_c$ starts decreasing as expected, and after going through the ionization minimum, a small increase is observed that corresponds to relativistic effects. An excellent website where $S_c$ data can be found for heavy ions in different materials can be found in [13]. In the case of electrons, [14] must be visited.

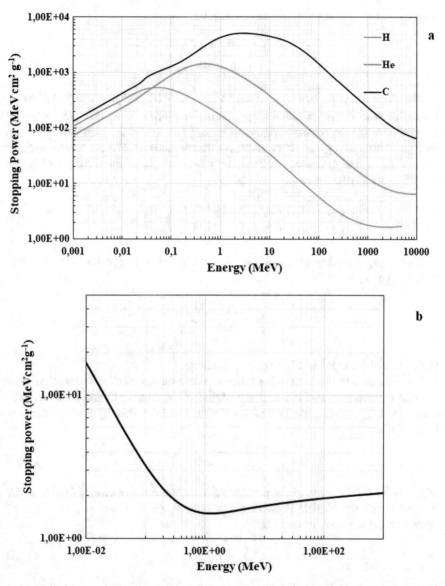

**Fig. 7.10** Mass stopping power in Si for **a** heavy particles and **b** electrons as a function of the energy

What we have presented thus far is the so-called linear stopping power. It is usually measured in MeVcm$^{-1}$. $S_c$ depends on the $\rho$ of the medium and hence on the aggregation state of the medium. It is convenient to define the mass stopping power to eliminate this dependency as follows

$$\frac{S_c}{\rho} = \frac{1}{\rho}\left(\frac{-dT}{dx}\right)_c \tag{7.31}$$

The *mass stopping power* is given in $MeVg^{-1}cm^2$ and is proportional to $\frac{Z}{A}$. Since this ratio is very constant along the periodic table, $\frac{S_c}{\rho}$ is very similar for all the elements [8]. In other words, particles deposit almost the same energy when they pass through a given mass thickness of any element. For the same reason, it is possible to find systematic relationships between $S_c$ of the different elements that allow us to deduce each other. Indeed:

$$\left(\frac{S_c}{\rho}\right)_2 = \frac{\left(\frac{Z}{A}\right)_2}{\left(\frac{Z}{A}\right)_1}\left(\frac{S_c}{\rho}\right)_1 \tag{7.32}$$

Moreover, for complex media, a Bragg addition rule applies, and the total stopping power is

$$\frac{S_c}{\rho} = \sum_{i=1}^{n} \varepsilon_i \left(\frac{S_c}{\rho}\right)_i \tag{7.33}$$

where $\varepsilon_i$ is the weight fraction of the $i$-th element.

The variables in the Bethe-Bloch equation can be grouped into those that characterize the medium and those of the particle. Thus, once the material has been determined, $S_c$ is just a function of the charge $z$ and the velocity $v$ of the particle. In particular:

$$S_c = z^2\lambda_0(v) \tag{7.34}$$

where $\lambda_0(v)$ is a function of the particle velocity and the properties of the material through which the particle is passing. For protons, $(S_c)_p = \lambda_0(v)$. Thus, for any other particle $x$, we will have:

$$(S_c)_x = z_x^2(S_c)_p \tag{7.35}$$

This is a scaling rule that can be extended to the kinetic energy, $T$, of the particle since $T = Mf(v)$. Then, the following happens.

$$S_c = z^2\lambda_1\left(\frac{T}{M}\right) \tag{7.36}$$

$\lambda_1$ is also determined by the properties of the material. For protons, once more, $(S_c)_p = \lambda_1(T)$. As before, the stopping power for other particles can be calculated in a straightforward way. Finally, we can develop a scaling rule for the loss of velocity of the particle. This is trivial to find using the above expression for $T$ and the first scaling rule. We obtain:

$$\frac{-dv}{dx} = \frac{z^2}{M}\lambda_2(v) \tag{7.37}$$

The fact that for a given medium $S_c$ depends only on the characteristics of the particle has interesting applications, as we will see in Chap. 12. We will describe an experimental device called $\Delta E$–$E$ telescope in which $S_c$ and $E$ can be measured at the same time, providing information on the particle charge or mass.

### 7.2.3 Bremsstrahlung

Bremsstrahlung consists of the emission of electromagnetic radiation during the deceleration of a charged particle when passing close to another charged particle of a different sign. It is a well-known effect, and while it is general for any charged particle, it is especially likely for electrons and positrons. The reason is their small mass. Indeed, the acceleration that a particle of mass m undergoes in the presence of a Coulomb field is the following

$$a = \frac{1}{m}\frac{Z_1 Z_2}{r^2}e^2 \tag{7.38}$$

$Z$ is the electric charge of the involved particles. Clearly, the smaller the mass is, the higher the acceleration.

Bremsstrahlung occurs when an electron passes close to a nucleus and deviates from its trajectory due to the attraction of the nuclear electric field, as shown in Fig. 7.11. The electron loses some energy that is released as electromagnetic radiation. The energy of the emitted photon is:

$$E_f = T_0 - T \tag{7.39}$$

**Fig. 7.11** Schematic view of the Bremsstrahlung effect

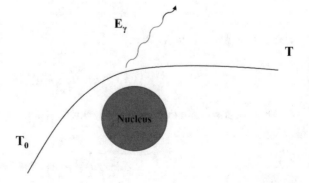

where $T_0$ and $T$ are the energies of the electron, or positron, before and after the interaction process with the nuclear electric field. Because the electrons are free after and before the collision, the Bremsstrahlung energy spectrum is continuous with energies ranging from $0 \leq E_f \leq T_0$. Qualitatively, the Bremsstrahlung cross section is known to depend on the charge of the medium as $Z^2$, considering only the effect of the nuclear electric field. In fact, there is an additional contribution to the Bremsstrahlung effect produced by collisions with atomic electrons, which depends on Z. Therefore, it is more correct to assume that the Bremsstrahlung cross section has the form $\frac{d\sigma(E_f)}{dE_f} \propto Z(Z+1)$ [15].

A Bremsstrahlung stopping power, $S_r$, can be defined as it leads to an electron loss of energy. The definition is identical to that of $S_c$, but now the energy loss refers only to that produced by Bremsstrahlung. The calculation is also similar once the cross section is known [16]. It can be found that

$$S_r \approx 4Z^2 r_0^2 \alpha \frac{N_A}{A} \rho T_0 \ell n \left( 183 Z^{-\frac{1}{3}} \right) \tag{7.40}$$

Similarly, a Bremsstrahlung mass stopping power can be defined as $\frac{S_r}{\rho}$. In the case of electrons, the energy lost by Bremsstrahlung can be relevant and, in many cases, larger than that lost by ionization or excitations. For that, it is more realistic to define a total stopping power as:

$$S = S_c + S_r \tag{7.41}$$

Their relative importance depends on the electron energy and Z of the medium. In Fig. 7.12, we give a comparison between both stopping powers for the case of water and W. For low energies, the collisional effects dominate. At higher energies, the Bremsstrahlung stopping power becomes more relevant. In general, $S_r$ is the most relevant energy loss at high energies, close to or above 10 meV, which is a relativistic zone. The energy at which $S_c = S_r$ is called the critical energy, $T_c$. In a good approach (Ref):

$$\frac{S_r}{S_c} \approx \frac{(Z+1.2)T}{800} \tag{7.42}$$

where $T$ is given in MeV. Hence,

$$T_c \approx \frac{800}{Z+1.2} \text{MeV} \tag{7.43}$$

The so-called radiation length, $X_0$, can be defined from (7.40) as

$$X_0 = \frac{1}{4Z^2 r_0^2 \alpha \frac{N_A}{A} \rho \ell n \left( 183 Z^{-\frac{1}{3}} \right)} \tag{7.44}$$

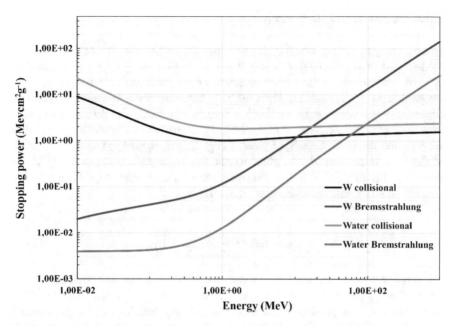

**Fig. 7.12** Comparison between collisional and Bremsstrahlung stopping powers

$X_0$ has length dimensions. In fact, $S_r$ can be written as:

$$S_r = \left(-\frac{dT}{dx}\right)_r = \frac{T_0}{X_0} \tag{7.45}$$

and it is easy to find:

$$T = T_0 e^{-\frac{x}{X_0}} \tag{7.46}$$

where $T$ is the remaining kinetic energy of the electrons after passing through a thickness $x$. Therefore, $X_0$ gives the distance at which the electron beam loses energy by a factor $\frac{1}{e}$.

It is advantageous to introduce this new concept. Note that $X_0$ contains all the information regarding the medium. $X_0$ characterizes the medium as a Bremsstrahlung producer. On the other hand, we can define the mass radiation length $\rho X_0$. There are empirical expressions for it [16]. One of them is

$$\rho X_0 (\text{gcm}^{-2}) = \frac{716.4 A}{Z(Z+1)\ell n\left(\frac{287}{\sqrt{Z}}\right)} \tag{7.47}$$

In the case of complex materials, a Bragg addition rule applies as those given previously.

### 7.2.4  Cherenkov Radiation

When a charged particle travels through a medium, it generates an electromagnetic field around its trajectory that propagates at a speed $c_i = \frac{c}{n}$, $c$ being the speed of light in vacuum and n the refractive index of the medium. If the velocity, $v$, of the particle is $v > c_i$, that is, when $n\beta > 1$, there will be a cooperative effect of electromagnetic waves such as shock waves in acoustics. The resulting coherent wavefront is called Cherenkov radiation [17]. Cherenkov radiation is located in a cone around the particle path, as seen in Fig. 7.13. It is easy to see that the angle of the cone is $\cos\theta = \frac{1}{n\beta}$. In other words, the angle reports the velocity of the particle and the radiation frequency since the refractive index depends on it. The threshold energy, $E_{\text{th}}$, for Cherenkov radiation is

$$E_{\text{th}} = m_0 c^2 \left[ \frac{n}{\sqrt{n^2 - 1}} - 1 \right] \tag{7.48}$$

this condition being accomplished more easily by electrons and positrons in the range of energies involved in this text. Indeed, at energies of some tenths MeV, only electrons and positrons can reach sufficiently high velocities to produce Cherenkov radiation. As we will see in Chap. 10, this is used for their detection. Nevertheless, Cherenkov radiation is very much used in High Energy Physics to detect charged particles.

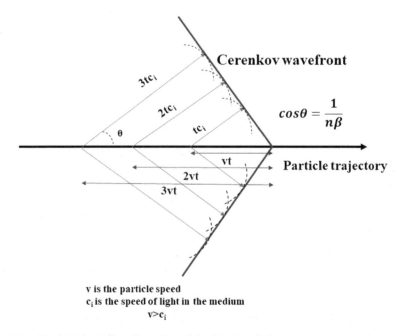

**Fig. 7.13** The Cerenkov effect. Formation of the Cerenkov light cone

Franck and Tamm demonstrated that the emission of Cherenkov radiation leads to a slight energy loss of the particle of charge $z$. Thus, a Cherenkov stopping power can be calculated, and it can be found that [18]

$$-\frac{dT}{dx} = z^2 \frac{\alpha^-}{c} \int v\,dv\left(1 - \frac{1}{\beta^2 n^2(v)}\right) \qquad (7.49)$$

which gives rise to an energy deposit of no more than a few keVcm$^{-1}$ in solid materials or some tenth keVcm$^{-1}$ in gases, which is practically negligible against the loss of energy by ionization. Franck and Tamm [18] also showed that the Cherenkov radiation spectrum has the form:

$$\frac{d^2N}{dx\,dv} = \frac{z^2\alpha}{c}\left(1 - \frac{1}{\beta^2 n^2(v)}\right) \qquad (7.50)$$

which gives the number of photons emitted per unit length and frequency. The spectrum is continuous and essentially uniform because n depends on $v$ very smoothly, especially on transparent media. In any case, the spectrum is grouped around the frequencies for which $\beta n(v) > 1$. This means that the light produced is concentrated essentially in the visible range or at most in the near-UV range. In addition, within the visible range in the blue zone, where n is greater. This is the reason why Cherenkov radiation presents a bluish color in transparent media.

## 7.2.5  Range, Specific Ionization, and Bragg Curves

In the case of charged particles, the parallel concept to the attenuation coefficient is that of range. The range serves to understand the behavior of a material as a charged-particle energy absorber and to evaluate its suitability as a radiation detector or, from another viewpoint, to develop radiation protection systems against charged particles.

Unlike photons, charged particles lose energy continuously along their path. Let us suppose a monochromatic beam of charged particles passing through a material. At a given distance within the material, it is expected that the number of particles will remain the same as that at the beginning. In addition, all the particles will have the same energy, although they are smaller than at the beginning. The particle beam will continue its path through the material until all of the original energy has been deposited. According to this picture, it is realistic to define a range of the particle beam that can be measured by performing a transmission experiment similar to that made before for $\gamma$ radiation (see Fig. 7.8).

Imagine a charged-particle beam moving through a certain material. The particle will undergo collisions along its path, transferring some energy to the medium in each of the collisions. According to the picture described before, if we measure the number of particles detected as a function of the material thickness, we should find the results depicted in Fig. 7.14a. All the particles will survive until reaching a thickness that we call the range of the particles in the material. Of course, it is expected that the range will depend on the energy of the particles. In real experiments, however, it can be observed that the geometric figure obtained is not an exact rectangle. In contrast, as observed in Fig. 7.14a, the curve turns out to be a rectangle but with smooth angles. In other words, part of the beam stops at a thickness lower than or higher than expected. The origin of this phenomenon is the statistical nature of the interaction of particles with matter. This causes a certain angular distribution of the beam particles around the path and a certain distribution of values of the individual range of each particle that forms the beam. It also causes fluctuations in the energy losses in each collision. This phenomenon is called straggling [19]. It can be found that the distribution of the particle beam range values is of Gaussian type. Thus, what the transmission experiment would truly give us is the average value of this distribution. Although the straggling effect is very small for heavy and fast particles, it exists and, therefore, modifies the expected transmission curve, making it difficult to find a characteristic value of the range. However, since the integrals of the ideal and real curves must be equal and the distribution of ranges is symmetric, the crossing point between both curves will be located at a thickness for which $\frac{N}{N_0} = 0.5$. It must represent the average value of the range distribution. For that, we call it R the range of the particles. Therefore, from an experimental point of view, the range of a particle beam in a medium is defined as the thickness of the medium for which the initial particle beam is reduced by 50%. Mass ranges can also be defined simply by multiplying the corresponding linear thicknesses by the mass density of the medium.

Electrons have a peculiar behavior, with possible high-energy transfers, sudden changes in their trajectory, and various competing energy-loss mechanisms. Thus, although the description made is basically valid, it is difficult to determine the range for electrons. In Fig. 7.14b, we have a typical transmission curve for a monochromatic electron beam. The appearance is clearly different from that obtained for heavy charged particles, and R cannot be defined as before. In the case of electrons, it is better to use the so-called extrapolated range, $R_e$, as seen in Fig. 7.14b.

The transmission curves for $\beta$ radiation have an exponential shape (see Fig. 7.14c). Thus, it is possible to define a linear, $\mu$, or mass, $\frac{\mu}{\rho}$, absorption coefficient as a result of the fitting of the transmission curve, which is:

$$N(x) = N(0)e^{-\mu x} = N(0)e^{-\frac{\mu}{\rho}(\rho x)} \tag{7.51}$$

In this case, the absorption coefficient has strict empirical meaning, although it must depend on the energy of the radiation and the medium through which the $\beta$-radiation passes. In fact, there are some empirical relationships between the value of the absorption coefficient and the $E_{max}$ of $\beta$ radiation for a given medium. Generally, they are potential-type functions [19], such as that for Al.

$$\frac{\mu}{\rho}\left(\frac{cm^2}{g}\right) = 17.0 E_{max}^{-1.14} (MeV) \tag{7.52}$$

**Fig. 7.14** A transmission curve for the case of **a** heavy charged particles, **b** monoenergetic electrons, and **c** β radiation

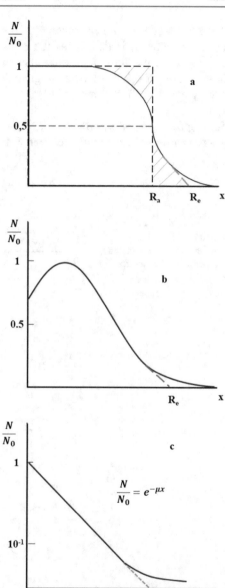

When studying the transmission of electrons, it is important to consider the effect of backscattering, as the electrons, especially those of low energy, could undergo large deviations from their path during a collision. This effect is more likely as the Z of the medium increases. This should be taken into account when designing a detector, as a nonnegligible number of electrons cannot penetrate into the active volume (see next chapters). To evaluate this effect, the backscatter coefficient, or albedo, is defined as the ratio between the number of electrons that impinge on a medium and the number of electrons scattered at an angle $> 90°$. Some values of this coefficient are given in Table 7.1 as a function of the electron energy and for different media [20].

*Scaling Laws for the Range*
There is a formal connection between $S$ and $R$ of charged particles, since

$$R = \int_0^R dx = -\int_0^{E_0} \left(\frac{dE}{dx}\right)^{-1} dE \qquad (7.53)$$

**Table 7.1** Albedo coefficients in some media as a function of the electron energies

| Energy (MeV) | Material | Albedo coefficient |
|---|---|---|
| 0.1 | C | 0.04 |
| | Al | 0.12 |
| | Cu | 0.28 |
| | Ag | 0.38 |
| | Au | 0.50 |
| 0.5 | C | 0.025 |
| | Al | 0.10 |
| | Cu | 0.26 |
| | Ag | 0.38 |
| | Au | 0.50 |
| 1.0 | C | 0.02 |
| | Al | 0.08 |
| | Cu | 0.23 |
| | Ag | 0.35 |
| | Au | 0.48 |
| 1.5 | C | 0.015 |
| | Al | 0.02 |
| | Cu | 0.1 |
| | Ag | 0.15 |
| | Au | 0.25 |
| 10 | C | 0.01 |
| | Al | 0.015 |
| | Cu | 0.04 |
| | Ag | 0.07 |
| | Au | 0.12 |

where $E_0$ is the energy of the particle at the beginning of its trajectory. Thus, some scale laws for $R$ can be derived from those for $S$. It is easy to see that for a particle of charge $z$,

mass $M$, velocity $v$, and energy $T$, we have

$$R = \frac{1}{z^2} f_T \left( \frac{T}{M} \right) \tag{7.54}$$

and

$$R = \frac{M}{z^2} f_v(v) \tag{7.55}$$

where the functions f characterize the medium. Thus, once we evaluate the range for a given particle, it is possible to deduce its value for another particle in the same medium.

The results obtained with these equations do not agree well with the experimental data, especially in the case of electrons. Therefore, it is necessary to use empirical formulas, of which there is a wide variety in the literature [19].

### Linear Energy Transfer (LET)

LET is a concept similar to S, although with some differences. Indeed, the charge-particle energy loss does not necessarily coincide with the energy deposited along its path. Indeed, the energy deposit may be nonlocal, for example, because of the production of the so-called δ rays: electrons that absorb enough energy to produce their own ionization track. This process is obviously included in the calculation of stopping power. However, δ rays transport the absorbed energy away from the path of the incident particle, as shown in Fig. 7.15. Furthermore, in the case of electrons, Bremsstrahlung certainly involves a loss of energy, but the deposit is not local. On many occasions, the parameter of interest is the local deposit of energy. For instance, for gas ionization detectors, $S$ does not allow a rigorous calculation of the ionization tracks of the charged particles through the gas; in biological materials, the radiation effect is related to the local deposit of energy. The concept of LET ('linear energy transfer'), $L_\Delta$, is introduced to evaluate all these specific cases. $L_\Delta$ is a restricted stopping power, since it excludes high-energy losses in the calculation. LET is defined as

$$L_\Delta = \left( -\frac{dT}{dx} \right)_c \quad E < \Delta \tag{7.56}$$

$\Delta$ being a cut-off parameter, which limits the transfer of energy considered in the calculation of $S$. The use of the cut-off parameter avoids inclusion in the calculation of nonlocal energy deposits, such as the case of δ rays. Its value depends on the medium, in which the calculation is intended.

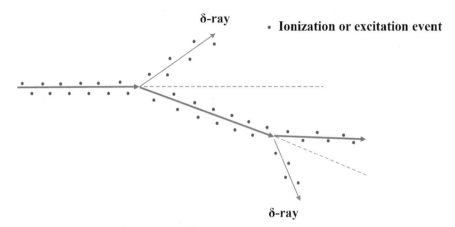

**Fig. 7.15** Production of δ rays

*Specific Ionization*

Related to LET is the so-called specific ionization, I, defined as the number of ions produced per unit length of the path of a charged particle [19]. Its average value $\bar{i}$ can be related to $S_c$ by

$$S_c = w\bar{i} \tag{7.57}$$

where $w$ is the average value of the energy needed to create an electron–ion pair in the medium through which the particle is passing. Usually, w is calculated as the ratio between the energy lost by the particle and the traveled distance. Since the particle can lose energy by excitation or other mechanisms different from ionization, w is usually greater than the ionization potentials of the material. Equation 7.57 applies for the first sections of the particle path, that is, for high speeds, where we can consider that $w$ is constant. This is not true for the last part of the trajectory, in which $v$ is low and $w$ is no longer constant.

*Bragg Curves for Specific Ionization*

Bragg curves [21] for specific ionization consist of the graphical representation of the residual range of the particle in the medium. In Fig. 7.16a, the expected Bragg curves for an individual heavy charged particle of a given energy and for a monochromatic beam of the same particles are given. At the beginning of the trajectory, the specific ionization is basically constant since the speed is still large and does not change. At the end of the trajectory, when the residual range is close to 0, $\bar{i}$ dramatically increases since the speed rapidly decreases. Now, $S$ (or LET) clearly increases. This $\bar{i}$ rise is called the ionization peak or Bragg peak. The straggling effect explains the widening of the ionization peak in the case of a particle beam. For electrons (Fig. 7.16b), the interpretation of the Bragg curves is more complicated. In this case, the shape of the curve is a

**Fig. 7.16** Specific Bragg ionization curves for **a** heavy particles and **b** electrons

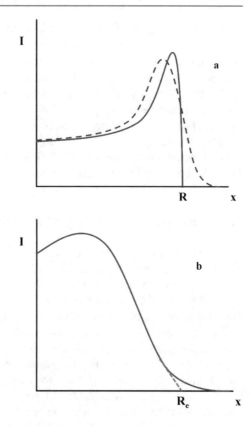

consequence of the appearance of the Bremsstrahlung effect that competes with ionization as a mechanism of energy loss.

### 7.2.6 Designing Charged-Particle Detectors

In general terms, gases, solids, and liquids could serve as detecting materials for charged radiation. The important thing is that the response of the detector can be easily interpretable. The Z of the medium must not be too high since the probability of backscattering increases with Z. This is especially important in the case of electrons. For the thickness of the chosen material, there are two possibilities. Should the goal be just particle counting, the thickness of the medium should be sufficient to absorb enough energy for the detector to give an analyzable response. If the objective is the measurement of the energy of the particle, the thickness of the material must be equal to or greater than R. This way, the particle will deposit all its energy. From this viewpoint, energy particle losses before entering the detector must be avoided.

## 7.3　Nuclear Reactions. Interaction of Neutrons with Matter

As we saw in Chap. 1, neutrons, together with protons, both called nucleons, are part of the nucleus. The discovery of the neutron is attributed to J. Chadwick (1932) [22], who, in 1933, made a first estimation of its mass. The mass of the neutron, 1.00866490 amu, is somewhat higher than that of the proton, which differs by approximately 1.3 meV/$c^2$. Bound in the nucleus is a stable particle, but free neutrons disintegrate by $\beta^-$ emission with $T_{1/2} = 12.4$ min. On the other hand, the neutron undergoes weak and strong interactions when passing through matter. The last one is used for its detection.

### 7.3.1　Nuclear Reactions with Neutrons

For a realistic design of neutron detectors, the only useful interaction is the strong interaction. We will not go into a detailed description of this interaction, a task that would correspond to a text on Nuclear Physics [23]. We will describe the different reactions the neutron can undergo, always with the objective of justifying the construction of neutron detectors.

Through strong interaction, many possible reaction channels for neutrons are opened. The probability of each depends on the neutron energy and the A of the target nucleus. In general, the excitation functions for neutrons, like many other nuclear projectiles, have two parts: One corresponds to a diffusive or potential-type function and the other to a resonant type. An example is shown in Fig. 7.17. The resonances in the excitation function can be interpreted on the basis of the compound nucleus model proposed by Bohr [24]. According to this model, a nuclear reaction occurs in two stages: After the collision, a conglomerate of the projectile and the target is formed, the compound nucleus, which subsequently decays into the reaction products with a certain probability. In the case of neutrons, a reaction would occur as follows:

$$n +^A X \rightarrow \left[ {}^{A+1}_{Z} X \right]^* \rightarrow b +^{A'}_{Z'} Y$$

Resonances take place when the neutron energy coincides with any of the energy levels of the compound nucleus. As a consequence, the probability that the compound nucleus will disintegrate through a certain channel will depend on the energy of the neutrons and the target nucleus. To analyze the different possibilities, it is convenient to classify the neutrons according to their energy [26]. In Table 7.2, we have a classification of neutrons according to their energy. The most characteristic reactions are presented schematically in Table 7.3.

The energies of the so-called thermal region correspond to those of thermal agitation of the surrounding environment. At an ambient temperature of 20 °C, the energy would be 0.025 eV assuming a Maxwell–Boltzmann neutron velocity distribution. As we will see in what follows, they are very interesting both from a scientific and a technological point of view.

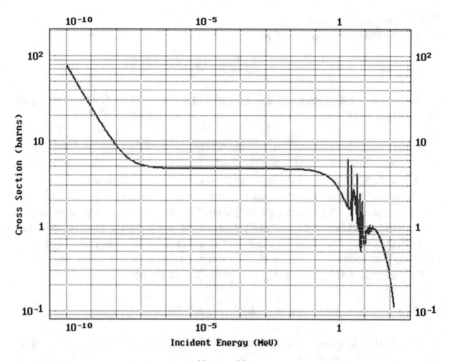

**Fig. 7.17** Neutron excitation function for $^{14}N(n, p)^{14}C$. Data were retrieved from the IAEA ENDF Database Retrieval System, 2004–2022: https://www-nds.iaea.org/endf/ [25]

**Table 7.2** Classification of neutrons according to their energies

| Neutron classification | Energy (eV) |
|---|---|
| Cold | 0.005 |
| Thermal | 0.025 |
| Epithermal | 1 |
| Slow | $10^2$ |
| Intermediates | $10^4$ |
| Rapid | $10^6$ |
| Ultrafast | $10^8$ |

**Table 7.3** Different types of neutron reactions

| Reaction type | Scheme |
|---|---|
| Elastic | $^A X(n, n)^A X$ |
| Inelastic | $^A X(n, n')^A X^*$ |
| Radiative capture | $^A X(n, \gamma)^{A+1} X$ |
| Particle emission | $^A_Z X(n, b)^{A'}_{Z'} Y$ |
| Fission | $^A X(n, vn)^{A'} Y, ^{A''} Z$ |

*Elastic and Inelastic Scattering*
Elastic scattering is represented by:

$$^AX(n,n)^AX$$

They occur with neutrons of any energy and are very likely for small $A$ nuclei. For example,

$$^1H(n,n)^1H$$

In the dispersion, the kinetic energy of the incident neutron is shared with the recoil nucleus and the scattered neutron. This leaves, consequently, with less kinetic energy. These reactions are very useful for shifting the energy spectrum of neutrons from the fast to the thermal range, i.e., to softening the neutron spectrum. This effect is used to design the moderator in a thermal fission nuclear reactor [27]. For that, $H_2O$ and $D_2O$, or even C, are used as moderators in nuclear reactors.

The inelastic scattering corresponds to the following processes:

$$^AX\left(n,n'\right)^AX^*$$

Now, part of the kinetic energy of n is used to populate the excited states of the recoil nucleus. For example,

$$^{26}Mg(n,\ n')^{26}Mg^*$$

*Radiative Capture*
This reaction can be represented by:

$$^AX(n,\gamma)^{A+1}X$$

The $\gamma$ radiation is emitted immediately by the compound nucleus after capture, either as a single photon or, more commonly, in a photon cascade. They are exoergic reactions and very likely with thermal or slow neutrons and large $A$ target nuclei. Usually, nucleus $^{A+1}X$ is a $\beta^-$ emitter since it is neutron-rich. On the other hand, it is possible to observe characteristic $\gamma$ radiation from the residual nucleus or its descendants. As we will see in Chap. 19, this leads to the development of a highly sensitive analytical method called neutron activation. Some examples are the following

$$n+^{103}Rh \to ^{104}Rh\left(T_{1/2} = 44\ s\right)+\gamma \quad \sigma = 144\,b$$

$$n+^{115}In \to ^{116}In\left(T_{1/2} = 54\,min\right)+\gamma \quad \sigma = 45\,b$$

A very important neutron radiation capture reaction, from a technological point of view, is the following

$$^{113}Cd(n, \gamma)^{114}Cd \quad \sigma = 20000\,b$$

$^{114}$Cd being stable. The high cross section of this reaction makes it advisable to use natural Cd (12.3% 113 Cd) as a component of control rods in nuclear reactors [27]. These rods serve to regulate the neutron flux inside the reactor and, consequently, the power produced. When necessary, they can stop the chain reaction and, therefore, the operation of the nuclear reactor. Other very interesting capture reactions, also from a technological point of view, are:

$$n +{}^{238}U(\sigma = 2.73b) \rightarrow{}^{239}U(\beta^-) \rightarrow{}^{239}Np(\beta^-) \rightarrow{}^{239}Pu$$

$$n +{}^{232}Th(\sigma = 7.4b) \rightarrow{}^{233}Th(\beta^-) \rightarrow{}^{233}Pa(\beta^-) \rightarrow{}^{233}U$$

Both $^{239}$Pu and $^{233}$U are fissile nuclei that can be used as nuclear fuel. Another interesting thermal neutron capture reaction is:

$$n +{}^1H \rightarrow{}^2H + \gamma \quad \sigma = 0.332\,b$$

which is, in fact, an exception in this group of reactions, since the target is very light. Shielding against neutrons is constructed very often with $H_2O$ for this reason.

*Emission of Particles*
The excitation energy of the compound nucleus dissipates here by emitting particles such as neutrons, protons, $\alpha$, and deuterons. They are represented by:

$$^A_Z X(n, b)^{A'}_{Z'}Y$$

These types of reactions occur for any A and, fundamentally, with fast neutrons. Indeed, they can provide the nucleons with enough energy to overcome nuclear binding. Interestingly, the most significant reactions in this group are the exceptions. Indeed, these reactions can occur in some light nuclei but with thermal neutrons. They are very useful for neutron detection.

$$n +{}^{10}B \rightarrow{}^7Li + \alpha \quad \sigma = 3837\,b$$

$$n +{}^3He \rightarrow{}^3H + p \quad \sigma = 5327\,b$$

$$n +{}^6Li \rightarrow H + \alpha \quad \sigma = 940\,b$$

$$n +{}^{14}N \rightarrow{}^{14}C + p \quad \sigma = 1.81\,b$$

From another point of view, the high reaction cross section for the case of $^{10}$B is why B is considered to be an excellent thermal neutron absorber.

*Fission*

Neutron-induced fission is a particular case of reactions in which particles are emitted. The general scheme of fission is as follows:

$$^{A}X(n, \upsilon n)^{A'}Y, ^{A''}Z$$

As we saw in Chap. 1, during spontaneous fission, the nuclear $^{A}X$ splits into two other nuclei of similar mass and close to half the mass of $^{A}X$. At the same time, $\upsilon$ neutrons are emitted. Fission can, in turn, be induced by neutron reactions. This is very likely for heavy nuclei and thermal neutrons [28]. However, fast neutron-induced fission is also important from many points of view, and for some heavy nuclei, it is the only possible neutron-induced fission process. The history of the discovery of neutron-induced fission is certainly very interesting, and its importance from a scientific and technological viewpoint is widely recognized [29]. Here, we will only give some data that will help to understand its utility for n detection. Nuclear fission is also very much important for producing energy.

$^{238}$U and $^{232}$Th undergo nuclear fission with fast neutrons, while $^{235}$U and $^{239}$Pu can do so with thermal and fast neutrons. However, the cross Sect. (577 and 741 b, respectively) for thermal neutrons is very high compared to that of the fast range. Neutron-induced fission is very interesting because of the production of radioactive nuclei. They are the so-called fission products (see Chap. 1). For example, $^{235}$U can undergo fission in approximately forty different ways, resulting in approximately eighty different fission products. They are neutron-rich and for that $\beta^{-}$ emitters. Their disintegration normally populates excited states of the daughter nuclei, and therefore, $\gamma$ radiation is also observed. In Fig. 3.4, the fission pathways and their yields for different fissile nuclides are presented. An example follows:

$$^{235}U(n, 3n) \, ^{134}Te \, ^{99}Zr \, \ldots$$

## 7.3.2   Path of Neutrons Through Matter

The path of neutrons through matter can be described in a global way, as we did for $\gamma$ radiation and charged particles. To judge how interesting a material is as a shield against neutrons or as a neutron detector, we can carry out a transmission experiment similar to that in Fig. 7.8 but with a neutron source [30]. Following our description of the different neutron reaction channels, it is very easy to see that, broadly speaking, neutrons behave in a very similar way as $\gamma$ radiation. Indeed, neutrons can deposit all or part of their energy, disappearing this way from the trajectory. Thus, after passing through the thickness x, the number of neutrons

$I(x)$ will be smaller than the original $I(0)$, but all the surviving neutrons will have the original energy. For that, it can be said that

$$I(x) = I(0)e^{-\Sigma x} \tag{7.58}$$

$\Sigma$ is the so-called macroscopic effective cross section, and it is

$$\Sigma = N\sigma \tag{7.59}$$

where $N$ is the number of targets (nuclei) per unit volume and $\sigma$ is the sum of all cross sections that account for all possible reaction channels for the neutron energy and A of the material. $\sum$ contains all the relevant information. Its dimensions are $L^{-1}$. This gives rise to the definition of the neutron mean free path, $\lambda$, $n$ the material that can be calculated as

$$\lambda = \frac{\int_0^\infty x e^{-\Sigma x} dx}{\int_0^\infty e^{-\Sigma x} dx} = \frac{1}{\Sigma} \tag{7.60}$$

$\lambda$ has dimensions of length and accounts for the average value of the distance traveled by the neutron before undergoing any possible reaction. Finally, note that it is possible to define a Bragg sum rule to calculate $\sum$ for complex materials.

## 7.3.3 Designing Neutron Detectors

It is not difficult to see that the problem of neutron detection is more complicated than the detection of charged particles or $\gamma$ radiation. It is not possible to detect neutrons directly because of the complexity of the interaction they undergo. Fortunately, charged particles are released during some neutron reactions. Consequently, neutron detection could be reduced to that of charged radiation, which we will see in the next chapters. Particle emission reactions are key when designing neutron detectors, and in fact, it is possible to apply the general principles of detection of charged particles. It is obvious, however, that the properties of the studied neutrons will have to be deduced indirectly through the properties of the nuclei released in the reaction, which complicates the interpretation of the results. Even elastic scattering is useful for neutron spectrometry. Indeed, the energy of the proton scattered in a reaction $^1H(n, n)^1H$, for example, informs us about the energy the neutron has deposited. Again, the problem is reduced to performing charged-particle spectrometry. Furthermore, neutron activation reactions and neutron-induced fission allow the design of neutron detectors, with the objective of counting rather than spectrometry, for somewhat more specialized applications.

**Exercises**

1. Find the stopping power in N for 5 MeV α-particles. Use (7.27) with $I = 85$ eV for N.
2. Find the stopping power in $H_2O$ for a 5 MeV α-particle. Use (7.27) assuming a Bragg rule for S. Take $I = 89$ and 19 eV for O and H, respectively.
3. It is known that the energy loss per unit length of 2.5 MeV protons in water is $13.8$ GeVm$^{-1}$. Estimate the energy loss per unit length for 10 meV α particles in water.
4. Estimate the ratio of Bremsstrahlung to collisional loss of energy per unit length in W for 1 and 10 meV electrons.
5. Estimate the W radiation length. Find the Bremsstrahlung energy loss of a 20 MeV electron beam after passing through a 10 cm thick W target.
6. Find the Al thickness needed to decrease by 20% a β radiation beam from a $^{204}$Tl source.
7. Find the ratio of the ranges of protons and α-particles of energy T in a medium. Suppose now that they have the same speed. What is the ratio under these conditions?
8. Demonstrate (7.4) and (7.5).
9. Let us have a $^{60}$Co γ source. Suppose that we are planning a Compton scattering experiment. Find the equivalent plot to Fig. 7.4 for the two photons emitted by the source.
10. Find the ratio between the pair-production cross section in Cu by 2.5 and 5 MeV γ-rays.
11. The attenuation coefficient for γ radiation is:
    (a) The probability that the photon interacts with matter per unit length.
    (b) The probability that the photon deposits a certain amount of energy in matter per unit length.
    (c) The thickness of the material necessary to attenuate the photon beam by 50%.
12. Estimate the γ-ray total interaction cross section of the above monochromatic source in Pb.
13. Use Fig. 7.9 to find the γ-ray total cross section for 100 and 1000 keV in lead.
14. Suppose that we have a $^{137}$Cs source emitting 662 keV photons. With a Ge detector, a spectrum is collected. 202500 counts are recorded in the photopeak (see Chap. 11). An Al foil of unknown thickness is introduced between the source and the detector, and the sample is counted again. Now, 165302 counts are collected for the same counting time. Find an expression to calculate the thickness of the foil on the basis of the attenuation coefficient. Discuss the precision of the method for determining thickness and the applicability range. If a 1 cm thick sheet of Pb had been introduced, what would then be the photopeak area of $^{137}$Cs?
15. Determine the thickness of a Si (Li) detector Be window (see Chap. 11) used in X-ray detection, knowing that the theoretical Fe $\frac{K_\alpha}{K_\beta}$. The ratio is 7.19 and that in a spectrum collected, the ratio is found to be 6.33.

16. An attenuation experiment through Pb is carried out with two $\gamma$ emitters. One of them is monochromatic. The results obtained are given in the following table. Plot both attenuation curves taking into account that the background was 523 counts in 4 min. Find which $\gamma$-emitter is monochromatic and its mass attenuation coefficient.

| Sample A | Thickness $(mg/cm^2)$ | Counts ($t =$ 4 min) | Sample B | Thickness $(mg/cm^2)$ | Counts ($t =$ 4 min) |
|----------|------------|------------|----------|------------|------------|
|          | 0          | 13,043     |          | 0          | 36,793     |
|          | 5          | 8221       |          | 5          | 14,651     |
|          | 10         | 5190       |          | 10         | 9306       |
|          | 15         | 3427       |          | 15         | 6280       |

17. Calculate the macroscopic cross section for thermal neutron capture by normal and heavy water if the corresponding microscopic cross sections are $\sigma_O = 2 \times 10^{-4}$ barn, $\sigma_H = 0.332$ barn, and $\sigma_D = 5.3 \times 10^{-4}$ barn. Calculate the mean free path of the neutrons in both cases.

18. An In sheet ($1$ $cm^2$ area 0.01 g mass) is irradiated for 10 s with a neutron flux from a reactor. Immediately after, the sheet was counted, and $3.8 \times 10^5$ counts were recorded over 6 min. The experiment was repeated with a new but identical In sheet surrounded by a thick Cd layer of Cd. The sheet is counted, and now $2.1 \times 10^5$ counts are recorded. Assume that the efficiency of the counter is 2% and that the half-life of radioactive indium is 54 min. Calculate the density of thermal neutrons in the flux from the reactor. The thermal neutron capture cross section follows a $\frac{1}{v}$ law, with $\sigma = 200$ barns for $v = 2200$ m/s.

19. The $^{207}$Pb(n, $\gamma$)$^{208}$Pb cross section is $10^{-24}$ $cm^2$. How much will the intensity of a neutron beam decrease from that energy as it passes through 10 cm of $^{207}$Pb ($\rho = 11$ g cm$^{-3}$)?

20. Assume a reactor with 3% enriched U nuclear fuel. Calculate how much $^{239}$Pu mass has been formed per 100 kg of U when 5% of $^{235}$U has been used.

21. A 100 mg Cu sample, activated with a cyclotron, contains $^{64}$Cu along with stable Cu (69% of $^{63}$Cu and 31% $^{65}$Cu). After irradiation, the sample has $10^9$ dps of activity. What is the relationship between the number of stable and radioactive atoms after 25.6 h?

# References

1. D.W. Anderson, *Absorption of Ionizing Radiation*, Chapters 6 and 7 (University Park Press, 1984)
2. E. Storm, H.I. Israel, Photon cross sections from 1 keV to 100 MeV for elements from $Z = 1$ to $Z = 100$. Nucl. Data Tables A **7**, 565–681 (1970)
3. D.R. White, An analysis of the Z dependence of photon and electron interactions. Phys. Med. Biol. **22**(1977), 219 (1977)
4. K.N. Klein, Y. Nishina, Über die Streuung von Strahlung durch freie Elektronen nach der neuen relativistischen Quantendynamik von Dirac. Z. Phys. **52**, 853–868 (1929)
5. H.A. Bethe, W. Heitler, On stopping fast particles and on the creation of positive electrons. Proc. R. Soc. Lond. A **146**, 83–112 (1934)
6. N. Tsoulfanidis, *Measurement and Detection of Radiation*, Chapter 4, Fifth Edition (CRC Press, 2021)
7. D.W. Anderson, *Absorption of Ionizing Radiation*, Chapter 5 (University Park Press, 1984)
8. D.W. Anderson, *Absorption of Ionizing Radiation*, Chapter 2 (University Park press, 1984)
9. N. Bohr, On the decrease of velocity of swiftly moving electrified particles in passing through matter. Phil. Mag. **30**, 581–682 (1915); H.A. Bethe, Bremsformel für Elektronen relativistischer Geschwindigkeit, Z. Phys. **76**, 293–299 (1932)
10. H.B. Gilbody, J.V. Ireland, Ionization of atomic hydrogen by protons in the energy range 60 to 400 keV. Proc. R. Soc. Lond. **277**, 137–141 (1964)
11. C. Möller, Zur Theorie des Durchgangs schneller Elektronen durch Materie. Ann. Phys. **14**, 531–585 (1932); N.F. Mott, The collision between two electrons. Proc. R. Soc. Lond. **126**, 259–267 (1930)
12. F. Bloch, Zur Bremsung rasch bewegter Teilchen beim Durchgang durch Materie. Ann. Phys. **16**, 285–320 (1933)
13. http://www.sr-niel.org/index.php/sr-niel-web-calculators/nuclear-stopping-power-calculator/protons-ions-nuclear-stopping-power-calculator
14. https://physics.nist.gov/PhysRefData/Star/Text/method.html
15. E. Segré, *Nuclei and Particles* (W. A. Benjamin, 1965), p. 57
16. D.W. Anderson, *Absorption of Ionizing Radiation*, Chapter 3 (University Park Press, 1984)
17. P.A. Cerenkov, Visible radiation produced by electrons moving in a medium with velocities exceeding that of light. Phys. Rev. **52**(4), 378 (1937)
18. Frank and Ig Tamm. Coherent visible radiation of fast electrons passing through matter. In Selected Papers (Springer, 1991), pp. 29–35
19. D.W. Anderson, *Absorption of Ionizing Radiation*, Chapter 4 (University Park Press, 1984)
20. T. Tabata, R. Ito, S. Okabe, An empirical equation for the backscattering coefficient of electrons. Nucl. Instrum. Methods **94**, 509–513 (1971)
21. W.H. Bragg, *Studies in Radioactivity* (Macmillan, 1912)
22. J. Chadwick, Existence of a neutron. Proc. R. Soc. A **136**, 692–708 (1932)
23. K.S. Krane, *Introductory Nuclear Physics* (Wiley & Sons, 1988)
24. N. Bohr, Transmutations of atomic nuclei. Science **86**, 161–165 (1937)
25. IAEA ENDF database retrieval system, 2004–2022: https://www-nds.iaea.org/endf/
26. K.S. Krane, *Introductory Nuclear Physics*, Chapter 12 (Wiley & Sons, 1988)
27. G.R. Choppin, J.-O. Liljenzin, J. Rydberg, C. Erkberg, *Radiochemistry and Nuclear Chemistry*, Chapter 20 (Elsevier Fourth Edition, 2013)
28. K.S. Krane, *Introductory Nuclear Physics*, Chapter 13 (Wiley & Sons, 1988)
29. O. Hahn, F. Strassman, Über den Nachweis und das Verhalten der bei der Bestrahlung des Urans mittels neutronen entstehenden Erdalkalimetalle. Naturwiss **27**, 11–15 (1939)
30. D.W. Anderson, *Absorption of Ionizing Radiation*, Chapter 9 (University Park Press, 1984)

# Principles of Radiation Detection: Counting and Spectrometry

**8**

**ABSTRACT**

The description of the main features of radiation detectors as well as the relevant parameters guiding their operation is the objective of the chapter. A general definition of radiation detectors based on the concepts of interaction of radiation with matter previously described is introduced. Then, the more relevant detector parameters, such as counting efficiency, dead time, energy spectrum, energy resolution, and detector background, are described in detail. The latter is of paramount importance for the detection of very low levels (or concentrations) of radioactivity in the environment.

## 8.1 Introduction

Human beings cannot detect, quantify, or characterize nuclear radiation. We can detect mechanical radiation, sound, at least in a wide range of frequencies, and electromagnetic radiation, within the so-called optical range, but not nuclear radiation.

A nuclear radiation detector comes to fill that gap. In general terms, it can be defined as a system in which nuclear radiation deposits energy through the interaction mechanisms we have studied in the previous chapters. In general, the energy deposit results in the direct or indirect appearance of an electric charge, $Q$. Usually, $Q$ is collected by the application of an electric field, and the electric current generated in this way informs about the passage of radiation and in many cases about the energy deposited, or even the physical nature of the radiation. The charge collection lasts for a collection time, $t_c$. Under these conditions, the problem of detecting nuclear radiation is reduced to the problem of processing electrical signals. Not every detector follows exactly this working model, but the concepts that we describe throughout this chapter, especially those of counting efficiency, energy resolution, and background, apply to any detector.

© The Author(s), under exclusive license to Springer Nature Switzerland AG 2022    259
M. García-León, *Detecting Environmental Radioactivity*, Graduate Texts
in Physics, https://doi.org/10.1007/978-3-031-09970-0_8

It is easily understood that our detector model, while ideal, is not universal, and it is not capable of detecting all types of particles. It will depend on the probability that the radiation we are interested in undergoes a given interaction mechanism in the material the detector has been constructed. In other words, detectors are more sensitive to one type of radiation than another, or more precisely, the signal produced by the detector is analyzable only for a certain type of radiation.

The information a detector gives can simply be the current intensity, $i(t)$, generated as the radiation passes through. Therefore, if the collection time is $t_c$, we have the following.

$$\int_0^{t_c} i(t)\mathrm{d}t = Q \tag{8.1}$$

In real experiments, a certain number of particles arrive at the detector per unit time, so a group of current pulses must be generated, each corresponding to a given $Q$ collected during a $t_c$. In other words, it can be expected that each particle produces a charge amount, which depends on the deposited energy, that will be recovered during a $t_c$. This is not the general behavior of the detectors. In many of them, $Q$ or $t_c$, or even both, is constant. Since $Q$ informs about the energy deposited by the particle, the distribution of $Q$ values gives the distribution of deposited energies, which is the principle of operation of radiation spectrometers. On the other hand, the variability of $t_c$ informs about the location where the particle deposited the energy in the detector. This is very useful for a good number of nuclear experiments. It is also interesting to emphasize that $t_c$ depends on the detector material since it is closely related to the mobility of charge in it.

In realistic experimental setups, the detectable signal from the detector is obtained using an ammeter. Since the ammeter has a finite response time, $T$, we observe the average value of the current intensity, $I(t)$, during the considered measurement time. That is,

$$I(t) = \frac{1}{T} \int_{t-T}^{T} i(t')\mathrm{d}t' \tag{8.2}$$

In this case, the detector is working in the current mode. In this mode of operation, the detector does not give details about the interaction events taking place inside and about what happens to each particle. Despite this, it is widely used for many problems, especially because of the simplicity of the experimental setup. The most popular applications of detectors that work in this way are related to personal and environmental dosimetry determination problems.

Details of each interaction event can be obtained with pulse-mode operation. In this operation mode, we can obtain a voltage pulse for each event. This is achieved by coupling an RC circuit to the detector. The detectable signal from the detector is now the time voltage pulse, $v(t)$, in the resistance, R, coupled to the circuit. The capacitor, C, integrates the charge released into the detector and discharges it through R until the potential drops to 0. If $\tau = RC$, the RC circuit time constant, when $\tau \ll t_c$ the current flowing through R is identical to that flowing through the detector. However, the common situation is that $\tau \gg t_c$. In this case, very little current will flow through R during $t_c$, and the detector current is integrated into C. Now, assuming that the time separation between pulses is large enough, the capacitor, C, will discharge through R until $v(t) = 0$, following an exponential law governed by $\tau$. In any case, the maximum pulse voltage $v_{max}$ is:

$$v_{max} = \frac{Q}{C} \tag{8.3}$$

This means that $v_{max}$, the height of the pulse, contains information about the energy deposited in the particle detector, as it depends on $Q$. If C is constant along a radiation measurement experiment, the set of $v_{max}$ gives the energy distribution deposited by the particles. Analysis of this pulse height distribution is the basis for radiation spectrometry. This mode of operation gives more detailed information than the current mode. Therefore, it is the most widely used in radiation detectors [1].

## 8.2 Counting Efficiency

### 8.2.1 Absolute Efficiency

As a result of measuring the activity of a sample, we obtain several pulses or counts during a counting time period. The number of pulses does not necessarily correspond to the number of particles emitted by the sample. To convert the number of counts into the real disintegrations taking place in the sample during such a time period, we need a counting efficiency, a counting yield. There are several definitions of counting efficiency, in many cases related to the specific counting problem in which the experimenter is involved. The most general definition, insofar as it includes all the effects that influence particle counting, is the absolute efficiency, $\varepsilon_{abs}$ [2].

Let us suppose an experiment in which we measure a sample of activity A during a counting time $\Delta t$ with a certain detector. In Fig. 8.1, we make a representation of such an experiment. In principle, we use an external counting technique, i.e., a counting technique in which the sample is outside the detector. We will see in the following chapters that it is possible to introduce the sample inside the detector in a so-called internal counting technique. If $N_p$ is the number of pulses (number of counts) observed in the detector counting scale during $\Delta t$ and $N$ is the number of decays that occur within the same time interval, the absolute efficiency will be as follows:

$$\varepsilon_{abs} = \frac{N_p - B}{N} \tag{8.4}$$

**Fig. 8.1** A schematic representation of a counting experiment of a sample with activity A

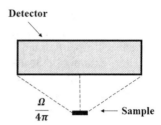

*B* being the detector background counts obtained during $\Delta t$. (We will study the background of detectors in the next section, and we will see that a certain number of counts appear in the detector counting scale even in the absence of radioactive samples. This effect must be subtracted from the total number of counts obtained with the sample.)

In general, $\varepsilon_{abs} \neq 1$ for several reasons. Indeed, first, the particle must be emitted within the solid angle, $\Omega$, which subtends the sample and the detector to be counted. In relative terms, this effect contributes to $\varepsilon_{abs}$ with a factor $\frac{\Omega}{4\pi}$ and accounts for the counting geometry of the experimental system. Second, the particle must overcome the effects of self-absorption produced by the thickness of the sample. The self-absorption factor, $K_{aut}$, depends on the energy of the particle and the composition of the sample. As expected, it is very important for $\alpha$ and $\beta$ radiation. It is less important for $\gamma$ radiation but significant for X-rays or soft $\gamma$-rays. Third, once it leaves the sample, the particle must overcome the absorption barrier that exists between the sample and the detector. Such a barrier is formed by the air and the detector window, or only the detector window, if the counting experiment is carried out under vacuum conditions. The absorption barrier absorbs energy from the particles, which can lead, in some cases, to prevent the particle from reaching the active volume of the detector. This absorption effect, $K_{abs}$, has the same physical origin as self-absorption, and it is more important for charged radiation than for electromagnetic radiation. Once the two barriers have been overcome, the particle enters the detector. This does not necessarily mean that it will be counted. There is an intrinsic detector efficiency, $\varepsilon_{int}$, defined as:

$$\varepsilon_{int} = \frac{N_p - B}{N_{in}} \tag{8.5}$$

Where $N_{in}$ is the number of particles that reach the active volume of the detector during the time interval $\Delta t$. In some detectors, $\varepsilon_{int} \approx 1$ for charged particles, specifically, but not for photons or neutrons, for which it can be <1. With regard to $\varepsilon_{int}$, there is not a general behavior of detectors. It depends on the type of detector and the particle being counted.

Just to illustrate the description we have made, we could say that

$$\varepsilon_{abs} = \varepsilon_{int} \frac{\Omega}{4\pi} K_{aut} K_{abs} I \tag{8.6}$$

where the factor $I$ is the radiation intensity (see Chap. 2), which gives the number of particles that we are counting that are emitted per decay. It could be said that the absolute efficiency is the product of all the described effects. Nevertheless, all such factors are not independent, so this expression is not exact. However, it helps to understand the processes that influence the value of $\varepsilon_{abs}$.

Thus, the activity of a source producing $N$ counts in the presence of a background $B$ (see next section), both during a $\Delta t$, would be:

$$A = \frac{N - B}{\Delta t \varepsilon_{abs}} \tag{8.7}$$

Usually, the sample and background are measured at different counting times. In these cases, it is convenient to use the counting rates, $n$ for the sample and b for the background, i.e., the number of counts per unit time, and now

$$A = \frac{n - b}{\varepsilon_{abs}} \tag{8.8}$$

The calculation of $\varepsilon_{abs}$ is more complicated than it can be deduced from (8.6), which, on the other hand, is not rigorous. There are enough mathematical methods [3] to calculate $\varepsilon_{abs}$. Nevertheless, an empirical approach is usually preferred. The basis of the empirical method is the use of radioactive samples of known activity, standard samples, with the same physical nature and identical geometry to that of the problem sample.

Should the standard sample have an activity $A_{std}$, $\varepsilon_{abs}$ will be:

$$\varepsilon_{abs} = \frac{n_{std} - b}{A_{std}} \tag{8.9}$$

where $n_{std}$ is the count rate produced by the standard sample and $b$ is the background count rate. As the absolute efficiency is a dimensionless quantity, given in many cases as percentages, the units of $A_{std}$, $n_{std}$, $b$ must be the same. Once $\varepsilon_{abs}$ is known, the activity of the problem sample is calculated by applying (8.5).

As we will see in Chap. 13, for environmental applications, it is necessary to calculate $\varepsilon_{abs}$ under real conditions or, in other words, it is necessary to calibrate the detector for counting efficiencies. In some cases, the calibration will be done for different radiation energies under identical geometry conditions, and in other cases for different thicknesses at a given energy, but always under identical geometry conditions. It is important to note that in the empirical approach to calculate $\varepsilon_{abs}$, the standard samples and the problem sample are at the same physical and geometric conditions.

## 8.2.2 Partial Efficiencies. Photopeak Efficiency

As said at the beginning, there are several definitions of efficiency [4], in many cases related to specific applications. As we will see in Chaps. 10, 11, and 13, it is not convenient to apply the concept of $\varepsilon_{abs}$ in environmental $\gamma$ spectrometry. We will see that the most interesting part of a $\gamma$ spectrum is the photopeak. It is associated with processes in which the $\gamma$ radiation deposits all its energy in the detector. Therefore, the photopeak identifies the radionuclide, so it can be used to unambiguously estimate its activity. For that, the photopeak efficiency is defined using the information contained in the photopeak. Although we will go in depth in the chapters mentioned above, we can advance a formulation of the photopeak efficiency $\varepsilon_{ph}$ as follows:

$$\varepsilon_{ph} = \frac{\Sigma_{ph} - B_{ph}}{A \Delta t I_\gamma} \tag{8.10}$$

where $\Sigma_{ph}$ and $B_{ph}$ are the number of counts and the background in the photopeak zone, respectively, and $I_\gamma$ is the $\gamma$ radiation intensity.

The photopeak efficiency is an example of the concept of partial efficiency that finds application for some specific problems. Basically, partial efficiency is defined in relation to part of the information provided by the detector and is used very frequently in spectrometry problems [5].

## 8.3    Background of Detectors

The detector background can be defined as the number of counts that appear in the counting scale of the system in the absence of a radioactive source. Obviously, it is crucial to determine and reduce the background of a detector when dealing with low- or very-low-activity samples. In fact, it is necessary to develop new counting techniques, as we will see in Chaps. 14 and 15, the Low-Level Counting Techniques (LLC), to address this problem.

### 8.3.1    Sources and Components

The origin of the detector background is varied [6], with several components of instrumental or fundamental nature. The relative importance of each component depends on the type of detector and the shielding technique used (see Chaps. 14 and 15). In what follows, we give a description of the different background components, although some more information can be found in the chapters mentioned above.

*Environmental Gamma Radiation Field*
The laboratory building materials, the air surrounding the detectors, the detector construction materials, the electronic chain, even the shielding (see Chap. 15), etc., contain radioactive $\gamma$ emitting elements. The main contributors are the radioactive elements belonging to the natural radioactive series (see Chap. 3) and other primordial radionuclides, such as $^{40}$K. All contribute to the detector background. In

Fig. 8.2, a Ge detector background $\gamma$ spectrum is presented. The exact interpretation of the figure will be possible after studying Chap. 11, but what it is interesting is now to see the number of peaks that represent the different $\gamma$ emitting elements that surround the Ge detector. As seen in Table 8.1, the peaks correspond mainly to the $^{238}$U and $^{232}$Th progeny, except for $^{40}$K. The $^{40}$K 1460.82 keV and the $^{208}$Tl 2615 keV, belonging to the $^{232}$Th series, dominate the background spectrum. From this point of view, a good selection of laboratory construction materials, for example, concrete, helps greatly to reduce the contribution of the environmental $\gamma$ field to the background of detectors.

In fact, some special cements can be found with concentrations of $^{40}$K and $^{232}$Th of 1200 and 1.4 ppm, respectively, in contrast to their usual concentrations in the Earth's crust of 2800 and 10.7 ppm [7].

*Radon in the Air*

The case of Rn must be studied individually within the environmental $\gamma$ radiation field component, since the techniques to suppress its contribution are very specific.

$^{222}$Rn and, to a lesser extent, $^{220}$Rn are very important sources of the $\gamma$ background. As gases, the rejection of this background component requires different techniques from those used for the general $\gamma$ field component, as we will see in Chap. 15. The contribution comes from the Rn progeny, $^{214}$Pb and $^{214}$Pb for $^{222}$Rn, as well as from $^{212}$Pb and $^{208}$Tl, which are descendants of $^{220}$Rn. Flushing of the air surrounding the detector and the counting chambers greatly helps to

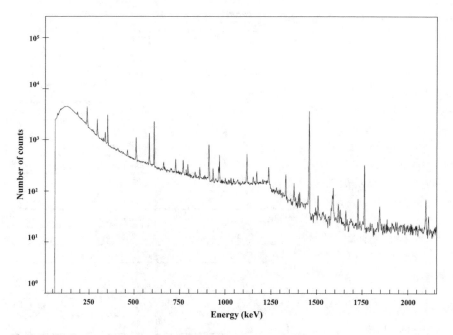

**Fig. 8.2**  Ge detector background $\gamma$-spectrum

**Table 8.1** Main $\gamma$-emitting elements in the background of a Ge detector

| Radionuclide | Energy (keV) |
|---|---|
| $^{238}U$ series | |
| $^{234}$Th | 63.29 |
| | 92.38 |
| | 92.8 |
| $^{226}$Ra | 186.21 |
| $^{214}$Pb | 241.98 |
| | 295.21 |
| | 351.92 |
| $^{214}$Bi | 609.31 |
| | 768.36 |
| | 934.06 |
| | 1120.29 |
| | 1238.11 |
| | 1377.65 |
| | 1408.01 |
| | 1509.23 |
| | 1729.65 |
| | 1764.49 |
| | 1847.44 |
| | 2204.22 |
| $^{232}Th$ series | |
| $^{228}$Ac | 129.08 |
| | 209.28 |
| | 270.23 |
| | 463 |
| | 794.7 |
| | 911.07 |
| | 964.6 |
| | 969.11 |
| | 1588 |
| $^{224}$Ra | 240.98 |
| $^{212}$Pb | 238.63 |
| | 300.09 |
| $^{212}$Bi | 727 |
| $^{208}$Tl | 278 |
| | 583 |
| | 860.37 |
| | 2614.66 |
| $^{40}$K | 1460.82 |

reduce the presence of Rn and thus decrease its contribution to the background. In other cases, the detector assembly is hermetically sealed inside a steel or plastic container, which avoids the introduction in the counting chamber of fresh air (Rn-rich) from the surrounding detector. Fortunately, both $^{222}$Rn and $^{220}$Rn are very short-lived radionuclides, 3.8 d and 55.6 s, respectively, and they rapidly decay.

### $\alpha$, $\beta$ or $\gamma$ Radioactivity in the Detector Substance

As commented before, the detector material contains impurities that also contribute to the background. The impurities consist mostly of primordial radionuclides, that is, $^{235}$U, $^{238}$U, $^{232}$Th, and their progeny, as well as $^{40}$K. In the case of the radioactive series, it is possible to find unsupported radionuclides, $^{226}$Ra in the case of the $^{238}$U series, or the whole family, depending on the detector material fabrication process. This is due, of course, to their different chemical behaviors. The fundamental contribution to this component of the background comes from charged radiation. Indeed, because it is not very penetrating, most of its energy is deposited in the detector. In Table 8.2, some information is given on the radioactive content of some common materials used for detector construction.

### Radioactivity $\alpha$, $\beta$, or $\gamma$ in the Shielding Materials

As we will see in Chap. 15, some materials are used to shield the detectors against external radiation, essentially against external $\gamma$ radiation. Despite their high contribution to background suppression, they can increase it at the same time because of the presence of radioactive impurities in the shielding material. Usually, Pb and, to a lesser extent, Fe are used as shielding materials. Cu is also used as an internal layer to attenuate the X-ray produced by $\mu$ in the main shield (see Chap. 15).

**Table 8.2** Potential radioimpurities in common detectors. In addition, surface contamination in the surroundings by natural radionuclides or $^{137}$Cs should be considered and removed by periodic cleaning

| Detector material | Radioimpurities |
|---|---|
| *Gas detectors* | |
| | $^{3}$H, $^{14}$C, and $^{85}$Kr in counting gases. $^{238}$U, $^{232}$Th, $^{40}$K, and $^{60}$Co in Fe cathodes. $^{238}$U, $^{232}$Th, $^{40}$K, and $^{226}$Ra in quartz |
| *Scintillation detectors* | |
| Organic | $^{3}$H, $^{14}$C, and natural radioactivity in additives. $^{40}$K, $^{238}$U, and $^{232}$Th series in photomultipliers. Glass vials in liquid scintillation counters contain more impurities than plastic or Teflon vials |
| Inorganic (NaI(Tl)) | $^{40}$K, $^{238}$U, and $^{232}$Th series in crystals and photomultipliers |
| *Semiconductor detectors* | |
| Si | Si is very pure. The contamination of the surroundings by natural radionuclides is the most important contribution. Recoil nuclei can contaminate the detector |
| Ge | $^{226}$Ra, $^{208}$Tl, and $^{40}$K in the components of the cryostat, molecular sieves, electronics around the detector, etc. Ge is generally very pure |

The usual manufacture of Fe makes it contain $^{60}$Co traces in addition to natural radioactivity [6]. For that, the iron Fe from old ships is preferred, as it was produced with older procedures and $^{60}$Co is not present. Cu is a clean material, although its neutron activation cross section is high, so it could become a secondary background source by the production of radionuclides near the detector. Fe has the same problem because of the high neutron activation cross section.

Pb has many advantages. One of them is the relatively low neutron cross sections and therefore the low probability of producing radionuclides by interaction with cosmic rays. However, Pb has intrinsic radioactive contamination due to the presence of $^{210}$Pb ($T_{1/2} = 22.2$ y) [8]. $^{210}$Pb decays by low-energy $\beta$-emission to $^{210}$Bi ($T_{1/2} = 5.01$ d). $^{210}$Bi is a high-energy $\beta$-emitter, $E_{max} = 1.16$ meV. While the radiation emitted by $^{210}$Pb is absorbed by the shielding, that from $^{210}$Bi can produce Bremsstrahlung and X-ray fluorescence in Pb that affect the background of the detectors.

The use of very old Pb, from seventeenth-century lead roofs or even from ancient Roman times, has been the solution to prevent the presence of $^{210}$Pb in Pb. Currently, it is possible to obtain modern Pb shielding with a certified low $^{210}$Pb content due to special fabrication processes [9].

*Cosmic Radiation*
A description of cosmic radiation that impinges on the Earth is given in Chap. 3. For detectors, the important components are electrons, photons, muons, and the nucleonic component, protons and neutrons. The flux of electrons and photons over the detectors can be reduced by Pb shielding, as we will see in Chap. 15. The nucleonic component essentially contributes by producing radionuclides through nuclear reactions. The main responsible is neutrons, since the proton flux is two orders of magnitude smaller. In Table 8.3, some examples of radionuclides produced by neutron-induced nuclear reactions with nuclei in materials close to the detector or forming part of the detector itself are given. The radiation emitted by these radionuclides may contribute to the background of the detector [10].

Muons contribute directly to the background of the detector, where they leave a track similar to that of $\beta$ radiation. Indeed, $\mu$ ionizes the matter as charged particles do and causes excitation. The photon and electron cascade can produce secondary reactions and release other particles and electromagnetic radiation. Direct Bremsstrahlung from $\mu$ is very unlikely due to its mass. However, the electrons released during ionization have enough energy to produce Bremsstrahlung when passing through the detector or the surrounding materials, relevantly through the shielding.

*Instrumental Noise*
The electronic components of the counter can introduce background noise. Either the voltage sources, the amplifiers, etc., introduce spurious electric pulses depending on parameters such as temperature and humidity in the detection laboratory. Improving the quality of electronic components and reducing their size is a way to decrease this influence. For laboratory ambient conditions, it is a key point to

**Table 8.3** Examples of radionuclides produced by n-induced nuclear reactions with nuclei in the materials surrounding or forming part of the detectors

| Energy (keV) | Radionuclide | Reaction |
|---|---|---|
| 10.37 | $^{68,71}$Ge | $^{70}$Ge(n, $\gamma$)$^{71}$Ge, $^{70}$Ge(n, 3n)$^{68}$Ge |
| 66.7 | $^{73m}$Ge | $^{72}$Ge(n, $\gamma$)$^{73}$Ge, $^{73}$Ge(n, n´)$^{73}$Ge |
| 122.1 | $^{57}$Fe | $^{57}$Fe(n, n´)$^{57}$Fe |
| 136.5 | $^{57}$Co | Neutron activations of Cu and Ni |
| 139.5 | $^{75m}$Ge | $^{74}$Ge(n, $\gamma$)$^{75}$Ge, $^{76}$Ge(n, 2n)$^{75}$Ge |
| 278.3 | $^{64}$Cu | $^{63}$Cu(n, $\gamma$)$^{64}$Cu, $^{65}$Cu(n, 2n)$^{64}$Cu |
| 558.4 | $^{114}$Cd | $^{113}$Cd(n, $\gamma$)$^{114}$Cd |
| 562.8 | $^{76}$Ge | $^{76}$Ge(n, n´)$^{76}$Ge |
| 579.2 | $^{207}$Pb | $^{207}$Pb(n, n´)$^{207}$Pb |
| 651.1 | $^{114}$Cd | $^{113}$Cd(n, $\gamma$)$^{114}$Cd |
| 669.6 | $^{63}$Cu | $^{63}$Cu(n, n´)$^{63}$Cu |
| 803.3 | $^{206}$Pb | $^{206}$Pb(n, n´)$^{206}$Pb |
| 1097.3 | $^{116}$In | $^{115}$In(n, $\gamma$)$^{116}$In |
| 1293.5 | $^{116}$In | $^{115}$In(n, $\gamma$)$^{116}$In |
| 2223 | $^{2}$H | $^{1}$H(n, $\gamma$) $^{2}$H |

keep them constant to avoid electronic fluctuations due to changes in humidity or temperature.

Furthermore, some detectors are sensitive to mechanical vibrations. This poses an additional condition to the laboratory design in which isolation against external vibrations is necessary [11].

## 8.3.2 Background Corrections

Background must be taken into account for a correct determination of activity. The background corrections become more necessary as the activity of the sample is lower. For low- or extremely low-activity samples, the detector background is a serious problem since the response of the detector to the sample radiation may be indistinguishable from the background itself. As we will see in Chap. 15, to avoid this problem, many counting techniques have been developed that include methods for the suppression or, better, the reduction of the different components of the background in detectors. Only with such techniques can very weak activity-counting experiments realistically be addressed. The relative influence of the background components in a detector depends on its sensitivity to the radiation involved. In this way, charged-particle detectors should be mostly affected by the muonic component or the $\alpha$ and $\beta$ radiation emitted by radionuclides that form part of the materials close to the detector or the detector substance itself. However, the relative importance of each component can change when using different shielding techniques. Shielding materials, as we have seen before and will see in Chap. 15, can be a source of background itself.

The following general equation can be used to discuss the influence of the background components in specific detectors. Indeed, if $B$ is the background number of counts per unit time, we could say that

$$B = B_\gamma + B_{Rn} + B_{det} + B_{sh} + B_{cos} + B_{nuc} + B_\mu + B_{inst} \qquad (8.11)$$

All the elements in the equation have an obvious meaning, and we can see that the cosmic component of the background has been divided into three subcomponents corresponding to cosmic electrons and photons, $B_{cos}$, nucleonic, $B_{nuc}$, and muonic, $B\mu$, subcomponents.

## 8.4   Dead Time

In what follows, it is convenient to distinguish two concepts: detector and counter. The counter is the detector plus the electronic modules needed to interpret the information it provides. One of the elements is the counter scale, where the counts produced by the detected particles are recorded. Thus, as happens for all the measurement instruments, in a counter, it is necessary to have a minimum time period between two events in the detector to distinguish them as different. This is the so-called dead time, $\tau$.

This concept is not very important in the determination of low radioactivity concentrations because it depends on the count rate, but it is convenient to take it into account for some applications.

### 8.4.1   Sources of Dead Time

$\tau$ depends on $t_c$ but also on the time resolution of the counter-associated electronics. In other words, there are physical and instrumental sources of dead time. The first ones are related to the physics behind the detection process in each detector, and the second ones are related to the time processing of the information (pulses) provided by the detector. Their relative contribution to $\tau$ depends on each counter. Dead-time losses appear in the count rate and obviously increase for very active samples. Therefore, an accurate measurement of the activity of a sample requires $\tau$ corrections.

### 8.4.2   Dead-Time Corrections

From an ideal point of view, we can imagine that the counters have two behavioral models [4] regarding dead time: the paralyzable, or extendible $\tau$ model, and the nonparalyzable, or not extendible $\tau$ model. They are ideal models, so a real counter usually follows a mixed behavior model depending on the count rate zone studied.

Let us suppose that the counter dead time is $\tau$. According to the nonextendible $\tau$ model, once one particle is counted, a period of dead time, $\tau$, is opened, and all the particles reaching the detector during such a time interval will not be counted. Once the time period $\tau$ passes, the counter recovers the possibility of counting a new particle. According to the extendible $\tau$ model, the situation is somewhat different. After counting the particle, a period of dead time, $\tau$, is opened in such a way that other particles arriving within that period will not be counted and will extend the dead time an additional $\tau$. In addition, so on. For that, this model is also known as paralyzable. To quantify the effect, we need some mathematical expressions.

Indeed, if we call $m$ the number of pulses that are counted per unit time in the counter, the experimental data, and $n$ the number of particles that reach the detector per unit time, which is directly related to the activity of the sample, we can find two mathematical expressions relating $m$ and $n$, one for each of the described models.

*Not Extendible $\tau$ Model*
In this case, $m\tau$ is the total fraction of lost dead time, and $nm\tau$ will be the number of lost particles per unit time. So:

$$n = m + nm\tau \tag{8.12}$$

or, in other words,

$$n = \frac{m}{1 - m\tau} \text{ or } m = \frac{n}{1 - n\tau} \tag{8.13}$$

In this way, n can be calculated on the basis of the experimental information, which is m. Of course, $\tau$ must be known.

*Extendible $\tau$ Model*
In this model, the periods of dead time vary, although $\tau$ is constant. It is known that the distribution of time intervals between random events that occur at a frequency n follows [4]

$$P(T)dT = ne^{-n\tau}dT \tag{8.14}$$

which gives us the probability $P(T)$ of observing a time interval ranging from $T$ to $T + dT$ between two random events. (We will see in Chap. 14 that radioactivity must be described as a random phenomenon.)

The probability of finding time intervals $>\tau$ will be

$$\int_{\tau}^{\infty} P(T)dT = e^{-n\tau} \tag{8.15}$$

Thus, the number of events detected by the counter, m, will be the frequency at which these intervals occur. Then,

$$m = ne^{-n\tau} \tag{8.16}$$

Clearly, for high n, $m \to \frac{n}{n\tau} = \frac{1}{\tau}$ in the not extendible model and $m \to 0$ for the extendible model. Nevertheless, for low n, both models are equivalent since (8.13 and 8.16) approaches to:

$$m \approx n(1 - n\tau) \tag{8.17}$$

This is the case for the determination of small amounts of radioactivity, which is very often the case for environmental applications.

## 8.5  Energy Spectra

A radiation detector not only serves to count particles. In contrast, it can provide additional information, such as the energy deposited by the particle in the detector itself. This information can be extracted from the response function of a detector, which, in the case of the pulse mode of operation, is a distribution of voltage pulse heights at the exit of the detector. In many cases, the distribution preserves information about the energy deposition processes in the detector. How the analysis of this distribution, or spectrum, provides such information is a matter of the following chapters. Here, we give a preliminary description of its structure.

### 8.5.1  Components

The pulse height distribution at the exit of a detector is a kind of picture reflecting the energy deposition processes that take place in the detector. It is also a picture of the energy spectrum of the emitted radiation from the sample. In the case, for example, of $\beta$ radiation, a continuum pulse height distribution should be expected to reflect the $\beta$ spectrum shape (see Fig. 8.3a). The absence of a structure in the pulse height distribution in these cases greatly complicates its analysis. In contrast, for discrete energy spectrum radiation, a certain structure is expected in the pulse height distribution, reflecting processes of total or partial energy absorption in the detector. In Fig. 8.3b, we present a picture of one such pulse height distribution. Discrete structures, or peaks, appear together with continuum areas. The first represents processes of total energy absorption, and the second represents partial deposits of energy. In general terms, the relative importance of the different zones depends on the specific cross section associated with the interaction processes. As we will see in Chaps. 10 and 11, the scheme represented in Fig. 8.3b corresponds to a $\gamma$ spectrum collected by a scintillation or semiconductor detector. Thus, the peaks represent photoelectric effects or total energy absorption processes in general, while the continuum represents the Compton effects. However, the reasoning that follows can also be applied to $\alpha$ spectra.

This pulse height distribution, or detector response function, or simply spectrum, is not but a frequency histogram of pulse heights, i.e., the number of pulses versus their heights. However, from a formal point of view, it is represented as a continuum mathematical function, $\frac{dn}{dc}[V^{-1}]$, the number of pulses per unit voltage, evaluated as a function of the pulse height $c$ [$V$], the channel number in what follows. Formally, the $c$ range extends from 0 to $\infty$.

The peaks in the spectrum appear as a response to total energy absorption processes. Thus, once the relationship between c and the energy is known, i.e., once the spectrometer is calibrated for energy (see the next chapters), the spectrum gives

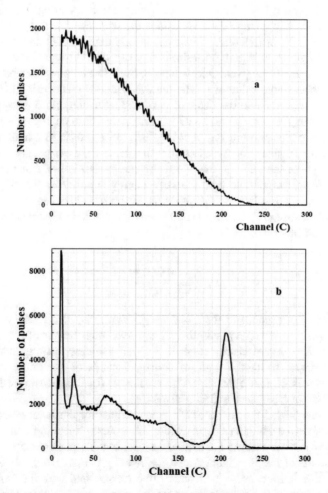

**Fig. 8.3**  Schematic representation of (**a**) pulse height distribution for continuum energy spectrum radiation and (**b**) for discrete energy spectrum radiation

a measurement of the radiation energy. When there are no discrete structures in the response function, as in the case of $\beta$ radiation, it is also possible to measure the energies by calibrating the spectrometer. However, the problem is not as readily solved as in the case of $\gamma$ or $\alpha$ radiation.


## 8.5.2   Energy Resolution

The capacity to measure energies or distinguish between close energies is given by the concept of energy resolution [5]. Indeed, as in other physical instrumentation, in which there is a resolution concept bound to the quantity the instruments are designed for, we can define an energy resolution for nuclear radiation spectrometers. It is expected that $c$ is constant for a given energy deposit. When the particle deposits all its energy, we should find the same $c$ for every event. In other words, it should happen that a Dirac delta appears in the response function; i.e., the peaks should be Dirac delta functions. However, in real experiments, the peaks have a certain width, which affects the discrimination capacity between close energies. It seems obvious that for a given energy deposit, there is a certain pulse height distribution and not a single value of $c$. The relative width of this distribution will give us an idea of how capable the detector is of distinguishing close energies. Obviously, thin peaks are convenient, so those corresponding to nearby energies do not overlap. This is illustrated in Fig. 8.4, in which the response to a total absorption energy process is given for two detectors that have different energy resolutions. Clearly, the wider the peak is, the worse the energy resolution, since many more pulse heights are compatible with the same energy deposit. To quantify this concept, we define the resolution in energies, $R$, as the ratio between the width of the peak at half-maximum in channels and the central channel of the peak or centroid channel, $c_0$. In other words:

$$R = \frac{FWHM}{c_0} \tag{8.18}$$

where $FWHM$ corresponds to '*Full Width Half-Maximum,*' that is, the width of the curve in a place that corresponds to half the maximum height of the peak. The definition of $R$ is purely instrumental. It is a dimensionless quantity and changes from one detector to another and, of course, depends on the energy. From the definition, it follows that a detector with resolution $R$ cannot distinguish energies in a range within $E \pm RE$ or within $E \pm FWHM$.

The reasons behind the variability in pulse heights for the same energy deposit are instrumental or fundamental in nature. Of course, fluctuations are expected due to electronic variations, fluctuations, noise, etc. Displacements of the electronic constants of the associated circuit for signal processing, such as gain of the amplifiers, instabilities, noise, and other effects associated with the experimental setup, are sources of fluctuations. Being of instrumental origin, they can be diminished or attenuated by using better instrumentation and laboratory conditioning at

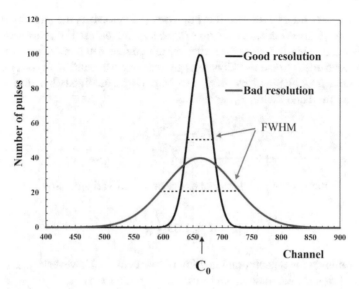

**Fig. 8.4** Energy resolution for two detectors at the same energy. The thinner peak reveals better resolution than the wider peak

temperature and humidity, for instance. However, the most important causes of peak broadening have a physical nature related to the detection process. In fact, they are a consequence of the statistical fluctuation in $Q$ released during the interaction process. In fact, the creation of $Q$ or, in general, the release of information carriers after the interaction in the detector is a statistical process. The number of information carriers for a given energy deposit fluctuates around an average value because the interaction of radiation with matter has a probabilistic nature (see Chap. 7). Assuming that the process is of a Poisson nature and that, on average, $N$ carriers are created for a given deposit of energy, the distribution of the number of information carriers will have a standard deviation of $N^{1/2}$. Should it be the only source of fluctuation, the distribution would be Gaussian, since for large $N$ the Poisson distribution tends to it, as we will see in Chap. 14 [12]. On the other hand, the relationship between the energy deposited and the height of the pulse, or, in other words, the number of information carriers and the channel associated with that event, is usually linear, almost proportional in many cases. This is a very plausible hypothesis, as we will see in the following chapters when studying the spectrometry procedures. Therefore, $c_0 = kN$, where $k$ is constant. Thus,

$$R = \frac{FWHM}{kN} = \frac{2.35}{\sqrt{N}}$$

(8.19)

where we have assumed that $FWHM = (2\ell n2)^{\frac{1}{2}}\sigma = 2.35kN^{\frac{1}{2}}$, since the distribution is of Gaussian type.

This is a very interesting result. R improves (decreases) as the amount of carrier information (charges) created per unit of energy increases: For $R$ to decrease, $N$ must be large. This is a condition that more suitable detectors have to meet for spectrometric measurements. However, in real experiments, $R$ is systematically smaller than that obtained by 8.19. U. Fano [13] quantified this difference by introducing the Fano factor, $F$, defined as:

$$F = \frac{\sigma_e}{\left(\sqrt{N}\right)^2} \qquad (8.20)$$

Where $\sigma_e$ is the experimental variance, the observed variance. So:

$$R = 2.35 \sqrt{\frac{F}{N}} \qquad (8.21)$$

$F$ depends on the detector and is difficult to calculate. However, it can be found that $F < 1$ for semiconductor and gas detectors and close to 1 for scintillation detectors.

A realistic expression of $R$ must include all the effects mentioned above. In fact, a correct formulation of $FWHM$ must contain the terms corresponding to the electronic instabilities, fluctuations of the constant characteristics of the detector, etc. These components are added quadratically, so.

$$FWHM_{tot}^2 = FWHM_{st}^2 + FWHM_{electro.}^2 + \dots \qquad (8.22)$$

Apart from the energy of the radiation emitted by a sample, the pulse height distribution can also give its activity. Indeed, it is apparent that in a given region of the distribution, within $c_1$ and $c_2$,

$$n = \frac{1}{\Delta t} \int_{c_1}^{c_2} \frac{dn}{dc} dc = \varepsilon_1 A \qquad (8.23)$$

$n$ is the total number of counts per unit of time recorded in this region, and $\Delta t$ is the counting time. Or, for the whole spectrum, or integral spectrum

$$n_0 = \frac{1}{\Delta t} \int_0^\infty \frac{dn}{dc} dc = \varepsilon_2 A \qquad (8.24)$$

$\varepsilon_1$ and $\varepsilon_2$ are counting efficiencies, the latter being $\varepsilon_{abs}$ while $\varepsilon_1$ is a partial efficiency as, for example, the photopeak efficiency.

As a last remark, when the detector gives an energy-independent response, as in the case of the Geiger–Müller detectors, see Chap. 9, the pulse height distribution has a structure not related to the deposited energy distribution. In such cases, it is better to directly use the whole integral of the spectrum, i.e., the total number of counts in the scale of the counter as the only information about the sample

(8.24). Now, it is important to know how this number changes with the only relevant parameter of the system, which is the electric field applied to the detector, or the applied voltage, $HV$. As we will see in Chap. 9, in these cases we must construct the so-called plateau of the detector, i.e., the curve $n_0$ versus $HV$ to find the optimum applied voltage for the experiment.

**Exercises**

1. Using a Geiger–Müller detector, the activity of a radioactive source emitting $\beta$ radiation is measured. In 10 min, 1000 counts are recorded. Assuming that the detector background was 10 counts in 10 min and that the counting efficiency was 40%, the activity of the sample in $Bq$. Assume that the intensity of the emitted radiation is 100%.

2. A given radionuclide ($T_{1/2} = 3.00$ d) emits $\beta^-$ ($I = 1.0\%$) and $\gamma$ ($I = 99.0\%$) radiations. A sample is counted with a GM detector, producing 3000 counts in 1 day. Calculate the activity of the sample at the beginning of the experiment assuming that the efficiency of the detector was 20.0% and that the background was measured, giving 10.000 counts in 10 d measurement time.

3. A sample containing 1 nCi of a $\beta$-emitter ($T_{1/2} = 60$ min) is measured with a Geiger–Müller detector for 15 min that produces 30.000 counts. If the background is 50 cpm, find the counting efficiency of the detector for such a $\beta$ energy. Note that the radionuclide will decay significantly during counting.

4. An experiment is carried out to measure the absolute efficiency of a Geiger–Müller detector as a function of the energy of the radiation emitted by a series of beta emitting reference samples of well-known activities. You find that the efficiency increases as the energy increases. However, from a certain energy, the efficiency reaches a constant value and no longer varies with increasing radiation energy. Explain why the absolute efficiency is energy dependent and why it does it that way. To what effect does the upper limit on the efficiency correspond?

5. The $^{253}$Cf emits, among other radiation sources, 5.979 meV $\alpha$-particles ($I = 0.2936\%$). A $^{253}$Cf sample is measured with a proportional detector, and 30,000 counts are obtained in 3 d of counting time. If the counting efficiency for that radiation is 40.0%, calculate the activity of the sample in $Bq$ at the time of starting the measurement. The detector background was 10,000 counts in 10 days.

6. In a geological survey, a sealed 250 m$^3$ air bag is found within a granitic base. This material contains some U. The $^{222}$Rn produced accumulates in the bag where a 0.5 m$^3$ sample is taken to isolate and measure the $^{222}$Rn present. After 14 h of preparation, the sample was measured for 12 h with a 20% efficiency detector. A total of 38,600 counts were recorded, with a 2 cpm background. If the mass of the granitic bases is estimated to be 500 kg, find the proportion of the U mass of the mineral.

7. Demonstrate (8.9) assuming that the carrier distribution is of the Poisson-derived Gaussian type.

8. Estimate the best energy resolution for a Si detector, providing a 1 meVelectron beam is totally stopped in the detector. Suppose a 100% carrier collection efficiency.

9. Make the same estimation for a 5 meV $\alpha$-particle beam that is totally stopped in the detector. Suppose a 100% carrier collection efficiency.

10. The Geiger–Müller dead time, $\tau$, can be described following the nonextendable model. In this case, the dead time can be found with the 'double source method,' in which two sources are measured under the same counting conditions as the Geiger–Müller detector. The method consists of measuring both samples alone and together. Source 1 and source 2 are individually measured, producing $m_1$ and $m_2$ counts per unit time on the counter, while the true counting rate should be $n_1$ and $n_2$ in the absence of dead time, respectively. Then, both sources are measured together, keeping the same conditions as they were measured individually. The result of the joint measurement is $m_{1+2}$ with the corresponding $n_{1+2}$. Assuming that $\tau$ is very small and that the background can be neglected, demonstrate that

$$\tau = \frac{m_1 + m_2 - m_{1+2}}{2m_1 m_2}$$

Note that $m_1 + m_2 > m_{1+2}$ but $n_1 + n_2 = n_{1+2}$. Hint: use (8.13).

11. Determine the dead time of a Geiger–Müller counter by using the following data based on the double source method: $m_1 = 10{,}000$ cpm, $m_2 = 9000$ cpm, and $m_{1+2} = 18{,}000$ cpm. Assume that the background is negligible.

12. The double source method is used to determine the dead time in a Geiger–Müller detector. Sample 1 gives 12,000 cpm, sample 2 gives 15,000, and sample 1 + 2 gives 26,000 cpm. Find $\tau$ and estimate the true counts for samples 1 and 2.

13. Sketch the ideal and real pulse height distribution expected for a monochromatic $\alpha$-emitter measured with a semiconductor detector.

14. Sketch the ideal and real pulse height distribution expected for a $\beta$-emitter measured also with a semiconductor detector. Suppose now that the daughter emits CE.

15. Sketch the integral pulse height distribution from the spectra of Exercises 12 and 13.

## References

1. G.F. Knoll, An excellent description of the introduction and the general concepts included in this chapter can be found, in *Radiation Detection and Measuremnt*, 4th ed. (Wiley, 2010)
2. N. Tsoulfanidis, *Measurement and Detection of Radiation*, Chapter 8, 5th edn. (CRC Press, 2021)
3. S. Hurtado, M. García-León, R. García-Tenorio, GEANT4 code for simulation of a germanium gamma-ray detector and its application to efficiency calibration. Nucl. Instrum. Meth. Phys. Res. A **518**, 764–774 (2004)
4. G.F. Knoll, *Radiation Detection and Measurement*, Chapter 4, 4th edn. (Wiley, 2010)

5. W.R. Leo, *Techniques for Nuclear and Particle Physics Experiments*, Chapter 5, 2nd edn. (Springer, Verlag, 1994)
6. P. *Theodórsson, Measurement of Weak Radioactivity*, Chapter 9, (World Sc. Pub. Co. 1996)
7. G. Heusser, Low-radioactivity background techniques. Annu. Rev. Nucl. Part. Sci. **45**, 543–590 (1995)
8. P. Povinec, Underground low-level counting, in *Low-Level Measurement of Radioactivity in the Environment: Techniques and Applications*, ed by M. García-León, R. García-Tenorio, (World Sci. Pub. Co. 1994), p. 113
9. P. Theodórsson, *Background Components of Low-level Beta- and Gamma-Detectors*, Ibid. p. 17.
10. G. Heusser, *Background in Ionising Radiation Detection-illustrated by Ge-spectrometry*, Ibid, p. 69
11. E. Armengaud et al., (60 auhtors), Performance of the EDELWEISS-III experiment for direct dark matter searches. J. Instrum. **12**, P08010 (2017)
12. N. Tsoulfanidis, *Measurement and Detection of Radiation*, Chapter 2, 5th edn. (CRC Press, 2021)
13. U. Fano, Ionisation yield of radiations. II. The fluctuations of the number of ions. Phys. Rev. **72**, 26 (1947)

# Gas Ionization Detectors

9

**ABSTRACT**

The principles of gas ionization detector (GID) operation are described in this chapter. GIDs are based on the ionization produced when charged particles pass through gases. Thus, they are especially efficient for the detection of $\alpha$- and $\beta$-particles, or electrons in general. The main components of the GID family are the Ionization Chamber, the proportional detector, and the Geiger–Müller detector, which operate under different working conditions. Applications for charged-particle counting or spectrometry are also described. The GID background components are analyzed, which helps to design detector shielding, as will be detailed in the following chapters.

## 9.1 Physics of Gas Ionization Detectors

Gas ionization detectors (GID in what follows) take advantage of the ionization produced by a charged particle when passing through a gas [1]. The collection of the released charge informs about the passage of the particle and, under certain conditions, about its energy and even its nature. A simple GID would consist of a container filled with an insulating gas, where an electric field is applied. The gas is ionized when irradiated by the charged particles, and the produced pairs of positive ions and electrons move under the applied electric field. The charge movement generates an electrical signal, which is the readable response of the detector. The filling gases are usually Ar, He, $N_2$, $H_2$, air, $O_2$, $CH_4$, or some mixtures of them. A standard mixture, very often used in conventional GID detectors, is the so-called P-10 or PR gas. It consists of 90% Ar + 10% $CH_4$. Another popular mixture is the compound of 99% He + 1% $CH_4H_{10}$ (isobutane), known as 'particle gas.' In what follows, we will see why the use of gas mixtures is better on some occasions than that of pure gases. The mixture and its proportions can be important in the case of

© The Author(s), under exclusive license to Springer Nature Switzerland AG 2022
M. García-León, *Detecting Environmental Radioactivity*, Graduate Texts
in Physics, https://doi.org/10.1007/978-3-031-09970-0_9

**Fig. 9.1** A schematic view
of the gas ionization detector

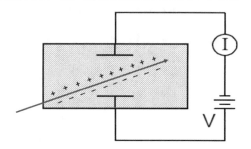

proportional and Geiger–Müller detectors since they determine the conditions of the so-called gas multiplication effect.

A schematic view of the GID design is presented in Fig. 9.1. It can be said that this scheme has basically been preserved all along the time from the first GID. Except for technical advances, it is a picture of current and primitive designs. Within the GID family, we include the Ionization Chamber, the proportional detector, and the Geiger–Müller (GM) detector. GIDs are very veteran detectors. From a historical point of view, they were, in fact, the first detectors whose measurable response was an electrical signal [2]. They were introduced into the young but intense and dynamic scientific panorama of nuclear radiation measurement in 1928 [3] when Geiger and Müller built their first GM tube and, since then, have rendered a very important service to this field of work. Today, GID has been surpassed in many Nuclear and Particle Physics applications by other types of detectors, such as scintillation or semiconductor detectors, as we will see in the next chapters. Even though this is true, they cannot be replaced in many works related to the effective counting of charged radiation. For environmental radioactivity, they are still a choice for charged-particle counting. The GID current designs do not differ from the original. It can be said that since the 1940s, its appearance and operation have not changed. However, it is worth recalling the new designs derived from the multianode proportional detector concept introduced by G. Charpak in the late 1960s [4]. These instruments are GIDs, despite the complexity of the information they provide, making them very particular. Today, they have found applications in Particle Physics and Nuclear Medicine.

## 9.1.1  Ionization in Gases

When a fast charged particle passes through a gas, it can excite or ionize, and it can also produce Bremsstrahlung, at least conceptually, since the low density and Z values of gases make this effect very unlikely. When the particle ionizes, it creates an ion–electron pair. This pair is the basis of the electrical signal that will serve to detect radiation. The ion-e pair can be created either directly or through δ rays (see Chap. 7). Regardless of the mechanism, the number of ion-e pairs formed is important. For the ion-e pair to appear, the particle must release at least the ionization energy of the electrons bound to the atoms or molecules forming the

**Table 9.1** Energy necessary to create an ion-e pair in eV for some gases commonly used as a counting gas

| Gas | Electrons | Alpha | Protons | C |
|---|---|---|---|---|
| Air | 33.8 | 35.1 | 36 | 34.7 eV |
| $H_2$ | 36.5 | 36.4 | 35.3 | |
| He | 41.3 | 42.7 | | |
| Ne | 29.7 | | | |
| Ar | 26.4 | 26.3 | 25.5 | |
| Xe | 23.6 | | | |
| $O_2$ | 34.8 | 36.4 | 33.6 | |
| $CO_2$ | 30.8 | 32.2 | 31.5 | |
| CO | 42 | 34.1 | | |
| *Its application the measurement* | | | | |
| $CH_4$ (methane) | 30 | 29.1 | | |
| $C_2H_2$ (acetylene) | 27.3 | 27.4 | | |
| $C_2H_4$ (ethylene) | 26.4 | 27.85 | | |
| $C_2H_6$ (ethane) | 24.8 | 26.4 | | |
| $C_3H_8$ (propane) | 25.7 | 26.1 | | |
| $C_4H_8$ (butylene) | 24.3 | 26.5 | | |

gas. In the gases of interest (Ar, He, $N_2$, $H_2$, air, $O_2$, $CH_4$), i.e., those commonly used as counting gases, this energy ranges from 10 to 20 eV [5]. Actually, as we saw in Chap. 7, when defining w, the particle must deposit more energy to create an ion-e pair, since part of the released energy is used to excite the gas. Thus, the energy required to create an ion-e pair is greater than the ionization energy. This is the reason why the energy required to create an ion-e pair ranges from 30 to 35 eV in the gases mentioned above, as seen in Table 9.1 [6].

## 9.1.2 Charge Transfer Reactions in Gases

To better understand the GID operation, it is important to know some relevant facts about the movement of charges in gases. In a gas, neutral molecules are in constant motion due to thermal agitation with an average free path of $10^{-6}$ to $10^{-8}$ m under normal conditions [7]. The ions created by the radiation take part in the thermal motion. Thus, they will diffuse from the densest to the least dense regions, and therefore they can collide with other molecules in their path. Some of these collisions could affect radiation detection processes. The most relevant collisions are those leading to charge transfer, which we describe in what follows.

*Electron Exchange*

This charge-exchange process occurs when a positive ion collides with a neutral molecule. The molecule loses one electron, which in turn is being captured by the positive ion. As a consequence, the former positive ion is neutralized, and the molecule is ionized. This effect is important in gas mixtures and causes the tendency to positively charge molecules with the lowest ionization potential. In principle, the charge created by the particle is preserved; thus, this effect should not affect the counting of the particle.

*Electron Capture*

Free electrons can collide with and be trapped by a neutral molecule, forming a negative ion. Such an ion behaves in the same way as positive ions but with opposite charge. This causes potential damage to the particle counting even though the total charge is the same. The reason is the lower mobility of the negative ion in the gas compared to that of the electron, which affects the collection time, $t_c$. This process is frequent for electronegative gases, as in the case of $O_2$. This does not occur for $N_2$, $H_2$, noble gases, gaseous hydrocarbons, and others.

*Recombination*

This occurs when electrons collide with positive ions. As a result, the positive ion is neutralized, and part of the charge created by the particle disappears. It also happens when the collision takes place between a positive and a negative ion. Now, part of the charge is lost as well. This is, in fact, the most likely process. In both cases, the recombination reduces the electric charge, which is a harmful effect for counting. Recombination is more important in the areas where the positive and negative charge densities, $n^+$ and $n^-$, are higher. In fact, it could be said that the recombination frequency, $\frac{dn}{dt}$, has the following form.

$$\frac{dn^+}{dt} = \frac{dn^-}{dt} = -\alpha n^+ n^- \tag{9.1}$$

that is, along the ionization track column (see Fig. 9.1).

On the other hand, when an electric field $E$ is applied to the gas, the charge will move, and this drift movement will add to the thermal diffusion movement. The drift speed, $v$, will have a different direction for electrons and positive ions. At a gas pressure $p$, it turns out that for the ions

$$v = \frac{\mu E}{p} \tag{9.2}$$

where $\mu$ is the charge mobility. This is largely constant for positive and negative ions over a wide range of $E$ and $p$, resulting in approximately $10^{-4}$ $m^2 atmV^{-1}$. For typical values of $p$ and $E$, the drift velocity for the ions turns out to be ~1 m/s. If the typical dimensions of the detector are ~cm, the transit time (collection) is ~10 ms. The mobility of the electrons is $10^3$ times higher, so the collection time is on the order of $\mu$s under these conditions.

### 9.1.3 Multiplication of Charge in Gases. Townsend Avalanche

As we will see, proportional and Geiger–Müller detectors take advantage of the phenomenon of gas multiplication. This effect causes amplification of the charge created in the detector as the charged particle passes and facilitates the analysis of the pulses provided by the detector. Gas multiplication originates from the secondary ionization produced by the electrons created in the first ionization track and depends on the intensity of the applied electric field. Due to this effect, an avalanche of ions appears, the so-called Townsend avalanche [8], which finalizes when all the released electrons are collected. In Fig. 9.2, the Townsend avalanche is represented. Under appropriate conditions, the number of ions produced in the avalanche can be kept proportional to the charge initially released so that the total pulse provided by the detector preserves the original information. This is not possible on other occasions. It all depends on the applied electric field. Gas multiplication amplifies the charge produced by the particle when passing through the gas. Thus, $v_{max}$ (see Chap. 8) will be significantly higher.

The existence of gas multiplication determines the type of GID we are dealing with. This depends on the applied electric field magnitude. The plot in Fig. 9.3 gives us the features of each of the GID operating regions, or, in other words, the differences among the detectors belonging to this family. We plot $v_{max}$ as a function of the applied voltage (HV), the electric field, to a general GID (see Fig. 9.1).

For lower HVs, recombination effects are important, and thus, the collection of the initial charge is not complete. Recombination starts to disappear as the applied voltage increases. When all the charge created in the primary ionization

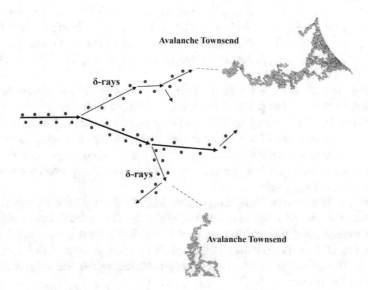

**Fig. 9.2** Ion avalanche during the gas multiplication process

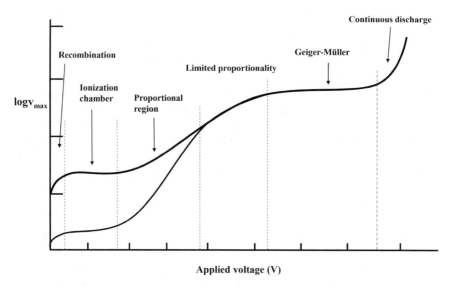

**Fig. 9.3**  Operating regions of a GID

track is collected, $v_{max}$ remains constant at a maximum value, and the GID works as an Ionization Chamber. This is the simplest GID detector and is based on the collection of the charge originally released by the particle that travels along the detector. This situation remains for a range of HV values until gas multiplication appears.

Gas multiplication tends to increase the value of $v_{max}$ since now the total collected charge increases because of secondary, or even further orders, ionization tracks, as seen in Fig. 9.2. Within a given HV range, the multiplication is linear: The amount of collected charge is proportional to that originally generated by the particle. Thus, $v_{max}$ increases linearly. In this region, the GID works as a proportional detector.

As the applied voltage continues increasing, nonlinear effects appear. The responsible is the cloud of positive ions produced, which moves very slowly toward the cathode. The presence of an ion cloud may locally decrease the electric field, affecting the production of ion–electron pairs. This region of limited proportionality is not useful for counting. Here, the resulting pulse no longer preserves the original information, although particles of different natures or energy could still produce different $v_{max}$ values.

When the HV is very high, the positive ion cloud can become large enough to prevent the charge avalanche from continuing. The created pulses all have the same height, and the information that the particle originally deposited in the detector is lost. It is the so-called Geiger–Müller region. A more detailed description of the Townsend avalanche in the Geiger–Müller region will be given in the corresponding section.

## 9.2    Ionization Chamber

The Ionization Chamber is the simplest detector within the GID family, although the information it provides on the radiation beam is far from being simple [9]. In fact, its response is spectrometric; i.e., the Ionization Chamber gives information on the energy deposited by the particle. Its operating scheme is shown in Fig. 9.1. Charged radiation ionizes the gas, and released charge is collected by the electrodes that are connected to a voltage source. The total charge collected is that directly released by the particle in the ionization track and is a measure of the deposited energy. The response from an Ionization Chamber provides information about the energy deposited by the particle in the detector. The movement of charge in the presence of an electric field is an electric current, $I$, which can be measured by an ammeter. In Fig. 9.4a, we plot the variation of $I$ as a function of the applied HV for an Ionization Chamber irradiated by a sample of activity A. The intensity increases as the HV increases until a plateau, a flat zone, is reached. This corresponds to the situation in which the maximum possible intensity current is obtained, i.e., when all the released charge is collected. For low voltages, the electric field is not high enough to avoid recombination effects, and the current intensity is lower than expected. Additionally, for that reason, as the voltage increases, the probability of recombination decreases, and the intensity increases. Of course, the magnitude of the current intensity depends on the activity of the sample since $I$ is in fact a measure of the average charge released during a given time (see Chap. 8). From this viewpoint, it is expected that the greater the activity is, the greater the intensity, as shown in Fig. 9.4a. From another viewpoint, since the charge amount released by the particles depends on the energy deposited, it is also possible to use $I$ as a measure of the energy of the particles. Ionization Chambers can be used as charged-particle spectrometers. They can also work in pulse mode (see Chap. 8). Now, the experimental information is provided by a group of pulses with heights $v_{max} = \frac{n_p e}{C}$, where $n_p$ is the number of ion–electron pairs, e is the electron charge, and $C$ is the capacitance associated with the RC circuit (see Chap. 8). The chamber response is now a distribution of $v_{max}$, each representing the energy deposited by the particle. In this operation mode, the plateau curve is presented in Fig. 9.4b. Here, we plot the total number of pulses (counts) per unit of time as a function of the applied voltage. The number of counts increases at the beginning for low voltages, since due to recombination, not all $n_p$ is collected, and the formed $v_{max}$ is not high enough to overcome the usual electronic threshold imposed in these experimental setups to avoid the counting of electronic noise. As the voltage increases, $v_{max}$ increases as the collection of $n_p$ increases. Finally, when all $n_p$ is recovered, we reach a plateau where all possible pulses are collected. On the other hand, as $v_{max}$ depends on the energy deposited by the particle, it is possible to construct the pulse spectrum, which informs the energy of the particles.

   Finally, it is interesting to note that typical $v_{max}$ values are low [7] since pulse formation depends only on the primary ionization track. Of course, some electronic

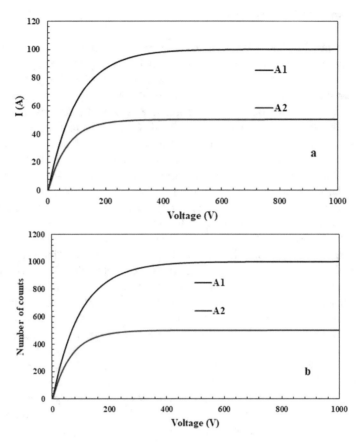

**Fig. 9.4** Response of an Ionization Chamber to irradiation of samples of activities $A_1$ and $A_2$; **a** variation of the intensity measured in the ionization chamber as a function of the applied voltage and **b** number of pulses per unit time from the ionization chamber

amplification steps are used to improve the interpretability of the response, but in some cases, it is better to use the gas amplification effect.

## 9.3   Proportional Counters

A proportional detector takes advantage of the gas multiplication effect. As said before, for gas multiplication to occur, high electric fields are needed [1]. In a simple parallel plate geometry, as shown in Fig. 9.1, very high voltages are necessary to obtain high electric fields. In contrast, with electrode cylindrical geometries, high fields can be obtained with relatively low applied voltages. For that reason, proportional and Geiger–Müller detectors are usually constructed in cylindrical geometries, as presented in Fig. 9.8. The cathode is the cylindrical container, whereas the anode is a wire situated along the symmetry axis of the cylinder. The electric field in this case depends on r, the distance to the anode, in the following way.

$$E(r) = \frac{V}{rln\left(\frac{b}{a}\right)} \tag{9.3}$$

where $V$ is the applied voltage and b and a are the cathode and anode radii, respectively. Clearly, $E$ is large when $r$ is small, so gas multiplication occurs essentially around the anode. As we shall see, this has important implications for charge collection, especially in the Geiger–Müller operating region. With this geometry, very high field values are obtained with relatively low voltage values. At least in comparison to the parallel geometry where the electric field is just $\frac{V}{d}$, d is the distance between plates. It is very easy to see that to obtain the same electric field with parallel plate geometry, it is necessary either to apply very high voltages or construct detectors with extremely short d. Both cases are not practical, and cylindrical geometry is preferred.

After gas multiplication, the total charge collected in a proportional detector, $Q$, is

$$Q = Mq \tag{9.4}$$

$q$ is the charge released in the primary ionization track, and $M$ is the so-called gas amplification or multiplication factor. $M$ depends on the different parameters conditioning the mobility of charge carriers in the gas and, of course, on the electric field or the applied voltage. In addition, for that, indirectly on the geometry of the detector. There are some empirical formulas for $M$ in the case of cylindrical geometry, and one is the following [10].

$$lnM = \frac{V}{ln\left(\frac{b}{a}\right)} \frac{ln2}{\Delta V} \left( ln\frac{V}{p\,a\,ln\left(\frac{b}{a}\right)} - lnK \right) \tag{9.5}$$

where $p$ is the gas pressure, while $K$ and $\Delta V$ are constant for each gas. $V$ is the applied voltage, of which $M$ depends exponentially. A consequence of (9.5) is the need for the applied voltage to be very stable throughout the experiment to have a well-determined value of $M$.

Since $M$ is constant, the information that a proportional detector provides is spectrometric, as in the case of Ionization Chambers. In fact, $Q$ will depend on $q$, that is, on the energy deposited by the particle. The advantage is that the response is physically amplified compared to that provided by an Ionization Chamber and, in principle, is readily usable. Typically, proportional detectors operate in pulse mode, so the plateau curve of Fig. 9.4 applies with a similar interpretation. Indeed, now $v_{max} = \frac{Mn_p e}{C}$. The interpretation of the curve shape is similar to that for Ionization Chambers, but not identical. Of course, for low voltages, the recombination effects are important, but they can now be compensated for by $M$, which depends on the applied voltage. This causes the shape of the plateau to change depending on the

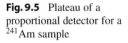

**Fig. 9.5** Plateau of a proportional detector for a $^{241}$Am sample

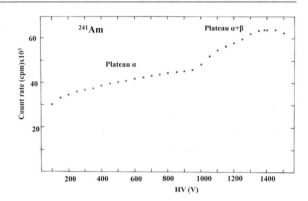

energy of the counted particles. Even the nature of the measured radiation. As an example, we can see the plateau curve obtained for a $^{241}$Am sample in Fig. 9.5. $^{241}$Am decays by several $\alpha$ emissions with different intensities feeding several excited states of its daughter, $^{237}$Np [11]. Therefore, $\alpha$ particles are observed from a $^{241}$Am sample, but soft $\gamma$-rays, X-rays, conversion electrons, and Auger electrons are also observed. A double plateau shape can be observed in Fig. 9.5, which is not easy to interpret. However, since the specific ionization for $\alpha$ particles is systematically higher than for electrons, it is expected that they do not need high $M$ to be counted, that is, to produce $v_{max}$ with heights above the threshold. For that, the first plateau, for low voltages (and hence low $M$), corresponds to the counting of $\alpha$ particles. However, it is necessary to increase $M$ to obtain pulses high enough to be counted in the case of electrons. We observe a second plateau for higher voltages, corresponding to the acceptance of the pulses produced by electrons. The first plateau is the so-called $\alpha$ plateau, and the second plateau is the $\alpha + \beta$ plateau since in this area, $\alpha$ particles are still counted as well. In this way, a proportional detector can distinguish the physical nature of the particles they count [12].

Similarly, the shapes of the plateaus for $\beta$ emitters can be explained. Figure 9.6 shows the plateaus obtained for some $\beta$ emitters with the same proportional detector as before [12]. The $v_{max}$ values depend on $E_{max}$ (the maximum energy of the $\beta$ radiation). Thus, different $M$ values will be needed to see the pulses counted for the different radionuclides. The lower the energy is, the lower the $n_p e$, and consequently, a higher $M$ will be needed to count the detected radiation. In this specific case, the shape of the plateaus was used to distinguish $\beta$ energies in a simple counting experiment.

Real $\alpha$- and $\beta$-spectrometry can be performed with proportional detectors as their response is spectrometric. Since the most widely used charged-particle spectrometers are semiconductor detectors, we will give some more details on the interpretation of $\alpha$- and $\beta$-spectra in Chap. 11.

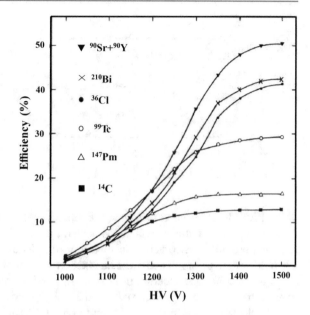

**Fig. 9.6** Plateaus of a proportional detector for several $\beta$ emitters. Figure reprinted from [12] in Chap. 14, with permission of Elsevier

## 9.4 Geiger–Müller Counters

As in the case of proportional detectors, Geiger–Müller detectors (GM in what follows) are based on gas multiplication effects. Thus, GM detectors are typically cylindrical in shape. However, to operate in the GM region, higher electric fields are needed, and the Townsend avalanche has special features [13].

Indeed, as the charged particle passes through the gas, ionization and excitation are produced. The deexcitation of the molecules usually leads to the emission of electromagnetic radiation, generally within the ultraviolet range of energies (see Fig. 9.7). The mean free path of the photons in the gas is typically high, so it can travel relatively long distances within the detector. The photon can undergo a photoelectric effect with some gas molecules and mostly with the detector housing. This phenomenon occurs independently of the GID operating region but becomes relevant within the GM region. Indeed, because the electric field is very high, the photoelectrons are sufficiently accelerated to gain enough energy and produce an ionization track. In this way, the avalanche propagates through the whole detector, and since the geometry is cylindrical, it takes place essentially along the anode. This situation is very specific for GM detectors. The avalanche stops when the electric field decreases. This occurs when the cloud of positive ions, which moves slowly in the gas, is large enough to screen the electric field. Under these conditions, it is easily understood that the response of a GM detector, the electric pulse, loses memory of the primary ionization event and thus does not keep information on the energy deposited by the particle. The pulse height is hence constant for all the detection events, since the dominant effect to stop the avalanche is the amount

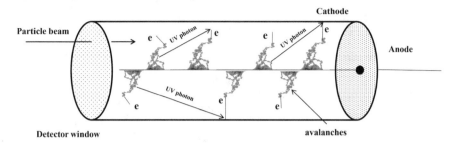

**Fig. 9.7**  Townsend avalanche in a GM detector

of positive charge needed to screen the electric field, and this is always approximately the same. In this way, the GM detector is useful only for counting particles. It cannot provide spectrometric information. Additionally, because of the high electric field, a new phenomenon must be taken into account when working with a GM detector. Despite the reduction of the electric field, the positive ions continue their way to the cathode. Once it arrives, the collision with the metallic walls may release electrons that, under a recovered electric field, could generate additional avalanches [9]. This would make it impossible to stop the avalanche in the detector. Initially, an external quenching technique was used to stop the avalanche by simply decreasing the applied voltage once detection occurred. Today, an internal quenching technique is used, which consists of adding a quenching agent to the filling counting gas. The agent is preferentially a halogen compound that transfers one electron to the positive ions when they collide on their way to the cathode. The positively charged halogen molecule arrives at the cathode. There, it is neutralized with high probability by capturing one electron. This process ended the avalanche.

An important drawback of GM detectors is their characteristic high dead time derived from the slow motion of the positive ions inside the gas. Because the electric field recovers high values only when the positive ions reach the cathode, the GM detector cannot detect particles during the time it takes them to arrive. Typically, this means dead times of some $100~\mu s$. This feature is not especially important when dealing with low radioactivity samples, although it should be considered when counting conventional samples with a GM detector.

As said before, GM does not give spectrometric information. In fact, the expected response function is presented in Fig. 9.8, where only a peak can be observed. This is the result of the single pulse height obtained independently of the energy deposited by the particle. Of course, the peak has a certain width derived from the natural statistical fluctuations of the energy deposits in matter. Moreover, a low pulse height tail is observed related to charge collection inhomogeneities, failures in the quenching effects, and others. This shape conditions that of the plateau of a GM detector, as shown in Fig. 9.9. A small slope in the plateau is observed to be related to such effects. In fact, the pulses progressively overcome the threshold as the applied voltage increases. Hence, a nonzero slope is observed. The information provided by GM detectors is poor but sufficient in many cases.

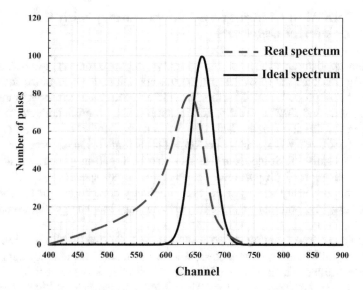

**Fig. 9.8** Pulse height spectrum in a GM detector

When the experimental need is just counting, they have great advantages compared to the other members of the family due to their simplicity.

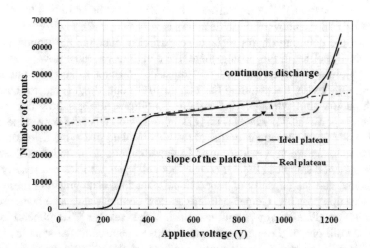

**Fig. 9.9** Plateau of a GM detector

## 9.5    Radiation Counting and Spectrometry with Gas Ionization Detectors

The main application of GID is the counting and spectrometry of charged particles. The reason is the Physics behind the operation of GID. Since Ionization Chambers and proportional detectors provide a spectrometric response, they can be used for $\alpha$- and $\beta$-spectrometry, and in some cases, proportional detectors have been used for CE and Auger electron spectrometry. Currently, however, charged-particle spectrometry is normally performed with semiconductor detectors, as we will see in Chap. 11. Some examples can also be found in the current literature on the application of proportional detectors for X-ray spectrometry. Of course, changes in gas density and pressure can sufficiently compensate for the low $Z$ of gases to meet the problem of detecting low- or very low-energy electromagnetic radiation.

Charged-particle counting experiments can be performed with the three members of the GID family; nevertheless, this is the main (and unique) application of Geiger–Müller detectors. Since they do not provide a spectrometric response, counting is the only possibility for Geiger–Müller in the field of charged-particle detection. In general, $\alpha$ and $\beta$ counting is carried out with Geiger–Müller, although if it is necessary to distinguish $\alpha$ from $\beta$ particles in the same sample and simultaneously, proportional detectors are the choice. In this case, the spectrometric response can be used to differentiate both particles.

An important problem when performing charged-particle counting or spectrometry measurements with GID is the absorption of energy in the detector window. This is a serious handicap for heavy or very low-energy charged particles. Gas-flow counters, essentially of proportional type, appear to be an option to solve this problem [9]. The window of these detectors is very thin, some few $\mu gcm^{-2}$, which allows for the counting of, for instance, $\alpha$ particles since the absorption is minimized. Nevertheless, in this arrangement, the counting gas must flow permanently during counting since the gas pressure must be constant to keep $M$ unchanged (see 9.6). Otherwise, the counting gas could diffuse outside the detector container through such a thin window. Under these conditions, the gas flow becomes an important experimental parameter. It is easily understood that the flow value conditions that of $M$ and consequently that of the electric field inside the detector. These gas-flow counters are very widely used in the simultaneous $\alpha$ and $\beta$ radioactivity in environmental samples [12]. Of course, both charged particles can readily cross the window, and the capacity of proportional detectors distinguishing these particles makes this possible. Geiger–Müller gas-flow counters have also been used for low- to intermediate-energy $\beta$ emitters. Such is the case for $^{99}Tc$, which has been measured with one of these experimental setups in environmental samples [14].

Additionally, to avoid the absorption effects by the detector window, the so-called internal counting technique or wall-less counters have been developed [12]. We will see them in Chap. 15 in more detail; now we can give some basic information. In principle, this technique is useful when the radioactive sample can be prepared in gaseous form at ambient temperature. This allows it to mix with the

**Fig. 9.10** Multiwire gas proportional detector scheme

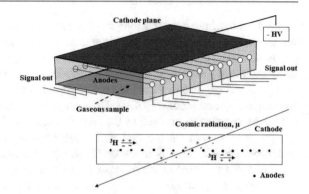

detector filling gas. Under these conditions, the absorption effects in the counting efficiency (see Chap. 8) can be overcome, and the geometry factor, the solid angle, can reach a 100% value, at least formally. This greatly helps in the counting and spectrometry of very low-energy radiation or heavy or very heavy charged parti-cles. When the chemical nature of the sample does not allow its preparation as a gas at ambient temperature, the alternative is to electroplate it on the inner walls of the detector, which after this process is filled with the counting gas [15].

Multielement (or multiwire) proportional detectors [4] is a design that takes advantage of wall-less counters. As shown in Fig. 9.10, instead of a counter with a single anode, the system is a compound of several anode–cathode arrangements all surrounded by a so-called ring counter, and the whole system is housed in a common carcase. The different single systems are connected in anticoincidence (see Chap. 15) among them. In this way, a selection of pulses can be done as a function of the range of the particles, which is a manner to select the counting of low-energy $\beta$ particles as those emitted by $^3$H. In fact, this concept was developed to allow mainly the counting of nonenriched $^3$H samples (see Chap. 13) [16]. On the other hand, all components are connected anticoincidence with the ring detector, which greatly reduces the background of the detector. We will return to this concept in Chap. 15 when studying Low-Level Counting Techniques.

GIDs are very widely used for the determination of radionuclides in the environ-ment. Together with liquid scintillation detectors (see Chap. 10), they are currently the main options for charge-particle counting measurements. The simplicity of the design of basic proportional or Geiger–Müller detectors makes possible the con-struction of multidetector systems, in which several detectors work at the same time. The advantage of this possibility is the increase in the experiment through-put. In contrast, the GID capacity for spectrometry is currently surpassed by semiconductor detectors.

## 9.6 Background in Gas Ionization Detectors

GIDs are mostly sensitive to charged radiation. Thus, the background component associated with the environmental $\gamma$ field, $B_\gamma$ (see 8.9), will have less relative importance than the contributions from $\alpha$ and $\beta$ radiation in the detector materials and the $\mu$ component of cosmic radiation, $B_{det}$, $B_{sh}$, and $B_\mu$. The nucleonic component $B_{nuc}$ could also be of interest considering the possibility of producing $\alpha$- or $\beta$-emitting radionuclides close to the detector. Usually, the filling gases are very pure and are presumed not to contain traces of natural radioactivity, including cosmogenic radionuclides. However, contamination during the filling process cannot be discarded. The housing of the GID detectors is mostly built with Cu, which is very clean except for the fact that $n$ can produce activation with a relatively high cross section. This is a source of background that can be relevant unless a shielding of $n$ is added to the detector (see Chap. 15).

The main contribution should come from $B_\mu$ and indirectly from $B_{cos}$, considering the secondary processes that $\mu$ can generate during its travel in the proximity of the detectors.

For the Rn component, $B_{Rn}$, the influence, as we saw in Chap. 8, could be important, provided the ubiquity of Rn is due to its gaseous nature. In this case, the relevant radiation is $\alpha$ and $\beta$ emitted by $^{222}$Rn, $^{220}$Rn, and their daughters; the $\gamma$ radiation is now not as important. As we saw in Chap. 8, it is convenient to remove the Rn from the detector surroundings by sealing the counter assembly with a plastic or steel container to prevent the contact of fresh air with the counter. As the different shielding techniques are implemented, the relative importance of each background component can change. For instance, in underground laboratories, once the cosmic-ray background is greatly reduced, the component associated with impurities in materials becomes the more relevant contribution [17] to the background in the GID. We will return to that matter in Chap. 15.

### Exercises

1. Use 9.2 to estimate the electron transit time within a gas detector of parallel plate geometry. The applied voltage is 300 V, and the plate distance is 5 cm. Suppose that the gas is at atmospheric pressure.
2. Suppose that there is a gas multiplication factor of $10^4$ in a gas proportional counter. What will be the maximum pulse height obtained when detecting a 5 meV $\alpha$-particle. Take the capacitance of the counting system as 100 pF and assume that 30 eV is needed to create an ion-e pair.
3. Make the same calculation for an Ionization Chamber. Compare the pulse heights obtained in both cases and explain the differences.
4. Let us suppose a cylindrical Geiger–Müller counter with an anode wire radius of 25 $\mu$m and an outer radius of 25 mm. If we apply 1000 V between the cathode and anode, estimate the distance from the center of the counter at which the electrons gain enough energy along 1 mm to ionize the He filling gas. Take 23 eV as the energy needed to create an ion–electron pair.
5. Suppose a parallel plate Geiger–Müller detector filled with the same gas as in the previous problem, where $d = 2.5$ cm. Estimate the voltage necessary to obtain the same electric field as that found in the anode surface in the previous cylindrical detector.

6. It is accepted that the consumption of 95% of the quenching gas in a Geiger–Müller is the life limit of the counter. Suppose that the counting gas of a 100 cm$^3$ volume GM detector is a compound of 90% Ar and 10% ethyl alcohol. If $10^9$ alcohol molecules are dissociated per Geiger discharge, estimate the lifetime of the counter as the maximum number of counts. (Note: prior to the use of halogens as quenching gases, alcohol molecules were the option for the first internal quenching GM counters).

7. Suppose that you measure the efficiency of a GM detector for a source that contains a certain activity of a $\beta$-emitter, $E_{max} = 1500$ keV. 20% is found. Explain why the efficiency value is <100%.

8. Suppose that you perform another efficiency measurement with a source of identical geometry and in the same counting geometry, but that contains an emitter $\beta$ of $E_{max} = 500$ keV. Will you get smaller or higher efficiency? Why?

9. Comment on the differences or similarities between proportional and Geiger–Müller detectors regarding:
   - Response function to $\alpha$- and $\beta$-radiation
   - Dead-time losses.
   - Ability to measure high-activity samples

10. Sketch the ideal and real pulse height distribution obtained with a proportional counter for
    - a $\alpha$-radiation source, emitting two $\alpha$ particles at 4 and 7 MeV.
    - a $\beta$-radiation source emitting $\beta$-radiation with $E_{max} = 1$ MeV.
    - Make the sketch assuming that the detector size and gas pressure are enough to stop both $\alpha$ and $\beta$ radiation.

11. Assume now that the detector dimensions are smaller than the $\alpha$-particle ranges. Sketch the spectra. Assume that the gas pressure cannot compensate for the detector size.

12. Sketch the spectra obtained for the previous radiation sources with a Geiger–Müller detector. Comment on the expected results for both cases considered above. Compare the differences found with respect to the proportional detector.

13. Sketch the response function of a proportional counter to the radiation emitted from a $^{55}$Fe source. Use https://www.nndc.bnl.gov/nudat2/ to find the $^{55}$Fe decay scheme.

14. Make the same study for the radiation emitted from a $^{137}$Cs source.

15. If we use a windowless gas proportional detector (internal gas counting technique) for $^{55}$Fe radiation spectrometry, should we expect differences in the spectra found compared to the experiment in Exercise 13?

16. A sealed $^{60}$Co source is measured with a proportional counter. Guess the spectrum obtained.

17. Discuss the detector features that could be optimized to efficiently detect high energy (>2 MeV) $\gamma$-radiation with a proportional counter.

18. In gas-flow detectors, the counting gas continuously flows through the detector. This technique is very useful when the detector window is very thin to

allow the counting of $\alpha$-particles. The gas cannot be confined under these conditions but will be lost through the detector window by osmosis. In the case where we want to perform a spectrometric experiment with a proportional detector, we discuss the critical counting variables based on 9.5.

19. Let us have a radioactive sample that contains two radiation emitters: one alpha and the other beta. We want to measure the alpha and beta activity of the sample, but with one detector. Suppose that we have two detectors in the laboratory: a GM and a proportional detector. Choose the appropriate detector for the measurement of both activities. Explain the selection.

20. Suppose that in the plateau area of the selected detector, counting efficiencies of 30% for $\alpha$ and 45% for $\beta$ are found. If the $\alpha$ and $\beta$ activities are 6 pCi and 30 pCi, respectively, what count number would you obtain when measuring $\alpha$ and $\beta$ activity, both during 30 min? Why?

# References

1. W.R. Leo, *Techniques for Nuclear and Particle Physics Experiments: A How-to Approach*, 2nd edn, Chapter 6. (Springer, 1994)
2. E. Rutherford, H. Geiger, An electrical method of counting the number of α particles from radioactive substances. Proc. Royal Soc. **81**, 141–161 (1908)
3. H. Geiger, W. Müller, Das Elektronenzählrohr. Physikalische Zeitschrift **29**, 839–841 (1928)
4. G. Charpak, R. Bouclier, T. Bressani, J. Favier, Č Zupančič, The use of multiwire proportional counters to select and localize charged particles. Nuc. Instrum. Method. **2**, 262–268 (1968)
5. D.W. Anderson, *Absorption of Ionizing Radiation*, Chapter 11. (University Park Press, 1984)
6. W. Binks, Energy per ion pair. Acta Radiologica **41**, 85–104 (1954); T.D. Strickler, 'Average energy loss per ion pair produced by alpha particels in binary gas mixtures', ORNL-3080. JC-34 Phys. (1961)
7. G.F. Knoll, *Radiation Detection and Measurements*, Chapter 5, 4th edn. Wiley (2010)
8. D. Xiao, Fundamental Theory of Townsend Discharge, in *Gas Discharge and Gas Insulation. Energy and Environment Research in China*, vol. 6. (Springer)
9. N. Tsoulfanidis, S. Landsberger, *Measurement and Detection of Radiation*, Chapter 5, 5th edn. (CRC Press, 2021)
10. G.F. Knoll, *Radiation Detection and Measurements*, Chapter 6, 4th edn. (Wiley, 2010)
11. https://www.nndc.bnl.gov/nudat3/
12. M. García-León, Ph. D. Thesis, University of Sevilla, 1983
13. G.F. Knoll, *Radiation Detection and Measurements*, Chapter 7, 4th edn. (Wiley, 2010)
14. M. García, G. Manjón, C.I. Sánchez-Angulo, $^{99}$Tc/$^{137}$Cs activity ratios in rainwater samples collected in the South of Spain. J. Environ. Radioact. **20**, 49–61 (1993)
15. L.D. Labeyrie, H.D. Livingston, A.G. Gordon, Measurement of 55Fe from nuclear fallout in marine sediments and seawater. Nucl. Instrum. Method. **128**, 575–580 (1975)
16. P. Povinec, Low-level gas counting, in *Low-Level Measurements of Man-Made Radionuclides in the Environment*, ed. by M. García-León, G. Madurga (World Scientific Pub. Cop. 1991), p. 38
17. P. Povinec, Underground low-level counting, in *Low-Level Measurements of Radioactivity in the Environment: Techniques and Applications*, ed. by. M. García-León, R. García-Tenorio (World Scientific Pub. Cop. 1994), p. 113

# Scintillation Detectors

10

**ABSTRACT**

The operating principles of scintillation detectors are described in this chapter. They take advantage of the excitation produced when charged particles pass through some materials. Scintillation detectors are, therefore, efficient for α- and β-particles, or electrons in general, as well as for γ-radiation, since high-Z scintillation detectors are available. Scintillation materials can be organic or inorganic in nature. Their response to the passage of radiation and range of applications are different. The application of NaI(Tl) to the measurement of γ-emitters is described in detail, which serves to introduce the principles of γ-spectrometry. Liquid scintillation counting is a technique widely used to determine environmental radioactivity, and special attention is given to it. The fundamental background components of scintillation detectors are also studied, which will guide the design of detector shielding for low-level counting or spectrometry applications, as we will see in the following chapters.

## 10.1 Physics of Scintillation Detectors

Scintillation detectors take advantage of the excitation that charged particles produce when passing through the matter. The subsequent deexcitation gives rise to the emission of light and scintillation, which provides information on the particle passage and the deposited energy. Scintillation detectors were introduced close to the beginning of radioactivity studies. In fact, they can be dated back to 1903 when *W*. Croke built the first scintillation ZnS detector [1]. They played an important role in the famous Rutherford experiment, which supports his planetary model of the atom [2]. It is interesting to note that by that time, it was possible to observe the scintillations only with the naked eye or a microscope. Interestingly, throughout the history of nuclear radiation detection, they gradually gave way to Gas Ionization Detectors (GID), since no further advances were made in the technique and

© The Author(s), under exclusive license to Springer Nature Switzerland AG 2022    299
M. García-León, *Detecting Environmental Radioactivity*, Graduate Texts
in Physics, https://doi.org/10.1007/978-3-031-09970-0_10

research of new scintillating materials. Scintillation detectors recover an important place in the field when photomultipliers were coupled with ZnS detectors [3], which facilitates the analysis of scintillations and, years later, when $\gamma$ spectrometry became possible with NaI(Tl) [4]. From that very moment, the development of scintillation detectors was very fast, with many new scintillating materials capable of meeting virtually all the problems in radiation detection.

Basically, a scintillation detector consists of a scintillating material coupled to a photomultiplier tube that converts the light intensity into electron intensity and consequently into an electric pulse. We will describe all of these elements in what follows. Regardless, the first condition for a material to be a good scintillation detector is the high probability of an excitation process occurring when a charged particle passes through. On the other hand, as we shall see, the photomultiplier tube is basically a photoelectric cell plus one electron multiplier. It is an alternative to the optical microscope used in the early days to observe the light emitted under the passage of radiation. The analysis of the generated current or voltage pulse informs about the passage of the radiation itself and its characteristics. A general scheme of scintillation detectors is given in Fig. 10.1.

A detailed study of the scintillation processes is far from the objectives of this text. It is a typical matter of an Atomic or Molecular Physics book. However, it is convenient to know some details to understand the operating principles of scintillation detectors.

The energy deposited by the particle in a scintillator is used to excite the material through different channels. All of them are characterized by an exponential deexcitation time process with a given time constant. Only the shortest time-constant emission is of interest for scintillation detectors [5]. This is the so-called prompt fluorescence radiation emission. In a good scintillator material, this must be the prominent process; i.e., most of the energy deposited by the particle should

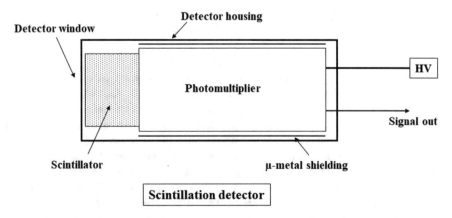

**Fig. 10.1**   A general scheme of a scintillation detector. As seen in the text and the following figures, the photomultiplier has to be polarized with a voltage source to make possible the electron multiplication process

be used to excite this specific process. Together with prompt fluorescence, other deexcitation mechanisms coexist in the material. They are the phosphorescence and delayed fluorescence radiation emission. In both cases, the time constant is longer, although the light is emitted at a different wavelength or the same wavelength. They are not convenient for counting.

It is important that the scintillating material is transparent to the light it emits [6]. This is related to the energy levels involved in the excitation-deexcitation process leading to scintillation. If the energy of the scintillation radiation coincides with the gap between levels of the molecule or of the crystal lattice, the probability of reabsorption will increase, and the signal could be lost. Doping of materials or crystals helps to mitigate this effect. The scintillation efficiency is also a parameter to take into account when selecting a good candidate as a scintillation detector [7]. Not all the energy deposited in the scintillation detector is used to generate light. There are several effects that dissipate part of the energy deposited in many cases in the form of heat. The scintillation efficiency is defined as the fraction of the deposited energy that is converted into scintillation light.

There are many scintillating materials in gas, liquid, or solid forms. Nevertheless, they are usually classified according to the physical process of light emission. In this way, we can discuss organic and inorganic scintillators. Gases deserve special mention. In organic scintillators, fluorescent emission is a property of the molecule, that is, the consequence of an excitation-deexcitation process of the molecule itself [7]. Thus, regardless of its aggregation state, the organic scintillator maintains its scintillating properties. In contrast, for inorganic scintillators, the fluorescence emission is associated with the crystalline structure in which it participates; that is, the light emission is the result of a deexcitation of the crystalline lattice [6]. Let us give some details.

## 10.1.1 Organic Scintillators

The fluorescence decay for organic scintillators is very fast, ~$10^{-9}$ s [8]. They consequently have a high time resolution, which makes them very suitable for many nuclear experiments. However, organic scintillators are generally transparent to their own fluorescent radiation. They are basically compounds of C, H, and O, and hence they have low effective Z. For that, organic scintillators are good electron detectors, since backscattering is unlikely. However, for the same reason, they are not effective for photons. Naphthalene, anthracene, and stilbene [9] were the most used organic scintillators at the beginning and today still form part of many scintillation detectors. Nevertheless, the most common organic scintillation detectors are today classified as plastics and liquid scintillators. Plastics are very versatile, as they can adopt different counting geometries. They basically consist of the dilution of an organic scintillator into a material that can be polymerized. Examples are polymethylmethacrylate and polyvinyltoluene [6]. They are especially useful for $\beta$ and electron spectrometry in general. The main features of the organic scintillators are given in Table 10.1. In general terms, their response to the deposited energy is linear, and thus, the intensity of scintillation light serves to determine the energy. The scintillation efficiency can vary and depends greatly on the purity of the material, since there are many different processes that could compete in the energy absorption and conversion into light. On the other hand, since different molecular energy levels can be populated by the energy deposited by the charged particle, two fluorescence components can be found in the scintillation process: prompt and slow fluorescence. The probability of excitation of a given molecular energy level has been found to depend on the nature of the charged particle [8]. In this way, a time analysis of the electric pulse, pulse shape analysis (PSA), serves to distinguish the type of particles being detected. This property is often boosted by mixing different scintillators and has the possibility of distinguishing particles [5]. In the case of liquid scintillators, as we will see, this provides a good advantage for low-level radioactivity counting [10].

Liquid scintillation is today the main contribution of scintillation detectors to the determination of radionuclides in the environment. At the beginning, its main application was oriented to the measurement of low $\beta$ energy emitters such as $^3H$ or $^{14}C$ due to the possibility of diluting the sample into the detector itself, which provided a high counting efficiency [10]. A liquid scintillator is built by diluting an organic scintillator into a solvent. In some cases, a third component is needed to shift the fluorescent radiation wavelengths toward those accepted by conventional photomultipliers. As we shall see later, liquid scintillation counting finds application in many problems related to the determination of radionuclides in the environment [10].

### 10.1.2 Inorganic Scintillators

As said before, in this case, fluorescent radiation is associated with an excitation-deexcitation process in the crystal lattice [5]. According to electronic band theory, electrons in solids fill two energy bands: the valence and conduction bands. Both are separated by a forbidden band in which the electrons cannot appear, since no energy states available exist there (see Fig. 10.2). The electrons in the valence band are bound to the structure of the crystal and cannot migrate in the presence of an electric field, whereas those in the conduction band are free and can move when an electric field is applied. When a charged particle passes through the scintillator crystal, it deposits some energy-producing electron–hole pairs; i.e., it provides the electrons in the valence band with enough energy to overcome the forbidden band. These pairs will migrate along the crystal with a given kinetic energy. The electrons that encounter holes along their way can subsequently be trapped, and fluorescent radiation is emitted in this process. The radiation energy corresponds to the gap between both bands.

This light informs about the radiation and the energy deposited. This is the basic operation of inorganic scintillators. In many cases, the energy gap between bands is too high, producing radiation wavelengths that do not match the sensitivity of conventional photomultipliers. This is a common situation that can be solved by

**Table 10.1**  Characteristics of some more common organic and inorganic scintillators

| Name | Compound | Wavelength of maximum emission(nm) | Decay time (ns) | Density $(gcm^{-3})$ | Index of Refraction | Light yield (photons/keV) |
|---|---|---|---|---|---|---|
| **Organic** | | | | | | |
| *Monocrystals* | | | | | | |
| Naphthalene | Naphthalene | 348 | 11 | 1.15 | 1.58 | |
| Anthracene | Anthracene | 448 | 30 | 1.25 | 1.59 | 0.5* |
| Trans-stilbene | Trans-stilbene | 384 | 4.5 | 1.16 | 1.58 | |
| p-terphenyl | p-terphenyl | 391 | 12-Jun | 1.23 | – | |
| *Plastics* | | | | | | |
| NE-102A | | 425 | 2.5 | 1.032 | 1.58 | |
| NE-104 | | 405 | 1.8 | 1.032 | 1.58 | |
| NE-110 | | 437 | 3.3 | 1.032 | 1.58 | |
| NE-111 | | 370 | 1.7 | 1.032 | 1.58 | |
| BC-400 | | 423 | 2.4 | 1.032 | 1.581 | |
| BC-404 | | 408 | 1.8 | 1.032 | 1.58 | |
| BC-408 | | 425 | 2.1 | 1.032 | 1.58 | |
| BC-412 | | 434 | 3.3 | 1.032 | 1.58 | |
| BC-414 | | 392 | 1.8 | 1.032 | 1.58 | |
| BC416 | | 434 | 4 | 1.032 | 1.58 | |
| BC-418 | | 391 | 1.4 | 1.032 | 1.58 | |
| BC-420 | | 391 | 1.5 | 1.032 | 1.58 | |
| BC-422 | | 370 | 1.6 | 1.032 | 1.58 | |
| BC-422Q | | 370 | 0.7 | 1.032 | 1.58 | |
| BC-428 | | 480 | 12.5 | 1.032 | 1.58 | |
| BC-430 | | 580 | 16.8 | 1.032 | 1.58 | |
| BC-434 | | 425 | 2.2 | 1.049 | 1.58 | |
| **Inorganic** | | | | | | |
| NaI(Tl) | NaI(Tl) | 415 | 250 | 3.67 | 1.85 | 40 |
| CsI(Tl) | CsI(Tl) | 540 | 1000 | 4.51 | 1.8 | 65 |
| CsI(Na) | CsI(Na) | 420 | 1000 | 4.51 | 1.84 | |
| CaI(Na) | CaI(Na) | 420 | 630 | 4.51 | 1.84 | 39 |
| KI(Tl) | KI(Tl) | 410 | 240/2500 | 3.13 | 1.71 | |
| $^6$LiI(Eu) | $^6$LiI(Eu) | 470 | 1400 | 4.06 | 1.96 | 14 |
| $CaF_2(Eu)$ | $CaF_2(Eu)$ | 435 | 900 | 3.19 | 1.44 | 24 |
| $BaF_2$ | $BaF_2$ | 310 | 630 | 4.88 | 1.49 | 2–6.5 |
| BGO | $Bi_4Ge_3O_{12}$ | 480 | 300 | 7.13 | 2.15 | 8.2 |

(continued)

**Table 10.1** (continued)

| Name | Compound | Wavelength of maximum emission(nm) | Decay time (ns) | Density $(gcm^{-3})$ | Index of Refraction | Light yield (photons/keV) |
|------|----------|-----------------------------------|-----------------|---------------------|---------------------|---------------------------|
| $CaWO_4$ | $CaWO_4$ | 430 | 2000 | 6.12 | 1.92 | |
| $ZnWO_4$ | $ZnWO_4$ | 480 | 5000 | 7.87 | 2.2 | |
| $CdWO_4$ | $CdWO_4$ | 470 | 5000 | 7.9 | 2.3 | 15 |
| $PbWO_4$ | $PbWO_4$ | 440 | 10 | 8.28 | 1.82 | 0.1 |
| CsF | CsF | 390 | 5 | 4.65 | 1.48 | |
| CsI | CsI | 305,450 | Several $\mu s$ | 4.51 | 1.8 | 2 |
| $CeF_3$ | $CeF_3$ | 310,340 | 10 | 6.16 | 1.68 | 4.4 |
| ZnS(Ag) | ZnS(Ag) | 450 | 200 | 4.09 | 2.35 | 90 |
| ZnO(Ga) | ZnO(Ga) | 385 | 0.4 | 5.61 | 2.02 | |
| GSO | $Gd_2O_2S$:Tb | 440 | 60 | 6.71 | 1.9 | 9 |
| YAG | $Y_3Al_5O_{12}$:Ce | 550 | 300 | 4.55 | 1.82 | 17 |
| YAP | $YAlO_3$:Ce | 370 | 27 | 5.37 | 1.95 | 18 |
| YAGG | $(Y, Ga)_3Al_5O_{12}$:Ce | 550 | 550 | 5.28 | – | |
| YAGGd | $(Y,Gd)_3Al_5O_{12}$:Ce | 570 | 106 | 4.69 | – | |
| LSO | $Lu_2SiO_5$:Ce | 420 | 40 | 7.4 | 1.82 | 25 |
| LSN | $La_3Si_6N_{11}$:Ce | 535 | 100 | – | – | |
| LYSN | $(La, Y)_3Si_6N_{11}$:Ce | 550 | 100 | – | – | |
| LuAG | $Lu_3Al_5O_{12}$:Ce | 540 | 71 | 6.7 | – | |
| LuAP | $LuAlO_3$:Ce | 365 | 17 | 8.54 | 1.94 | 17 |

[*] The light output is 50% of that of NaI(Tl). Anthracene provides the maximum light yield among the organic compounds

**Fig. 10.2** Electronic band structure of crystals and activator energy levels

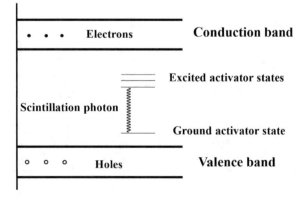

doping the solid with activators. Dopants introduce energy levels and trap energy levels within the forbidden band that are available to electrons. These activators can also be ionized as the charged particles pass along the crystal. Thus, the electrons can be trapped by a hole at one of these activator levels, which deexcites into the ground state by emitting photons of lower energy or larger wavelengths. A first consequence of the use of activators is the transparency of the crystal to the light emitted. For pure crystals, the probability of reabsorption of the emitting light during scintillation is high, since the amount of energy needed to create an electron-hole pair is close to that of the emitted photons. This is not the case when activators are used. There are, however, nonactivated inorganic scintillators with features that make them very suitable for radiation detection even in comparison to organic scintillators. This can be seen in Table 10.1, where a complete account of some of the most common inorganic scintillators is presented [11]. In general terms, considering the different processes, the energy needed to create an electron-hole pair is higher than that of the forbidden gap, three times higher in general. This means approximately 20 eV in many cases [11], which limits the energy resolution of these devices, according to the results presented in Chap. 8.

The fluorescence decay time is typically longer than that in organic scintillators. Therefore, they are somewhat slower. However, some modern materials have timing features that are competitive with organic scintillators (see Table 10.1) [12]. In contrast, high-Z crystals can be obtained, which makes them especially useful for $\gamma$-radiation. As in the case of organic materials, it is possible for some inorganic scintillators to perform a pulse shape analysis to differentiate among the counted particles, and this is because the scintillation efficiency depends on their mass and charge [6].

Among the many possible inorganic scintillators in Table 10.1, NaI(Tl) is the most remarkable. In fact, despite being one of the oldest members of the family [4], it is still the most used today in gamma spectrometry. This is mainly due to its high Z, good light yield, energy linearity, and acceptable timing features. In addition, it can be constructed in different sizes and geometries. However, it is highly hygroscopic, which is its main drawback from a practical point of view. For that, it is necessary to adequately seal the NaI(Tl) crystals to avoid water absorption, which will affect the light transmission within the crystal.

### 10.1.3 Gas Scintillators

The use of gases as scintillation detectors has received attention as an alternative to gas ionization detectors [13]. In particular, the scintillation properties of noble gases such as Xe, Ar, and He have been investigated and applied in several problems related to particle physics. Compared to inorganic scintillators, light production is much simpler because it is just an atomic excitation-deexcitation process. However, they produce ultraviolet light, so they need photomultipliers with enough sensitivity to these wavelength ranges. Alternatively, we can use wavelength shifters to match the sensitivity of conventional photomultipliers. A mixture

**Table 10.2**  Characteristics of common gas scintillators

| Element | Form | Wavelength of maximum emission(nm) | Decay time (ns) | Density (gcm$^{-3}$) | Index of Refraction | Light yield (photons/keV) |
|---------|--------|------------------|------------------|---------|---------|------|
| Xe | Gas | 325 | | | | 0.8 |
| Kr | Gas | 318 | | | | 0.4 |
| Ar | Gas | 250 | | | | 0.2 |
| He | Gas | 390 | | | | 0.2 |
| Xe | Liquid | 120–170 | 5(fast)–860(slow) | 1.4 | 1.29 | 40 |
| Kr | Liquid | 120–170 | Feb-85 | 2.41 | 1.4 | |
| Ar | Liquid | 120–170 | 22-Mar | 3.06 | 1.6 | |

with other gases, such as $N$, is enough to do this. In addition, there exist several competing processes that make the scintillation efficiency clearly lower than in the case of inorganic scintillators. However, they are very fast, even more so than organic scintillators, with response times of a few or fractions of nanoseconds. Some features of the most commonly used scintillation gases are given in Table 10.2.

## 10.1.4 Photomultipliers

Photomultipliers convert the intensity of scintillation light into an electronic current [7]. This provides an electric signal that can be analyzed. A basic photomultiplier diagram can be seen in Fig. 10.3, while some realistic geometries of commercially available photomultipliers can be found in [14]. They consist of two main parts: the photocathode and the dynodes. The photocathode is essentially a photoelectric cell, while each of the dynodes is connected to the previous one with an increasing positive voltage.

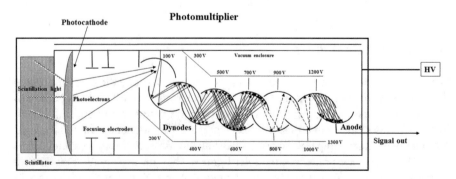

**Fig. 10.3**  Basic scheme of a photomultiplier

In the photocathode, the scintillation light is absorbed by the photoelectric effect, and a certain number of photoelectrons are released. The primary photoelectron current is introduced into the dynode chamber, where it is amplified. Indeed, photoelectrons are attracted by the first dynode, where they collide, releasing a certain number of electrons. These electrons are at the same time attracted by the subsequent dynode from which some more electrons are released after the collision. This process is repeated depending on the number of dynodes. The total current is collected in the final anode, which provides an electric signal that can be analyzed. Due to successive collisions, the primary photoelectron current is multiplied by a factor ranging from $10^7$ to $10^{10}$. Normally, the charge amplification in the photomultiplier is linear. Thus, the total charge collected in the final anode contains information on the number of electrons produced in the photocathode and ultimately on the light released in the scintillator. This means that the response of the photomultiplier preserves the information on the energy deposited by the particle in the scintillator. Nonlinearity effects can appear for very high current pulses. In this case, the effect of space charge can affect the electron trajectories, breaking the linear relationship between the light produced in the scintillator and the final electron current collected in the anode. However, it can be said that under normal operation, the linearity is preserved. In addition, the temporal information is also preserved since the response of the photomultiplication process is fast and only delayed by a few nanoseconds with regard to light production in the scintillator [6].

Of course, the whole setup must be kept under vacuum conditions; otherwise, the electron trajectories will be affected. The light spectrum sensitivity of photomultipliers is very wide. An example can be seen in Fig. 10.4, where the sensitivity of some commercial photomultipliers is presented and compared to the emission spectrum of some scintillators [11]. Photomultipliers with appropriate sensitivity to ultraviolet, visible, or near-infrared light can be found on the market [14]. In fact, they are certainly a very versatile instrument that finds application in many fields, apart from what we are studying in this chapter.

An alternative to the conventional dynode geometry presented in Fig. 10.3 is that called a continuous channel type. This geometry is very versatile and permits, in some cases, ad hoc experimental setups. A very widely used photomultiplication instrument is the microchannel plate (Fig. 10.5) [11]. Perhaps the main drawback of these instruments is the possible appearance of positive ions that travel in the opposite direction to that of electrons. They can produce secondary electrons when they collide with the walls of the channel. This is more likely as the electron current increases. The timing response of microchannel plates is even faster than that of conventional photomultipliers.

A major disadvantage of photomultipliers is the thermionic effect by which photoelectrons can be spontaneously released from the photocathode and give rise to dark currents that cause typical electronic noise of photomultipliers [6]. This can be partly solved by lowering the temperature of the photomultiplier. For this, it is sufficient to keep the laboratory temperature constant. In some cases, especially for older photomultipliers, they must be surrounded by an $LN_2$ stream. It

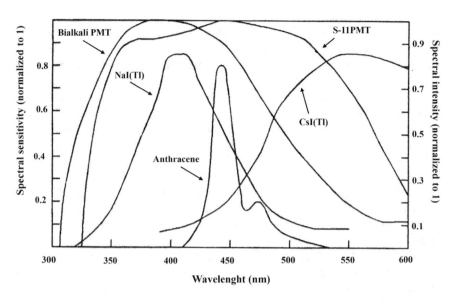

**Fig. 10.4**   Spectrum sensitivity of photomultipliers

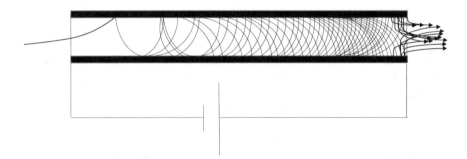

**Fig. 10.5**   Continuous channel photomultipliers

is also recommended to select the photomultiplier depending on the experiment to be carried out. As can be understood, in the case one is interested in the measurement of low-energy events (low photoelectron currents), the relevancy of the thermionic effect is high. Since the relative importance of this effect depends on the area of the photocathode and the material it is constructed, careful planning of the experiment should be performed. From this viewpoint, the bialkali photocathodes are very much advisable and, in fact, used because of the relatively low dark currents they produce by the thermionic effect. In addition, the existence of dark currents in the absence of an applied voltage is considered. Indeed, it is possible to observe a dark current increase after exposure of the photomultipliers to room light, especially to fluorescent tubes. For that, it is highly recommended to store the photomultiplier under dark conditions. However, the photomultiplier must also operate under dark conditions, as ambient light can generate an electron avalanche in the dynodes, with catastrophic consequences for the device. The reason is the large electron current caused by the tremendous photon flux over the photomultiplier from the ambient light. Dark currents from natural radioactivity contamination of photomultiplier components have also been documented and are very important in Low-Level Counting [15]. The presence of $^{40}$K in the photocathode or the glass envelope is very frequent and is the main reason for such an effect. Obviously, it should be avoided by selecting the photomultiplier construction material.

External magnetic fields affect the electron trajectories in the photomultipliers. This breaks the linearity of the photomultiplier response. Even the Earth's magnetic field could affect them, since the electron kinetic energies are very low (typically approximately 100 eV). To avoid that, photomultipliers are shielded against magnetic fields with a high magnetic susceptibility covering the metal layer, such as $\mu$-metal, an Fe alloy [6]. Related to that is the recent use of semiconductor photodiodes as an alternative to conventional photomultipliers for special applications [11]. These devices are less sensitive to magnetic fields. They are based on the production of electron-hole pairs in the photodiode by scintillation light. It can be said that the number of pairs per absorbed photon is higher than the number of photoelectrons in a conventional photocathode. Therefore, the so-called quantum efficiency is higher, and consequently, it is expected that the energy resolution is better (see Chap. 8). In conventional environmental radionuclide measurements, this alternative is not commonly used. For that, we will not discuss such a device in this text.

## 10.2 Counting and Spectrometry with Scintillation Detectors

The wide variety of scintillation detectors makes it possible to address most of the radiation counting and spectrometry problems. We will give a brief description of the different applications in what follows, since there is ample literature on this topic. In separate sections, we will pay more attention to $\gamma$ spectrometry and liquid scintillation counting. The first topic has historically been the main contribution of scintillation detectors to the general field of radiation counting and spectrometry. Today, the second topic is the main application of scintillation detectors to the field of environmental radioactivity.

*$\beta$ and CE Counting and Spectrometry*
$\beta$ and CE counting and spectrometry are possible with organic scintillators, since the low-Z value of scintillators makes electron backscattering and Bremsstrahlung unlikely. In addition, plastic scintillators are virtually transparent for $\gamma$-radiation. The use of liquid scintillators for $\beta$ radiation counting will be discussed later. Now, it is interesting to highlight the important contribution that plastic scintillators have made to the field of $\beta$ radiation spectrometry [16]. In fact, before the

introduction of semiconductor detectors (see Chap. 11) as charged-particle spectrometers, plastic scintillators were used for the determination of many of the $\beta$
spectra, allowing the development and testing of the first theories of $\beta$ radiation.
Their acceptable energy resolution, on the other hand, allowed one to carry out a
good determination of CE spectra as well [17].

*α and Other Heavy-charged Particle Counting and Spectrometry*
Scintillation detectors cannot successfully solve the problem of $\alpha$ and heavier
charged-particle spectrometry. However, they can comfortably address the problem of $\alpha$ radiation counting or even of heavier particles with SZn(Ag) scintillators
[18]. The features of this interesting inorganic scintillator can be found in Table
10.1. SZn(Ag) has a high scintillation efficiency. In fact, its scintillation yield is
higher than that of some other scintillators, including NaI(Tl). The effective atomic
number is high, which would make it useful for $\gamma$ spectrometry. However, it is not
transparent to its own light. Therefore, it is not realistic to use large volumes, which
would be necessary for efficient $\gamma$ detection. It is produced as thin polycrystalline
films, which is sufficient for short-range charged-particle counting. New SZn(Ag)
detector designs can give counting efficiencies of up to 90% [19].

*Special Applications*
The time response of scintillators makes them very useful for special applications
[20]. Being fast detectors, organic scintillators can be used to study the time intervals between nuclear events. Its application to the measurement of $T_{1/2}$ or $\tau$ of
nuclear levels, typically ~ns, is very well known [21]. The technique used is that
of delayed coincidences, whose experimental setup is presented in Fig. 10.6. The
same setup serves for general radiation coincidence experiments and thus to study
genetic relationships among radiations and angular correlations between emitted
radiations from nuclei. As is known, this kind of experiment allows one to obtain
fundamental information on the nucleus decay and energy level scheme, including
level quantum numbers [22]. The configuration of Fig. 10.6 requires fast time-
response detectors since very short times are involved in the studied transitions.
Thus, plastic scintillators are very useful in these experiments. Of course, when
necessary, one of the detectors can be replaced by an inorganic scintillator to detect
$\gamma$ radiation. In this way, particle–particle coincidence or particle-$\gamma$ coincidence
experiments are possible [23].

This setup has also been used in Low-Level Counting (see Chaps. 14 and 15)
because it permits one to select the radiation of interest, lowering the background
of the experiment. However, the counting efficiency is greatly reduced.

As we previously observed, the time constant of the fast fluorescence is different
for each scintillator (see Table 10.1). This is used to discriminate particles that can
penetrate more or less into a detector. An example is the 'phoswich' configuration
formed by a combination of a NaI(Tl) detector ($\tau = 0.23$ μs) and a CsI ($\tau =$
1 μs) detector. In Fig. 10.7, we have an example of such a configuration. Fewer
penetrating particles will be detected by NaI(Tl), giving pulses of shorter duration
than the more penetrating ones that will reach CsI, giving longer duration pulses.

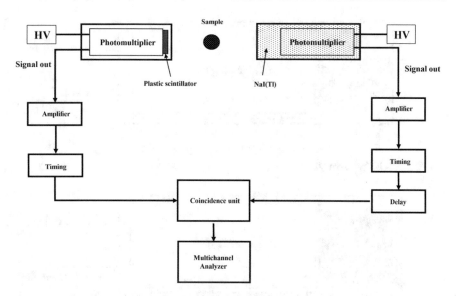

**Fig. 10.6** Experimental setup for coincidence experiments. The use of plastic or inorganic scintillation detectors allows us to carry out a variety of experiments

A more complex design is also presented in Fig. 10.7, in which a combination of scintillators can simultaneously measure $\alpha$, $\beta$, and $\gamma$ radiations in the sample by analyzing the pulse shape [24].

Even the fluorescence time constant depends on the type of particle that produces the scintillation. This is used to discriminate between particles by studying the shape of the pulse with a technique known as pulse shape analysis which is very much used in liquid scintillation counters [25].

## 10.3  Gamma-Ray Spectrometry with Scintillation Detectors

Scintillation detectors make $\gamma$ spectrometry available to many laboratories. The discovery of the scintillation properties of NaI doped with Tl [4] was crucial for that. Today, they have been replaced by semiconductor detectors (see Chap. 11) due to their better energy resolution. However, the experience gained along the history of $\gamma$ spectrometry comes mostly from the use of scintillation detectors, mainly NaI(Tl), and for that, we will present in this section the principles of $\gamma$ spectrometry. Of course, these principles can be applied to semiconductor $\gamma$ spectrometry with some differences that will be explained in the corresponding section of Chap. 11.

The suitability of NaI(Tl) for $\gamma$ spectrometry and that of many other inorganic scintillators (see Table 10.1) derives from its high Z. This makes the photoelectric effect and pair production, when possible, a very likely process in NaI(Tl).

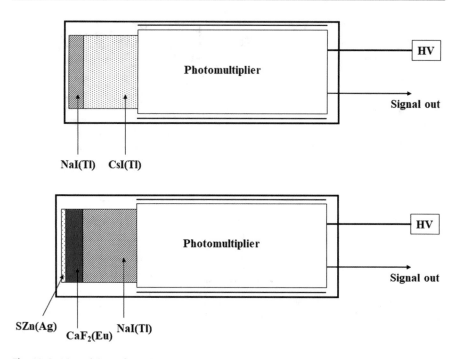

**Fig. 10.7** Phoswich configurations

However, $\gamma$ radiation is not directly responsible for the scintillation in the crystal. They are the electrons released by photoelectric, Compton, or pair-production effects that produce the emission of light as a consequence of its interaction with the scintillator. The intensity of the scintillation depends on the energy deposited by the electrons and indirectly by the $\gamma$ radiation. A scheme of this process is given in Fig. 10.8. The electric pulse provided by the photomultiplier preserves the information on the energy deposited by the $\gamma$ photons since the multiplication factor is constant for a given voltage applied to the dynodes. This is the basis of $\gamma$ spectrometry. Inasmuch as this is true, the pulse height frequency histogram (see Chap. 8) will be a picture of the different $\gamma$ interaction processes taking place at the crystal. Therefore, part of the histogram will represent the total absorption of $\gamma$ energy, which will allow its measurement.

In Fig. 10.9, we give a general scheme of a $\gamma$ spectrometer based on NaI(Tl). Actually, this is a very general scheme since it contains all the elements necessary to produce, collect, and analyze the electric pulse and to construct the pulse height spectrum. We will see that this scheme can be applied for $\alpha$ and $\beta$ spectrometry, the only difference being the detector. A high-voltage source serves to polarize the photomultiplier in this case. The preamplifier converts the electric pulse from the photomultiplier into a voltage pulse and slightly amplifies it. The amplifier shapes and amplifies the pulse. Normally, the selected shape is a Gaussian-type pulse,

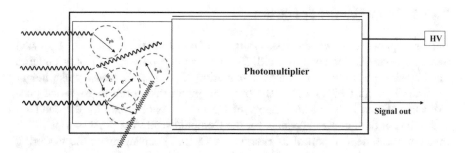

**Fig. 10.8**   Gamma spectrometry with scintillation detectors. Formation of the pulse

which is readily analyzed by the following multichannel analyzer that classifies the pulses according to their heights.

For conventional $\gamma$ spectrometry applications, NaI(Tl) detectors have a cylindrical geometry. The dimensions of the cylinder characterize the detector. Therefore, when we talk of a $2 \times 1$ NaI (Tl) detector, we mean that the diameter of the circle is 2 in. and the height of the cylinder is 1 in. (see Fig. 10.10). Usually, the $3' \times 3'$ NaI(Tl) is taken as the reference detector when evaluating the features of other detectors. We will see that the size of the detector is very important for the analysis of the realistic response of NaI(Tl) spectrometers. Today, inches are substituted for cm (1 in. = 2.54 cm) when describing the dimensions of the NaI(Tl) cylinder.

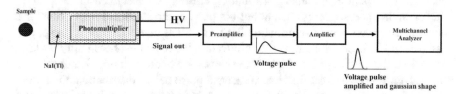
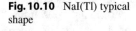

**Fig. 10.9**   Basic scheme of a NaI(Tl)-based $\gamma$-spectrometer

**Fig. 10.10**   NaI(Tl) typical shape

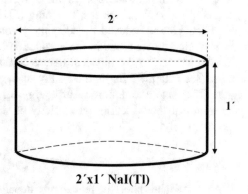

## 10.3.1 Pulse Height Spectrum

Let us suppose that we are measuring a monochromatic $\gamma$ emitter, $E_\gamma$, well below the pair-production threshold. In that case, the photons can only undergo photoelectric absorption or Compton scattering. Then, the expected pulse height spectrum is that of Fig. 10.11a. The peak, or photopeak, corresponds to the photoelectric effects in which the energy of the photon is absorbed by the detector, while the continuum area accounts for the Compton scattering processes in which the photon can deposit part of its energy (see Chap. 7) depending on the scattering angle. However, it is important to recall that we are actually observing the energy released by the electrons produced by the $\gamma$ interaction in the detector. For that, there will be no pulses between the peak, the photoelectric peak, and the Compton area since the maximum energy of the Compton electrons is $\frac{4E_\gamma^2}{1+4E_\gamma}$, which corresponds to backscattered photons, is $<E_\gamma$. On the other hand, the probability distribution of Compton effects in the crystal is responsible for the Compton area shape (see Chap. 7). Thus, pulse grouping is observed at high energy, since for large angles scattering the Compton electrons have similar energies. This is the Compton edge, as described in Chap. 7.

Figure 10.11a represents the ideal response of a NaI(Tl) spectrometer to the monochromatic $\gamma$ emitter. The real response is presented in Fig. 10.11b. The general shape of the spectrum remains, but the peak wides, the Compton edge is smoothed, and the empty area between the Compton edge and the peak is filled with some pulses. The smoothing of the Compton edge and the widening of the peak are the result of the energy resolution effects, as we studied in Chap. 8. In this case, this is essentially related, but not only, to the statistical fluctuations that occur in the production of scintillation light, as well as photoelectrons in the photocathode and to fluctuations in the multiplication factor in the photomultiplier [26]. Therefore, the voltage pulse height at the exit of the photomultiplier is not constant for a given energy deposit. Instead, we have a pulse height distribution. Of course, electronic instabilities, gain shifts, and refraction index gradients in the scintillator also have a certain influence on the widening of the peaks. Changes in the multiplication factor in photomultipliers due to thermal or magnetic effects, dark currents, etc., could also influence the widening [5]. However, while all these instrumental effects can be reduced by adequately controlling the experimental conditions, the statistical fluctuations in the production and collection of charge cannot, since they are related to the physical nature of these processes. Peak widening, as we know (see Chap. 8), limits the ability of the detector to distinguish close $\gamma$ energies. Of course, the definition of energy resolution of (8.16) applies here, and if the origin of the widening was strictly of statistical type, it is expected that $R$ is approximately

$$R \propto E_\gamma^{-\frac{1}{2}} \tag{10.1}$$

As we will see later, the real energy dependence of $R$ is more complex, although similar to 10.1. This is a very logical result, which does not always apply for

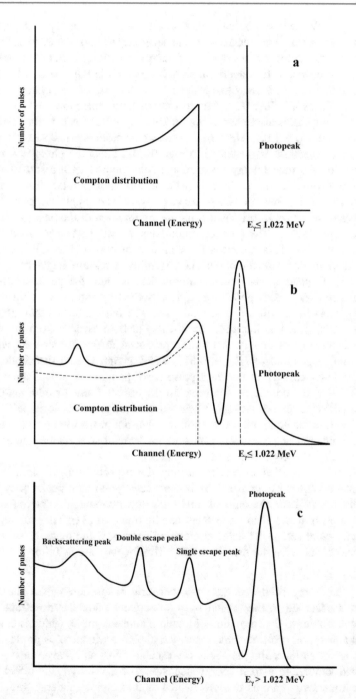

**Fig. 10.11** **a** Ideal $\gamma$ spectrum; **b** real $\gamma$ spectrum obtained with a NaI(Tl) spectrometer for a monochromatic $\gamma$ emitter with $E_\gamma \leq 1.022$ MeV; and **c** real spectrum obtained for $E_\gamma > 1.022$ MeV

all detectors. We will see below that $R$ can be used to identify real peaks in a complex $\gamma$ spectrum. This means that it is necessary to find a more refined energy dependence of $R$. A reference value of the energy resolution is that found for a 3' × 3' NaI(Tl) at 662 keV. It is approximately 7%, which is the best $R$ value for this energy and that size. It is clear that $R$ depends on the size of the crystal for a given energy, insofar as it influences the uniformity of light collection.

The detector size is also responsible for some specific effects in the response to $\gamma$ photons with $E_\gamma > 1.022$ MeV. In this case, pair production is possible, and the expected real spectrum is presented in Fig. 10.11c. Now, the photopeak includes the effects of the total energy absorption of the photon as a result of the pair production. From this viewpoint, it is better to talk about a total absorption peak. However, in addition, two satellite peaks can be observed: single escape and double escape peaks. Such peaks originate from the annihilation of the positron produced during pair production. The positron annihilates in matter when it encounters one electron in its way. This produces the emission of two 0.511 meV energy photons emitted in opposite directions, i.e., forming a relative angle of 180° (see Fig. 10.12). Depending on the size of the detector and the position where the positron annihilates, both photons can be absorbed by the detector through, for instance, photoelectric effects. The total pulse will fall within the total absorption peak. However, it is possible that one of them or both escapes from the detector without depositing its energy. All pulses produced in the former case will simulate a total deposition of $(E_\gamma - 0.511)$ MeV, giving rise to the single escape peak. The latter case gives rise to a pulse corresponding to a deposition of $(E_\gamma - 1.022)$ MeV, the double escape peak. In the case that one or both annihilation photons will suffer one or multiple Compton scattering, the resulting pulse will be included within the continuum Compton area but not in the total peak. Obviously, the appearance of escape peaks depends on the detector size. As the size increases, the probability of escape effects must decrease.

There are some other effects modifying the expected pulse height spectrum that must be known. Apart from the backscattering and sum peaks, they are not especially relevant for the case of environmental radionuclide determination. A full and more in-depth account of them can be found in [27]. They all depend on the experimental setup, the detector size and geometry, the material surrounding the detectors, and even the sample activity. Briefly, they are as follows:

*X-ray Escape Peaks*

Usually, the X-ray produced after photoelectric absorption within the NaI(Tl) detector is absorbed in the crystal by a subsequent photoelectric effect. In this way, it contributes to the formation of a total photopeak pulse, which is counted in the total absorption peak. In some cases, the photoelectric effect is produced close to the detector surface, and the X-ray can escape. Thus, an X-ray escape peak can be formed simulating a total energy deposition of $E_\gamma - E_x$. This effect is relevant for low-energy gamma photons, which tend to rapidly undergo photoelectric effects, as well as for small detectors.

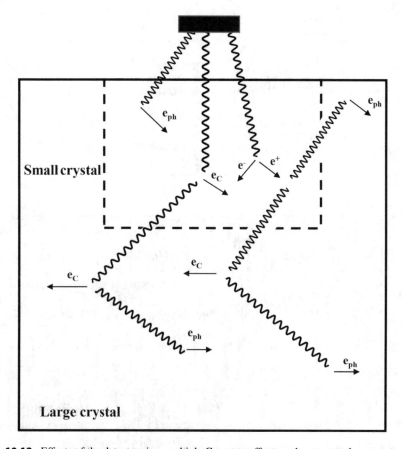

**Fig. 10.12**  Effects of the detector size: multiple Compton effects and escape peaks

*Electron Escape*
Especially for high energy $\gamma$, the electrons released by the photoelectric effect, Compton or pair production, can escape from the crystal without completely depositing its energy. These events produced pulses that will be located within the Compton zone and, in fact, tend to deform it. This effect is relevant for high-energy photons and small detectors.

*Bremsstrahlung Escape*
The electrons released by gamma radiation can dissipate energy in the crystal in the form of Bremsstrahlung. Part of this energy can escape the crystal depending on its size and the energy of the $\gamma$ radiation. The result is once a deformation of the Compton area. This effect is relevant for high-energy photons.

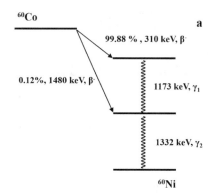

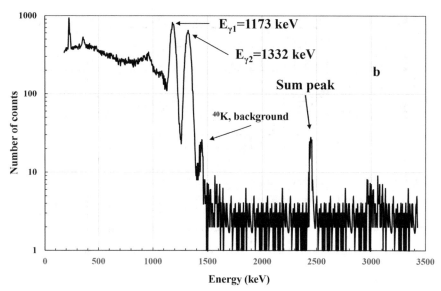

**Fig. 10.13** **a** $^{60}$Co decay scheme and **b** $^{60}$Co spectrum obtained with a NaI(Tl) detector. The sum peak corresponds to the simultaneous counting of the 1.173 and 1.332 meV photons emitted in the cascade are observed

### Annihilation Radiation

When studying $\beta^+$ emitters or $\gamma$ emitters with $E_\gamma > 1.022$ MeV, it is possible to observe the so-called annihilation peak in the pulse height spectrum, corresponding to 0.511 MeV or less likely to 1.022 MeV. Both correspond to the annihilation of $\beta^+$ emitted by the sample or produced by the high-energy $\gamma$ photons in the surroundings. The annihilation could occur in the sample itself or in the surrounding materials.

*Bremsstrahlung*

When studying $\beta$ emitters, it is possible to observe the Bremsstrahlung radiation produced by the electrons either in the sample or in the surrounding materials. This effect may be important for high-energy beta emitters and in the case of high activities.

*X-ray Peaks*

Photons from the sample are emitted in all directions. Many of them impinge on the materials surrounding the detector and undergo a photoelectric effect. The subsequent emitted X-rays can reach the detector and be detected. Peaks corresponding to these X-rays can appear in the spectrum. They come not from the sample but from the surrounding materials. These peaks could be important when counting low-activity samples. Logically, they modify the response in the low-energy zone of the spectrum. For that, it is very convenient to know in depth the experimental setup by taking a precise account of the materials surrounding the detectors, especially the shielding, as we will see in Chap. 15. Regarding this problem, an important effect is the X-ray fluorescence that occurs in the Pb shielding of Low-Level Spectrometers (see Chap. 15) caused by cosmic muons (see Chap. 8).

*Backscattering Peak*

Some photons emitted from the sample are not directed to the detector but rather to the surrounding material. There, they can undergo a $\theta > 90^0$ Compton scattering coming back to the detector where they can be detected. The resulting pulse is located within the Compton area, but since the energy of the backscattered photons is very similar (see Chap. 7), a peak-shaped pulse grouping appears, called the backscattering peak. In many cases, this false peak modifies the Compton area in such a way that makes it difficult to analyze real peaks. In addition, in some cases, they can be taken as real photopeaks. Thus, a careful analysis of the spectrum must be performed.

*Sum Peaks*

This effect is associated with the decay scheme of the sample. When photons are emitted in a cascade, the probability of coincidence peaks appears as a consequence of simultaneous photoelectric effects, i.e., effects occurring within the resolving time of the detector. A very popular example is that presented in Fig. 10.13, where we see the decay scheme of $^{60}$Co, with two photons emitted in cascade. The spectrum collected with NaI(Tl) contains the individual photopeaks and a sum peak simulating a photon of $(1.173 + 1.332)$ MeV energy. Sum peaks are a serious source of errors when analyzing $\gamma$ spectra. First, because they can be taken as real peaks and second, because part of the original pulses that might be counted in the corresponding photopeak is included in this false peak. We will return to this problem in Chap. 13 when describing $\gamma$ spectrometry efficiency calibration.

## 10.3.2 Identification of Radionuclides and Activity Calculation

As described in Chap. 8, the pulse height spectrum contains much information about the measured sample. From the point of view of the measurement of radioactivity in the environment, the most important information is the identification of the radionuclides in the sample and the quantification of its activity. However, some other parameters and quantities can be derived from a $\gamma$ spectrum, such as nuclear reaction cross sections, $\gamma$ line intensities, and nuclear level characterization. Of course, it is necessary to translate the information in the spectrum into the quantities of interest.

To identify the radionuclides that are present in the sample, we must measure the energy of the detected $\gamma$ radiation, since it is very specific for each radionuclide and characterizes its decay scheme. For that, the spectrometer must be calibrated in energy. In other words, we must find the relationship between the pulse height, the channel (see Chap. 8), and the energy deposited. This information can be extracted from the photopeaks because they are related to total energy absorption processes. As commented before, the response of the NaI(Tl) scintillator is very linear. Ideally, the number of photons, $L$, produced per unit length, $x$, in the scintillator is expected to be proportional to the amount of energy, $E$, deposited per unit length [5]. Or in a mathematical way.

$$\frac{dL}{dx} = k\frac{dE}{dx} \tag{10.2}$$

$k$ is a proportionality constant depending on the scintillator. In real measurements, the relationship can be found to be linear or very linear. From this point of view, the relationship between the energy $E_\gamma$ and the channel C should be a straight line

$$E_\gamma = mC + n \tag{10.3}$$

The slope, m, and the y-intercept, n, can be found by fitting experimental values of C corresponding to the well-known $E_\gamma$ of standard radionuclides. However, for a more precise determination, second- or third-order polynomials are sometimes used. For example,

$$E_\gamma = a + bC + cC^2 + dC^3 \tag{10.4}$$

Normally, c and d are very small, which demonstrates that a linear approach is good enough. In Fig. 10.14, we present an energy calibration curve and a parabolic fitting of the data. The second-order coefficient is very small, which reveals that a straight-line fitting would not have been a bad choice. Energy calibration curves were used to determine the energy of $\gamma$ radiation emitted from unknown radionuclides. Once the photopeak is identified in a pulse height spectrum, its central channel is determined, and the corresponding energy is obtained by interpolation in the calibration curve. Such an energy helps to identify the radionuclide.

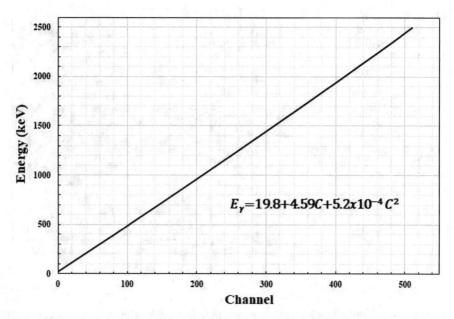

**Fig. 10.14** Energy calibration curve for a $3' \times 3'$ NaI(Tl) spectrometer

The photopeaks are easily identifiable in $\gamma$ spectra provided that they are well-defined statistically. However, for low radionuclide concentration samples, the definition of the photopeak is not as clear. Moreover, for these cases, pulse groupings can be found in the spectra that could seem similar to photopeaks. Nevertheless, as presented in (10.1), we know that the energy resolution experimentally measured follows a well-defined energy dependence that can be used as a criterion to distinguish real photopeaks from such pulse groupings. Therefore, a calibration of the energy resolution of the spectrometer is highly convenient to confirm that a presumed photopeak is truly a photopeak.

An example of resolution calibration is presented in Fig. 10.15, in which the energy dependence of $R$ is close to

$$R = \frac{\left(a + bE_\gamma\right)^{\frac{1}{2}}}{E_\gamma} \tag{10.5}$$

a result close to that expected from (10.1) but not exactly the same. The conclusion is that a clear candidate photopeak should have an energy resolution compatible with that foreseen by (10.5).

To quantify the sample activity or, more specifically, the radionuclide activity in the sample, it is necessary to calibrate the detector for efficiencies or, more precisely, to calibrate for photopeak efficiencies as defined in (8.8). The photopeak efficiency $\varepsilon_{ph}$ depends on $E_\gamma$ but also on the sample and sample detector

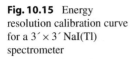

**Fig. 10.15** Energy resolution calibration curve for a 3′ × 3′ NaI(Tl) spectrometer

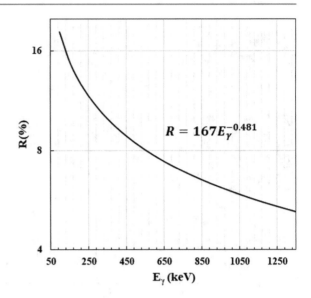

$$R = 167E_\gamma^{-0.481}$$

geometries, sample density, detector size, etc. The calibration of the $\gamma$ spectrometer is usually carried out by empirical methods, constructing calibration curves with radionuclides of known activities and samples with the same geometry and composition as the sample problem. In some cases, this is difficult to achieve, so standard samples of a similar composition can be used. As we will see in Chap. 11, on some occasions, the problem is addressed by using radiation transport Monte Carlo simulations in the sample. This method avoids the experimental problem of preparing a sample of identical composition. In general, the dependence of $\varepsilon_{ph}$ on $E_\gamma$ for a specific sample geometry and composition can be found by fitting the experimental data to polynomials of the following type.

$$ln\varepsilon_{ph} = a + bln E_\gamma + c(ln E_\gamma)^2 \qquad (10.6)$$

An example can be seen in Fig. 10.16. Once the efficiency for a given energy is known, the activity, $A$, of the radionuclide emitting such energy in a real sample will be

$$A = \frac{\Sigma_{ph} - B_{ph}}{\varepsilon_{ph}\Delta t I_\gamma} \qquad (10.7)$$

$\Sigma_{ph}$ and $B_{ph}$ are the peak area and the background under the peak, respectively, both collected during a time interval $\Delta t$. $I_\gamma$ is the intensity of the $\gamma$ line.

The calculation of $\Sigma_{ph}$ and $B_{ph}$ is not easy in many cases. This is due to the difficulties in the determination of the photopeak limits in NaI(Tl) and other inorganic scintillators as a result of their relatively poor energy resolution. In the case of low-activity samples, this problem becomes very difficult to solve. As we will

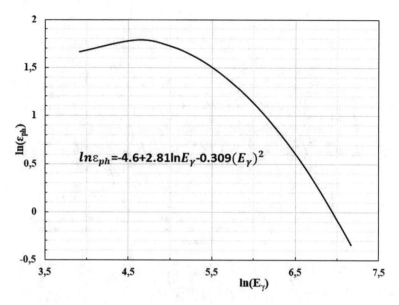

$$ln\varepsilon_{ph} = -4.6 + 2.81\ln E_\gamma - 0.309(E_\gamma)^2$$

**Fig. 10.16**  $3' \times 3'$ NaI(Tl) photopeak efficiency calibration curve

see in Chap. 11, that is not as important for Ge $\gamma$ spectrometers since the starting and ending channels of the photopeak are sharply defined in general. The determination of the photopeak limits follows statistical arguments. An example can be found in Fig. 10.17a. The photopeak limits can be calculated by analyzing the statistical significance of the differences in the number of counts (pulses) between adjacent channels around the peak. Thus, in general, the number of counts in channel $c_i$ is expected to be statistically equivalent to the number of counts in channel $c_i + 1$ when approaching the beginning of the peak. In other words, $n_i$ is normally within the interval $n_{i+1} \pm \sigma_{ni+1}$. Only when $n_i < n_{i+1} - \sigma_{ni+1}$ we can guess that the peak starts at channel $c_i$. Applying a similar argument to find the end of the peak, we could say that the ending channel is $c_f$ if $n_f < n_{f-1} - \sigma_{nf-1}$ and $n_f$ is within $n_{f+1} \pm \sigma_{nif+1}$. In some cases, the spectrum needs mathematical smoothing before applying this kind of statistical analysis [28]. Once the limits are known, $\Sigma_{ph}$ is simply the sum of the number of counts from channel $c_i$ to channel $c_f$. The determination of $B_{ph}$ is very relevant for low-activity samples, which is the usual case in environmental measurements. Now, the photopeak is superimposed onto a continuum that contains the Compton from higher energy photons. It can also be superimposed on the peaks of the $\gamma$ emitters that make up part of the $\gamma$ radiation field (see Chap. 8 and 15). In Fig. 10.17b and c, we can see a representation of these contributions to $B_{ph}$. It can be said that

$$B_{ph} = B_C + B_\gamma \tag{10.8}$$

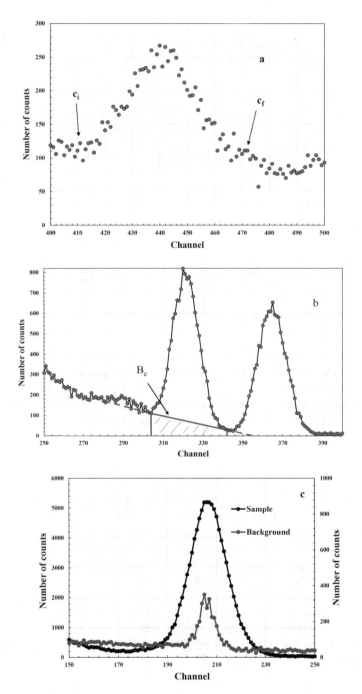

**Fig. 10.17**  **a** poorly defined limits in a photopeak obtained with a NaI(Tl) $\gamma$ spectrometer; **b** $B_C$ contribution to $B_{ph}$; and **c** $B_\gamma$ contribution to $B_{ph}$

$B_C$ is the contribution of the continuum Compton from higher energy photons and $B_\gamma$ is that of the $\gamma$ radiation field background. $B_C$ is calculated by assuming that the shape of the Compton continuum is a straight line connecting $c_i$ and $c_f$. Thus, the area under this line will be

$$B_C = \frac{n_i - n_f}{2}\left(c_f - c_i\right) \tag{10.9}$$

with an obvious meaning of the notation. $B_\gamma$ is found by measuring the background spectrum of the spectrometer to determine the net area of the possible photo-peak in this zone. Notably, $B_\gamma$ must be the net area of the background photopeak. Otherwise, we could extract twice the Compton contribution in (10.7).

The relatively poor energy resolution makes it difficult to distinguish close energies in NaI(Tl) spectrometers (see Fig. 10.18 as an example). Especially for low-activity samples, this can be a very important drawback. In fact, today, low-activity sample spectrometry is carried out with Ge spectrometers (see Chap. 11). For such cases, it is difficult to find clearly isolated peaks that are readily ana-lyzable. In fact, before the arrival of semiconductor detectors, it was necessary to develop mathematical methods to resolve complex $\gamma$ spectra based on the applica-tion of a superposition principle that assumed that the total spectrum was the sum of individual $\gamma$ spectra of previously identified radionuclides [28]. Today, scin-tillation $\gamma$ spectrometry has been surpassed by semiconductor detectors. A more detailed account of modern $\gamma$ spectrometry will be given in Chap. 11.

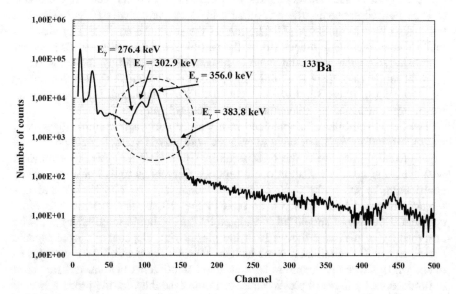

**Fig. 10.18** An example of poorly resolved peaks in a NaI(Tl) spectrometer, measuring the $^{133}Ba$ $\gamma$ spectrum

## 10.4  Counting and Spectrometry with Liquid Scintillation Detectors

### 10.4.1 Technical Aspects

Indeed, as mentioned before, liquid scintillation counting is nowadays the main contribution of scintillation detectors to the field of measurements of environmental radioactivity. The most important reason is the possibility of internal counting and thus maximizing the counting efficiency to values close to 100% [10]. The sample can be diluted in, or mixed with, the scintillator, forming the scintillation cocktail, in such a way that a very tight contact is produced between the sample and detector. This feature makes liquid scintillation counters (LSCs) very competitive with regard to GID, where internal gas counting is possible as well. On the other hand, the automatic sample change in actual LSC designs greatly increases their throughput. Pulse shape analysis (PSA) is also possible with LSC, allowing, for example, the simultaneous determination of $\alpha$ and $\beta$ emitters [29]. Similar to the other scintillation counters, LSC is basically a compound of a scintillator, a liquid contained in a vial in this case, and a photomultiplier, following the scheme of Fig. 10.1. In actual setups, the vial is surrounded by two photomultipliers (see Fig. 10.19), and only coincident electric pulses from the two phototubes are accepted as real counts. Pulses produced only in one of them are supposed to be background or noise contributions. This greatly enhances the sensitivity of the technique [30]. The liquid scintillator is an organic scintillator. See Table 10.2 for an account of their characteristics. However, the scintillator generally contains at least two components, the scintillant and the solvent material, with the first being dissolved in the latter [7]. The role of the solvent is to efficiently absorb a major part of the energy deposited by the particles in the scintillator. This energy is transferred to the scintillant, who is responsible for emitting the scintillating light. In some cases, there is a third component, the wavelength shifter, that moves the wavelength of the scintillation light to longer values and thus matches the wavelengths accepted by conventional photomultipliers [6].

The liquid scintillator is contained in a vial. In early applications, glass vials were used, but today, there is an alternative with plastic vials. Depending on the experimental needs, one is the most suitable. Glass vials are often contaminated with $^{40}$K, and cosmic radiation could produce Cherenkov radiation in the walls, contributing to the background of the measurement [31]. The background differences between both vials can be seen in Fig. 10.20. The difference is quite apparent. However, it should be considered that many liquid scintillators could be absorbed by or penetrate the walls of the plastic vial. Depending on the organic component of the scintillator, some authors consider the possibility of dissolution of the walls. Currently, there are several liquid scintillators that are not supposed to produce such effects in plastic vials, or in the case they do, the process is very slow, and counting time measurements can be carried out with absolute confidence [32].

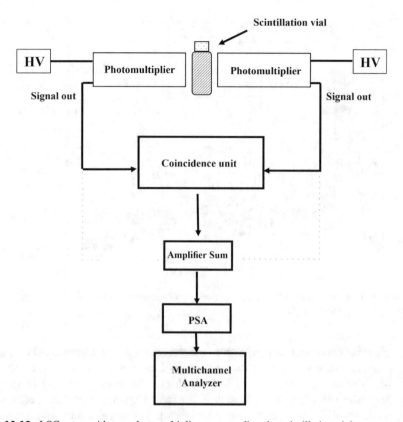

**Fig. 10.19** LSC setup with two photomultipliers surrounding the scintillation vial

The main advantage of plastic vials is the diffuse surface they have compared to glass. This reduces the internal reflection of the light, increasing the light intensity that arrives at the photomultipliers. Conventionally, 20 ml volume vials are mostly used, although there are special systems that allow 100 ml vials, which is convenient for special applications.

Quenching effects greatly modify the counting in an LSC. In principle, quenching causes a decrease in the number of scintillation photons arriving at the photomultiplier photocathode.

The reasons are of chemical and physical nature and, of course, related to the energy and type of the counted radiation as well as to the scintillator, the sample history, specifically its preparation, and others. The influence of quenching effects can be seen in Fig. 10.21. Independent of the quenching origin, the result is a decrease in the number of photons arriving at the photocathode. This decreases the expected pulse height corresponding to a given deposited energy. The modification of the spectrum shape is presented in the figure, where a drift to lower channel values is observed. There are several types of quenching. Physical quenching due

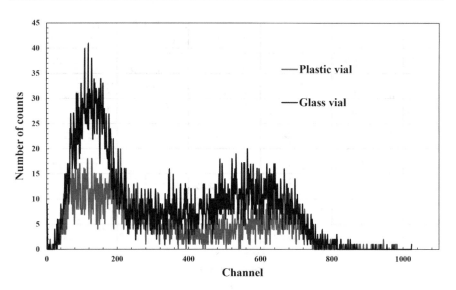

**Fig. 10.20** Background spectra of a liquid scintillation counter for 20 ml plastic and glass vials. Courtesy of J.-L- García-León

to precipitation and microcrystallization processes in the scintillation cocktail; ionization quenching due to a large concentration of excited molecules in the solvent, very likely when counting $\alpha$ particles, which prevents energy transfer to scintillant molecules and others [32]. However, the most important quenching effects in real environmental applications are those known as chemical and color quenching [10]. Chemical quenching appears when the energy absorbed by the solvent is transmitted to neighboring molecules that do not emit light during deexcitation. Therefore, this produces a lost signal. In many cases, molecular oxygen in the scintillation cocktail may cause this quenching effect. Color quenching is derived from the presence of colored substances in the scintillation cocktail, commonly associated with the sample under analysis. These substances absorb some wavelengths of the scintillating light, preventing their detection by the photocathode and thus diminishing the number of photoelectrons produced. As a consequence, part of the signal is also lost. All these possible effects require special precautions when preparing the sample. We will provide more information on this topic in Chap. 13. However, for instance, to reduce color effects, the presence of organic matter in the scintillation cocktail should be avoided, which is mainly responsible for color quenching. In the case of chemical quenching, it is important to reduce the presence of oxygen, chlorides, alcohols, and others [33].

Despite the precautions that must be taken during sample preparation, quenching effects can occur in real measurements. Thus, they must be evaluated to give a precise determination of the activity of the sample. Several methods can be studied in depth in the specialized literature [34]. The external standard method or

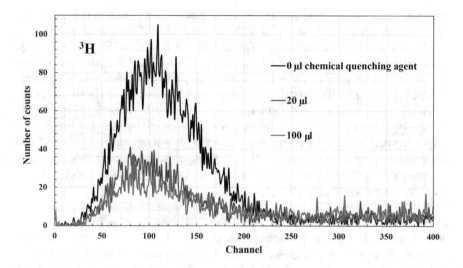

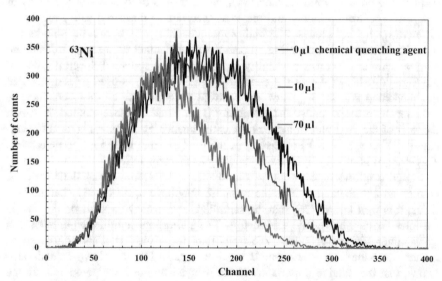

**Fig. 10.21** Some $\beta$ spectra taken with an LSC affected by quenching effects. In the case of $^3$H, some amounts of a chemical agent have been diluted in the scintillation cocktail, and the effect is observed in the spectra. In the case of $^{63}$Ni, a color quenching agent was used. Courtesy of J. L. García-León

external quench monitoring consists of the use of an external $\gamma$ radioactive sample, normally incorporated into the counting system, which produces Compton electrons in the scintillation cocktail, whose spectrum variation due to quenching effects is assumed to represent that of the sample itself. The $^{133}$Ba or $^{152}$Eu sources are commonly used in popular LSC systems [35]. The end point of the Compton spectrum is known, and the variations in real measurements due to quenching are quantified as a parameter known as SQP. The variation of the counting efficiency can be estimated as a function of SQP, and in the literature [36], there is some fitting of the efficiency to the quadratic polynomials of SQP. An alternative is the use of internal standards in which a known activity of the measured radionuclide is added to the scintillation cocktail once the problem sample has been measured. Of course, the evaluation of the efficiency is very realistic since all the quenching effects apply in the same way as in the real sample. However, the problem sample cannot be used in other measurements since a high activity has to be poured into the scintillation cocktail to measure the quenching effects with precision. The internal method is applied mainly in high-activity metrology [37]. It is not of practical use in environmental applications.

The most popular method is based on the parametrization of the spectrum shape changes under quenching-controlled conditions. The main advantage of these methods lies in the possibility that they can be applied at the same time as the measurement of a problem sample. There are several possibilities. One of them is the channel ratio method [38], in which the ratio between the areas of a defined spectrum window and the total spectrum is measured as a function of different quenching grades. The higher the quenching effects are, the smaller the ratio. The fitting of the counting efficiency to the channel ratio serves to calibrate the real samples. Another possibility is the use of the so-called Spectral Index of the Sample (SIS) in which the centroid of the spectrum is used to parameterize the quenching effects [38]. Now, the centroid drifts toward the lower channels as the quenching increases. A last possibility is the Final Channel Method, in which the drift of the final part of a spectrum as the quenching increases is used to evaluate it. This method lacks precision, as the final channel is difficult to determine. Therefore, it is of very little use in environmental radioactivity.

Liquid scintillators are organic scintillators and therefore share their features. One important characteristic is the response they have to differently charged particles. It is well known [39] that the scintillation efficiency depends on the charge particles for certain organic scintillators. The anthracene scintillation efficiency is seven times higher for 1 meV electrons than for 1 meV protons or four times higher for 2 meV electrons than 2 meV protons in NE102. However, it is also known that the relative importance of the delayed fluorescence component in the scintillation process depends on the charged particles. Although the characteristic time constant for prompt fluorescence is some nanoseconds, that for delayed or slow fluorescence is some 100 ns. This can be used to distinguish $\alpha$ from $\beta$ particles in an LSC. In Fig. 10.22, a typical voltage pulse is given for $\alpha$ and $\beta$ particles. Obviously, the time shapes are different. Thus, they can be differentiated electronically, and their simultaneous counting is possible [40].

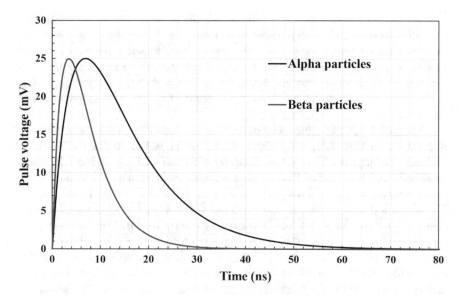

**Fig. 10.22** Shape of the voltage pulse for $\alpha$ and $\beta$ particles useful to distinguish them in a sample measured by LSC

## 10.4.2 Applications

The use of LSC started around the 1950s with applications to Fundamental Physics and Biology. Thus, pioneering measurements can be found in the search for neutrino and fission products in the human body [41]. However, it was the possibility of using internal counting for the measurement of very low-energy $\beta$ emitters, such as $^3$H and $^{14}$C, that boosted the development of LSC [42]. This possibility led LSC to play a very important role in radiocarbon dating [43]. Today, LSC is still used for $^{14}$C dating, although Accelerator Mass Spectrometry has surpassed it, as we will see in Chap. 18.

The use of LSC in environmental measurements is wide. Many $\beta$ emitter radionuclides are currently measured in the environment using liquid scintillation. Examples are $^3$H and $^{14}$C but also $^{90}$Sr, $^{99}$Tc, $^{210}$Pb, $^{241}$Pu, and others [10]. Of course, when dealing with real environmental samples, quenching effects must be carefully taken into account. Nevertheless, methods exist to make the appropriate corrections, and as we will see in Chap. 13, radiochemical methods are developed to reduce their influence.

LSC is also used for the determination of $\alpha$ emitters. Of course, the poor energy resolution of these detectors does not justify their use as an alternative to semiconductor detectors (see Chap. 11). Nevertheless, internal counting provides higher counting efficiencies and, above all, the possibility of simultaneous determination of $\alpha$ and $\beta$ emitters by pulse shape analysis, making them very useful for several applications [44]. It can be found in current literature measurements of $^{210}$Po,

$^{226}$Ra, $^{222}$Rn, and U isotopes by liquid scintillation counting among natural $\alpha$ emitters [45]. However, it is also possible to measure the $^{241}$Am and Pu isotopes [46]. Internal counting makes the use of LSC possible for the determination of electron capture emitters in the environment. From this viewpoint, the measurement of $^{55}$Fe [47] has received much attention during the last times. For these cases, LSC helps in the measurement of X-rays emitted or alternatively the corresponding Auger electrons.

Liquid scintillation counters can be used as Cherenkov counters for high-energy $\beta$ emitters. In this case, the sample is diluted in water, where the threshold Cherenkov energy is 263 keV (see Chap. 7). $^{90}$Sr has an $E_{max} = 546$ keV, above the threshold. Thus, part of the electrons it emits could produce Cherenkov radiation. However, only part. Therefore, the efficiency is low. Fortunately, its daughter $^{90}$Y is also a $\beta$ emitter with an $E_{max} = 2283$ keV that can be used for Cherenkov counting of $^{90}$Sr since they are in secular equilibrium [36]. This technique can also be used to determine $^{210}$Pb by measuring its daughter $^{210}$Bi, which emits $\beta$ radiation with $E_{max} = 1161$ keV [36, 48]. The technical condition is that the photomultiplier must be sensitive to the Cherenkov radiation wavelength, but this is readily achieved since the Cherenkov light in water is well within the sensitivity of conventional photomultipliers.

Of course, LSC is used very much in Biology and Health Sciences in general. These are probably the fields where LSC currently plays a more important role. There is a huge amount of information on this in the current literature. However, this application is not within the scope of this book.

## 10.5    Background in Scintillation Detectors

Scintillation detectors are sensitive to different background sources since they can be used for charged particles, as well as $\gamma$- and X-rays. Thus, by analyzing (8.9), it seems that organic scintillators will be more sensitive to $B_{cos}$ and $B_{nuc}$ and $B_{\mu}$, while inorganic detectors will be mainly sensitive to $B_{\gamma}$. $B_{Rn}$ could affect both types of detectors since this component produces charged particles and $\gamma$ emitters. Of course, a more specific analysis would give a more precise evaluation of the relative importance of each component. For example, NaI(Tl) detectors, depending on their size, are also very sensitive to $B_{\mu}$ since they can detect Bremsstrahlung produced by cosmic $\mu$ in the surroundings of the detector and the detector itself [31]. The purity of the material is certainly an important drawback in NaI(Tl) detectors. Fortunately, it is currently possible to produce large high-purity NaI(Tl) crystals with U and Th contents below a few ppt and $^{40}$K contents around ppb.

The contamination traces in photomultipliers significantly affect the background of scintillation detectors. In fact, they can become the main source of background for the determination of very low radionuclide concentrations. Progress has been made in reducing the U, Th, and $^{40}$K contamination of glass and ceramic parts of photomultipliers to ppt and ppb levels. The use of bent light guides allows one to isolate the detector from the photomultiplier in some cases, reducing the influence of its contamination in the measurement background. For very specific applications, i.e., very low radioactivity measurements, the use of photodiodes instead of conventional photomultipliers greatly helps reduce the background [31].

In the case of LSC, as commented on before, the contamination traces in the vial are important. Depending on the application and the final chemical form of the sample, the use of glass or plastic vials is better. Electrostatic effects are also important when manipulating plastic vials because of the accumulation of electric charge in their walls. Usually, real LSC counters have systems to compensate for that charge and avoid electric discharges that could influence the background of the measurement [10, 49]. Chemiluminescence is a specific process in LSCs that affects their background. This corresponds to the emission of light by the interaction of the radiation in the sample with the vial walls. As commented before, usual LSC setups contain at least two photomultipliers connected in coincidence to recover as much as possible of the light scintillation flux produced. Normally, the light produced by chemiluminescence excites only one photomultiplier, and therefore, the pulse produced is easily rejected [31]. Photoluminescence of some substances present in the sample is also possible in an LSC experiment. This is derived from the excitation of such substances by ambient light during the manipulation of the sample. Since photoluminescence gives rise to a delayed emission of light, it is convenient to store the scintillation cocktail in darkness some 24 h before counting. In this way, it is possible to suppress this background source [32].

**Exercises**

1. The energy-light conversion efficiency is approximately 13% in NaI(Tl). Suppose a 2 MeV photon beam is totally absorbed in the crystal. Estimate the number of light photons produced.
2. Estimate the scintillation efficiency of anthracene if the 2 meV energy deposited produces 40,000 photons.
3. Compare inorganic and organic scintillators regarding the following properties:
   - time response
   - $\gamma$-ray spectrometry ability
   - charged-particle spectrometry ability
4. Select a scintillation detector to carry out the following experiments. Explain the choice.
   - $\beta$ counting
   - $\alpha$ counting
   - $\beta$ spectrometry
   - $\alpha$ spectrometry
   - $\gamma$ spectrometry
5. We have a $^{207}$Bi source that emits $\gamma$ radiation of energies of 570 keV ($I = 98\%$) and 1064 keV ($I = 74\%$). After passing through an Al sheet, it is observed with NaI(Tl) that 1% of both $\gamma$ photons were scattered at an angle of 30°. Sketch the $\gamma$ spectrum obtained and describe the different parts; give the exact relationship

between the areas of the photopeaks that appear, assuming efficiencies of 2.24% and 1.26%, respectively.

6. If the area of the photopeak with the highest energy is 20.000 counts obtained during a time of 10 min, find the activity of the sample in $Bq$.

7. We have a radioactive sample that contains two emitters: one alpha and one beta. We also plan to measure the alpha and beta activities of the sample. Suppose that you have only one detector in the laboratory, a SZn (Ag) scintillation detector. Could you measure both activities?

8. Suppose that the energy resolution of a NaI(Tl) detector at 662 keV is 7%. Sketch the spectrum obtained for a $^{60}$Co sample with the same detector.

9. Find the Compton edge energy for the spectra obtained in the previous exercise as well as the backscattering peak energy. Place both events in the sketched spectrum.

10. Suppose that a $\gamma$-emitter has the following decay scheme.
    Sketch the spectrum that can be obtained with a NaI(Tl) detector and explain each part. Take $E_{\gamma 1}$ and $E_{\gamma 2} < 1.022$ MeV and consider it a possible CE decay.

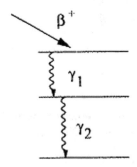

11. Do you expect to find sum peaks in the spectrum? Design a method to mitigate the sum effects.

12. Suppose now that the radionuclide emits $\gamma$-radiation with energies $E_\gamma > 1.022$ MeV. If you measure the sample with a $3' \times 3'$ NaI(Tl), sketch the expected spectrum explaining each part.

13. Suppose that the previous sample is measured with a $2' \times 2'$ NaI(Tl). Will you expect qualitative or quantitative differences between both spectra?

14. Explain why a GM detector is more efficient than a NaI(Tl) for detecting $\beta$ radiation. In contrast, explain why a NaI(Tl) detector is more efficient for $\gamma$ radiation than a GM detector. Is it possible to make $\beta$ spectrometry with a GM detector? Why is it possible to carry out $\gamma$- spectrometry with a NaI(Tl) detector?

15. Briefly explain why a CsI(Tl) detector is more useful than a proportional detector for $\gamma$-spectrometry.

16. Under which conditions could a proportional detector be a good electromagnetic radiation spectrometer?

17. Suppose a radionuclide emitting $\gamma$-radiation with energies $E_1 > E_2 > E_3$. All the emitted energies are below 1.022 MeV. Sketch the expected spectrum obtained with a NaI(Tl) detector, explaining each part.

18. It is necessary to determine the activity of a $^3$H sample. Design the experiment and choose the appropriate detector. Use https://www.nndc.bnl.gov/nudat2/ to find its decay scheme.

19. Discuss the experimental needs for the activity determination of a $^{55}$Fe sample with a scintillation detector. Use https://www.nndc.bnl.gov/nudat2/ to find its decay scheme.

20. Suppose that you are expected to determine the $\beta$-radiation maximum energy of $^3$H and $^{14}$C and you have a plastic scintillation detector and a liquid scintillation detector in the laboratory. Design the experiment and discuss the reasons for selecting the detector.

## References

1. W. Crookes, The emanations of radium. Proc. Roy. Soc. London **71**, 405–408 (1903); W. Crokes, Certain properties of the emanations of radium. Chem. News **87**, 241 (1903)
2. E. Rutherford, The scattering of $\alpha$ and $\beta$ particles by matter and the structure of the atom. Phil. Mag. **21**, 669–688 (1911)
3. S.C. Curran, W.R. Baker, *United States Atomic Energy Commission Report MDDC*, vol. 1296 (1944)
4. R. Hofstadter, Detection of gamma rays with thallium-activated sodium iodide crystals. Phys. Rev. **75**, 796–809 (1949)
5. W.R. Leo, *Techniques for Nuclear and Particle Physics Experiments: A How-to Approach*, 2nd edn. Chapter 7. (Springer, 1994)
6. G.F. Knoll, *Radiation Detection and Measurements*, Chapter 8, 5th edn. Wiley, 2010)
7. N. Tsoulfanidis, S. Landsberger, *Measurement and Detection of Radiation*, Chapter 6, 5th edn. (CRC Press, 2021)
8. J.B. Birks, *The Theory and Practice of Scintillation Counting*, Chapters 3 and 6. (Pwergamon Press, 1964)
9. S. Niese, The discovery of organic solid and liquid scintillators by H. Kallmann and L. Herforth 50 years ago. J. Radioanal. Nucl. Chem. **241**, 499–501 (1999)
10. X. Hou, X. Dai, Environmental liquid scintillation analysis, in *Handbook of Radioactivity Analysis*, vol. 2. (Elsevier 2020), pp. 41–135
11. P. Križan, Photon detectors, in *Handbook of Particle Detection and Imaging*, vol. 2, Chapter 13, ed. by C. Grupen, I. Buvat (Springer, 2012)
12. T. Yanagida, Inorganic scintillating materials and scintillation detectors. Proc. Jpn. Acad. Ser. B **94**, 75–97 (2018)
13. M. Mutterer, Gas scintillation nuclear particle detectors. Nucl. Instrum. Method. **196**, 73–81 (1982)
14. https://www.hamamatsu.com/eu/en.html
15. P. Theodorsson, *Measurement of Weak Radioactivity*. World Scientific Publishing Co. (1996), p. 226
16. L.S. Ring, D.J. Zaffarano, $\beta$ spectra of some short-lived nuclides, in *ISC-648 Ames Laboratory report for the US Atomic Energy Commission* (1955)
17. D.G. Gardner, W.W. Meinke, $\beta$-ray spectroscopy using a hollow plastic scintillator. Int. J. Appl. Radiat. Isot. **3**, 232 (1958)

18. S. Yamamoto, H. Tomita, Comparison of light outputs, decay times, and imaging performance of a ZnS(Ag) scintillator for alpha particles, beta particles, and gamma photons. Appl. Radiat. Isot. **168**, 109527 (2021)
19. E. Bodewits, D. Cester, M. Lunardon, S. Moretto, P. Schotanus, L. Stevanato, G. Viesti, Characterization of a large area ZnS(Ag) detector for gross alpha and beta activity measurements in tap water plants. IEEE Trans. Nucl. Sci. **63**, 1565–1569 (2016)
20. G. Consolati, D. Franco, C. Jollet, A. Meregaglia, A. Minotti, S. Perasso, A. Tonazzo, A new antineutrino detection technique based on positronium tagging with plastic scintillators. Nucl. Instrum. Methods. **795**, 364–369 (2015)
21. M.A. Caprio, N.V. Zamfir, E.A. McCutchan, R.F. Casten, Z. Berant, H. Amro, C.J. Barton, C.W. Beausang, D.S. Brenner, J.R. Cooper, R.L. Gill, G. Gürdal1, A.A. Hecht, C. Hutter, R. Krücken, D.A. Meyer, J.R. Novak, N. Pietralla, P.H. Regan, J.J. Ressler, Measurement of $2^{+1}$ level lifetimes in $^{162}$Yb and $^{162}$Er by fast electronic scintillation timing. Eur. Phys. J. A **16**, 177–180 (2003)
22. Angular correlations in nuclear disintegration, in *Proceedings of the International Conference on Angular Correlations in Nuclear Disintegration*, ed. by H. van Krugten, B. van Nooijen. (Delft, The Netherlands, Springer, 1970)
23. K. Furutaka, S. Nakamura, H. Harada, T. Katoh, Evaluation of β-γ coincidence measurement system using plastic scintillation β-ray detector developed for the determination of γ-ray emission probabilities of short-lived nuclides. J. Nucl. Sci. Technol. **37**, 832–839 (2000)
24. S. Min, B. Seo, C. Roh, S. Hong, J. Cheong, Phoswich detectors in sensing applications. Sensors **21**, 4047 (2021)
25. G.F. Knoll, *Radiation Detection and Measurements*, Chapter 17, 4th edn. (Wiley, 2010)
26. N. Tsoulfanidis, S. Landsberger, *Measurement and Detection of Radiation*, Chapter 12, 5th edn. (CRC Press, 2021)
27. G.F. Knoll, *Radiation Detection and Measurements*, Chapter 10, 4th edn. (Wiley, 2010)
28. L. Salmon, Analysis o gamma-ray scintillation spectra by the method of least-squares. Nucl. Instrum. Methods. **14**, 193–199 (1961)
29. D.L. Horrocks, *Applications of liquid scintillation counting* (Academic Press, London, 1974), p.346
30. M. Villa, G. Manjón, M. García-León, Study of color quenching effects in the calibration of liquid scintillation counters: the case of $^{210}$Pb'. Nucl. Instrum. Method. Phys. Res. A **496**, 413–424 (2003)
31. P. Theodorsson, *Measurement of Weak Radioactivity*, Chapter 9, World Scientific Publishing Co. (1996)
32. M. Villa, Ph.D. Thesis, University of Seville (2004)
33. G. Manjón, A. Absi, M. Villa, H.P. Moreno, R. García-Tenorio, Efficiency, background and interference in colored samples using a LSC Quantulus 1220, LSC 2001 Adv. Liq. Scintill. Spectrom. Radiocarb. 83–91 (2002)
34. L. Kahiola, Cosmic particle spectrum as a quench monitor in low-level liquid scintillation spectrometry. Nucl. Instrum. Method. A **339**, 295–296 (1994)
35. P.J. Meechan, V.A. Bhatt, Color quenching in environmentally friendly cocktails. Health Phys. **73**, 808–813 (1997)
36. F. Vaca, G. Manjón, M. García-León, Efficiency calibration of a liquid scintillation counter for $^{90}$Y Cherenkov counting. Nucl. Instrum. Method. Phys. Res. A **406**, 267–275 (1998)
37. C. Grau, E. Gunther, E. García, A. Grau-Malonda, Ionization quenching in LSC. Appl. Radiat. Isot. **60**, 447–451 (2004)
38. H. Fujii, M. Takiue, Radioassay of dual-labeled samples by sequential Cerenkov counting and liquid efficiency tracing technique. Nucl. Inst. Method. Phys. Res. A **273**, 377–380 (1988)
39. R.L. Craun, D.L. Smith, Analysis of response data for several organic scintillators. Nucl. Inst. Method. 239–244 (1970)
40. L. Salonen, H. Hukkanen, Advantages of low-background liquid scintillation alpha spectrometry and pulse shape analysis in measuring $^{222}$Rn, uranium, and $^{226}$Ra in groundwater simples. J. Radioanal. Nucl. Chem. **226**, 67–74 (1977)

41. C.T. Peng, Liquid scintillation counting: past and present. LSC 2001 Adv. Liq. Scintill. Spectrom. Radiocarb. 1–10 (2002)
42. F. Schönhoffer, E. Henrik, Recent progress and applications of low-level liquid scintillation counting. J. Radioanal. Nucl. Chem. **115**, 317–335 (1987)
43. A. Hogg, G. Cook, Liquid scintillation counting (LSC)-past, present, and future. Radiocarb. 1–14 (2021)
44. G.Ö. Çakal, R. Güven, H. Yücel, An application of LSC method for the measurement of gross alpha and beta activities in spiked water and drinking water samples. Nukleonika **60**, 637–642 (2015)
45. G. Manjón, I. Vioque, H.P. Moreno, R. García-Tenorio, M. García-León, Determination of $^{226}$Ra and $^{224}$Ra in drinking Waters by liquid scintillation counting. Appl. Radiat. Isot. **48**, 535–540 (1997)
46. X.L. Hou, P. Roos, Critical comparison of radiometric and mass spectrometric methods for the determination of radionuclides in environmental, biological and nuclear waste samples. Anal. Chim. Acta **606**, 105–139 (2008)
47. X.L. Hou, L. Togneri, M. Olsson, S. Englund, O. Gottfridsson, M. Forsstrom, H. Hirvonen, Standardization of radioanalytical methods for determination of $^{63}$Ni and $^{55}$Fe in waste and environmental samples, NKS-356. Nord. Nucl. Saf. Res. (2015)
48. M.S. Al-Masri, A. Hamwi, H. Mikhlallaty, Radiochemical determination of $^{210}$Pb in the enviroment using Cerenkov counting. J. Radioanal. Nucl. Chem. **219**(1), 73 (1997)
49. D.L. Horroks, Studies of background sources in liquid scintillation counting. Int. J. Appl. Radiat. Isot. **36**, 609–617 (1985)

# Semiconductor Detectors

# 11

**ABSTRACT**

The operating principles of semiconductor detectors are described in this chapter. They are based on the production of electron-hole pairs in semiconductors when charged particles pass through. Semiconductor detectors are efficient for charged particles, and therefore for $\alpha$- and $\beta$-particles, or electrons in general but also for $\gamma$-radiation since high $Z$ semiconductor detectors can be constructed. Their energy resolution is much better than that of the detector families studied previously, making them the best option for $\alpha$-, $\beta$-, and $\gamma$-spectrometry. The principles of $\gamma$-spectrometry already explained in the previous chapter are applicable for semiconductors. Those for charged particles and specifically for $\alpha$-spectrometry are introduced in this chapter. The semiconductor detector background components are analyzed to help design detector shielding, as will be described in the following chapters. In the case of Si detectors, shielding is not necessary, which makes them very competitive, as no special facilities are necessary to measure low radioactivity concentration samples.

## 11.1 Physics of Semiconductor Detectors

At first, semiconductor detectors were known as solid-state ionization chambers. This is due to the similarity of its operation to GID. Indeed, they take advantage of the ionization produced by charged particles along its track in matter. However, in semiconductors, electron-hole pairs are produced instead of ion–electron pairs. Semiconductor detectors have helped to solve several pending problems within the radiation detection field. First, they provided better energy resolution than GID and scintillation detectors for $\gamma$ and charged particle radiation, and second, their density made it possible to construct smaller detectors for charged particles with enough counting efficiency. To understand the operation of semiconductor detectors, it is necessary to recall the band theory of electrons in solids [1].

© The Author(s), under exclusive license to Springer Nature Switzerland AG 2022
M. García-León, *Detecting Environmental Radioactivity*, Graduate Texts
in Physics, https://doi.org/10.1007/978-3-031-09970-0_11

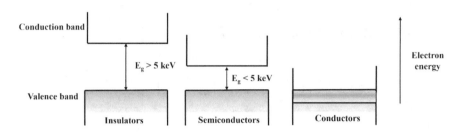

**Fig. 11.1**   Electron energy distribution within bands of energy in a solid

When individual atoms form a lattice, the electrons reorganize into energy bands. The energy band extends along the lattice in such a way that the electrons can occupy different energy states within the band. Two bands are formed, the valence and conduction bands (see Fig. 11.1). The electrons in the valence band are bound to the crystal structure and cannot move along the lattice. The electrons in the conduction band are free and can move under the action of an electric field or simply by thermal agitation. Between the valence and conduction bands, there exists a forbidden gap in which no electrons can be found.

According to this picture, the solids can be classified with regard to electrical conductivity. Thus, a solid is a conductor if the conduction band contains electrons that can move under an electric field. Therefore, in a conductor, it is possible to observe an electric current when a field is applied. This does not occur for insulating and semiconductor materials where no electrons are present in the conduction band. The difference between insulators and semiconductors lies in the width of the forbidden gap (see Table 11.1), which is large in the case of insulating materials and moderate for semiconductors [2–4]. To observe electric current in such materials, it is necessary to promote electrons from the valence to the conduction band. This is easier in semiconductors than in insulating materials because less energy is needed. Of course, thermal excitation could contribute to breaking this situation and bringing electrons from the valence into the conduction band. Should this occur, a vacancy is created in the valence band, a hole, which in fact behaves as a positive charge under the action of an electric field. The produced electron and hole are diffused away by thermal agitation, as in the case of the ion–electron pair in gases. Once again, the production of an electron-hole pair by thermal excitation is more likely in semiconductors than in insulating materials, since the probability, $p(T)$, of this to occur depends on the temperature $T$ as [2]

$$p(T) = CT^{\frac{3}{2}}e^{-\frac{E_g}{2kT}} \tag{11.1}$$

where $E_g$ is the width of the forbidden gap, $k$ is the Boltzmann constant, and $C$ is a constant characteristic of the material. As we will see below, this fact makes it advisable to cool down the semiconductor detector in some cases to decrease such probability as much as possible.

**Table 11.1** Forbidden gap of energy for different materials

| Material | Energy gap (eV) | |
|---|---|---|
| | 0 K | 300 K |
| Si | 1.17 | 1.11 |
| Ge | 0.74 | 0.66 |
| InSb | 0.23 | 0.17 |
| InAS | 0.43 | 0.36 |
| InP | 1.42 | 1.27 |
| GaP | 2.32 | 2.25 |
| GaAs | 1.52 | 1.43 |
| GaSb | 0.81 | 0.68 |
| CdSe | 1.84 | 1.74 |
| CdTe | 1.61 | 1.44 |
| ZnO | 3.44 | 3.2 |
| ZnS | 3.91 | 3.6 |
| Natural diamond | | 5.47 |
| $SiO_2$ | | 9 |
| AlN | | 6 |

## 11.1.1 Electron-hole Production

As the charged particle passes through a solid, electrons and holes are created along its pathway, since the particle provides the necessary energy for that before stopping down in the solid. Under the action of an electric field, electrons and holes move in the solid in opposite directions. In the case of holes, the movement is physically different from that of electrons. Indeed, the hole, being a charge vacancy, can be filled by one electron in the valence band, producing a new hole that can be filled in turn by another electron, producing a new hole, and so on. This phenomenon can be interpreted as the movement of a hole along the lattice or, in other words, as that of a positive charge. Similar to (9.2) for GID, the drift velocity of electrons, $v_e$, and holes, $v_h$, can be formulated as

$$v_e = \mu_e E \tag{11.2}$$

and

$$v_h = \mu_h E \tag{11.3}$$

for low to moderate electric fields, $E$; $\mu$ is the mobility of electrons and holes in the solid. Interestingly, the mobility of electrons and holes is of the same order of magnitude [1]. This means that holes also contribute to the formation of the electric pulse, a different situation from that of GID, where positive ions practically do not contribute to the electric pulse. The carrier mobility, together with other

**Table 11.2** Some characteristics of semiconductor materials

| Parameter | Si | Ge | 4H-SiC | InP | GaAs | CdTe |
|---|---|---|---|---|---|---|
| Density (gcm$^{-3}$) | 2.33 | 5.32 | 3.21 | 4.78 | 5.72 | 5.85 |
| Bandgap (eV) | 1.11 | 0.66 | 3.26 | 1.35 | 1.43 | 1.44 |
| Electron mobility (cm$^2$V$^{-1}$ s$^{-1}$) | 1350 | 3900 | 1000 | 4600 | 8000 | 1100 |
| Hole mobility (cm$^2$V$^{-1}$ s$^{-1}$) | 480 | 1900 | 115 | 150 | 400 | 100 |
| Dielectric constant (relative to vacuum) | 12 | 16 | 9.7 | 12.4 | 12.8 | 10.9 |
| Resistivity ($\Omega$ cm) | <10$^4$ | 50 | >10$^5$ | 10$^6$ | 10$^7$ | 10$^9$ |
| Energy per electron-hole pair (eV) | 3.62 | 2.96 | 7.8 | 4.2 | 4.2 | 4.43 |

characteristics of some semiconductor materials, is given in Table 11.2 [5, 6]. On the other hand, according to the literature, the expected drift velocities in semiconductors are some $10^7$ cms$^{-1}$. This means collection times of some 10 ns for normal detector dimensions, ~$10^{-1}$ cm. In other words, semiconductor detectors are relatively fast detectors compared to GIDs.

## 11.1.2 Energy Resolution

As mentioned above, the production of electron-hole pairs is less energy-consuming in semiconductors. Therefore, they are a very good choice as radiation detectors. The reason is the small value of the forbidden energy gap (see Tables 11.1 and 11.2). Nevertheless, it is important to realize that the energy needed to produce an electron-hole pair is higher than the gap width, since part of the energy deposited by the particle along its track is used to excite phonons in the solid and collective excitation modes, in general [5]. The average energy needed to create an electron-hole pair is approximately 2.9 eV in Ge and 3.7 eV in Si (see Table 11.2). state that easily provides the of the particle energy and of its physical nature to be considered a constant [6]. Of course, it is clearly lower than the energy needed to create an electron–ion pair in a gas (approximately 30 eV; see Chaps. 9) or a scintillation photon in an organic or inorganic material (see Table 10.1). Consequently, according to (8.17), the energy resolution is expected to be better than that for those detectors. The reason is that the number of charge carriers per unit of deposited energy is significantly higher.

A basic scheme of a semiconductor detector is presented in Fig. 11.2. A semiconductor is connected to a high-voltage source, which generates an electric field inside the detector. The electron-hole pairs created by the particles irradiating the detector are collected by the electric field in the electrodes and the electric pulse formed. The analysis of the pulses is carried out in the same way as already described in previous chapters.

This would be a simple description of the operation of a pure semiconductor detector in which the observed current corresponds to the movement of the electron-hole pairs produced along the track of the charged particle. Indeed, in

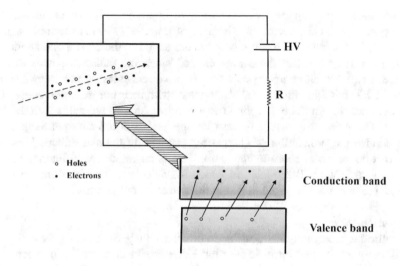

**Fig. 11.2**   Basic scheme of a semiconductor detector

a pure semiconductor material, there are no free electrons or holes, but those produced by thermal agitation (11.1). Nevertheless, in practice, the situation is somewhat different, since impurities exist in the semiconductor that is responsible for the presence of free electrons or holes in the material. This affects or even dominates the electrical conductivity of the semiconductor [7].

### 11.1.3  Types of Semiconductor Detectors

The presence of free-charge carriers due to impurities makes it very difficult or impossible to detect radiation. In fact, in real semiconductors, the charge current obtained by the application of an electric field in the material will be mostly due to impurity-derived free-charge carriers. Important efforts have been made in recent years to obtain high-purity semiconductor materials, and important progress has been achieved in this area. Today, it is possible to produce high-purity Ge that can be used directly as a semiconductor detector. However, this is not the case for Si, the other semiconductor material conventionally used in radiation detection in which the purer material available today still has impurity concentrations dominating its conductivity [2]. Because the absence of free-charge carriers is essential for detecting radiation, the main goal when constructing a semiconductor detector is to obtain an area in the material where no free charge is present. This can be achieved by different techniques. A brief account of them is given in what follows.

*Intrinsic Detectors*
When the semiconductor material is virtually free of impurities, we have an intrinsic semiconductor detector. This is only possible today with Ge, which can be

manufactured with impurity concentrations of approximately $10^9$ atoms/cm$^3$ [7], in practice a pure Ge detector. In this case, the detector follows the simple scheme of Fig. 11.2 in which two electric contacts are added to the crystal and an electric field is applied to collect the charge created when the radiation passes through the Ge. Large Ge ultrapure crystals can be produced today, which allows one to increase the counting efficiency. This, together with their intrinsic high-energy resolution, makes them the best option for $\gamma$-spectrometry, as we will see in the next section. Ge high-purity detectors must be operated at liquid nitrogen temperature to minimize the production of charge-free carriers by thermal effects. However, they can be stored at room temperature. This is an important difference regarding the Ge(Li) we will see later, in which the detector has to be stored at liquid nitrogen temperature to avoid breaking the detector configuration.

*The p-n Junction*

In contrast, it is currently not possible to produce a large Si volume pure enough to realistically use it as an intrinsic detector. However, the conductivity of a semiconductor can be modified by adding impurities as they introduce free-charge carriers into the crystal [1]. This is well known and can be used to fabricate realistic Si semiconductor detectors with the help of the concept of a p–n junction, which has been extensively studied in electronics. Before going into detail, let us revise some basic concepts in Solid-State Physics. In an intrinsic semiconductor material, the density of free-charge carriers is zero, except when electrons are promoted from the valence band to the conduction band due to thermal effects. In this case, the density of electrons in the conduction band (n) is equal to the density of holes in the valence band (p) since each hole appeared because of the electron promoted to the conduction band. The presence of impurities causes an excess of electrons or holes, and therefore, both densities are different. When $n > p$, the semiconductor material is said to be n-type and, on the contrary, when $p > n$, , p-type. Both types of ligands can be artificially obtained by controlling the process of addition of impurities, doping, to the semiconductor. Indeed, if we replace a Si atom (4 electronic valence) in the lattice with a 5 electronic valence element (As or P, for instance), an unpaired electron will appear. This electron is very weakly bound to the lattice because the binding energy is reduced by the effect of the dielectric constant of the material. The result is the appearance of an energy state very close to the conduction band (see Fig. 11.3a), a donor state that easily provides the conduction band with electrons by thermal effects. This increases n, producing an n-type semiconductor. In contrast, if we replace a Si atom with a 3 electronic valence element (e.g., B, Al, Ga, and In), a Si electron valence remains unpaired in the lattice. Now, B tends to capture one electron from the lattice to saturate the covalent bond. However, the binding energy of this electron is lower because the B nuclear charge is smaller. The result is the appearance of an energy level close to the valence band (see Fig. 11.3b), an acceptor level, which easily accepts electrons from the valence band by thermal effects, increasing p and producing a p-type semiconductor.

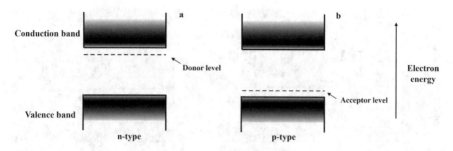

**Fig. 11.3** Donor (**a**) and acceptor (**b**) energy levels introduced in a Si lattice by doping it with pentavalent or trivalent impurities. See the text for easier comprehension

Both n- and p-type semiconductors are useless as detectors since the n- or p-density dominates the conductivity of the material. However, a p–n junction makes it possible to build a detector, since it helps to remove the free charge. Let us deposit an n-type semiconductor on one of the surfaces of a p-type semiconductor (see Fig. 11.4). In this way, we are forming a p–n junction in which an excess of electrons exists in the n-side and of holes in the p-side. Due to thermal effects, the electrons and holes from both sides will migrate across the junction such that the holes in the p-side will be occupied by the electrons. Similarly, it can be said that holes will migrate toward the n-side as the electrons leave. At the beginning, both sides were electrically neutral, but due to the migration of electrons, a net negative space charge was formed on the p side, and correspondingly, a net positive space charge was formed on the n side (see Fig. 11.4). This is because the migration of electrons leaves positively ionized donor impurities on the n-side and negatively ionized acceptor impurities on the p-side. The space charge created induces an electric potential across the junction, the so-called contact potential. It will progressively slow down the charge diffusion processes until a charge-state equilibrium can be reached when no more migration can take place. At equilibrium, such contact potential is found to reach a value close to 50–90% of the band gap in the semiconductor [8]. As a result, a region appears in the semiconductor with no free-charge carriers, the depletion region, limited by n- and p-type semiconductor regions, as shown in Fig. 11.4. The extension of this region on the p- or n-side will depend on the original concentrations of n and p.

The depletion region this way generated is a good candidate for acting as a radiation detector, since no free-charge carriers are present. Thus, a particle passing through the region will produce some electron-hole pairs along its pathway. The contact electric field will move the electrons toward the n-type region and the holes in the opposite direction. This electrical current can be used to detect particles. However, the contact potential is too low, generally ~1 V, to collect all the charged created and therefore avoid recombination and trapping processes [8]. Furthermore, the thickness of the depleted region is too small compared to the radiation range in these materials. Therefore, the performance of such simple p-n junctions as detectors is not acceptable. It is necessary to increase the electric field

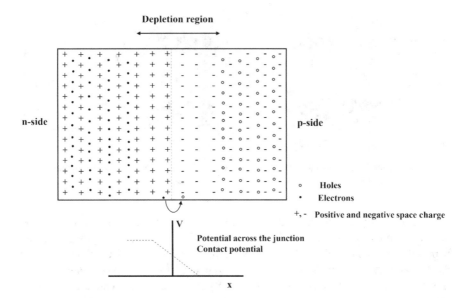

**Fig. 11.4** p–n junction

inside the depletion region and its length. This can be achieved by applying an external voltage to the junction. However, the selection of its polarity is crucial since the junction acts as a conventional diode [2]. Indeed, if the junction is biased in the forward direction, it will conduct the electrical current very easily, while if it is biased in the reverse direction, the conductivity will decrease dramatically. In our case, if we apply a positive voltage to the p-side of the junction, the potential barrier in the depletion region will decrease, and therefore, the electrons from the n-side will move toward the p-side. The holes will consequently move to the n-side, and the current will increase (Fig. 11.5a). This forward biased junction is not useful as a detector at all, since the movement of the majority carriers under the electric field will mask that of the electron-hole pairs eventually created by a particle crossing the junction. However, if we apply a positive potential to the n-side of the junction, the opposite situation will appear (Fig. 11.5b). The potential barrier increases in the depletion region, avoiding the movement of the majority charge carriers from the n-side to the p-side and vice versa. Moreover, the minority carriers on the p-side, the electrons, will move to the n-side, occupying the holes still empty on the n-side. Under this situation, the electric field in the depletion region increases as does its thickness. This reverse-biased junction can act as a realistic detector since the magnitude of the electric field ensures a complete collection of charge, and the thickness of the depletion region is large enough compared to the range of charged particles in the semiconductor.

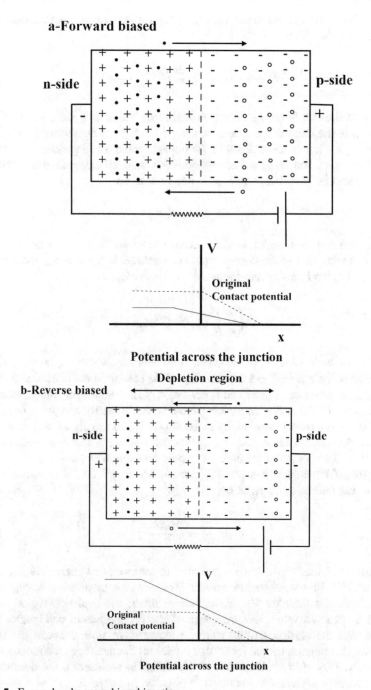

**Fig. 11.5** Forward and reverse-biased junctions

By solving the Poisson equation [2], it can be found that the thickness of the depletion region, $W$, is.

$$W = x_n + x_p = (\frac{2\varepsilon V_b}{q_e} \frac{N_a + N_d}{N_a N_d})^{\frac{1}{2}} \tag{11.4}$$

where $x$ is the length of the depletion region on the n or p side, $V_b$ is the bias voltage, $\varepsilon$ is the dielectric constant of the medium, $q_e$ is the electron charge, and $N_d$ and $N_a$ are the concentrations of the dopant on the n and p sides, respectively. Therefore, $N_d q_e$ and $N_a q_e$ are charge densities on each side. Since the depletion region should be electrically neutral, it turns out to be

$$N_d x_n = N_a x_p \tag{11.5}$$

Thus, the depletion region extends more or less into the n- or p-side, depending on the concentration of the dopant type. For instance, if $N_a > > N_d$, the depletion region will extend mostly into the n-side. In this case,

$$W \approx \left(\frac{2\varepsilon V_b}{q_e N_d}\right)^{\frac{1}{2}} = (2\varepsilon \mu_n \rho_n V_b)^{\frac{1}{2}} \tag{11.6}$$

where it has been taken into account that $\rho_n = (\mu_n q_e N_d)^{-1}$, $\rho_n$ and $\mu_n$ are the resistivity of the material and the mobility of the carrier on the n side, respectively (a similar equation can be derived for $N_a < < N_d$). As the bias voltage increases, the thickness of the depletion region increases. Additionally, to have a large $W$, it is necessary to have high-resistivity materials. This is normally associated with the purity of the semiconductor used as the junction wafer. For that, the semiconductor material used to construct the detector should be as pure as possible.

The depletion region has no free charge. Thus, it forms a capacitor with a capacity per unit area, $C$, given by

$$C = \frac{\varepsilon}{W} = (\frac{q_e \varepsilon N_d}{2V_b})^{\frac{1}{2}} \tag{11.7}$$

A small detector capacitance is essential to achieve good energy resolution [7]. For that, high bias voltages are needed. However, the breakdown voltage of the material poses a limit for $V_b$. If the voltage inside the depletion region reaches such a breakdown value, the properties of the reverse-biased configuration will be lost, and the electric current starts to increase. In several books [9, 10], an interesting nomogram can be found that relates all the quantities mentioned above.

Usually, $W < d$ is the thickness of the wafer. In such cases, the detectors are called partially depleted. However, it is possible to arrive at a $W = d$ condition simply by increasing $V_b$ to a value, $V_d$, called the depletion voltage, which can be found from (11.6) as

$$V_d = \frac{q_e N_d d^2}{2\varepsilon} \tag{11.8}$$

with an obvious meaning of the notation. These Fully Depleted Detectors have several advantages over those partially depleted, although the latter are of the most conventional use in the determination of environmental radionuclides, specifically $\alpha$-emitting radionuclides.

There are several methods for fabricating Si detectors based on p-n junctions [1, 2, 7]. The Diffused Junction Detectors are the older configuration. In this case, an n-type impurity is deposited on one side of a p-type Si wafer. This is done by exposing this side to a vapor of the impurity, usually p. The junction is formed at a given distance from the surface of the wafer. In practical terms, this means that a dead layer appears at the wafer surface, acting as a detector window where part of the particle energy can be absorbed. This situation is not convenient for spectrometric applications.

Si Surface Barrier Detectors help to overcome this drawback. In this case, the junction is formed by creating a high density of electron traps on the surface of an n-type Si wafer. The traps act as p-type impurities. A very thin Au layer is deposited on the surface and acts as an ohmic contact. The thickness of the layer is clearly smaller than those obtained for Diffused Junction Detectors, which is its main comparative advantage. However, Surface Barrier Detectors are very sensitive to ambient light since the Au layer thickness is sufficient to allow light in the optical range to pass through and excite the creation of electron-hole pairs. For that, they must be operated in darkness. On the other hand, the Au layer is very fragile and can be affected by the deposition of ambient dust or oil from the vacuum pumps needed for $\alpha$-spectrometry measurements. From this viewpoint, the Diffused Junction Detectors are more rugged and, therefore, still find room for some applications.

Today, the most common method for the fabrication of p-n junctions is the implantation of impurities into Si wafers using accelerators. Ion Implanted Detectors are obtained by bombarding the surface of the wafer by a $p^+$ or $n^+$ ion beam, $P$ or $B$, for instance. Because the energy and beam intensity can be controlled in these experiments, the depth of implantation and the abundance of impurities can be regulated. This is a clear advantage over previous fabrication methods, which are based mostly on empirical recipes. Ion Implanted Detectors are less sensitive to ambient conditions than Surface Barrier Detectors. Thus, they have been substituted in most applications.

*Li-Drifted Detectors*
Under the best purity conditions, W can reach values from 100 to 2000 μm for p-n junctions based on Si wafers. This limits its applications to low-range charged particles, i.e., $\alpha$ particles and $\beta$ radiation from medium to intermediate energies. Of course, even with the highest possible W, these detectors are not efficient for

$\gamma$ since the photoelectric cross section is too low for the atomic number of Si, ($Z = 14$), as seen in Chap. 7. As we will study in the next section, Ge ($Z = 32$) detectors are the best option for $\gamma$-spectrometry since $Z$ in this case is high enough to make probable the photoelectric effect over a wide range of $\gamma$ radiation energies. However, in the case that we could wide W or construct detectors with adequate length depletion regions (some few mm), it would be possible to efficiently detect X-rays or low-energy $\gamma$-rays [11]. The possibility of constructing Li drifted detectors allows one to address this problem.

Li drifted detectors were originally developed to construct Ge detectors of adequate size for $\gamma$-spectrometry by the time when it was not possible to obtain large high-purity Ge wafers. The so-called Ge(Li) detectors are today obsolete, but the technique is of application to Si, and Si(Li) detectors are currently the most popular option to carry out X-ray or Low-Energy Photon Spectrometry (LEPS) in general.

In general terms, the construction of Si(Li) detectors is based on the diffusion of p-type Li into n-type Si (or Ge as originally done) [2]. Li is deposited onto one of the Si wafer faces, and diffusion takes place. This creates a p–n junction at the Si surface. The junction is reverse biased, and the temperature increased. This helps to move the Li ions under the action of the electric field, which diffuse through the Si lattice. The concentration of donor impurities increases along the p-type Si, reaching values similar to those of acceptor impurities. As a consequence, there is a charge compensation process that gives place to an intrinsic region in which virtually there are no free-charge carriers. The electrical configuration of this new system, the p-i-n configuration, is given in Fig. 11.6. The i-region serves as a detector since it does not contain free-charge carriers. The n-and p-regions, in which an excess or defect of Li exists, can act as electrical contacts through which an electric field is applied, which collects the electron-hole pairs created by the radiation. The intrinsic region in Si extends up to approximately 10 mm (15 mm in the case of Ge). The external high voltage applied to this type of detector must be high enough to make for efficient charge collection in such a long-depleted region. For that, it typically reaches values from 500 to few thousands of $V$. This should be compared to the $V_b$ values, which are typically lower than or close to 100 V [1].

The Si(Li) detectors must be operated at liquid nitrogen temperature (77 K) to diminish the probability of current noise of thermal origin. It is also advisable, although not mandatory, to store the detectors at 77 K. The reason is the instability of the p-i-n configuration because of the Li mobility in the lattice. The mobility is not high since the dimensions of the Li ion compared to those of the Si lattice are comparable. Nevertheless, it is convenient to store the detector at liquid nitrogen temperature. This will extend its lifetime for longer. In the case of Ge(Li), storing at 77 K is crucial, since now the Li mobility is very high at room temperature, which can break the p-i-n configuration very rapidly.

**Fig. 11.6** p-i-n configuration

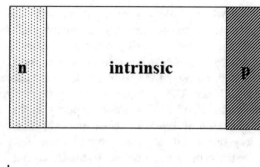

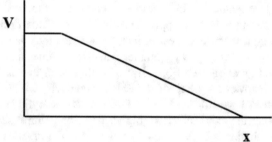

## 11.2  Gamma-Ray Spectrometry with Semiconductor Detectors

Detection and, more specifically, $\gamma$-spectrometry require high $Z$ detector materials and high-energy resolution configurations. Both conditions are met by Ge.
Indeed, at $Z = 32$, the photoelectric cross section is high enough for a wide $\gamma$-ray
energy range (see Chap. 7). The good energy resolution of semiconductor detectors has already been discussed. Furthermore, thick enough depleted regions are
necessary to allow $\gamma$-radiation to deposit its energy fully or partially, as seen in
(7.19). Ge(Li) detectors achieved all these conditions, since approximately 10 mm
'intrinsic' regions could be obtained with p-i-n configurations. Nevertheless, as
already mentioned, they were difficult to operate, mainly because of the need for
storage at liquid nitrogen temperature. For that, they have been substituted by high-
purity Ge detectors. As seen in (11.6), $W$ is inversely proportional to the impurity
density. Thus, low $N$ can provide long $W$ at a reasonable $V_b$. At present, it is possible to obtain Ge of sufficient size with impurity concentrations of approximately
$10^9$ atoms/cm$^3$ or relative abundances of $10^{-12}$ in Ge [7]. This means virtually
pure Ge. Under these conditions, depleted regions of 10–15 mm can be obtained.
This, together with the fact that Ge has a high $Z$, makes intrinsic or high-purity
Ge detectors the best option to carry out $\gamma$-spectrometry studies.

There are several Ge detector configurations, all of which are related to the n-
or p-type character of the Ge crystal. Indeed, although highly pure, Ge still has
a very low impurity concentration. Thus, if the impurity is p- or n-type, the Ge

crystal will be mildly p- or n-type. In any case, the impurity concentrations are very small, so the conductivity or resistivity of Ge is low or high, respectively [2].

The more conventional geometry is the so-called planar geometry (Fig. 11.7a). In this case, the detector volume is a p-type Ge cylinder (it is also possible to start with an n-type Ge, although less conventional) with two electrical contacts at both bases. They are constructed by depositing a $n^+$ donor impurity in one of the surfaces and a $p^+$ acceptor impurity in the other. Deposits are made by the implantation of ions with accelerators [2]. The reverse voltage is applied in such a way that the detector is operated under fully depleted conditions, allowing a more uniform electric field across the Ge crystal. Under these conditions, depletion region lengths up to 20 mm are obtained, but the detector volume is relatively low, reaching values of approximately 30 $cm^3$. This limits the efficiency for medium to high $\gamma$-ray energies. In fact, planar geometry is mostly applied to the detection of low-energy $\gamma$-rays. Higher detector volumes can be obtained with the coaxial configuration (see Fig. 11.7b). In this case, the external surface of the cylinder is recovered by one of the electric contacts, while the other contact recovers an internal well opened along the axis of the cylinder. Since the crystal can grow along the axial direction, higher detector volumes, up to 1000 $cm^3$ or more, can be obtained. External contact is $n^+$ for Ge crystals of type p and $p^+$ for Ge of type n [7]. The former is the conventional configuration in which the outer layer n, usually Li, has a thickness of approximately 0.5 mm. This limits the range of utility of the detector to energies beyond 40 keV. However, it is also possible to use a configuration in which the outer contact is p-type, usually B. In this case, outer layer thicknesses close to $10^{-1}$ $\mu$m can be obtained, which allows the measurement of lower, close to 10 keV, energy $\gamma$-rays. There is an additional limitation to the range of energies that can be measured with Ge detectors. As stated before, Ge detectors must be operated at liquid nitrogen temperature. To do that, the detector volume must be in contact with a Dewar cryostat filled with $LN_2$ through a cold Cu finger (see Fig. 11.8). At the same time, the detector must be isolated from the environment to keep its temperature constant and to avoid moisture or dust, which could affect its operation. For this, the detector is encapsulated under vacuum conditions, which means the appearance of an additional detector window that can attenuate the $\gamma$-radiation impinging on the detector. The old Ge detectors were encapsulated in Al containers, with a thickness that makes it difficult to detect low-energy $\gamma$-radiation. Modern designs include thin Be or polymer windows in the container, which facilitates efficient detection of energies close to a few keV [2]. In this way, the traditional advantages of coaxial detectors are maintained, although the energy response is extended in such a way that $\gamma$-radiation from approximately 5 keV to 10 meV can be detected efficiently. This is the energy range of interest in the determination of environmental radioactivity. In fact, most laboratories currently have this type of detector for environmental work involving $\gamma$-radiation measurements.

Ge detectors are usually characterized by two parameters: the relative efficiency and the energy resolution. The relative efficiency helps to compare the efficiencies of the Ge detectors with those of NaI(Tl), which is expected to be more efficient

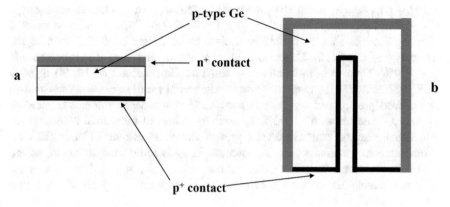

**Fig. 11.7** High-purity Ge detectors in planar (**a**) and coaxial (**b**) configurations

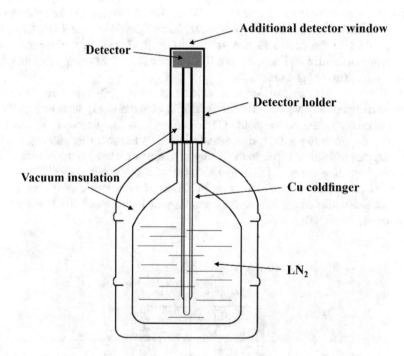

**Fig. 11.8** Conventional Dewar cryostat

for $\gamma$-radiation since its average atomic number is higher. Relative efficiency is defined as

$$\varepsilon_r(\%) = \frac{\varepsilon_{Ge}}{\varepsilon_{NaI(Tl)}} 100 \tag{11.9}$$

with $\varepsilon_{Ge}$ and $\varepsilon_{NaI(Tl)}$ being the photopeak efficiencies of the Ge detector and a $3' \times 3'$ NaI(Tl) detector, respectively, for the 1332 keV $^{60}$Co $\gamma$-radiation emitted from a point-like source situated at a distance of 25 cm along the axis of the cylinders (see Chap. 10). Due to the different atomic numbers, it is expected that $\varepsilon_r < 100\%$. Thus, conventionally, one can find Ge detectors of 10, 30 or 40% relative efficiency in the market. However, the technique of pure Ge crystal growth has evolved greatly, and today, it is possible to obtain large volumes capable of producing detectors with $\varepsilon_r > 100\%$, reaching values of more than 200%, which will increase as the pure Ge crystal growth technique evolves [12]. In the case of environmental radioactivity, the increase in $\varepsilon_r$ is more than welcome, as the main purpose is the quantification of radionuclides. This, in combination with the Ge energy resolution, as we see in what follows, converts $\gamma$-spectrometry into a powerful analytical tool.

In fact, energy resolution is also a parameter that characterizes the detectors, and the manufacturers provide these data together with $\varepsilon_r$. Conventionally, the energy resolution is given as FWHM (see Chap. 8), specifically for the 122 keV photopeak from $^{57}$Co and for the 1332 keV photopeak from $^{60}$Co. Values of 1.00–1.25 keV can be found in the first case and from 1.81 to 2.13 keV in the second case, depending on the size and geometry of the detector [12]. This energy resolution is excellent for many applications.

Finally, it is interesting to describe the well geometry, which is a very efficient detector and certainly very useful for specific applications. As seen in Fig. 11.9, a well-like hole in the center of the Ge detector allows the placement of samples in a quasi $4\pi$ geometry, which significantly increases the counting efficiency. This geometry has wide application in the dating of marine or lake sediments with $^{210}$Pb (see the pioneering paper in [13]), in which the 46.5 keV emitted by $^{210}$Pb is used for its quantification in sediment samples. The main drawback of the technique is the obvious size limitation of the sample, which decreases the FOM of the experiment, as we will see in Chap. 14.

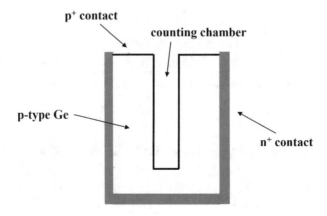

**Fig. 11.9** Geometry of the well detector

## 11.2.1 Pulse Height Spectrum

A typical Ge $\gamma$-spectrometer has the same structure as that presented in Fig. 10.9, substituting the NaI(Tl) detector for a Ge detector. The response of a Ge detector to $\gamma$ radiation is similar to that of a NaI(Tl) detector. Thus, all the concepts described in Chap. 10 apply in the present case. The relevant difference lies in the better energy resolution that the Ge detectors have compared to NaI(Tl). From this viewpoint, Ge detectors are able to distinguish between close energies significantly better than NaI(Tl). This makes them the most popular (or better, the only) option to solve the problem of environmental radioactivity determination, in which complex $\gamma$-spectra are very often managed. In Fig. 11.10, examples are given to compare the $\gamma$-spectra obtained with Ge and with NaI(Tl). Thinner photopeaks were obtained with Ge as a result of its better energy resolution. Thus, radionuclides are distinguished more clearly in the spectrum, and multiplets are better resolved. Of course, since the Ge atomic number is smaller than that of NaI(Tl), the relevance of the Compton continuum in the whole spectrum is higher. For the same reason, the valley between the Compton edge and the photopeak is more relevant since the probability of multiple Compton scattering is higher. Moreover, for small-sized Ge detectors, the photopeak is formed by pulses from photoelectric absorptions but mostly by photon Compton scattering followed by photoelectric absorptions. In fact, the peak-to-Compton ratio is a Ge detector feature that manufacturers generally provide to fully describe it. It is defined as the ratio between the count number at the photopeak centroid and the count number at the Compton area associated with the peak [7]. As the size of the detector increases, this ratio and $\varepsilon_r$ increase.

Its lower atomic number also explains the relevance of escape peaks in Ge detectors. Indeed, the probability of photoelectric absorption of the photons created inside the detector is lower, so it is more likely to find the single or double escape peaks associated with the pair production in the detector by high-energy $\gamma$-rays or the X-ray escape photopeak, in this case coming from the Ge characteristic X-ray. Again, the detector size tends to compensate for these effects. In contrast, the much better energy resolution of Ge makes it possible to clearly identify photopeaks in statistically poorly defined $\gamma$-spectra, which is a common situation in environmental applications. An example is given in Fig. 11.11, which demonstrates that Ge detectors are the only option for these problems, despite their lower counting efficiency.

## 11.2.2 Identification of Radionuclides and Activity Calculation

As in the case of NaI(Tl), Ge $\gamma$-spectrometers must be calibrated to obtain the information contained in the spectra. First, to identify the radionuclides in the sample, an energy calibration must be performed. For that, a set of well-known energy $\gamma$-emitters are selected, and their spectra are obtained under the same counting

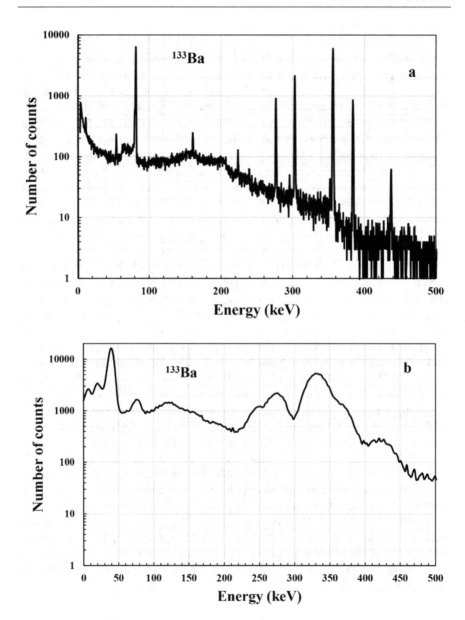

**Fig. 11.10** Comparison between **a** Ge and **b** NaI(Tl) $\gamma$-spectrum obtained for $^{133}$Ba

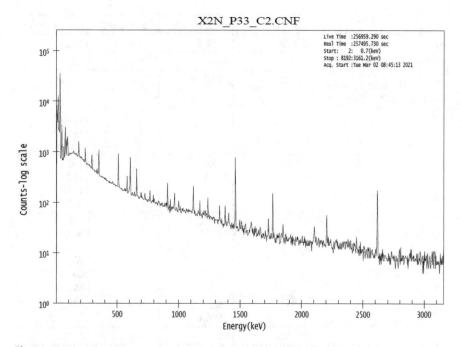

**Fig. 11.11** A real environmental gamma spectrum of a marine sediment sample

conditions as the real sample. In Table 11.3, an example of the commonly used standards is given.

Of course, a calibration of energy resolution as in the case of NaI(Tl) must be done by using the same standards. The determination of the peak centroid can be done manually by identifying the channel at which the number of counts is maximum in the photopeak. The other possibility is to fit the data to a Gaussian-shaped peak [2]. The fit provides the centroid and the FWHM. This, in turn, helps to ensure that the peak is a photopeak. Since the available range of energies in Ge detectors is longer than in NaI(Tl), the fitting functions for $E_\gamma$ and FWHM are more complex than for NaI(Tl). Some of them are given in Table 11.4 [15, 16].

Once the radionuclide is identified by the $\gamma$-ray energy or energies, its activity is calculated by applying (10.7). A photopeak efficiency calibration is also necessary for determining $\varepsilon_{ph}$. The calibration of the Ge spectrometer is carried out once again by using standard samples of well-known activities covering the energy range of the detector. The most common approach is to fit the set of obtained values to a curve that can take different shapes, as observed in the literature. We give some examples in Table 11.4, and in Fig. 11.12, we plot some real efficiency data at two standard counting geometries.

Efficiency calibration is not a trivial task. There are different effects that must be considered when using an empirical approach. For instance, sum-coincidence or self-absorption effects are sample-dependent. Thus, the applicability of the

**Table 11.3** Common γ-emitters used for energy calibration. Other radionuclides are possible. One interesting example is [116m]Ho [14], which provides 58 defined photon energies

| Radionuclide | Energy (keV) | Radionuclide | Energy (keV) |
|---|---|---|---|
| [22]Na | 1274.6 | [141]Ce | 145.4 |
| [24]Na | 2754.1 | [144]Ce | 133.5 |
| [54]Mn | 834.8 | [152]Eu | 121.8 |
|  |  |  | 344.3 |
|  |  |  | 778.9 |
|  |  |  | 964.1 |
|  |  |  | 1112.1 |
|  |  |  | 1408.0 |
| [57]Co | 122.0 | [182]Ta | 100.1 |
|  | 136.5 |  | 152.4 |
|  |  |  | 179.4 |
|  |  |  | 222.1 |
| [60]Co | 1173.2 | [192]Ir | 295.9 |
|  | 1332.5 |  | 308.4 |
|  |  |  | 316.5 |
|  |  |  | 468.1 |
|  |  |  | 604.4 |
|  |  |  | 612.4 |
| [88]Y | 898.0 | [198]Au | 411.8 |
|  | 1836.1 |  |  |
| [109]Cd | 88.0 | [203]Hg | 279.2 |
| [124]Sb | 1691.0 | [207]Bi | 569.7 |
|  |  |  | 1063.7 |
| [131]I | 364.5 | [208]Tl | 583.1 |
|  |  |  | 2614.7 |
| [133]Ba | 53.1 | [212]Pb | 238.6 |
|  | 81.0 |  |  |
|  | 276.4 |  |  |
|  | 302.8 |  |  |
|  | 356.0 |  |  |
|  | 383.8 |  |  |
| [137]Cs | 661.6 | [241]Am | 59.5 |
| [139]Ce | 165.9 | Annihilation | 511.0 |
| [140]La | 1596.2 |  |  |

efficiency curve to the real situation must be ensured. This is a key issue in environmental applications, as we will see in Chap. 13. However, it can be said that the application of Monte Carlo calculations has made great progress in this area [17]. There, the photon trajectories and the interactions are simulated within the sample, the surrounding materials, and the detector. The use of a Monte Carlo approach requires the knowledge of several important parameters, such as the detector dimensions, exact composition, including that of the detector window (many times protected by proprietary regulations), and the geometry and composition of the sample. Many of these data are not easy to obtain. In some cases, X-ray

**Table 11.4** Fitting functions for energy, FWHM, and efficiency calibrations

| Parameter | Function |
|---|---|
| Energy (E) | $E = a + b(ch)$ |
| | $E = a + bx + c\,(ch)^2$ |
| Full Width Half Maximum (FWHM) | $FWHM = a + bE$ |
| | $FWM = \sqrt{a + bE}$ |
| | $FWHM = \sqrt{a + bE + cE^2}$ |
| Photopeak efficiency ($\varepsilon$) | $\ln\varepsilon = \sum_i a_i(\ln(E))^i$  i = 0–5 or 6 |
| | $\ln\varepsilon = a_1 - (a_2 + a_3 e^{-a_4 E}))e^{-a_5 E}\ln E$ |
| | $\ln\varepsilon + k =$ |
| | $(a_1 + a_2\ln E + a_3(\ln E)^2)\arctan(e^{(a_4 + a_5\ln E + a_6(\ln E)^2)})$ |
| | $\varepsilon = a_1 E^{-a_2} + \sum_{i=2}^{4} a_{2i-1}e^{-a_{2i}E}$ |
| | $\varepsilon = \frac{1}{E}\sum_i a_i(\ln E)^{i-1}$  i = 1–8 ($\neq$5,7) |
| | $\varepsilon = a_1(e^{-a_2 E^{a_3}} + e^{-a_a E^{a_5}})(1 - e^{-a_6 E^{a_7}})$ |

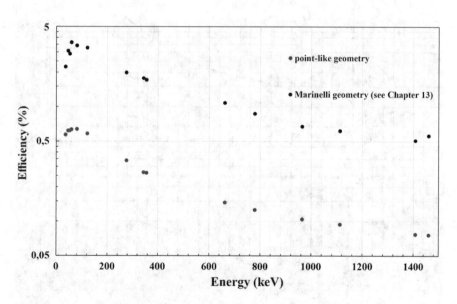

**Fig. 11.12** Photopeak efficiency curves for two standard counting geometries

radiography of the detector is needed to obtain the exact dimensions. Regardless, the Monte Carlo approach allows us to better simulate the effects previously mentioned, and its use is increasing.

## 11.3   Alpha- and Beta-Spectrometry with Semiconductor Detectors

Charged particle spectrometry is the main application of Si junction detectors. Specifically, $\alpha$-spectrometry is of wide use in the determination of environmental radioactivity. Their energy resolution makes Si detectors superior to any other alternative. On the other hand, it is possible to obtain sufficient thicknesses of the depletion region compared to the $\alpha$-particle range in Si [1]. The same can be said for $\beta$-particles of low to medium energies (see Fig. 11.13) [18].

The identification of Si detectors follows certain rules. Thus, the characteristics of the detector are given by a set of figures. For instance, a PD50-15–100

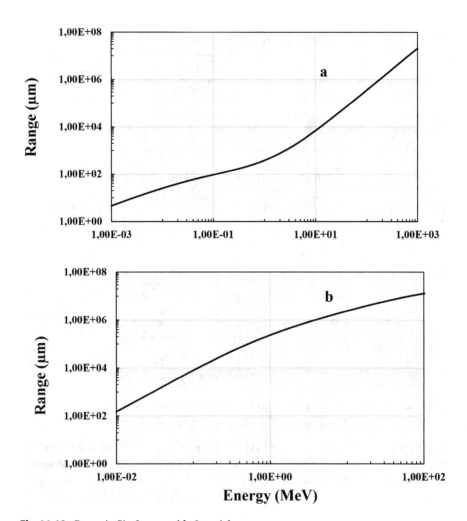

**Fig. 11.13**  Range in Si of **a** $\alpha$- and **b** $\beta$-particles

Si detector is a partially depleted type with a 50 mm$^2$ active area, an FWHM of 15 keV for the $^{241}$Am $\alpha$-emission of 5.485 meV (see Fig. 11.14c), and a depletion region 100 µm long at the advised bias. For activity determinations, higher active areas are better. Nevertheless, it is important to see that as the area increases, the energy resolution tends to worsen because of the increase in the detector-associated capacity (see comments to 11.7). This situation is currently being overcome by improving the purity of Si wafers. Thus, it is possible to find detectors of high active area, preserving an optimum energy resolution (see Table 11.5). Generally, the junction configuration used in environmental radioactivity is of partially depleted type at the $V_b$ advised by the manufacturer. However, FD detectors can also be an option for these applications [2].

Si detectors must be operated under vacuum conditions that minimize the continuous loss of energy that charged particles undergo along their trajectory in matter (see Chap. 7). For that, the sample detector assembly is enclosed in a vacuum chamber. However, the vacuum level has to be adequately optimized, since an extremely good vacuum could facilitate the implantation of recoil nuclei from the disintegration process into the detector. This is especially important for the case of $\alpha$-particles emitting radionuclides. Normally, vacuums of 10$^{-2}$ to 10$^{-3}$ mm Hg are used, although this figure can be modified according to the experiment. The vacuum chamber also serves to prevent ambient light from reaching the detector. Indeed, as commented before, some Si detectors are sensitive to light in the optical range, since the forbidden gap width and the energy necessary to create an electron-hole pair are small enough compared to the light energy in the optical range.

For the same reason, the sample must be as thin as possible to avoid energy losses by self-absorption processes. There are several methods to achieve that. Electrodeposition, vacuum sublimation, controlled evaporation, and others help to obtain thin or very thin samples and thus to take advantage of the excellent energy resolution of Si detectors. We will return to this problem in Chap. 13.

## 11.3.1 Pulse Height Spectrum

A typical Si p–n junction $\alpha$- or $\beta$-spectrometer has the same structure as that presented in Fig. 10.9, now substituting the NaI(Tl) detector with a Si detector placed in a vacuum chamber.

The main application of this scheme in environmental radioactivity is the determination of $\alpha$-emitting radionuclides by $\alpha$-spectrometry. The discrete nature of the $\alpha$-spectrum facilitates its use, and since these detectors are not very sensitive to the different background sources (see Chap. 8 and the comments in the final section of this chapter), the scheme of Fig. 10.9 can be used without special modifications. For a monochromatic $\alpha$-emitting nuclide, the expected spectrum is a simple Dirac $\delta$ function centered at the $\alpha$-particle energy, as seen in Fig. 11.14a. Under realistic conditions, a widening of the peak is observed (Fig. 11.14b), as in the previously studied cases of NaI(Tl) and Ge detectors. The reasons for such a

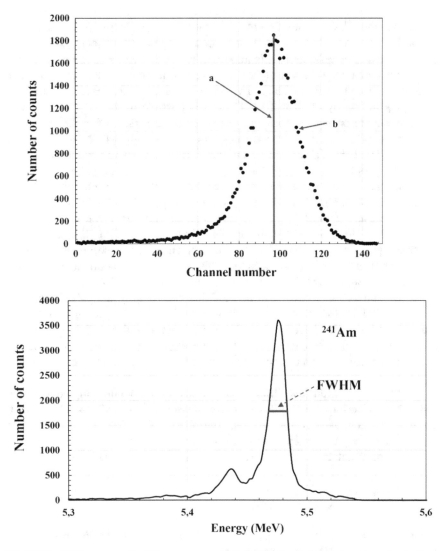

**Fig. 11.14** Ideal response (**a**) and real response (**b**) of a Si α-spectrometer to a monochromatic α source. The figure includes a real $^{241}$Am α-spectrum indicating how the FWHM is determined

widening are similar and, therefore, related to the statistical nature of the charge carrier production per unit of energy. However, in this case, a certain asymmetry at the low-energy side appears. This is because α-particles continuously lose energy along their trajectory toward the detector. In fact, part of the energy can be lost in the sample by self-absorption, part in the residual air between the sample and the detector, and part in the detector window.

**Table 11.5** FWHM for several Si detectors of different sizes

| Active area (mm$^2$) | FWHM (keV) |
|---|---|
| 25 | 12 |
| 50 | 12 |
| 150 | 14 |
| 300 | 16 |
| 450 | 17 |
| 600 | 22 |
| 900 | 27 |
| 1200 | 35 |
| 2000 | 35 |
| 3000 | 50 |
| 5000 | 75 |

In other words, the $\alpha$-particles arriving at the detector have a certain energy distribution with a maximum energy at the original energy of the particle. For these reasons, the thickness of the sample and the vacuum conditions in the counting chamber must be optimized to minimize energy losses. Regarding the influence of the detector window, it is interesting to note that its absorbing effect can be minimized if the $\alpha$-particle trajectories are normal to the detector window plane. As shown in Fig. 11.15, $\alpha$-particle trajectories with angles $>90^0$ will explore longer dead distances in the window, depositing more energy before entering the active detector volume.

In real experiments, increasing the distance sample detector is a natural way to select normal trajectories and thus improve the instrumental energy resolution by decreasing the asymmetric deformation, which contributes to peak widening. Nevertheless, this, in turn, diminishes the counting efficiency, which is not convenient in low-level radioactivity measurements. Conventionally, in $\alpha$-spectrometry, the energy resolution is estimated directly from the FWHM value. It is calculated as explained in Chap. 8. The FWHM is fairly constant in Si p–n junction detectors

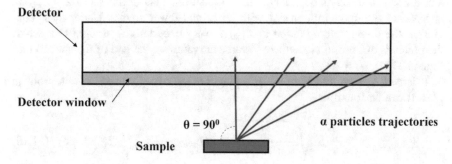

**Fig. 11.15** $\alpha$-particle trajectories that impinge at different angles on the detector window

for $\alpha$-particle energies ranging from 4 to 9 MeV [7]. This is the range of interest in the determination of environmental radioactivity. The energy calibration of Si $\alpha$-spectrometers can easily be performed. The energy-channel relationship is just a straight line similar to (10.3). It can be found by calibrating the spectrometer with well-known energy $\alpha$-emitters as $^{241}$Am and $^{210}$Po. In many cases, the use of a pulse generator helps to widen the validity of the straight line obtained. In metrological applications, the determination of the peak centroid is usually performed by fitting the peak to a given curve assumed to represent the peak shape in the spectrometer. It is possible to find different peak shapes in the literature [19]. For environmental applications, it is sufficient to take the channel with the maximum count number as the peak centroid. Actually, for practical purposes, energy calibration is not strictly necessary, provided that radiochemistry must be used to extract the radionuclide of interest from the sample (see Chap. 13). However, it is convenient to have a calibration curve to identify unknown peaks that could appear in complex $\alpha$-spectra.

The application of $\beta$-spectrometry with Si detectors for environmental radionuclide determinations is very scarce. The main reason is the continuum nature of the $\beta$-spectrum that makes it difficult to interpret the results, especially when treating low-level radioactivity samples. However, some examples can be found in the literature, such as the measurement of $^{99}$Tc in seaweed samples [20]. As in the case of $\alpha$-spectrometry, radiochemistry is compulsory to extract the radionuclide of interest from the sample. Otherwise, the interpretation of the results is simply impossible.

Generally, the energy calibration of the $\beta$-spectrometer is performed using electron conversion emissions, for example, from $^{137}$Cs and $^{207}$Bi. The discrete nature of the recorded spectra allows one to associate a given energy (see Fig. 11.16) with a specific channel. An energy-channel linear relationship is sufficient to calibrate the spectrometer for energies. In real applications, after radiochemical separation, it is expected that the radionuclide of interest is the only radioactive species present in the sample.

However, there may also be interferences. Hence, it must be identified to ensure that we are measuring what we want to measure. The energy calibration is useful for that, although in $\beta$-spectra we have no peaks, but a continuum spectrum with a shape which depends on the type of $\beta$ transition. This means that the spectrum approaches the maximum energy with fewer and fewer counts. Under these conditions, the direct determination of $E_{\mathrm{max}}$ is very uncertain. A straight line across the points is not useful since there is some curvature at the end of the spectra (see Fig. 11.17).

It is well known that Fermi–Kurie plots are very useful to solve this problem [21]. It can be found that

$$\sqrt{\frac{N}{p^2 F(Z', p)}} = k(E_{max} - E) \qquad (11.10)$$

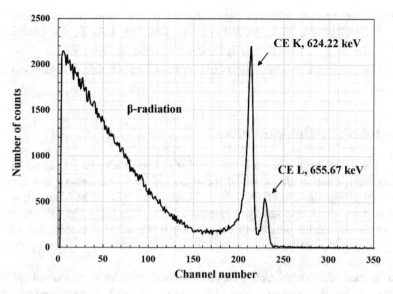

**Fig. 11.16** $^{137}$Cs conversion electron (CE) spectrum obtained with a Si detector. The continuum $\beta$-spectrum emitted by the radionuclide also appears in the recorded spectrum

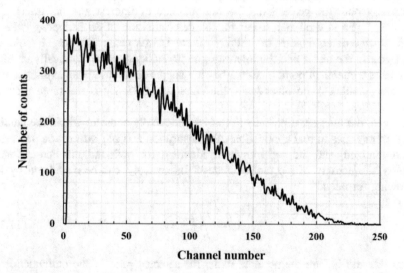

**Fig. 11.17** $^{204}$Tl $\beta$-spectrum obtained with a Si detector

$N$ is the number of counts at a given energy E, p is the corresponding momentum of the electrons, $F(Z', p)$ is the Fermi function, $E_{max}$ is the maximum $\beta$ energy, and $k$ is a constant that is energy independent. By plotting $\sqrt{\frac{N}{p^2 F(Z',p)}}$ against the energy, a straight line should be found that intercepts the channel axis at $E_{max}$.

## 11.3.2  Activity Determination

Either for $\alpha$- or $\beta$-spectrometry, the activity of the sample is calculated by integrating the $\alpha$-peak or the whole $\beta$-spectrum and using the appropriate counting efficiency as explained in Chap. 8. In addition, as we will see in Chap. 13, when a radiochemical extraction has been carried out, the chemical yield must be known to properly convert the peak or spectrum integral into activity. In this case, a chemical yield tracer must be used to monitor extraction efficiency. We will return to this problem in Chap. 13.

In the case of $\alpha$-spectrometry, the most popular application, the sample of interest is spiked with a known amount of yield tracer prior to chemical extraction. Usually, the tracer is also an $\alpha$-emitter, so it is expected that the corresponding $\alpha$-peaks appear in the spectrum. In Fig. 11.18, we have an example of an $\alpha$-spectrum containing both the peaks from the radionuclide of interest and the tracer. For instance, $^{242}$Pu is used as a tracer for the determination of Pu isotopes, $^{208}$Po or $^{209}$Po for the determination of $^{210}$Po, $^{232}$U for U isotopes, $^{229}$Th for Th isotopes, etc. Because the tracer and the interesting radionuclide behave identically or similarly from a chemical point of view, they are supposed to have been extracted with the same yield. On the other hand, the counting efficiency is the same for both radionuclides.

Indeed, the intrinsic efficiency in (8.6) is practically constant within the studied range of energies in real applications, the absorption and self-absorption processes are minimized, and the solid angle is identical for both radionuclides. Consequently, the activity of the radionuclide of interest, $A_s$, can be calculated by the following formulas.

$$A_s = \frac{I_y}{I_s} \frac{\Sigma_s}{\Sigma_y} A_y \qquad (11.11)$$

where $\Sigma_s$ and $\Sigma_y$ are the net area under the selected peak of the radionuclide of interest and the yield tracer, respectively; $I_s$ and $I_y$ are the corresponding radiation intensities of the measured $\alpha$ peaks; and $A_y$ is the activity of the tracer, which in turn is known. Of course, the integration of $\alpha$-peaks is not trivial since the low-energy tail theoretically extends up to the zero channel. In real low-level radioactivity $\alpha$-spectra, it is easy to define the integration limits since the number of counts on the low-energy side of the peak is very low. Certainly, the integration

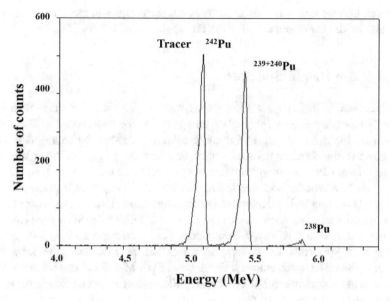

**Fig. 11.18**  Example of an $\alpha$-spectra from a Pu isotopes sample with a $^{242}$Pu spike

criteria should be kept fixed for all experiments to obtain reliable results. The procedure presented here is called the isotope dilution method and will be detailed in Chap. 13.

## 11.4   X-ray Spectrometry with Semiconductor Detectors

Ge detector configurations are capable of detecting a wide range of photon energies, including those within the X-ray range. They are very useful in environmental applications, since with only one measurement, it is possible to have information on an ample set of radionuclides, including those emitting X-rays after conversion electron events. The measurement of such X-rays reveals or confirms the presence of these radionuclides in the sample. As previously described, these spectrometers have to be calibrated in energy, FWHM, and efficiency following the standard rules. With regard to the detection of X-rays, the main drawback of this technique is the Compton continuum from high-energy $\gamma$-rays. It can mask the low-energy X-ray peaks that in environmental samples are usually not very well defined from a statistical viewpoint. For that, Si(Li) spectrometers are preferable for X-ray spectrometry. In fact, due to its atomic number, it is basically transparent for energies greater than 50 keV [22]. For this energy and below, the photoelectric effect is the predominant photon interaction mechanism, since virtually there is no Compton scattering. Consequently, the pulse height spectrum is basically the compound of a set of photopeaks. Despite this, Si(Li) is not widely used in environmental radionuclide determinations. There are only a few examples in the current scientific literature for the measurement of [55]Fe, Pu isotopes, and others [23].

### 11.4.1 Pulse Height Spectrum

The photopeak in Si(Li) spectra has a Gaussian or near-Gaussian shape with low-energy tailing provoked by incomplete charge collection in the crystal. This effect depends on the detector size and is more relevant because the energy deposit is done close to the detector window, which acts as a dead layer. In other words, it is more relevant for low energies when the photoelectric effect takes place just by entering the detector. This is also the reason why the X-ray escape peaks are very relevant when using Si(Li) detectors. The fact that photoelectric effects occur close to the detector window increases the escape probability of a Si X-ray (1.8 keV) from the detector. In this way, a peak situated at an energy $(E_x-1.8)$ keV usually appears in the spectrum. The calibration of a Si(Li) spectrometer follows the same steps as in the other spectrometers. Energy and FWHM calibrations are performed with the same objectives as before: the identification of radionuclides and the discrimination of false peaks. In many cases, this calibration is not strictly necessary since the peaks appearing in the spectrum correspond to expected radionuclides. As explained above, the application of Si(Li) to environmental radionuclide determination is usually limited to the measurement of specific radionuclides that could have been previously chemically extracted. However, for the sake of comparison, it is interesting to know that the FWHM for 5.9 keV [55]Fe X-ray can range from 140 to 180 eV, reaching a value of 130 eV, close to the statistical limit, in some cases. This makes Si(Li) superior to other X-ray spectrometer options, such as the use of proportional or scintillation detectors [23].

### 11.4.2 Activity Determination

The determination of activity requires calibration of the efficiency of the spectrometer. There is a scarcity of activity standards that makes calibration difficult. However, it is possible to find some of them. [241]Am emits a relatively ample range of low-energy photons and is an option for efficiency calibration. Monte Carlo simulations, as previously described for $\gamma$-spectrometry, are a very useful tool considering that we now deal with very low energies. This makes it crucial to reproduce the composition and geometry of the real samples when using an empirical approach to obtain the efficiency calibration curve. As expected, the detector window clearly affects the efficiency calibration curve. Some examples are given in Fig. 11.19. Thinner windows are welcome, and the use of *Be* or polymers is an excellent choice [24]. However, it is interesting to note that when using very thin polymer detector windows, the counting must be performed in a dark ambient. The reason is the nonnegligible probability that the ambient light passes through the window and reaches the detector, generating electron-hole pairs.

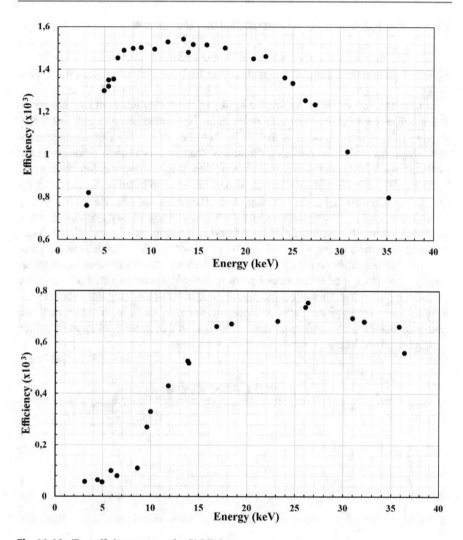

**Fig. 11.19** Two efficiency curves for Si(Li) detectors

Once the efficiency is known, the activity can be calculated by using an expression such as (8.7). For that, an appropriate peak integration must be done to avoid the tailing effects. This means making an assumption about the peak shape and fitting the data to it. As commented before, Gaussian or near-Gaussian shapes are used. Some of them can be found in [25].

## 11.5    Background in Semiconductor Detectors

For the determination of radionuclides in the environment, the most used are the p–
n junction Si and Ge high-purity detectors. For that, we only give here information
on them.

Si detectors are virtually insensitive to any of the background sources described
in Chap. 8, except for the electronic noise, $B_{inst}$. The reasons are varied. First, they
are very thin, and the $Z$ of Si is low. Both features make them almost transparent
to $B_\gamma$, $B_{cos}$, $B_\mu$ and to the radiation eventually produced by nuclear reactions in the
surroundings, giving rise to $B_{\mathrm{nucl}}$. However, they work under vacuum conditions.
For that, $B_{Rn}$ is irrelevant. Moreover, Si and Ge are the cleanest materials available today. Therefore, $B_{\mathrm{det}}$ is not important. However, the materials surrounding
the detector, including the electrical connections and the vacuum chamber, could
contain some impurities, mostly of the natural $U$ isotopes and Th series, which can
be easily identified in the recorded spectrum. However, the amount is usually not
relevant. Furthermore, continuous use of the detector can result in its contamination by implantation of recoil nuclei from $\alpha$ disintegrations in measured samples.
As commented above, this effect can be minimized by optimizing the vacuum conditions. As a consequence, these detectors do not need passive shielding as those
described in Chap. 15 for normal operation. This makes them readily useful since
no special infrastructure is needed to work with them.

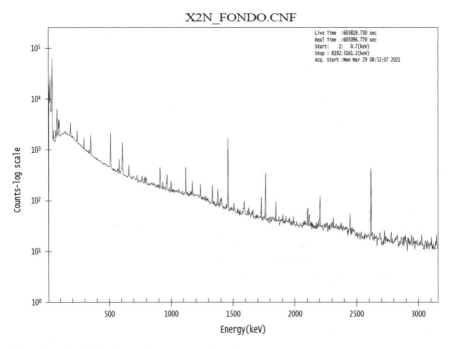

**Fig. 11.20** Typical background $\gamma$-spectrum in a Ge detector

The case of Ge is very different [26]. In general terms, they are sensitive to $B_\gamma$, $B_{Rn}$, and $B_\mu$. In the latter case, the Bremsstrahlung radiation created by the $\mu$ passing through the detector. This contributes to the continuum of the spectrum. Of course, since the Ge atomic number is lower than that of NaI(Tl), the contribution of this component is comparatively not as important. The relative importance of each background component depends on the detector size. Since they need passive or active shielding to decrease the background (see Chap. 15), $B_{sh}$ is important, as well as $B_{cos}$ and $B_{nucl}$, since the probability of nuclear reactions around the detector is relevant. The detector material is very pure, but not necessarily the electrical connections and, significantly. For that, $B_{det}$ can be of importance in the determination of very low levels of activity. In Fig. 11.20, we give an example of the background of Ge spectrometers. More information on this topic is presented in Chap. 15.

**Exercises**

1. Estimate the temperature needed for an electron to gain enough thermal energy to overcome the gap in Si and Ge.
2. In the figure, we have the decay scheme of the nucleus $^A$X has a very long half-life.

    Suppose that you have a Si (Li) detector. Describe an experiment to measure the activity of a sample containing $^A$X. Set up the equations of the method assuming that the necessary parameters are known.

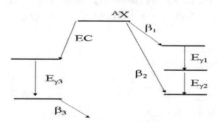

3. Sketch the spectrum obtained with coaxial HPGe. Consider $E_{\gamma 1} < E_{\gamma 2} < E_{\gamma 3}$ with $E_{\gamma 3} > 1.022$ meV.
4. Suppose that you carry out the same experiment with a much smaller active volume detector. To what extent will the response function obtained vary?
5. Suppose now that the Ge detector is of the extended energy range type. How will the spectrum change?
6. In the figure, we have the decay scheme of the nucleus $X$ that decays into the nucleus $Y$, emitting the corresponding radiation.

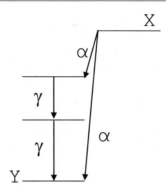

Propose two methods to measure the activity of a sample that contains a certain mass of nucleus $X$. Compare them and comment on the advantages and disadvantages of the methods.

7. Of the detectors you know, choose the best to measure the energy of the gamma radiation emitted by the source. Justify the selection.

8. Of the detectors you know, choose the best to measure the energy of the charged radiation emitted by the source. Justify the selection.

9. According to the decay scheme, is electron emission possible? And what about X-rays? Justify the answer.

10. $^{210}$Pb is extremely useful for sediment dating. According to its decay scheme, there are at least three methods to measure its activity. Propose two of them and briefly explain their basis. Select the appropriate detectors, justifying the choice.

11. We have a $^{213}$Bi sample whose decay scheme can be found at https://www.nndc.bnl.gov/nudat2/. Sketch and describe the spectrum obtained with the following detectors:
    (a) Geiger–Müller.
    (b) Proportional detector.
    (c) NaI(Tl).
    (d) HPGe.
    (e) Si ion, PD 100–50-15.

12. Suppose that we have a $^{64}$Cu simple whose decay scheme is

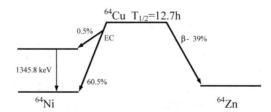

Sketch the $\gamma$-spectrum that would be obtained with a Ge detector. Explain each part.

13. See the next decay scheme

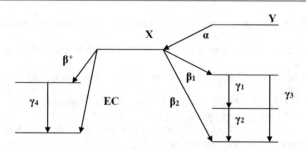

We have a sample with a certain number of $Y$ nuclei and would like to measure its activity. However, we do not have a suitable detector for $\alpha$ particles. Consequently, it must be measured through the radiation emitted by $X$. Propose a method to measure it

14. Find an expression for the activity of $Y$ as a function of the $X$ activity of $X$. Suppose all the necessary data are known. Note that there are no reasons to apply equilibrium concepts.

15. Choose the most suitable detector to measure the energies of the $\gamma$ radiation emitted from the source. Sketch the expected spectrum explaining each part. Suppose $E_{\gamma 1} < E_{\gamma 2} < E_{\gamma 3} < E_{\gamma 4}$ with $E_{\gamma 4} > 1.022$ MeV.

16. Propose an experimental method to show the existence of $\gamma 3$ emission.

17. A point-like $^{60}$Co sample is measured with a $3' \times 3'$ NaI(Tl) and a 30% relative efficiency Ge detector, sketch the $\gamma$-spectra obtained with both detectors and estimate the photopeak ratios if the activity of the sample is $A$.

18. Sketch the spectrum obtained with an Ion Implanted Si detector for a $^{137}$Cs source. Explain each of the spectrum regions.

19. Sketch the spectrum obtained with a PD100-50-15 Si detector for a $^{210}$Po sample. Describe and explain the shape and different parts. Suppose now that the spectrum is obtained without applying the reverse bias voltage. Explain the differences.

20. Suppose now the spectrum is taken with a PD100-300-20. Sketch the spectrum and describe the differences from those found in the previous problem.

21. Sketch the spectrum obtained with a PD100-100-15 Si detector for a $^{232}$U sample. Identify each of the obtained peaks.

## References

1. W.R. Leo, *Techniques for Nuclear and Particle Physics Experiments: A How-To Approach*, 2nd edn. Chapter 10. (Springer, 1994)
2. G.F. Knoll, *Radiation Detection and Measurements*, Chapter 11, 4th edn. (Wiley, 2010)
3. E. Vella, F. Messina, M. Cannas, R. Boscaino, Unraveling exciton dynamics in amorphous silicon dioxide: interpretation of the optical features from 8 to 11 eV. Phys. Rev. **B83**, 174201 (2011)
4. M. Feneberg, R. Leute, B. Neuschl, K. Thonke, M. Bickermann, High-excitation and high-resolution photoluminescence spectra of bulk AlN. Phys. Rev. **B82**, 075208 (2010)

5. C.A. Klein, Bandgap dependence and related features of radiation ionization energies in semiconductors. J. Appl. Phys. **39**, 2029 (1968)
6. G. Lutz, Semiconductors as detectors, in *Semiconductor Radiation Detectors*. (Springer, 2007), pp. 79–93
7. N. Tsoulfanidis, S. Landsberger, *Measurement and Detection of Radiation*, Chapter 7, 5th edn. (CRC Press, 2021)
8. H. Spieler, *Semiconductor Detector Systems*. (Oxford University Press, 2005). https://www-physics.lbl.gov/~spieler/physics_198_notes/
9. G.D.O´ Killey, *Detection and Measurement of Nuclear Radiation*. US Atomic Energy Commission report NAS-NS-3105 (1962), p. 64.
10. J. L. Blankenship, C.J. Borkowski, ´Improved techniques for making P+-i-N+ diode detectors´. I.R.E. Trans. Nucl. Sci. **NS-7**(2–3), 190 (I960); ibid., Performance of silicon surface barrier detectors with charge sensitive amplifiers. **NS-8**(1), 17 (1961)
11. K. Debertin, R.G. Helmer, *Gamma and X-ray Spectrometry with Semiconductor Detectors*. (North-Holland, 1988)
12. See, for instance, https://www.mirion.com/products/standard-high-purity-germanium-detectors; https://www.ortec-online.com/products/radiation-detectors
13. P.G. Appleby, P.J. Nolan, F. Oldfield, N. Richardson, S.R. Higgitt, [210]Pb dating of lake sediments and ombrotrophic peats by gamma essay. Sci. Total Environ. **69**, 157–177 (1988)
14. E.M.O. Bernardesa, J.U. Delgado, L. Tauhata, C.J. da Silva, A. Iwahara, R. Poledna, A.S. Paschoa, [166m]Ho: a multi-$\gamma$ standard for the calibration of Ge spectrometers. Appl. Radiat. Isot. **56**, 157–161 (2002)
15. A.F. Sánchez-Reyes, M-L. Febrian, J. Bar, J. Tejada, Absolute efficiency calibration function for the energy range 63–3054 keV for a coaxial Ge (Li) detector. Nucl. Instrum. Method. Phys. Res. **B28**, 123–127 (1987)
16. S. Hurtado, M. García-León, R. García-Tenorio, A fitting algorithm based on simulated annealing techniques for efficiency calibration of HPGe detectors using different mathematical functions. Nucl. Inst. Method. Phys. Res. **A594**, 362–367 (2008)
17. S. Hurtado, M. García-León, R. García-Tenorio, GEANT4 code for simulation of a germanium gamma-ray detector and its application to efficiency calibration. Nucl. Inst. Method. Phys. Res. **A518**, 764–774 (2004)
18. https://physics.nist.gov/PhysRefData/Star/Text/ASTAR.html; https://physics.nist.gov/PhysRefData/Star/Text/ESTAR-t.html.
19. Giuseppe A. Marzo, A comparison of different peak shapes for deconvolution of alpha-particle spectra. Nucl. Inst. Method. Phys. Res. **A832**, 191–201 (2016)
20. E. Holm, J. Rioseco, R.B.R. Persson, Surface barrier detectors for the determination of [99]Tc by beta spectrometry, in *Methods of Low-Level Counting and Spectrometry, IAEA Proceedings Series*. (1981), STI/PUB/592, pp. 277–281
21. K.S. Krane, *Introductory Nuclear Physics*, Chapter 9. (Wiley, 1988)
22. D.W. Anderson, *Absorption of Ionizing Radiation*, Chapter 7. (University Park Press, 1984). https://physics.nist.gov/PhysRefData/XrayMassCoef/tab3.html
23. L. Mikelic, V. Orescanin, S. Lulic, Determination of [55]Fe in the waste waters of the Krsko nuclear power plant measured simultaneously by a liquid scintillation spectrometer (LSC) and an X-ray spectrometer (XRS). Nucl. Inst. Method. Phys. Res. B **263**, 95–98 (2007); S. Salminen-Paatero, P. Vanninen, J. Paatero, Identification of the isotopic composition of Pu and U and its applications in environmental and CBRN research. Def. Technol. **17**, 728–740 (2021)
24. P. Yalçın, A. Sülün, A. Baştuğ, Y. Kurucu, Y. Şahin, The efficiency determination for Si(Li) Detector in the 3–723 keV energy range. Can. J. Anal. Sci. Spectrosc. **50**, 108–115 (2005)
25. C.S. Rossington, R.D. Giauque, J.M. Jaklevic, A direct comparison of Ge and Si(Li) Detectors in the 2–20 keV Range. IEEE Trans. Nucl. Sci. **39**, 570–576 (1992)
26. G. Heusser, Background in ionizing radiation detection illustrated by Ge spectrometry, in *Low-Level Measurement of Radioactivity in the Environment: Techniques and Applications*, ed. by M. García-León, R. García-Tenorio. (World Sci. Pub. Co., 1994), p. 69

# Dosimeters, Other Detectors, and Specific Designs

# 12

**ABSTRACT**

Radiation dosimeters are presented in the chapter to determine the main procedures for determining the absorbed radiation dose. Nuclear track detectors are also described, as they are used for environmental radioactivity, mostly for $^{222}$Rn determinations. $\Delta E$–$E$ telescopes and time-of-flight detectors are very much used in conventional or accelerator-based mass spectrometry. They are presented in this chapter as well. The operating principles of Cherenkov detectors are described as they give a very interesting information on how to use Cherenkov counting with liquid scintillation counters.

## 12.1  Dosimeters

The principles, definitions, magnitudes, and units of radiation dosimetry are already given in Chap. 6. It is well established that the effects of nuclear radiation on biological or any other material systems are measured in terms of deposited energy per unit mass. As the radiation detectors exactly inform the energy lost by the particles along their path in matter, they are the ideal devices to measure radiation doses. There are two types of dosimeters: active and passive. The former provides information on the radiation doses delivered at the time of measurement. The latter records the doses received during a given time period and is read out afterward.

Active dosimeters are very useful for field dosimetry, while passive dosimeters are mainly used for personal dosimetry of exposed individuals. In what follows, some examples of both types of dosimeters are given. For a deeper understanding of the matter, the reader is referred to more specialized books [1].

© The Author(s), under exclusive license to Springer Nature Switzerland AG 2022
M. García-León, *Detecting Environmental Radioactivity*, Graduate Texts
in Physics, https://doi.org/10.1007/978-3-031-09970-0_12

## 12.1.1 Active Dosimeters

Active dosimeters are very useful in field dosimetry or as survey meters. Gas ionization, scintillation, or semiconductor detectors are used for that. Air-filled ionization chambers provide a direct measurement of $\gamma$-ray radiation exposure since the response of the detector is directly related to ionization in air (see Chap. 6). A very popular design for these devices is given in Fig. 12.1, where a free-air ionization chamber is presented. The $\gamma$-ray beam is collimated to interact in the central part of the chamber, which, in turn, consists of just two parallel plates separated by a certain distance that limits a region in free air. The exposure rate is deduced from the measured current intensity, $I$, and the expected air mass between the plates. Thus, $R = \frac{I}{M}$. This can be easily deduced from the air density, which is well known as a function of the temperature.

As shown in Chap. 6, the radiation exposure is a very specific magnitude since it is defined as the energy deposited by electromagnetic radiation, either X- or $\gamma$-rays, in air. The absorbed dose is, on the contrary, a more general magnitude, defined as the energy deposited per unit mass of the irradiated material. Gas-filled ionization chambers can be used to measure the absorbed dose produced in generic materials compared to that measured in the filling gas. For that, the ionization chamber is irradiated by the radiation source studied. The absorbed dose, $D_m$, in the investigated material for the same irradiation conditions is given by the Bragg–Gray principle [2]. Thus, $D_m$ is

$$D_m = w S_m P \tag{12.1}$$

$w$ is the average energy needed to create an ion pair in the gas (see 7.57), $S_m$ is the relative mass stopping power of the material to that of the gas, and $P$ is the number of ion pairs per unit mass produced in the filling gas. Alternatively, gas ionization detectors working in the proportional or Geiger–Müller (GM) regions

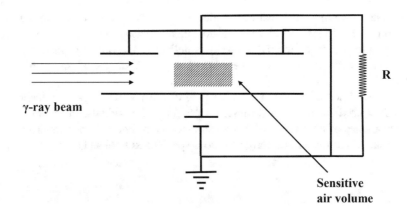

**Fig. 12.1** Schematic presentation of an air-filled ionization chamber scheme

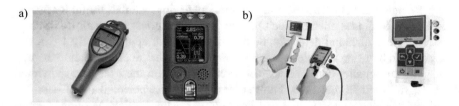

**Fig. 12.2** Some examples of portable survey meters. **a** Courtesy of Tracerco Ltd: Tracerco™ T401 contamination monitor and Tracerco™ Personal Electronic Dosimeter (PED); **b** Courtesy of Bertin Technologies ©Bertin: SaphyRAD C contamination monitor

can be used for dosimetry measurements, although the most used design is based on ionization chambers.

Portable devices are very important in this field, where there is a need to determine doses or dose rates from radiation sources that cannot be transported to the laboratory or have been accidentally dispersed in the environment. Survey meters can be constructed with gas ionization detectors, especially GM, due to the simplicity of their operation but also with scintillation detectors such as NaI(Tl), CsI (Tl), and SZn(Ag) or semiconductor detectors such as Si, Ge, CdS, or CdSe [3]. Depending on the expected radiation emitted at the site, it will be more convenient to use one or another. Furthermore, the expected strength of the radiation source is also an important criterion for selecting the appropriate survey meter. For instance, scintillation detectors are more convenient than GM for very active sources since they are faster. From this viewpoint, it is interesting to note that despite the simplicity of the operation of the GM counters, their long dead time makes them inappropriate for active samples. Some examples of portable dosimeters are presented in Fig. 12.2.

To convert the response of the survey meter into absorbed doses, it is necessary to calibrate the dosimeter, taking a well-calibrated ionization chamber as a reference [2]. Thus, the response of the survey meter to the well-known absorbed dose as measured by the reference ionization chamber is determined, and a calibration curve is constructed. Additionally, belonging to this family of portable survey meters are portable area dosimeters that allow the measurement of doses from contaminated places in the laboratory. The working principles are the same, and the main feature of these devices is the large active area of the detector. The detector response is usually given in $Bqm^{-2}$ or $Gym^{-2}$.

A recent development in dosimetry is the use of semiconductor detectors as dosimeters in human radiation therapy [4]. Semiconductors can be constructed in very small sizes and operated in active or passive modes, and their response to radiation is very easily read out. All these features are very well suited to the needs in human radiation therapy, where the spatial distribution of the dose received by the individual must be perfectly known [5]. Furthermore, the spatial resolution required in many treatments makes it impossible to use conventional dosimeters. However, semiconductors are more sensitive than ionization chambers as a result of their higher density. There are several semiconductor materials useful for micro-dosimetry: Si diodes, metal oxide semiconductor field effect transistor (MOSFET) detectors, and diamond detectors. Spatial resolutions of approximately 1 μm can be obtained by using such devices, which is extremely helpful for radiation therapy planning and quality control since the radiation dose on the irradiated individual can be followed in real time. Furthermore, the contact potential (see Chap. 11) that spontaneously appears in a p–n junction makes the application of an external electric potential unnecessary.

### 12.1.2 Passive Dosimeters

In Detector Physics, a very important goal is to obtain a fast response of the detector to radiation. However, for some applications, it is more interesting to deal with materials that can preserve the radiation effects for reasonably long times. This is the case for passive dosimeters, in which, by different physical processes, the material can accumulate irradiation effects, which are later evaluated by different means. One of these passive dosimeters is a chemical dosimeter based on the radiolysis produced by radiation in aqueous solutions. The Fricke dosimeter [6] is a good example of these devices, in which the dose is measured by analyzing $Fe^{3+}$ produced in an Fe aqueous solution after irradiation. This dosimeter is very useful for high doses with a dynamic range of 1–500 Gy.

However, the main application of passive dosimeters is personal dosimetry, in which the important goal is the measurement of the radiation dose received by the individual during a long time period, for instance, during the working time. The most popular approaches are described below.

*Film Badge Dosimetry*
Film badge dosimetry is based on the use of a photographic emulsion as a means of recording the dose received by the individual during a fixed time period. As is known, the photographic emulsion is formed by AgBr grains in gelatine. As in the case of conventional photography, the grains are activated as the radiation passes through the emulsion. Information on the radiation passage is preserved in this way almost indefinitely. The photographic emulsion is developed afterward, which reveals not only the pass of radiation but also the amount of radiation that has crossed the material during the studied time period, or, in other words, the radiation dose received by the individual. The dose accumulated in the badge is evaluated by comparing the observed darkening of the film with that produced in an identical film irradiated by a well-known radiation dose [3].

*Thermoluminescence TLD Dosimetry*
A TLD dosimeter (see Fig. 12.2) consists of a doped inorganic crystal in which the forbidden gap is populated with a high concentration of trapping levels according to the scheme presented in Fig. 12.3a. When radiation passes through the material, some electron–hole pairs are produced. Many of the electrons will promote from

the valence to the conduction band, but some of them are captured by the trapping centers. In the ideal case, a good number of trapping centers will be situated at a sufficient distance from the conduction band to make its promotion to the conduction band at normal room temperature very unlikely. In other words, these excited states are thought to have a long mean life. In this way, the amount of trapped electrons preserves information on the energy deposited during irradiation or, in other words, on the absorbed dose. At the same time, trapping centers close to the valence band may capture some of the holes produced. For the same reason, the mean life of these states can be long. It is apparent that the number of trapped electrons and holes is a measure of the dose absorbed by the TLD dosimeter. To evaluate this quantity, the TLD dosimeter is heated. In this way, some energy is communicated to the electrons that can be promoted from the trapping centers to the conduction band. Then, they can be recaptured by a hole in the material, which results in the emission of a deexcitation photon. At the same time, the energy provided by raising the temperature is enough for the electrons in the valence band to be captured by the neighboring holes. Now, the new hole in the valence band can be occupied by one of the electrons. In this process, a deexcitation photon is emitted as well. In many cases, the photon energies involved are approximately 3–4 eV, i.e., in the optical range.

The photon flux can be measured with the help of a photomultiplier, as in the case of scintillation detectors (see Chap. 10). The TLD glow curve gives the

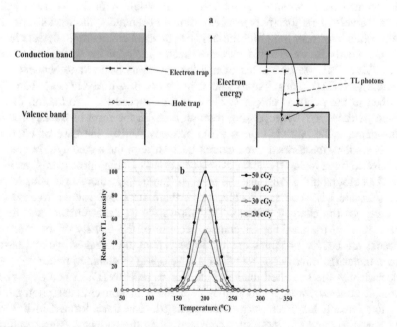

**Fig. 12.3** **a** Energy levels in a TLD and processes of TL photon production and **b** TLD glow curve showing the photon emission intensity as a function of temperature

thermoluminescent photon flux as a function of temperature. An example is presented in Fig. 12.3b. The area under the glow curve is a measure of the absorbed dose. A simple calibration to known doses allows the measurement of the radiation dose [7]. These materials are so that they can be reutilized, since it is expected that after heating, the trapping centers will be empty. Typical materials for TLD dosimetry are $CaSO_4$ (Mg), $CaF_2$ (Mg), LiF (Mg), or LiF (Ti). The doping material is in parentheses. The most widely used TLD dosimeter is based on LiF with a sensitivity as low as 0.2 $\mu$Gy [7].

## 12.2    Track Detectors

As described in Chap. 7, charged particles ionize or excite the matter, resulting in the formation of a track when crossing a material. From the beginning, this has been used for the detection of charged particles, as it is the working principle of bubble or cloud chambers [8] of crucial importance in the pioneering studies in Particle Physics and Cosmic Ray research. As expected, the length and thickness of the observed track depend on the mass, the charge, and, of course, the energy of the particle. This principle is useful in the measurement of environmental radioactivity, although not specifically with the use of bubble or cloud chambers. Indeed, the background of such devices makes it very difficult to distinguish the interesting from the superfluous tracks and, overall, because of the extremely short lifetime of the formed track, which is absolutely incompatible with the measurement of very low levels of radioactivity concentrations. Fortunately, the tracks formed in certain solids have very long lifetimes and can be used efficiently to detect charged particles, mostly heavy charged particles, such as $\alpha$ particles or fission products [9]. In Table 12.1, we give a list of materials used as Solid-State Nuclear Track Detectors (SSNTD). These passive detectors work in a different way than those described in the previous chapters. They, in fact, record the information that the charged particles left when passing through them in the form of tracks produced by the damage the particle causes in its pathway. Unlike the case of bubble or cloud chambers, the formed track cannot be seen with the naked eye. In contrast, they have to be revealed by chemical processes and etching procedures, as in the case of photographic emulsions. As seen in Table 12.1, there is a wide variety of useful materials, such as SSNTD. All of them are dense enough to absorb all the energy of the charged particle. This allows for its determination and for the identification of the particle. Of course, because of the density of the materials, the tracks are not long or thick enough to be visible to the naked eye and have to be seen under the microscope and after a developing or etching process.

Specifically, the so-called nuclear emulsion is an SSNTD formed by AgBr in gelatin at concentrations much higher than those in conventional photographic emulsions, and it has been very much used [9]. The track formed in this case corresponds to the AgBr molecules activated along the pathway. After the film is developed, the obtained image reveals a track that informs the stopping power of

**Table 12.1** Common materials used for SSNTD

| Material | Composition |
|---|---|
| *Organic* | |
| Cellulose nitrate (Daicell,...) | $C_6H_8O_9N_2$ |
| Cellulose triacetate (Cellit, Triafol-T,...) | $C_3H_4O_2$ |
| Bisphenol A polycarbonate (Lexan, Makrofol,...) | $C_{16}H_{14}O_3$ |
| Allyldiglycol-carbonate (CR-39, MA-AND,...) | $C_{12}H_{18}O_7$ |
| Polymethylmethacrylate (Plexiglass, Lucite, Perspex,...) | $C_5H_8O_2$ |
| Polyethylene terephthalate (Cronar, Melinex, ...) | $C_5H_4O_2$ |
| *Glasses* | |
| Soda lime | $74SiO_2:13Na_2O:10.5CaO:1.3Al_2O_3:K_2O,MgO...$ |
| Silica | $SiO_2$ |
| Flint | $18SiO_2:4PbO:1.5Na_2O:K_2O:...$ |
| *Inorganic* | |
| Quartz | $SiO_2$ |
| Phlogopite mica | $KMg_2Al_2Si_3O_{10}(OH)_2$ |
| Muscovite mica | $KAl_3Si_3O_{10}(OH)_2$ |

the particle. Thus, the length of the track and grain density on the track give the energy of the particle and even its identity.

As seen in Table 12.1, there are many other materials that can be used as SSNTDs. In these cases, the tracks appear after an etching process normally carried out with strong alkali or acid solutions. The chemical treatment causes the appearance of very characteristic pits in the regions damaged by the particle. A typical plot of tracks obtained with an SSNTD is given in Fig. 12.5. Once the tracks are revealed independently, they have to be counted. This is done manually with the help of a microscope or automatically with specific software to drive a computer-assisted counting and identification process.

SSNTD is used in Cosmic Ray, Astrophysics, and Nuclear Reaction studies. For instance, the contribution of SSNTD to understanding the basic mechanism in heavy-ion reactions or pion-induced nuclear reactions is well recognized. Its role in the characterization of exotic radioactivity (see Chap. 1) is also relevant. There are also many examples in the current literature on the application of SSNTD in Geology. For instance, the so-called fission-track dating method is based on the counting of tracks produced by fission products of U isotopes, mostly $^{235}U$ and $^{238}U$. From a technical point of view, the method is only available to those groups with a nuclear reactor that is used to generate the fission of such nuclei. The fission-track record informs about the composition of U isotopes, and so on the sample age. Moreover, the use of SSNTD for mineral exploration or seismology is well known [10]. In both cases, the method consists of determining $^{222}$Rn in the interesting sample. Normally, the measurement is carried out by depositing the SSNTD in situ, which is collected after some exposure time to the $^{222}$Rn present at the site. SSNTD also finds room for applications in Radiobiology or Material Science (Fig. 12.4).

However, with respect to environmental radioactivity, the main, although not the only, application of SSNTD is the determination of $^{222}$Rn in the environment [11], either at work or at home (Rn indoors). The importance of knowledge of the Rn level in houses or workplaces has already been studied in previous chapters since its contribution to the radiation dose to the population and specifically to the lungs is very relevant. SSNTDs, usually sealed in a container, are left for some weeks in houses. After such an exposure time, the SSNTD is etched, and the tracks are counted. Data must be refined by subtracting the background and adequately identifying the track origin. The $^{222}$Rn emanation rate, $E$ $(Bqm^{-2}$ $h^{-1})$, from the surface studied in a house or workplace can be calculated from the accumulated

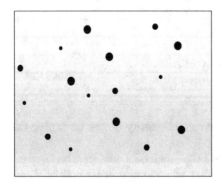

**Fig. 12.4** Tracks in SSNTD. Typical plots for $^{222}$Rn $\alpha$-particles and $^{238}$U spontaneous fission products

**Fig. 12.5** Schematic view of a $\Delta E$–$E$ telescope

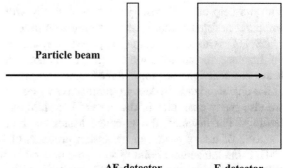

Particle beam

$\Delta$E-detector                    E-detector

activity, C $(Bqm^{-3} h^{-1})$, measured by the SSNTD as

$$E = \frac{CV\lambda}{AT} \tag{12.2}$$

V is the volume of the container, A is the surface of the explored area, $\lambda$ is the $^{222}$Rn decay constant, and T is the effective exposure time, which is related to the measuring time t, as in [11]

$$T = \left[ t - \frac{1}{\lambda}\left(1 - e^{-\lambda t}\right) \right] \tag{12.3}$$

## 12.3   ΔE–E Telescopes

ΔE–E telescopes allow the determination of the charge or mass or both of the components of an ion beam [12]. This is, as we will see in Chap. 18, of crucial importance in Accelerator Mass Spectrometry (AMS). In fact, such telescopes are very useful for discriminating among isobars in the ion beam; this increasing order of magnitude increases the sensitivity of the AMS system and, therefore, facilitates the precise measurement of extremely low levels of radioactivity in the environment. ΔE–E telescopes were developed in early experiments in Nuclear Physics as a unique tool for the determination of the charge of nuclear reaction products. ΔE–E telescopes are based on the Bethe-Bloch (7.28) for the stopping power. The strongest dependence on the speed of the particles resides in the $\frac{1}{\beta^2}$ term. The logarithmic contribution is very small. From this viewpoint, with a very good approximation, it can be said that

$$S_c \approx 4\pi z^2 r_0^2 \frac{m_0 c^2}{\beta^2} \rho \frac{N_A}{A} Z \tag{12.4}$$

Should we know $\rho$, A, and Z, i.e., the characteristics of the medium the particle is passing through, it turns out.

$$S_c = -\frac{dT}{dx} \approx k\frac{1}{\beta^2} = k\frac{Mz^2}{T} \tag{12.5}$$

T and M are the kinetic energy and mass of the particle with charge z. The constant k includes all the features of the material through which the particle is passing.

Thus, let us suppose that a beam of charged particles is passing through a detector. Since k is known, the simultaneous determination of T and $\frac{dT}{dx}$ of the particles will inform, $Mz^2$, i.e., on their mass or charge or both together. (12.5) is only valid under the assumptions leading to (7.28). In other words, ΔE–E telescopes can be used only for fast charged particles. This fits perfectly within the AMS working conditions. As we will discuss in Chap. 18, $\Delta E–E$ telescopes help very much to discriminate among isobars in an ion beam since their slight differences in $S_c$ can be detected with them. In this way, it is easy to quantify the desired mass without the interference of isobaric species. Usually $\Delta E–E$ telescopes consist of two detectors connected in coincidence, and the particle beam is made to cross both detectors. One of them is thin enough so that the particles release only a small amount of energy, while the second detector is thick enough so that the particles deposit all their energy. In Fig. 12.7, we present a schematic view of one of these systems. In many cases, the $\Delta E$ and E detectors are of the Si semiconductor type [12]. The excellent energy resolution they provide makes them perfect candidates as $\Delta E–E$ telescopes.

Moreover, very thin, some tenths μm, Si detectors can be constructed keeping the energy resolution performances of the more standard sizes. In another version of this system, the radiation beam passes through a single multianode gas detector (Fig. 12.6). Now, the energy deposited by the particle is observed along its path by the electrical signal produced in each anode. The signal produced in the first sections of the path corresponds to the stopping power, and the sum of all the signals produced gives the total energy deposited. The gas detector can work in both the ionization chamber and the proportional region, i.e., in a working region in which the response of the detector is spectrometric. $\Delta E–E$ telescopes based on gas ionization detectors are very useful for low-energy ion beams or very heavy ions that can be easily stopped in a Si $\Delta E$ detector.

The use of gas ionization detectors and specifically ionization chambers is the best option for heavy ions. The main reason is that they are easy to construct and practically do not become damaged during irradiation. Of course, the position of the anodes regarding the chamber window is an important issue, and some previous calculations of the involved isobar stopping power in the counting gas are convenient before constructing the detector. For example, in Fig. 12.7 taken

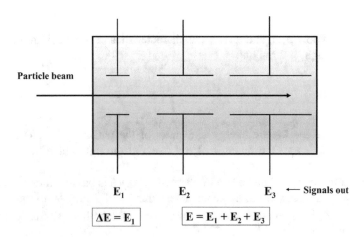

**Fig. 12.6**  Multianode $\Delta E–E$ telescope

**Fig. 12.7** Range of $^{36}$Cl (dotted line) and $^{36}$S in the gas detector in Fig. 12.6. The difference allows isobar discrimination with the $\Delta E$–$E$ telescope of Fig. 12.6. Courtesy of Dr. F. J. Santos

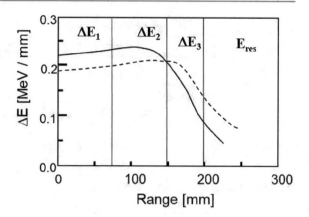

from [13], the stopping powers of $^{36}$Cl and $^{36}$S in Ar at 69 Torr are presented. It seems clear that $^{36}$Cl deposits more energy per unit length than $^{36}$S in the first 15 cm. Beyond this distance, the $^{36}$S stopping power is higher. Consequently, for the measurement of $^{36}$Cl, the first anode or group of anodes should be situated in the first 15 cm of the detector. This will help identify $^{36}$Cl since its signal should be more intense in this part of the ionization chamber. In contrast, the anodes located beyond 15 cm will give a more intense signal for $^{36}$S. Of course, the same ionization chamber can be used for different isobaric groups. To adapt the same design to new conditions, the gas pressure inside the chamber, the beam energy, or both can be changed.

Very thin scintillator films also exist and are of practical use in these measurements, but the better energy resolution of the options described above makes them not very widely used. The response of a $\Delta E$–$E$ telescope consists of a $\Delta E$–$E$ plot, as presented in Fig. 18.5. The different points identify the measured $Mz^2$. It is very easy to see that, in the case where z is known, each spot corresponds to a given mass. Since in AMS the charge state of the ion beam is selected at the high-energy side of the spectrometer, the $\Delta E$–$E$ plot serves to identify the mass of the studied radionuclide. The size of the spot depends on the ion beam energy distribution and on the detector (or detectors) energy resolution. It can be said that in the ideal case, when the ion beam energy is constant, the mass resolution of the method is limited by the energy resolution of the detectors.

## 12.4  Time-Of-Flight Spectrometers

Time-of-flight (TOF) spectrometers are also very useful in AMS since they can be used for mass discrimination. TOF systems measure the time it takes a particle to travel a known distance, $L$. In other words, a TOF spectrometer can determine the speed of the particles and consequently their kinetic energies. In fact, this method is used extensively to measure neutron energies [12].

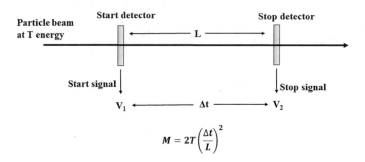

**Fig. 12.8** A schematic view of a TOF system

A schematic TOF spectrometer is presented in Fig. 12.8. A typical system consists of two detectors, the start detector and the stop detector, located at a distance $L$. The time elapsed $\Delta t$ between the electrical signal produced at the start and stop detectors when the beam passes through them is measured electronically. The velocity is trivially deduced from the time since $v = \frac{L}{\Delta t}$. The kinetic energy, T, of the particle follows rapidly once the mass, $M$, is known from the well-known equation $T = \frac{1}{2}Mv^2$ in the nonrelativistic approach, which is the normal case for AMS.

However, as mentioned before, it is also very useful for distinguishing masses between the components of a particle beam. Indeed, once v is measured and $T$ is known, as happens in AMS, the difference in $v$ among the components of the particle beam gives the differences in $M$, which is the main goal in AMS. In fact, this is another method for rejecting isobars that interfere with the component of interest in the ion beam (see Chap. 18). A measure of the mass resolution of these systems can be obtained from the mass uncertainty calculated from the kinetic energy equation. Since at the nonrelativistic approach $M = \frac{2T\Delta t^2}{L^2}$, , the relative uncertainty in the mass (see Chap. 14) is

$$\frac{\sigma_M}{M} = \sqrt{(\frac{\sigma_T}{T})^2 + (\frac{\sigma_L}{L})^2 + (\frac{\sigma_{\Delta t}}{\Delta t})^2} \tag{12.6}$$

with an obvious meaning for the notation. The uncertainty in length is normally very small compared to the other uncertainties and can be neglected. This means that electronic time resolution is the key issue in these systems. In AMS, the energies usually range from 0.1 to 2 MeV/amu, which means velocities of ~cm/ns. Since the typical time resolution provided by these spectrometers is a few tenths of nanoseconds, TOFs with distances of one or several meters are sufficient to distinguish between ion masses. Normally, the TOF systems used in AMS are made up of two detectors, generally very thin carbon sheets that release electrons when the ion beam crosses them (see Fig. 12.9). The electron current is amplified by a microchannel plate (see Fig. 10.5 and Chap. 10), and an electrical signal is produced. The time difference between both electrical signals is measured

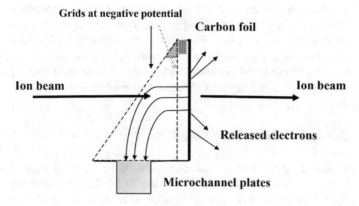

**Fig. 12.9** Typical $C$ foil-based microchannel plate device (see Chap. 10)

electronically, and a TOF spectrum is the response of the detector assembly (see Fig. 18.6). The masses of the interfering species can be identified from such a spectrum since the time of flight must be different once the energy is well determined. Interestingly, when measuring heavy masses in AMS, the use of TOF systems might be the best approach to discriminate between isobars. Indeed, for heavy ions, the stopping power is not exactly proportional to $z^2$ but to $z_{eff}^2$, an effective charge that is defined as [14]

$$z_{eff} \approx z^{\frac{1}{3}} \frac{v}{v_0} \tag{12.7}$$

where $v$ is the ion speed and $v_0 = \frac{e^2}{h}$. This effect was already commented on in Chap. 7 and is due to the screening of the electrons on the nuclear charge and the change in charge state the ions suffer when passing through the gas in a detector. Both effects diminish the ability to distinguish masses with a $\Delta E$–$E$ telescope for heavy ions.

## 12.5 Cherenkov Detectors

As shown in Chap. 7, the emission of Cherenkov light occurs when a charged particle travels through a medium at a speed higher than the speed of light in such a medium. Cherenkov light can be used to detect radiation and, in some cases, to give its energy. Detectors based on Cherenkov radiation find application mainly in Particle Physics due to the necessary energies for Cherenkov light production. Indeed, the need to overcome a threshold energy or speed (see Chap. 7) makes them essentially useful for high-energy charged-particle beam detection and spectrometry. The operation of Cherenkov detectors is, however, similar to that of scintillation detectors with the use of photomultipliers to convert the Cherenkov photon current into an electron current. Of course, the wavelength sensitivity of the photomultiplier should fit the wavelength Cherenkov light spectrum produced in the medium. In many cases, wavelength shifters are necessary [15]. Although there is little application to environmental radioactivity determinations, it is interesting to know the designs currently available in this field of Detector Physics [16].

There are basically three types of Cherenkov detectors.

### 12.5.1 Cherenkov Threshold Counters

It is the simplest design. Threshold detectors provide an electrical signal just when the particle velocity is higher than the Cherenkov threshold velocity in the detector material. As can be understood, the application of these detectors is very limited: They can only distinguish those particles in the beam that have a linear momentum above the thresholds from the rest of the components. In contrast, the design is very simple.

### 12.5.2 Cherenkov Differential Detectors

In this case, the Cherenkov detector can discriminate speed ranges within the particle beam. This provides spectrometric information about it once the mass of the particle is known. As already described in Chap. 7, see Fig. 7.13, the angle of the Cherenkov cone wavefront is a measure of the particle speed. For that, its determination gives the particle energy. An application of the most common principles of Geometrical Optics allows the measurement of such an angle and for that of the energy of the particle crossing the medium.

### 12.5.3 Cherenkov Circular Image Detectors

In this case, the detector material is covered by a spherical mirror that focuses the light cone along the trajectory of the particle to form a ring. The radius of the ring depends on the angle of the cone and can therefore be used to measure the particle velocity and hence its energy.

The application of Cherenkov detectors to many problems in particle or Astroparticle Physics is very popular, since the energies involved in the interesting processes make the production of Cherenkov light very likely in common media such as water or in the air. As we will see in Chaps. 14 and 15, Cherenkov detectors play a very important role in various problems requiring low-level counting [17]: Many underground laboratories contain several such detectors for the search of very rare physical processes. However, as mentioned before, direct application to environmental radioactivity of Cherenkov detectors has thus far been scarce. It was already described in Chap. 10 how liquid scintillation counters (LSC) can be used as Cherenkov detectors since the energy of some interesting β-emitters is above the Cherenkov threshold in distilled water. This has been exploited to detect $^{90}Sr$-$^{90}Y$ or $^{210}Pb$, through its daughter $^{210}Bi$, in environmental samples [18]. In these cases, the experimental setup is identical to that of a common LSC since the Cherenkov light wavelength in water fits into the conventional photomultiplier sensitivity range.

**Exercises**

1. An air-based ionization chamber dosemeter determines a 15 Gy absorbed dose for 5 min at 5 cm depth in a pool from a 25 MeV $\beta$-radiation beam. Estimate the dose absorbed in water at the same depth.

2. The factor G is used in Fricke dosimetry as the number of chemical species produced by 100 eV of energy absorbed. Suppose a Fricke dosemeter based on ferrous sulfate in which the $Fe^{++}$ concentration is $10^{-3}$ M. Calculate the radiation dose required to decrease the ferrous concentration to $10^{-4}$ M. Take G as 15.5.

3. Demonstrate (12.6).

4. Estimate the flight path in a TOF experiment to obtain a mass resolution of 0.1% for a $3.57 \pm 0.01$ MeV $^{129}I$ beam in an AMS experiment. Take $\Delta t = 1$ ns and consider $\frac{\Delta L}{L}$ negligible as usual in these experimental setups.

5. TOF systems are often used to perform neutron spectrometry. Find the length of a TOF fast neutron spectrometer with an energy resolution of 0.1% and $\Delta t = 1$ ns. (12.6) must be adapted to the situation in which M is known.

6. Based on Fig. 18.6, estimate the mass resolution of the TOF spectrometer for the $^{129}I$ region. As usual, the length resolution can be neglected, and suppose that the beam is accelerated to $30.0 \pm 0.1$ MeV.

7. Find the Cherenkov threshold energy for electrons in quartz (n = 1.47) and water.

8. Using https://www.nndc.bnl.gov/nudat2/ find at least three $\beta$- or CE emitters that could be measured in a quartz or water-based Cherenkov detector.

9. Is it possible to determine the $^3H$ activity of a sample by detecting Cherenkov radiation with a liquid scintillation counter? Use just distilled water as the scintillation cocktail. Discuss the possibility of measuring $^{14}C$, $^{90}Sr$, or $^{210}Pb$ with the same method.

10. In the so-called exotic radioactivity, a heavy nucleus decays, emitting another nucleus heavier than $\alpha$ particles. For example: $^{226}Ra \rightarrow {}^{14}C + {}^{212}Pb$ with a branching ratio $\frac{\lambda_{14_C}}{\lambda_\alpha} = 3.2 \times 10^{-11}$

    From an experimental point of view, it is very difficult to measure such a branching ratio. Explain why and propose a method to determine it.

11. Consider now the following decay.

$$^{232}U \rightarrow {}^{208}Pb + {}^{24}Ne$$

with a branching ratio

$$\frac{\lambda_{24_{Ne}}}{\lambda_\alpha} = (2.0 \pm 0.5) \times 10^{-12}$$

Propose a method to measure the branching ratio.

12. Estimate the energies of $^{14}$C and $^{24}$Ne in the decays previously shown. Discuss how realistic the measurement of such heavy ions is with a conventional detector. Which detector family would be suitable for measurement?

13. In Fig. 18.5, a few $\Delta E$–$E$ plots are depicted. Discuss why we find spots instead of well-defined points in the plots.

14. Check how (12.7) can modify (12.5) and discuss to what extent it could affect the method for the determination of the mass of a nuclear species.

# References

1. M.G. Stabin, *Radiation Protection and Dosimetry: An Introduction to Health Physics.* (Springer, 2010)
2. G.F. Knoll, *Radiation Detection and Measurements*, Chapter 5, 4th edn. (Wiley, 2010)
3. N. Tsoulfanidis, S. Landsberger, *Measurement and Radiation Detection*, Chapter 16, 4th edn. (CRC Press, 2015)
4. A.B. Rosenfeld, Advanced semiconductor dosimetry in radiation therapy, in *Concepts and Trends in Medical Radiation Dosimetry*, ed. by A. Rozenfeld. (American Institute of Physics, 2011), pp. 48–74
5. N. Nedev, E. Manolov, D. Nesheva, K. Krezhov, R. Nedvev, M. Curiel, B. Valdez, A. Mladenov, Z. Levi, Radiation dosimeter based on metal oxide semiconductor structures containing silicon nanocrystals. Key Eng. Mater. **495**, 120–123 (2012)
6. G.R. Choppin, J.-O. Liljenzin, J. Rydberg, C. Erkberg, *Radiochemistry and Nuclear Chemistry*, Chapter 7, 4th edn. (Elsevier, 2013)
7. G.F. Knoll, *Radiation Detection and Measurements*, Chapter 19, 4th edn. Wiley, 2010)
8. C.T.R. Wilson, On a method of making visible the paths of ionising particles through a gas. Proc. Royal Soc. Lond. A: Math. Phys. Eng. Sci. **85**, 285–288 (1911)
9. R. Ilic, S.A. Durrani, Solid state nuclear track detectors, in *Handbook of Radioactivity Analysis*, 2nd edn. (Elsevier Science, 2003), p. 179
10. H.A. Khan, I.E. Qureshi, SSNTD applications in science and technology: a brief review. Radiat. Meas. **31**, 25–36 (1999)
11. R.M. Amin, A study of radon emitted from building materials using solid-state nuclear track detectors. J. Radiat. Res. Appl. Sci. **8**, 516–522 (2015)
12. N. Tsoulfanidis, S. Landsberger, *Measurement and Detection of Radiation*, Chapter 13, 4th edn. (CRC Press, 2015).
13. F.J. Santos, Ph. D. Thesis, University of Sevilla (2004)
14. D.W. Anderson, *Absorption of Ionizing Radiation*, Chapter 2. (University Park Press, 1984)
15. G.F. Knoll, *Radiation Detection and Measurements*, Chapter 9, 4th edn. (Wiley, 2010)
16. B. Ratcliff, J. Schwiening, Cherenkov counters, in *Handbook of Particle Detection and Imaging*, vol. 2(19), ed. by C. Grupen, I. Buvat (Springer, 2012), pp. 453–471
17. Hyper-Kamiokande deteector. http://www.hyperk.org/en/index.html
18. N. Todorovi, Stojkovi, J. Nikolov, B. Tenjovi, "$^{90}$Sr determination in water samples using Čerenkov radiation. J. Environ. Radioact. **169**, 197–202 (2017)

# Radiochemistry for Environmental Samples

**13**

**ABSTRACT**

The determination of environmental radioactivity requires the use of radiochemistry methods. In this chapter, the principles of radiochemistry are introduced with the aim of understanding the methodology used in the literature for the determination of important environmental radionuclides. Radiochemical methods start, however, with sampling. In fact, sampling, as well as sample transport and storage, are very important components of the procedure for environmental radioactivity quantification. A general description of the different methods used in laboratories is given in the chapter. The different steps in radiochemistry procedures for environmental radionuclide separations are also described, including the determination of the chemical recovery, or yield, of the procedure. The behavior, dynamics, and fate of radionuclides in nature depend on their physical and chemical forms. Speciation studies are also introduced. The general detector efficiency calibration methods are introduced in this chapter, as they are directly connected to the final radioactive source form and, in turn, related to the radiochemical procedure. Analytical quality control is necessary in an environmental radioactivity laboratory. The main principles of analytical quality control practices have been included in the chapter since they are the only way to ensure the quality and reproducibility of the methodology used in laboratories.

## 13.1 Sampling Techniques

The determination of radioactivity in the environment starts with sampling. It is a very important part of the procedure and must be considered when designing the experiment. In other words, it is necessary to pay adequate attention to sampling works since they will determine and ensure the quality of the experiment we are trying to perform. The sample should be representative of the problem we are

© The Author(s), under exclusive license to Springer Nature Switzerland AG 2022
M. García-León, *Detecting Environmental Radioactivity*, Graduate Texts
in Physics, https://doi.org/10.1007/978-3-031-09970-0_13

investigating and useful for determining the radionuclide or radionuclides we need to measure. Representativeness also encompasses the absence of contamination or cross-contamination during sampling, with the sampling instruments or with other samples. Therefore, the material used for the sampling should be sufficiently clean to carry out the work with quality. This, in many cases, means precleaning and purification of the sampling instruments. The homogeneity of the sample is also an important issue because the radionuclide or radionuclides under investigation may not be homogeneously distributed in the sample. This is quite common in soils or sediments affected by radioactivity inputs from nuclear detonations or nuclear reactor accidents. The presence of radioactive (hot) particles can bias the results and provide a nonrepresentative activity concentration in the soil or sediment. In seawater or lacustrine water, stratification effects produced by temperature or salinity gradients, or the appearance of different redox conditions along the depth, may give different radioactive concentrations in the water columns. These differences can also appear in shallow marine waters, mainly in estuarine areas, where the mixing of marine and fresh waters creates environmental changes associated with changes in salinity. This may cause the preferential redissolution of the radionuclides from the sediment in some places, producing changes in concentrations within small areas. Tidal conditions affect radioactivity concentrations in estuaries and sometimes in rivers where tidal waves enter upstream.

Seaweed sampling is another example where representativity can be lost if the origin of the sample is not clear. Now, it is necessary to ensure that seaweed grows in the marine environment of interest. In other words, sampling of algae on beaches is not recommended since it is not possible to assign the real origin of the collected sample.

In short, a deep understanding of the environment under study is necessary to adequately cope with the sampling work.

The sample size is a very important parameter. In fact, as we will see in the next chapter, it affects the generalized $FOM_g$ of the experiment (see 14.15). As the sample size increases, the availability of the radionuclide or radionuclides under study increases, and therefore, the uncertainty in its determination decreases. Indeed, if a water sample contains approximately 1 mBql$^{-1}$ of the radionuclide, it will be better to measure 100 L of water instead of 1 L, since the uncertainty of the measurement will be smaller (see next chapter). Thus, the sample size must be established prior to sampling work. It can be estimated on the basis of the expected radionuclide concentration in the environmental compartment. Nevertheless, as we will see in the next sections, as the sample size increases, some more losses are expected during the extraction of the radionuclide of interest; i.e., a decrease in the radiochemical chemical recovery could take place. Therefore, a balance of the size needed for a good measurement precision and the capacity of the radiochemical procedure to give good recoveries must be done.

The containers used for sample collection must be cleaned before use. This means that bags, bottles, and, of course, the instrumentation need for sampling, corers, knives, scissors, etc. The samples must be rinsed with acid and distilled water to remove any possible element that could contaminate the sample.

Last but not least, the collected sample must be conveniently identified following the standards adopted in the laboratory. The place must be identified, as well as the location, date, and any other circumstance that helps to understand the foreseen results. This might seem a trivial comment, but in many cases not having a well-formulated identification protocol is at the origin of misunderstandings and misinterpretation of results.

The importance of sampling for environmental radioactivity is crucial, and several protocols exist that should be followed to carry out reproducible experiments. Some of them are described in the following.

## 13.1.1 Solid Samples

*Soils*

The selection of the sampling area is a key point in the planning of soil sample collection. The sampling area should correspond to flat open areas, without trees, bushes, or other objects that could shadow the deposition of radionuclides. It must be far from ditches and roads. When sampling is carried out within a radioactivity monitoring program, the selected area must be kept throughout the duration of the program to guarantee the comparability of the results. Usually, also in monitoring programs, vegetation and soils are collected at the same site.

Once the area is selected, the size of the collection site must be fixed, and the samples must be taken from a place with a soil corer of well-defined dimensions. Various samples are usually taken during sampling work from the fixed collection site. For normal applications, it is enough to take 5 cm deep soil cores. However, when radionuclide migration or mobility studies are necessary, it is also advisable to take deeper soil cores. For that, 30–50 cm deep cores are taken that are sliced every 2–5 cm. Large stones, vegetation, and roots must be removed from the site. The soil samples are transported to the laboratory in well-identifiable plastic bags. The pretreatment of the samples depends on the planned analysis. Conventionally, they are dried at room temperature or at 105 °C, depending on whether volatile radionuclides are presumed to be present. After drying, the soil sample was sieved and ground before $\gamma$-spectrometry. Should the soil be used for other specific analyses, pretreatment must vary depending on the radionuclide investigated [1]. The activity concentration can be given in Bqkg$^{-1}$ or in Bqm$^{-2}$, which is more convenient for evaluating the atmospheric deposition of radionuclides in accidental releases. In this case, the area covered by the soil sample must be known. An important issue in soil sampling is the presence of radioactive (hot) particles, as described in Chaps. 4 and 5. If they are suspected to be present, they must be identified by using different procedures that are described in detail in Chap. 20. For example, sample splitting and subsequent analysis by $\gamma$-spectrometry can hint at the presence of radioactive particles. Indeed, a clear difference in the activity between the subsamples reveals the presence of particles. Then, they can be located more precisely using autoradiography or other techniques [2].

*Sediments*

The sampling of sediments is a difficult technique strongly dependent on the studies to be carried out. Analysis of sediment surface contamination requires simple dredges, which usually collect the first few cm of the water system bed. In Fig. 13.1a, one of the most commonly used dredges is presented, and the adequate model must be selected depending on the water system, currents, and depth of the collection site. Of course, the dredge should penetrate the sediment sufficiently and collect the same amount of sediment each time. In addition, it must be completely closed after taking the sample and should not disturb the surrounding area of the sediment. From this perspective, the descent speed of the dredge should be carefully controlled to avoid dispersion of the material on the sediment surface. The use of dredges for deep-water sediment collections is possible, but it must be ensured that the cable is long enough to reach the river, sea, or lake waterbed. For that, previous depth measurements are advisable, or an electronic deep measuring system should be used. From a scientific viewpoint, the collection of sediment cores is more interesting, as the construction of radioactivity depth profiles can provide very rich information on the studied environment, including its time evolution. Sampling is very complicated in this case and requires specialized personnel [3]. There are several models of sediment core samplers. One example can be seen in Fig. 13.1b. The corer is introduced into the water and falls under its own weight, reaching the bed. Once this happens, there are several mechanisms to introduce the sampler into the sediment. Very often, this is a PVC cylinder with a length that depends on the depth to be sampled. The vertical structure of the sediment core must be preserved, and any sediment disturbance must be avoided. In fact, this is the key issue that makes sample collection difficult. In some devices, it can be performed by retaining the sample and creating a vacuum inside the corer. Alternatively, it is possible to inject air and thus create a buoyancy condition that helps to lift the core to the water surface. In both cases, the purpose is to minimize disturbances to the sediment sampled. Once the core is collected, the PVC cylinder should be sealed with tight fitting caps on both sides.

The use of freeze-coring is especially advisable to maintain the original sediment structure. This means the use of stainless-steel corers that are equipped with a cryogenic circuit surrounding the sampler [4]. The freezing liquid is mostly dry ice mixed with ethanol and, less frequently, liquid nitrogen. The frozen samples must be transported to the laboratory under cooling conditions. This makes sampling difficult in remote areas. Once collected, the sediment core is transported to the laboratory, where it is cut into slices, usually 1 or 2 cm thick, depending on the chosen spatial resolution. The samples are then treated as soil samples.

For many applications, it is important to study the dynamics of sediment formation in a system. This is of special interest for marine studies where deep waters are involved and the vertical transport of sediments can trace the movement of relevant radionuclides. The use of sediment traps is the choice in this case. Sediment traps have a cylindrical or funnel shape and are submerged in water at the selected depth. The sediment trap (see Fig. 13.1c) can stay at the sampling point for hours

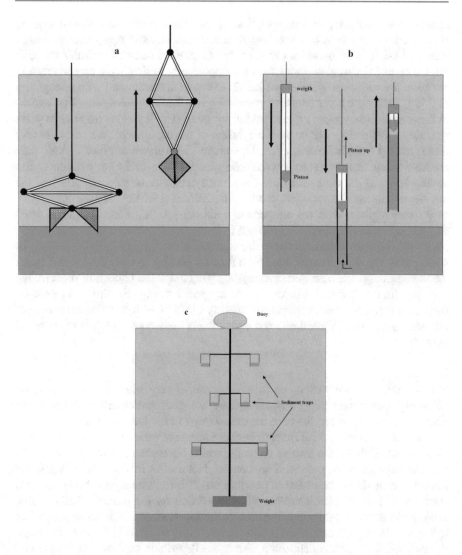

**Fig. 13.1** Sediment sampling with **a** dredges **b** sediment corers, and **c** sediment traps

or years, depending on the involved study [5]. The trapped material in this way is analyzed for radioactivity as a soil or sediment.

*Ice Cores*

Ice is formed by compressing the accumulated snow layers in cool areas. In regions with permanent ices, the ice column formed in this way preserves very interesting information on the local and global environment. The determination of radionuclides in ice layers is a very active research area that includes international efforts, as the sampling of ice cores is a difficult task. In fact, sampling is usually performed in very remote areas and requires sophisticated equipment and high technical skills. It is interesting to note that there exist national facilities for storing ice cores [6] that give researchers the opportunity to deal with very interesting samples. Ice cores are cylinders of ice, 5–15 cm in diameter, and with a length up to 3 km. Each core layer contains particles, gases, radionuclides, etc., that remain in the ice since they were trapped by falling snow. The layers correspond to years or seasons, depending on the time resolution of the experiment. Ice cores are collected mainly from polar areas, i.e., Greenland and Antarctica. Nevertheless, they are also taken from mid- to low-latitude glaciers. The drills are cylinders with cutting edges at the ends. They are introduced manually or mechanically into the ice, depending on the sampled depth or the hardness of the ice (see Fig. 13.2). In warmer regions, with ice temperatures above −10 °C, a hand auger is often enough to obtain the sample. The depth in these cases reaches approximately 20–30 m as much. However, in the polar regions, a mechanical drill is needed. In some cases, the drill is surrounded by a thermal heating annulus device that melts the water around the core during sampling. This helps the collection process. Ice cores are transported to the laboratory or to special storage facilities that must be thermalized at very low temperatures, usually close to −36 °C. The cores are cut into slices, some a few cm thick, and each slice is treated according to the planned experiment.

*Vegetation*
The sampling of vegetation, pasture, or grass is very often included in environmental monitoring programs. First, they are good indicators of radioactivity contamination. However, they are important steps in the human food chain.

There are several general rules to perform adequate sampling work. It is convenient to sample the same area as the soils, which allows the comparison of results and the calculation of radionuclide transfer factors. As in the case of soils, the sampling area should be as flat as possible and free from trees and shrubs that can interfere with atmospheric radionuclide deposition on the pasture. Additionally, roads and tunnels must be far from the collection area, which is usually 1 $m^2$ or a few $m^2$. In any case, the collection surface must be fixed to be able to assign the radionuclide surface deposition. For gross deposition studies, the sample can include stems and grasses and is usually collected 1–5 cm from the ground with scissors. Samples are packaged in plastic bags and transported to the laboratory. There, they were weighed and dried at room temperature or at 105 °C and then ground. It is convenient to weigh the sample before and after drying and give the results in $Bqkg^{-1}$ dry weight or fresh weight [1, 7]. Differences between the radioactivity content of the leaves, stems, and roots may be important. This should be taken into account when sampling vegetables for human consumption or when a study is planned on the distribution of radionuclides throughout the plant. For this, the different parts of the plant must be collected separately. Samples are packaged in plastic bags and transported to the laboratory.

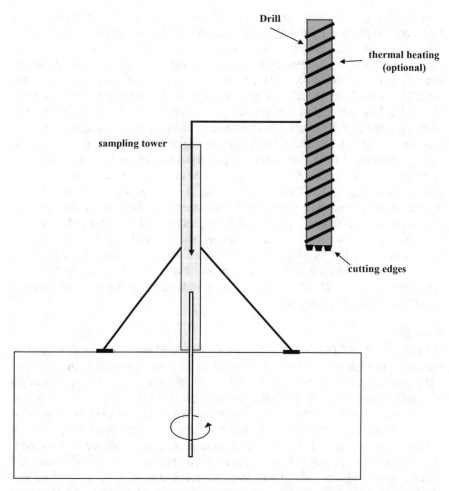

**Fig. 13.2** Ice corer devices. The drill can be introduced in the ice manually or mechanically depending on the hardness of the ice

In all cases, the collected samples must be transported to the laboratory as soon as possible. Otherwise, microorganisms could grow in the wet material. If the shipment to the laboratory is not immediate, air drying of the sample will be convenient to make it lose as much moisture as possible prior to introduction into the bag.

*Seaweed*

Seaweeds are excellent bioindicators of the presence of radionuclides in the marine environment [8]. They can be found in intertidal coastal areas or subtidal coastal areas where light is sufficient for photosynthesis. They can also be found on beaches, but it is not convenient to collect them, as it is not possible to know their origin [9]. Collection in intertidal areas must be performed during low tides. This must be planned following the information of local tidal tables. It is important to have an exact record of the location of the collected algae. Some groups plan the collection by systematically classifying the collection area into quadrants and extending the sampling work in lines perpendicular to the coast. On other occasions, a random collection is good enough. It can be done by hand or using a knife for those specimens that are strongly adhered to the bed or rock. The seaweed collected must be cleaned from the soil, sediments, and other materials. They must be weighed on site and stored in plastic containers. If the samples cannot be rapidly transported to the laboratory, a preservative such as formaldehyde must be added to the container. Subtidal algae can be collected by snorkeling or scuba diving, depending on the depth of the water. As before, algae can be collected by hand or using knives or other cutting instruments for those specimens strongly adhered to the marine bed. The collected samples must be weighed and preserved, if necessary, as before. Once in the laboratory, the seaweed must be dried at room temperature or at 105 °C and then ground and sieved to prepare it for analysis. It is convenient to know the fresh-to-dry weight ratio to find where the radionuclide is preferentially present. This is of paramount importance in the case that seaweed is used for human consumption.

*Foodstuff*

Food represents a direct way through which radionuclides can have an impact on humans. Therefore, the analysis of radioactivity in foods is present in the radiological protection programs of countries. Of course, the collection of the food sample must follow some rules to make the results significant [7].

It is more convenient to analyze individual foodstuffs rather than mixed diet samples. However, in the latter case, it is compulsory to distinguish between adult and baby diet samples. This will provide a more realistic assessment of radionuclide ingestion by the population. In general terms, food products must be collected directly from the producer and then from the retail market. Samples must be prepared as is usually done at home, and the edible part must be analyzed. From this point of view, bones must be separated from meat and fish. This can be easily achieved by heating the sample at 150 °C [7]. Of course, the mass of the sample must be referred to the original value, excluding the inedible parts.

Milk must be collected from the producer or directly from the market. Preservation of milk is a problem, although for short periods, the sample can be stored in a refrigerator prior to analysis. In the opposite case, some preservative products must be added to the milk sample as formaline or $NaN_3$ [7]. In general, the composition of the diet is not constant throughout the world. This means that significant foodstuffs vary between countries. Eggs, mollusks, leafy vegetables, root or bulb vegetables, grain and rice, and even seaweed are also significant or very significant for human consumption, depending on the country. To do this, diet analysis must be planned according to statistical studies on the consumption behavior of specific societies.

## 13.1.2 Liquid Samples

*River Water*

River water can be collected with simple polyethylene bottles [10]. Samples are usually taken at half the depth of the river. Stagnant water areas or sites close to the shoreline must be avoided. The amount of sample will depend on the expected radionuclide concentrations. Once collected, the water sample must be filtered immediately after sampling, preferably on site, to remove the suspended matter. For that, a filtering pump is used, and the water is passed through a 0.4 μm Nucleopore or Micropore filters. The water sample is retrieved in a clean plastic bottle. It is convenient to acidify water up to pH = 2 with $HNO_3$ or HCl to avoid microorganism growth or radionuclide deposition on the walls of the bottle. Additionally, it is also convenient to store the sample in the refrigerator until analysis. This storage method is fairly adequate when simple $\gamma$-spectrometry measurements are planned. However, acidulation must affect the chemical state of some radionuclides. Therefore, in the case that the radionuclides of interest must be extracted by using radiochemical procedures or speciation studies are planned (see next sections), the acidulation is not convenient, and other methods of storage before analysis have to be used. In these cases, the analysis should be carried out as soon as possible. Meanwhile, the sample can be stored in a refrigerator but not for a long period of time. The filter containing suspended matter must be stored in a Petri dish and shipped to the laboratory. There, it must be dried and weighed. Since the weight of the clean filter is known, the amount of suspended matter loaded can be easily calculated. For speciation studies (see specific section), filtration (0.45 μm, 0.2 μm, 0.1 μm pore size) and ultrafiltration techniques (~nm) are needed and must be performed on site [11]. As we will see, the separation of size species in waters (dissolved, colloids, and particulate material, see Chaps. 5 and 20) is essential to know the dynamics of the radionuclides in the aquatic system. This is applicable to seawater and lake water sampling works.

*Seawater*

Surface seawater samples can be collected following the previous procedure. However, in deep waters, vertical transport is relevant for the dynamics of radionuclides in marine systems. For that, it is important to collect samples at different well-defined depths. This can be done with special samplers. The most popular are the so-called Niskin bottles [12]. They consist of Teflon cylinders with two stoppers on both sides of the cylinder. The bottle is submerged in seawater attached to a cable with the two stoppers open. When the desired depth is reached, a messenger (small weight) is sent through the cable that activates an elastic mechanism that closes both stoppers and thus traps the water at the selected depth. In Fig. 13.3, we can see some Niskin bottles and the sampling mechanism. The Niskin bottles can be grouped into rosettes containing several bottles, each with an individual messenger. The rosette is submerged in water, and each messenger hits the bottle at a given depth. In this way, seawater samples can be taken at different depths with the same device. As in the case of river water, seawater samples must be filtered to separate suspended matter. This is done on site, usually on oceanographic vessels. In many vessels, there are laboratories for sample storage, processing, and measurement [5]. All the principles described before with respect to the preservation of water samples apply in this case.

*Lake Water*

Lake water sampling is very similar to that in rivers, and all the recommendations previously described also apply here. For deep-water sampling, Niskin bottles were used.

*Tap Water*

Tap water is collected in plastic bottles that have previously been rinsed and cleaned. Collection can take place directly at the water processing plant. When sampling is carried out at home, the pipes must be flushed 5–10 min before sample

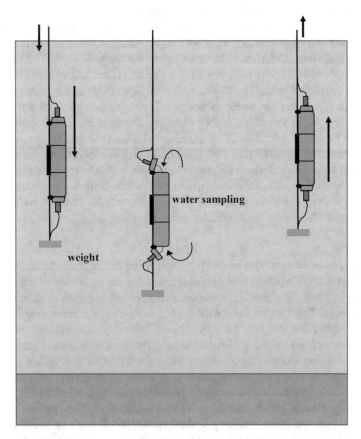

**Fig. 13.3** Niskin bottles for water sampling. On some occasions, the bottles can be grouped into rosettes

collection. In the case where the sampling is done after night, the flushing time should be longer, close to 30 min [7].

## 13.1.3 Atmospheric Samples

*Atmospheric Deposition*

Fallout radionuclides are usually collected with stainless-steel funnels of well-known collection surface (see Fig. 13.4a). The funnels must be placed in open areas or building roofs, always overhanging trees, bushes, or buildings to avoid shadowing. The funnel collection surface is normally 1 $m^2$, but other surfaces are also adequate. In any case, the collection surface must be well known, since in many cases the relevant information is given in $Bqm^{-2}$. A plastic bag is connected to the outlet of the funnel outlet to collect rainwater. Usually, sample collection is carried out monthly. Before the bottle is removed, the funnel walls must be cleaned with distilled water and a slightly acidified solution. In this way, the plastic bottle contains not only the collected rainwater but also the dry component of the fallout [13]. In some cases, the acidified solution is not used to avoid possible changes in the chemical state of key radionuclides before analysis. The total deposition sample obtained in this way is shipped to the laboratory and stored until analysis. As in the case of water, the suspended matter in the rainwater is removed by filtration. In this way, it is possible to distinguish between the soluble and insoluble phases in rainwater. However, for a perfect discrimination between dry and wet fallout, a more complicated device is used (Fig. 13.4b). Now, a double funnel system is deployed in the collection area. One funnel was utilized for wet fallout collection, and the other was utilized for dry fallout. The wet collector is open, and the dry collector is covered during rainy periods. In contrast, during dry periods, the wet collector is covered, and the dry collector is open. This can be done with a movable cover that automatically changes from the wet to the dry collector at the start of the rain and vice versa [13]. Once the sampling period has ended, the dry collector walls are rinsed several times with distilled water and a slightly acid solution to remove all the sampled particles. The obtained slurry is retrieved in a plastic bottle at the collector outlet. The wet collector is treated as shown above. In some cases, an ionic exchange column (see next sections) is attached to the funnel outlet (see Fig. 13.4c) [1, 13]. Rainwater flows throughout the ionic exchange resin, and the relevant radionuclide is retained in the resin. Once the collecting period has ended, the funnel is washed with water that is also passed through the resin. It is important to keep the resin wet throughout the sampling period. Otherwise, the retention efficiency can decrease.

*Aerosols*

Airborne radionuclides are collected using a high-volume air sampler. Air is passed through polypropylene filters with the help of a high flow pump, and aerosols are expected to remain in the filter [13]. Some high-flux stations allow one to sample

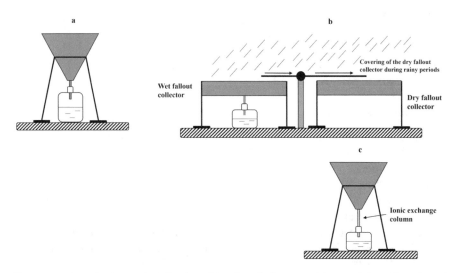

**Fig. 13.4** Atmospheric sample collection with **a** funnels, **b** automatic wet and dry collectors, and **c** ionic exchange collector coupled to a funnel

some 15,000 m$^3$ of air per day, which makes possible the determination of very low levels of radioactivity in air. This high flux also permits a higher time resolution in radioactivity determination, since it is possible to prepare weekly and, in some cases, daily samples [14]. The air filter is measured directly by $\gamma$-spectrometry or chemically treated for specific radionuclides. The described method allows for total particle collection. However, in some cases, for instance, when radiological studies are necessary, the aerosol particle size is significant. In fact, depending on its size, the particle can be incorporated into the human respiratory tract (see Chap. 6). The collection of aerosols of different particle sizes can be performed with devices called cascade impactor collectors. In this case, air is passed through different stages in such a way that heavier particles deposit in the first stage, while lighter particles deposit in the subsequent stage. The final result is the selective collection of the different particle sizes contained in the aerosols. In this way, it is possible to determine the radioactivity concentration for each particle size. This sampling procedure is used extensively for speciation studies in which knowledge of the association of radionuclides with aerosol particle size is of paramount importance [15]. Air sampling in the lower stratosphere is also possible, although this is not usually available in many laboratories. In this case, air filters are collected by devices installed in aircraft, and the collection is carried out along a flight tract of constant height and increasing latitudes [16].

The collection of radioactive gases is carried out by using the same filtering devices but with special filters. For the collection of I isotopes, air is passed through a try-ethylene diamide (TEDA)-activated charcoal, which traps gaseous iodine chemical species with a high efficiency. In real applications, air is passed before through a conventional air filter that traps I in the aerosol. Air is driven into the activated charcoal cartridge where gaseous species I is trapped [14]. Similar systems have been used for radioactive noble gas sampling, although the efficiency is clearly lower than that of iodine [17].

### 13.1.4 Biological Samples

Samples of human tissues are not included in environmental radioactivity protection programs, with the exception of severe accidental situations. However, routine bioassay monitoring of workers exposed to radiation includes in vitro analysis of urine and, on some occasions, of blood and feces [18]. Normally, urine is collected during a 24 h sampling period, although on some occasions, a one-spot sample is sufficient. In both cases, the sample is acidified with 1% $HNO_3$ for storage, although it is very convenient to start the analysis soon after collection. Feces collection also involves a 24 h sampling period. In this case, the samples are dried directly and stored in a refrigerator in the event that the analysis is not performed soon after collection. However, it is more convenient to analyze the sample as soon as possible. For this, the sample is ashed, and the ashes are analyzed for $\gamma$-spectrometry or used for radiochemical extraction of specific radionuclides.

## 13.2 Sample Transport and Storage

As a general rule, sample analysis must be carried out soon after collection. In general, terms, this is difficult to accomplish, as the sample must be transported to the laboratory and stored before analysis [18, 19]. The transport of samples is not a trivial task, provided that all the properties and representativeness of the sample must be preserved before the analysis. Therefore, sample contamination, losses, or cross-contamination must be avoided. For this, the use of adequate containers is of paramount importance. Plastic bags are a very good option for sediments, soils, or seaweed. The transport of water samples is carried out in plastic bottles instead of the classic glass bottles. The reason is that in glasses, the probability of radionuclide absorption in the walls is considerably higher. As commented before, it is very convenient to acidify the water after collection once the suspended matter has been separated in the case where radionuclide determination in the liquid and solid phases is planned. For that, $HNO_3$ is conventionally used until the water pH reaches a value of 1.7 or 2. However, acidification could cause changes in the chemical species of some radionuclides, which is not convenient for speciation studies. Petri dishes are a good option for the transport of air or suspended matter filter samples.

In any case, bags, bottles, or Petri dishes must be tightly closed to avoid loss of volatile species. In addition, transport should be performed at a cool temperature in a dark environment to minimize heat- or light-boosted chemical reactions. High-altitude plane transport must be carefully planned, as the interaction of cosmic rays during flight can produce cosmogenic radionuclides in the samples.

In the laboratory, the samples must be stored under controlled conditions to avoid degradation, spoilage, etc. If the analysis is planned to occur upon arrival or within a short time period after arrival, storing the sample in a refrigerator under dark conditions is sufficient. In some cases, for biological samples, for example, some preserving agents, such as formaline, alcohol, and $NaHSO_3$, must be used.

For long-term storage, it is convenient to pretreat the sample and obtain it in a more stable form. Sediments, soils, seaweed, and biological material should be dried at 70–80 °C or room temperature to avoid loss of interesting radionuclides. The dry residue is then ground and stored in tightly sealed (if possible under vacuum conditions) very clean plastic bags. The resulting sample must be stored in a refrigerator or at least in a cool and dark environment. Samples must be weighed before and after drying, as the radioactivity concentration can be reported as dry weight $Bqkg^{-1}$ or fresh weight.

On many occasions, the plastic bottles used for water transport can be stored directly under dark and cool conditions if the sample was acidified upon collection. However, in some cases, a volume reduction is necessary before the sample is stored. The simpler method is evaporation, but this must be done very carefully to avoid loss of volatile species or due to projections from the sample during heating. For specific applications, it can be necessary to precipitate the interesting radionuclides in the water sample and thus reduce the high-volume water to a few grams of precipitate. In the next section, we give some more details.

## 13.3  Chemical Procedures

Ge semiconductor $\gamma$-spectrometry has made radiochemistry unnecessary in many applications. Their excellent energy resolution provides a very powerful tool to identify and quantify radionuclide concentrations in environmental samples just by determining the net area under the photopeak of interest (see Chap. 11). The crucial problem is now the construction of good efficiency calibration curves, as we will see in the next sections [20].

However, radiochemistry is still necessary, as there are many $\alpha$- and $\beta$-emitters that are relevant in environmental studies. Furthermore, the extremely low environmental levels of some $\gamma$-emitting nuclides make it necessary, or at least advisable, to preconcentrate the sample before $\gamma$-spectrometry. This means carrying out a radiochemical procedure that can be more or less complicated depending on the expected radionuclide concentration.

Radiochemistry procedures are similar but not identical to conventional chemical methods. First, safety considerations compel us to modify working protocols to avoid unnecessary radiation exposure. Second, the half-life of some radionuclides (or tracers used to measure recoveries) makes it compulsory to speed procedures that should last well-fixed time periods under conventional conditions.

From a basic viewpoint, the number of atoms involved in chemical reactions is extremely low, and hence, the mass concentrations involved are also very low. It is known that the kinetics and thermodynamics of chemical reactions depend on

**Table 13.1** Number of atoms and mass of several interesting radionuclides for environmental levels of activity, 1 mBq. The number of atoms is calculated as $N = \frac{\pi 10^4}{\ln 2} T_{\frac{1}{2}}$ (y) and the corresponding mass as $m = \frac{N P_m}{N_A}$

| Radionuclide | $T_{1/2}$ (y) | Number of atoms | Mass (fg) |
|---|---|---|---|
| $^3$H | 12.32 | $5.58 \times 10^5$ | $2.78 \times 10^{-3}$ |
| $^{14}$C | 5700 | $2.58 \times 10^8$ | 6.00 |
| $^{90}$Sr | 28.90 | $1.31 \times 10^6$ | $1.96 \times 10^{-1}$ |
| $^{99}$Tc | $2.11 \times 105$ | $9.56 \times 10^9$ | $1.57 \times 10^3$ |
| $^{129}$I | $1.57 \times 107$ | $7.11 \times 10^{11}$ | $1.52 \times 10^5$ |
| $^{137}$Cs | 30.08 | $1.36 \times 10^6$ | $3.10 \times 10^{-1}$ |
| $^{239}$Pu | 24,110 | $1.09 \times 10^9$ | 4.34 |
| $^{241}$Am | 432.6 | $1.96 \times 10^7$ | 7.85 |

the mass concentrations of the involved compound. In other words, the procedures applied in conventional chemistry for macroscopic amounts of the counterpart stable isotopes might not be directly applicable to the environmental radionuclide [21]. In Table 13.1, the number of atoms and the corresponding mass of environmentally interesting radionuclides, which are equivalent to the usual levels of radioactivity concentrations, are presented. It is very clear that the amounts involved are so small that classical procedures may fail.

Moreover, radioactive disintegrations can cause chemical transmutations. This is the case for $\alpha$ particle emission, where the recoil energy of the daughter radionuclide is greater than the chemical bond energies. On the other hand, from a practical viewpoint, some problems appear. As commented in the previous section, it is convenient to collect and store liquid samples in plastic bottles because radionuclides can be absorbed by the glass walls. For that, it is convenient to acidify the sample to provide enough $H^+$ ions, which prevent adsorption. The probability of this effect occurring increases as concentration decreases [22]. For that, it is very relevant in environmental radioactivity. Additionally, for that reason, the use of traditional glassware in chemistry laboratories must be avoided or done with great care.

An interesting advantage of radiochemistry with regard to conventional chemistry is the fact that the final product of the chemical separation is identified by the emitted radiation. This means that the purity of the extraction is measured with respect to other radioactive species that could interfere with the measurement. The concept of chemical purity of the extracted compound, which is a key issue in conventional analytical chemistry, does not apply here. The so-called decontamination factors [22] are measured here with respect to potentially interfering radioactive species and not to chemical species in general. In most cases, this simplifies the procedure, as the radiation emitted by the radionuclide of interest can be determined without ambiguity with high-energy resolution $\alpha$- or $\gamma$-spectrometers. In the case of $\beta$-emitters, the separation must follow more conventional criteria since the energy spectrum is a continuum, and therefore, it is difficult to distinguish different $\beta$-emitters with a $\beta$-spectrometer [23].

There are many radiochemical procedures for radionuclides of interest. Some schemes are shown in Fig. 13.5. Of course, we will not describe all of them in

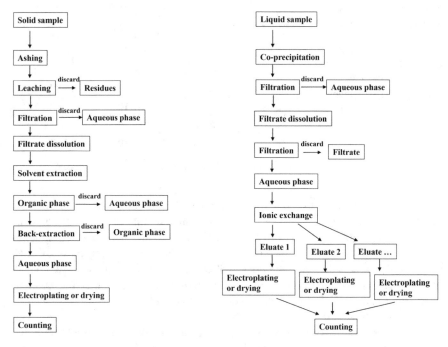

**Fig. 13.5** Flowchart of typical radiochemical procedures for solid and liquid samples

detail. Instead, we will discuss their common aspects in the next section, and the reader is advised to read the specialized literature [24].

Three common steps can be distinguished among the radiochemical procedures: preconcentration, chemical separation, and radiation source preparation. We will describe them in the following paragraphs.

## 13.3.1 Preconcentration Processes

Preconcentration of the sample, and more specifically of the radionuclide of interest, is necessary in most cases to reduce the sample volume and thus make its handling easier. In other cases, it is necessary to perform a previous selection of the radionuclide or radionuclides of interest to facilitate the extraction. Of course, losses must be avoided during preconcentration, and the adopted procedure must be compatible with the rest of the extraction method [25].

*Evaporation*
Evaporation is a very simple procedure for reducing the volume of water samples. Evaporation must take place at room temperature or at 70–80 °C. In this way, boiling is avoided, so the probability of losses is reduced. Because in many cases the water sample has been acidified, evaporation must be performed in a fume

hood. The sample can be evaporated to dryness, and then the residue is solubilized using different methods (see the following paragraphs). Alternatively, the sample can be evaporated to reach the volume necessary to fill standard geometry beakers (Marinelli beakers, for instance; see the next paragraphs) for $\gamma$-spectrometry.

*Precipitation and Coprecipitation*

Precipitation or coprecipitation is frequently used for large volume water samples. It consists of the formation of an insoluble phase in a solution and the subsequent precipitation of solids that can be collected in different manners. This method allows for substituting the handling of hundreds, or thousands, of liters of water by some grams or milligrams of precipitate, which greatly facilitates the chemical procedure. Precipitation of such small amounts of radionuclides is not easy. For that, a macroscopic amount of a carrier is added to the sample to improve the precipitation process. The carrier is a stable isotope of the radionuclide of interest, presumably in the same chemical state. It is expected to carry the radionuclide of interest during precipitation. However, the more conventional method to produce radionuclide precipitation does not involve the use of stable isotopes but rather of the so-called nonisotopic carriers or scavengers. They carry indiscriminately different chemical species. A very popular method is based on the precipitation of $Fe(OH)_3$ at basic pH. For that, some given mass of $Fe^{2+}$ is dissolved in the water sample at acidic pH. The addition of NaOH or $NH_4OH$ to the solution causes the formation of hydroxides and their precipitation at basic pH. Coprecipitation with $Fe(OH)_3$ is quantitative for many radionuclides [26]. The precipitation process is usually performed in a hot plate, and the precipitate is washed and redissolved. On many occasions, a new coprecipitation is performed to recover as much radionuclide as possible. Finally, the precipitate is recovered by filtration or centrifugation and then dissolved once more to continue the chemical process. The use of iron hydroxide precipitation is the most popular procedure, but for some applications, other hydroxides such as $Mn(OH)_3$, $Al(OH)_3$, $La(OH)_3$, $Zr(OH)_3$, $Th(OH)_3$, $Mg(OH)_2$, and $Ni(OH)_2$ can be used. They also help to coprecipitate a good number of radionuclides [24, 26]. Other compounds cannot be precipitated using these scavengers. This is the case for Cs, whose isotopes, $^{137}Cs$, $^{134}Cs$, and, to a lesser extent, $^{135}Cs$, are relevant from an environmental point of view. In this case, ammonium phosphomolybdate (AMP) is the choice [27].

*Wet and Dry Ashing*

Wet or dry ashing is used to remove organic matter from the sample because it makes it difficult to perform chemical reactions. Wet or dry ashing is often used to treat soils, sediments, seaweed, etc. Dry ashing is usually carried out at 300–550 °C in a laboratory furnace. This seems high enough to remove all the organic phases. However, depending on the sample, this temperature can be increased. The main drawback of dry ashing is the probability of losing volatile radionuclides. The choice in this case is wet ashing of the sample, which is a way to oxidize the organic matter. Now, the sample is attacked with a mixture of mineral acids ($HNO_3$, $H_2SO_4$, and $HCLO_4$). The use of $H_2O_2$ greatly helps in the total elimination of the sample.

*Solubilization*

Solid samples (sediments, soils, seaweed, rocks, etc.) must be solubilized to perform adequate radionuclide extraction. The leaching or solubilization of the sample was carried out with a mixture of mineral acids ($HNO_3$, $H_2SO_4$, $HCLO_4$, and HF). The process can be improved with the help of a microwave oven where the pressure and temperature of solubilization can be controlled [28]. Total dissolution of the sample is not always possible. For these cases, a fusion technique is necessary to obtain the total decomposition of the sample. Now, a salt (flux) is thoroughly mixed with the sample, and the mixture is heated to a temperature above the flux melting point (800–1000 °C). The components of the mixture react during heating. The molten mixture is cooled and solubilized following the previous procedures.

The preconcentration process described here can be combined according to the needs of the entire radiochemical procedure.

## 13.3.2 Separation and Purification Procedures

The next step, once the sample has been pretreated, is chemical separation. There are many different procedures. Their main goal is the purification of the sample or the rejection of interferences that can affect the determination of radionuclides. From this point of view, the selected method will depend on the final determination procedure. If the radionuclide is measured by a radiometric method, the method must ensure that the final radioactive source is free from radioactive interference. Radioactive interference can be defined as radioactive species that emit radiation that cannot be distinguished from that emitted by the radionuclide of interest. From this viewpoint, the interferences are not necessarily isotopes of the radionuclide of interest. In contrast, they can be different radioactive species. Thus, a careful study of the potential radioactive interferences must be performed before developing the chemical separation method. It is very interesting to note that the radiation detector is a tool for rejecting radioactive interferences. For example, if the interfering radionuclide emits $\gamma$-radiation and the radionuclide of interest $\alpha$-radiation, it seems very clear that a Si $\alpha$-spectrometer is the best option. However, if the radionuclide is determined by a mass spectrometry method, the problem is substantially different. It is now important to design a chemical method capable of rejecting isobars or molecules with the same mass as the radionuclide of interest. It is important to note that the radionuclide is usually at extremely low concentrations in the sample. This means that the potential formation of interfering molecules by the combination of elements in the sample is relevant although their concentrations are very low. In Table 13.2, a list of potential interfering molecules in ICP–MS mass spectrometry is given [29]. In Chap. 18, information on AMS interferences can be found.

Despite the very low concentrations of the potentially interfering molecules, they are still orders of magnitude higher than those of the radionuclide of interest. To avoid interferences in mass spectrometry, it is sometimes more convenient to inject a molecule of the radionuclide of interest into the spectrometer. This is the

**Table 13.2**  Molecules interfering with radionuclide measurements in mass spectrometry

| Radionuclide | Interference |
|---|---|
| $^{90}$Sr | $^{90}$Zr$^+$, $^{90}$Y$^+$, $^{180}$W$^{++}$, $^{180}$Hf$^{++}$, $^{180}$Ta$^{++}$ <br> $^{38}$Ar$^{40}$Ar$^{12}$C$^+$, $^{50}$Cr$^{40}$Ar$^+$, $^{53}$Cr$^{37}$Cl$^+$, $^{52}$Cr$^{38}$Ar$^+$ <br> $^{54}$Cr$^{36}$Ar$^+$, $^{54}$Fe$^{36}$Ar$^+$, $^{56}$Fe$^{16}$O(H$_2$O)$^+$, $^{72}$Ge$^{18}$O$^+$ <br> $^{74}$Ge$^{16}$O$^+$, $^{76}$Ge$^{14}$N$^+$, $^{78}$Kr$^{12}$C$^+$, $^{55}$Mn$^{35}$Cl$^+$, $^{58}$Ni$^{16}$O$_2^+$, <br> $^{60}$Ni$^{16}$O$^{14}$N$^+$, $^{45}$Sc$^{45}$Sc$^+$, $^{74}$Se$^{16}$O$^+$, $^{76}$Se$^{14}$N$^+$, $^{78}$Se$^{12}$C$^+$, <br> $^{50}$Ti$^{40}$Ar$^+$, $^{50}$V$^{40}$Ar$^+$ |
| $^{99}$Tc | $^{98}$Mo$^+$, $^{98}$Ru$^+$, $^{99}$Ru$^+$, $^{100}$Mo$^+$, $^{100}$Ru$^+$ <br> $^{83}$Kr$^{16}$O$^+$, $^{59}$Ti$^{40}$Ti$^+$, $^{98}$Mo$^1$H$^+$, $^{59}$Co$^{40}$Ar$^+$, $^{62}$Ni$^{37}$Cl$^+$, <br> $^{81}$Br$^{18}$O$^+$, $^{98}$Ru$^1$H$^+$, $^{63}$Cu$^{36}$Ar$^+$, $^{64}$Zn$^{35}$Cl$^+$, $^{67}$Zn$^{16}$O$_2^+$, $^{51}$V$^{16}$O$_3^+$, $^{198}$Pt$^{++}$, <br> $^{198}$Hg$^{++}$ |
| $^{129}$I | $^{129}$Xe$^+$, $^{97}$MoO$_2^+$, $^{113}$CdO$^+$, $^{127}$IH$_2^+$, $^{127}$I$^2$H$^+$ |
| $^{135}$Cs | $^{135}$Ba$^+$, $^{134}$Ba$^1$H$^+$, $^{119}$Sn$^{16}$O$^+$, $^{95}$Mo$^{40}$Ar$^+$, $^{98}$Mo$^{37}$Cl$^+$, $^{134}$Xe$^1$H$^+$ |
| $^{137}$Cs | $^{137}$Ba$^+$, $^{97}$Mo$^{40}$Ar$^+$, $^{121}$Sb$^{16}$O$^+$, $^{100}$Mo$^{37}$Cl$^+$ |
| $^{237}$Np | $^{238}$U$^+$, $^{197}$Au$^{40}$Ar$^+$, $^{203}$Tl$^{34}$S$^+$, $^{205}$Tl$^{32}$S$^+$, $^{200}$Hg$^{37}$Cl$^+$, $^{202}$Hg$^{35}$Cl$^+$, $^{206}$Pb$^{31}$P$^+$ |
| $^{239,240}$Pu | $^{238}$UH$^+$, $^{238}$UH$_2^+$, $^{207}$Pb$^{16}$O$_2^+$, $^{199}$Hg$^{40}$Ar$^+$, $^{160}$Dy$^{79}$Br$^+$, $^{160}$Gd$^{79}$Br$^+$, <br> $^{158}$Gd$^{81}$Br$^+$, $^{204}$Pb$^{35}$Cl$^+$, $^{208}$Pb$^{16}$O$_2^+$, $^{200}$Hg$^{40}$Ar$^+$, $^{161}$Dy$^{79}$Br$^+$ |
| $^{234,235,236,238}$U | $^{194}$Pt$^{40}$Ar$^+$, $^{206}$Pb$^{14}$N$_2^+$, $^{204}$Pb$^{14}$N$^{16}$O$^+$, $^{195}$Pt$^{40}$Ar$^+$, $^{207}$Pb$^{14}$N$_2^+$, $^{196}$Pt$^{40}$Ar$^+$, <br> $^{204}$Pb$^{16}$O$_2^+$, $^{208}$Pb$^{14}$N$_2^+$, $^{196}$Hg$^{40}$Ar$^+$, $^{205}$Tl$^{31}$P$^+$, $^{204}$Pb$^{32}$S$^+$, $^{199}$Hg$^{37}$Cl$^+$. <br> $^{206}$Pb$^{14}$N$^{16}$O$^+$, $^{198}$Pt$^{40}$Ar$^+$, $^{206}$Pb$^{16}$O$_2^+$, $^{208}$Pb$^{14}$N$^{16}$O$^+$, $^{234}$UH$^+$, $^{235}$UH$^+$ |

case, for example, for the $^{10}$Be or $^{239}$Pu and $^{240}$Pu AMS determinations. BeO or BeF$_2$ and PuO$_2$ are injected into the accelerator instead of Be and Pu to better reject interferences from $^{10}$B and $^{238}$UH [30, 31].

Some general separation techniques follow. They can be used in combination to make a more selective extraction. The procedures involve the use of reagents, labware, etc. Because the radionuclide concentrations are extremely low, the reagents must be very pure. Additionally, it is highly recommended to carefully clean the labware that will be used. In many cases, it is convenient to rinse the materials with acids and distilled water prior to analysis. In fact, most of the tools are single-use materials. This greatly decreases the probability of cross-contamination [7, 10].

*Volatilization and Distillation*

Volatilization and distillation are very simple techniques for extracting volatile species from a sample. It is based on the vapor pressure differences among the elements or their compounds. Some examples of interesting species and their volatile chemical forms are given in Table 13.3 [24]. In a simple version of this procedure, the sample is heated to a given temperature. The emanating volatile species are swept to a low-temperature region with the help of an inert gas, such as N$_2$ or He. There, the gas condenses, and the chemical separation can continue according to the expected compound. In Fig. 13.6, a simple distillation apparatus is presented. The sample, a soil or sediment, is heated to 500–600 °C, and the gases are condensed on wetted glass wool that is at a sufficiently low temperature [32]. A more sophisticated version of this procedure allows fractional separation of different radionuclides by programming the heating temperature. In this way, the radionuclides are selectively volatilized. Furthermore, in some cases, the sample is pretreated to form volatile radionuclide compounds (see Table 13.3). This improves selectivity. Volatilization techniques have not been widely applied to inorganic matrices, but there are several methods in the literature for the separation of $^{210}$Po, $^{99}$Tc, and $^{129}$I by volatilization techniques [32–34]. A very popular application of distillation is the determination of $^3$H in water or soils. In this case, the sample is pretreated with NaOH and KMnO$_4$ to limit the volatilization of other elements during distillation [35].

Volatilization is a very selective technique and conceptually simple. This is its main advantage. However, the range of applicability to environmental radioactivity is limited, since not all interesting radionuclides form volatile forms at reasonable temperatures that could be adequately controlled [24].

*Solvent Extraction*

Solvent or liquid–liquid extraction is a separation technique that is very much used because it is simple and selective. It consists of the transfer of radionuclides between two liquid phases, normally an aqueous solution and an organic liquid. Both phases must be immiscible or, at least, very little miscible [24]. The transfer of the radionuclide from the aqueous phase to the organic phase is a consequence of the increase in solubility in the organic liquid and the subsequent decrease in solubility in water. This can be achieved by complexing the species to obtain a neutral, nonhydrated compound. A detailed and fundamental explanation of this process is beyond the objectives of this text. However, the reader is referred to [24] for a deeper description of the chemical transformation of solvent extraction.

**Table 13.3** Chemical form of volatile species of radionuclides

| Radionuclide | Volatile chemical form |
|---|---|
| $^3$H | Element, hydrides, oxides, halides |
| $^{14}$C | Hydrides, oxides, halides |
| $^{39}$Ar | Element |
| $^{55}$Fe | Halides |
| $^{81,85}$Kr | Element, halides |
| $^{89,90}$Sr | Halides |
| $^{95}$Zr | Halides |
| $^{99}$Tc | Oxides, halides |
| $^{129,131}$I | Element, hydride, halides |
| $^{133}$Xe | Element, halides |
| $^{134,135,136,137}$Cs | Element |
| $^{210}$Po | Element, halides |
| $^{220,222}$Rn | Element, halides |
| $^{234,235,236,238}$U | Halides |
| $^{237}$Np | Halides |

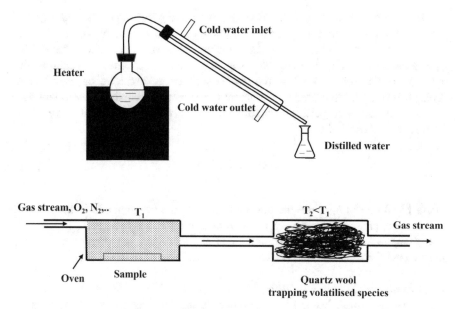

**Fig. 13.6**  Examples of distillation apparatus

There are three kinds of organic compounds that can produce this change in solubility and subsequent transfer among phases: polydentate organic anions $A^-$, neutral organic molecules B, and large organic cations $RB^+$.

Polydentate anions form chelate structures. When these structures are in contact with water, they dissolve. The chelating agent introduced in this way into the water ionizes and complexes the radionuclide. The radionuclide chelate in turn dissolves in the organic phase. Polydentate organic anions are acetylacetonate, pyrazolones, benzoylacetonate, thenoyltrifluoroacetone (TTA), diethylexylphosphoric acid (HDEPH), and dibutylphosphoric acid (HDBP).

Uncharged extracting compounds or neutral molecules are historically the most important extracting agents. Diethyl ether was used to purify U in the Manhattan Project, and tributylphosphate (TBP), the esther of an organophosphorus compound, is a key extracting agent in the PUREX process for the separation of Pu from spent nuclear fuel [36]. The chemical process is now more complicated, although complexation processes are carried out by bonding the radionuclide with O in the extractant structure. Complexation is improved by introducing a salting agent ($HNO_3$, $H_2SO_4$, etc.) into the aqueous phase. Neutral molecules used in these separations are also other alcohols, ethers, or ketones.

Cationic extracting compounds are strong bases that form complexes with radionuclides, essentially actinides. They are tertiary or quaternary amines, with the most used being trioctylamine (TOA) and triisooctylamine (TIOA). The complexing reaction can now be described as an ion exchange.

Liquid-liquid extraction is carried out by mixing the aqueous and organic phases. The water solution contains radionuclides. This is done by vigorous shaking of the mixture with hands or using ultrasonic devices. After shaking, the phases were allowed to separate. Once this happens, the organic phase containing the radionuclide is collected, while the aqueous phase is discarded or mixed once again with an additional organic extractant to recover the remaining radionuclides.

The distribution or partition coefficient, $K_d$, serves to estimate the extraction efficiency. It is defined as

$$K_d = \frac{[A_{org}]}{[A_{aq}]}$$  (13.1)

where $[A_{org}]$ and $[A_{aq}]$ are the concentrations of the radionuclide in the organic and aqueous phases, respectively, once equilibrium has been reached.

In Table 13.4 [24], some radionuclides together with the organic extracting agents used for separation are listed.

*Ion Exchange*

Ion exchange is probably the most popular separation technique. It is highly specific, and the decontamination factor is very high. The technique is very simple, fast, and in most cases the only way to separate elements of very similar chemical behavior [24]. This technique received a significant boost when synthetic resins were discovered to have ion-exchange properties [37].

In general terms, an aqueous solution containing radionuclide ions is passed through a synthetic insoluble, although permeable, organic resin with fixed charged functional groups. These groups are associated with mobile counterions of opposite charge. The ions in the aqueous solution are exchanged with the counterions, and in turn, they remain bound into the resin. This binding can be broken afterward by passing another solution through the resin.

Resins are synthetic polymers cross-linked with divinylbenzene (DVB) with functional groups that can exchange cations or anions. In the first case, the

**Table 13.4** Usual liquid–liquid extractants for radionuclide extraction

| Radionuclide | Extractant |
|---|---|
| $^{89,90}$Sr | dicyclohexano-18-crown-6 in trichloromethane |
| $^{99}$Tc | Tri-n-octylamine (TnOA), tributylphosphate (TBP) |
| $^{210}$Pb | Aliquat-336® |
| $^{224,228,226}$Ra | Aliquat-336® in xylene |
| Pu isotopes | Trioctylphosphine oxide (TOPO), trisooltylamine (TIOA), TnOA, TBP |
| Am isotopes | Trioctylphosphine oxide (TOPO), trisooltylamine (TIOA), TnOA, TBP |
| Np isotopes | Trioctylphosphine oxide (TOPO), trisooltylamine (TIOA), TnOA, TBP |
| U isotopes | Ethyl acetate, TIOA |
| Th isotopes | Thenoyltrifluoroacetonate (TTA), TIOA, Aliquat-336® |

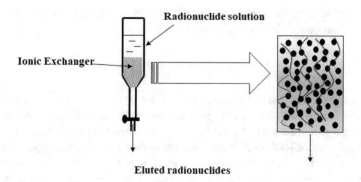

**Fig. 13.7** Usual ionic exchange resin setup

group is typically formed by sulfonate, $R_{resin}SO_3^{-1}H^{+1}$, or carboxylate anions, $R_{resin}COO^{-1}H^{+1}$, and the cation displaces the H ion. Examples are as follows: Amberlite IR-100, Dowex 30, Dowex 50, Zerolit 225, and Amberlite IR-120. In anionic exchangers, the resin typically contains quaternary amine groups, such as $R_{resin}N(CH_3)_3^{+1}OH^{-}1$, and the anion displaces the OH ion. Examples of this type of exchanger are Amberlite IRA-400 and Dowex 1.

The ion-exchange resins are prepared in columns, and the aqueous solution passes at a flux that must be determined to obtain the optimum retention of the radionuclide of interest. The length and diameter of the prepared column must also be determined for specific separation. A typical experimental setup using ionic exchange resins is presented in Fig. 13.7. The retention capacity or absorbability of the resin is given by the distribution coefficient D. An ion $X$ is defined as

$$D = \frac{W_X}{W_{X,s}} \qquad (13.2)$$

$W_x$ is the weight of retained $X$ ions per kg of dry resin, and $W_{x,s}$ is the weight of $X$ ions in solution per liter of the solution. An alternative definition is the volume distribution coefficient, $D_v$, which is

$$D_v = \frac{W_{X,r}}{W_{X,s}} \qquad (13.3)$$

$W_{x,r}$ is the weight of the retained $X$ ions per liter of resin. If r is the resin density, then

$$D_v = rD \qquad (13.4)$$

To estimate the resin separation capacity for different ions, for example, ions 1 and 2, we can define the separation factor $\alpha$ as

$$\alpha = \frac{D_1}{D_2} \tag{13.5}$$

The properties of ion exchangers depend not only on the number and type of functional groups but also on the cross-linking configuration. Manufacturers usually indicate with a number after an '$X$' the percentage of cross-linking of the resin. For example, an AG-X8 resin is an 8% cross-linked resin. As this figure increases, the density of the functional group increases, which is supposed to improve the capacity for ionic exchange and the selectivity. However, as this figure increases, the diffusion rate in the resin decreases. Because the capacity of exchange is strongly dependent on the diffusion rate of the solution, a compromise must be achieved to obtain the optimum cross-linking value. The size of the resin particles is also an important parameter. As the particle diameter increases, the flow rate will increase. Typically, particle diameters ranging from 0.04 to 0.3 mm (400-50 mesh) are used. Smaller diameters of 200–400 mesh improve the separation but decrease the flow rate.

There also exist inorganic ion exchangers, natural as clays and zeolites, synthetic as molecular sieves, heteropoly acid salts as molydophosphates, and others. Unlike organic exchangers, they can be used in high-temperature or high-radiation environments. They have been used for some specific applications in environmental radioactivity [27].

The absorbed ions can be eluted from the resin by using complexing agents or acid solutions. By alternating the pH of the solution, the elements can be eluted selectively. A set of typical ion-exchange resins used for some significant radionuclides is presented in Table 13.5 [24].

*Extraction chromatography*
Extraction chromatography is closely related to solvent extraction techniques. However, the organic phase is now made stationary by adsorption onto a solid support. The liquid phase remains mobile. The solid phase is packed into columns or disks, the liquid phase passes through the column, and the ions remain in it. From this viewpoint, extraction chromatography has the advantages of solvent and ion-exchange extraction techniques. Extraction chromatography resins are widely available today. The extractant compound can vary according to the application (see Table 13.6) and is absorbed on an inert polymeric substrate. Different extractants are very selective for specific radionuclides. This avoids many repetitive chemical steps in extraction procedures.

## 13.3.3 Source Preparation for Counting and Spectrometry

The final step of a radiochemical procedure is the preparation of the source for radiation counting or spectrometry. The selected method will depend both on the

**Table 13.5** Ionic resins conventionally used for radionuclide separation

| Functional groups | Resins | Radionuclides |
|---|---|---|
| *Cation exchangers* | | |
| -SO$_3$H<br>-COOH<br>-OH<br>-SH | Dowex, AG, or Eichrom 50 W- X4<br>Dowex, AG, or Eichrom 50 W- X8<br>Amberlite IR-120<br>Duolite. GT73<br>Amberlite IRA-743A<br>Amberlite IRC-718<br>Chelex® 100<br>Eichrom Diphonix® | $^{26}$Al<br>$^{41}$Ca<br>$^{55}$Fe<br>$^{60}$Co<br>$^{63}$Ni<br>$^{89,90}$Sr<br>$^{95}$Zr<br>$^{90}$Y<br>$^{137}$Cs<br>Ra isotopes<br>U isotopes<br>Th isotopes<br>Pu isotopes<br>$^{241}$Am |
| *Anion exchangers* | | |
| -NH$_2$<br>-NHR<br>-NR$_2$<br>-NR$_3$$^+$ | Dowex., AG, or Eichrom. 1- X4<br>Dowex, AG, or Eichrom 1- X8<br>Bio-Rex. 5<br>Dowex or AG 4- X4<br>Amberlite. IRA-68 | $^{55}$Fe<br>$^{60}$Co<br>$^{95}$Zr<br>$^{99}$Tc<br>$^{210}$Pb<br>U isotopes |

**Table 13.6** Chromatography extractants used for significant radionuclides

| Extractant | Radionuclides |
|---|---|
| Aliphatic quaternary amine (TEVA®) | Pu, Np, Th, U, and Am isotopes, $^{99}$Tc |
| Diamyl, amylphosphonate (DAAP, U TEVA®) | Pu, Np, Th, U, and Am isotopes |
| Octylphenyl-N, N-di-isobutyl carbamoylphosphine oxide (CMPO) dissolved in tri-n-butyl phosphate (TBP) (TRU®) | $^{89,90}$Sr, Ra isotopes, Pu, Np, Th, U, and Am isotopes |
| Dimethylglyoxime | $^{59,63}$Ni |
| 4,4'(5')-bis(t-butyl-cyclohexano)-18-crown-6 in n-octanol | $^{90}$Sr, $^{90}$Y |
| 4,4'(5')-bis(t-butyl-cyclohexano)-18-crown-6 in isodecanol | $^{210}$Pb |
| Aliquat-336 N | $^{99}$Tc |
| Diamyl, amylphosphonate | Pu, Np, Th, U, and Am isotopes |
| TOPO | Pu, Np, Th, U, and Am isotopes |

radiation to be measured and the available detector. The approach for mass spectrometry methods is different because the goal is atom counting. The most common procedures are followed.

*Evaporation*

Evaporation is a common method of preparing the source, although of limited application [19]. Usually, some evaporation steps are needed to reduce the final solution volume containing the radionuclide of interest after separation. Evaporation must be performed to avoid losses caused by violent boiling that could produce material projection from the evaporation container. Reduced volume can be used for $\gamma$-spectrometry at well-established geometries.

On many occasions, it is necessary to evaporate the solution to dryness and redissolve the residue with an appropriate solvent before introduction into the counting chamber. For instance, in the case where a liquid scintillation counter will be used, the final solution must be conditioned to dissolve it in the scintillation cocktail (see Chap. 10).

For gross $\alpha$- or $\beta$-counting, the dry residue is often counted directly with a gas proportional or Geiger–Müller counter [38]. Nevertheless, self-absorption effects limit this approach to a great extent. In fact, the thickness and inhomogeneity of the sample make this simple procedure not advisable.

Sample evaporation was used for $\alpha$-spectrometry measurements. However, the final sample is inhomogeneous. The use of spreading agents during evaporation decreases the inhomogeneity. However, this method is practically obsolete, as electrodeposition produces thinner and more homogeneous sources [39].

*Filtered Precipitate*

The use of precipitation as a concentration (and separation) procedure allows direct counting of the precipitate once it is filtered and dried. This has been used for measuring Sr carbonates or Ra sulfate precipitates [40] with gas counters. If the precipitate thickness is expected to be the same for all separations, it is not necessary to construct a thickness-efficiency calibration curve.

$^{137}$Cs precipitates with ammonium phosphomolybdate (AMP), and this possibility is used to separate Cs from high-volume waters. The precipitate was filtered, dried, and measured directly by $\gamma$-spectrometry [41]. In this case, there is no need to calibrate the counter for the sample thickness due to the penetrability of $\gamma$-radiation.

*Electrodeposition*

Electrodeposition is a redox process and allows the preparation of very thin and homogeneously distributed radiation sources. This is an ideal option for $\alpha$-spectrometry and, of course, for $\beta$-spectrometry, although in this case its use is less frequent. Unfortunately, it cannot be used for all of the interesting radionuclides. It is very sensitive to the presence of impurities or organic matter in the final solution and is not as fast as precipitation procedures. However, it is the method of choice for most $\alpha$-emitting radionuclides. During electrodeposition, radionuclides present in solutions as ions are deposited on an electrode by electrochemical reactions [24]. In Fig. 13.8, an electrodeposition device is depicted. The electrodeposition cell is usually a Teflon cylindrical container where the solution is incorporated. A metallic planchet at the bottom of the cylinder is the cathode, and a platinum wire is introduced in the solution as an anode. A voltage is applied between the anode and the cathode, and the cations tend to travel toward the cathode where they are deposited. Because the electric field is supposed to be homogeneous along the cathode, the ion deposit will also be homogeneous. Cathode disks are usually made of well-polished Ag or stainless steel, which is the common option [39]. Electrodeposition can be applied for the preparation of many radionuclide sources. U and Th isotopes, transuranic radionuclides, Ra isotopes, Tc, and many others can be prepared by electrodeposition [42–44]. All of them are environmentally significant. The electrodeposition yield depends on many variables, of course, the acid or basic condition of the solution, its pH, the applied voltage, or, more specifically, the current intensity across the solution. All these parameters should be optimized specifically for each case and expected concentrations in the solution, as electrodeposition behavior has been found to depend on element concentration [24]. Electrodeposition can be spontaneous. In this case, the reduction occurs spontaneously, and it is better to talk about self-deposition. To allow self-deposition to occur, an appropriate potential must appear between the cations and the electrode. $^{210}$Po is normally self-deposited on Ni, Ag, or stainless-steel planchets [45].

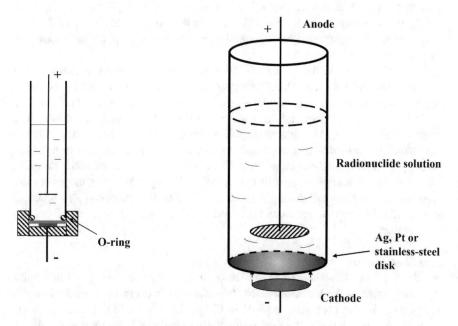

**Fig. 13.8** Usual setup of an electrolytic cell

Once electrodeposition has taken place, the cathode is removed, washed, and dried. Occasionally, it is covered by a very thin mylar film to avoid losses.

*Gas Counting*

In some specific cases, the preparation of gas sources for counting is possible. In the so-called gas internal counting techniques (see Chap. 15), radionuclides can be introduced directly into the gas detector. Of course, this is possible only for radionuclides that form gaseous compounds at room temperature. $^{14}C$ as CO, $CO_2$, or $CH_4$, $^3H$ as elemental H, or other hydrogenated gaseous species has been measured very frequently using this technique at environmental levels [46, 47]. $^{222}Rn$ emanating from $^{226}Ra$ samples was measured with ionization chambers for the determination of $^{226}Ra$ [48]. For that, the Rn emanating from the sample is swept with $N_2$ and introduced into the detector. The gas is conducted through a quartz tube system containing cool traps and zeolite traps for sample purification. Very similar systems have been used for the determination of radioactive noble gases in environmental samples [17].

*Direct $\gamma$-spectrometry*

Generally, radiochemistry is not needed for the determination of $\gamma$-emitting radionuclides. The mean free path of gamma radiation in matter is long enough to allow for direct measurement without previous separation. In some cases, pre-concentration is needed, but it is very frequent that the sample can be directly measured without chemical extraction. Of course, as we will see in the next section, the counting geometry must be fixed. In this way, an efficiency calibration curve previously constructed can be applied to convert the net photopeak area (see Chaps. 10 and 11) into activity.

For liquid samples, the procedure is very simple. A known volume of the sample is poured into a well-defined geometry container and directly counted. Containers are normally cylindrical-shaped or Marinelli geometry beakers (see Fig. 13.14) [49]. It is important to ensure that the radionuclides in solution will not be absorbed during measurements, which can usually take several days. On the other hand, the selection of the sample volume is also a key question to decrease the experiment limit of detection (see Chap. 15). In the case of solid samples, such as soils, sediments, seaweed, and rocks, it is important to homogenize the sample by powdering and sieving if necessary. Marinelli beakers can also be used, although the geometries are more varied since flasks, Petri dishes, and others can be an option.

*Mass Spectrometry*

The form of the final source in the case of mass spectrometry determination depends greatly on the technique used. It is closely related to the ion source used. More details of this problem are given in Chaps. 16 and 18. In the case of Accelerator Mass Spectrometry, the final product after chemical separation is usually in solid form. It is dried, powdered, and mixed with a material that facilitates heat and electricity conduction at the ion source. The mixture is pelletized to form the so-called cathodes for the Cs-sputtering ion source [30].

In the case of ICP–MS, the final sample undergoes slight acid or basic dissolution. The liquid is introduced into a nebulizer, where fine-particle aerosols are produced. They, in turn, are introduced into the ion source where the particles are vaporized, and the radionuclides of interest are ionized. Details can be found in Chap. 16. ICP–MS also accepts solid samples. Now, the samples are irradiated by a laser beam, and the particles in this way are introduced into the ion source. This laser ablation method is described in Chap. 16.

## 13.4  Yield Determination

The radionuclide recovered after a radiochemical method is not always 100% of that present in the sample. In contrast, very often the recovery is significantly lower than 100%. The reason is that the procedures are complicated, and the radionuclide concentrations involved are very low. In any case, it is necessary to know the recovery yield to evaluate the activity concentration in the sample. The definition of recovery yield, $Y$, is simple [50].

$$Y = \frac{Q_r}{Q_s} \qquad (13.6)$$

Here, $Q_r$ is the amount of radionuclide recovered after separation, and $Q_s$ is the amount of radionuclide present in the sample before separation. Once extraction has been performed, the sample is counted, and the activity concentration, $A_c$, can be found as

$$A_c = \frac{n - f}{Y I \varepsilon k} \qquad (13.7)$$

n and f are the sample and background count rates, respectively, $I$ is the radiation intensity, $\varepsilon$ is the counting efficiency, and $Y$ is the recovery yield. The factor k can be the mass of the analyzed sample or its volume. Thus, $A_c$ would be alternatively given in units of activity per gram or liter.

The recovery yield, $Y$, can be estimated in different ways. The first approach consists of the analysis of known radionuclide amounts in matrices identical to or very similar to those that will be analyzed. Of course, the chemical procedure must be identical, and the recovered radionuclide fraction is taken as $Y$. This value is then applied to the analysis of unknown samples. For this approach, reference matrix materials can be used in which the radionuclide concentration is previously known. When the radionuclide of interest is not present in the sample or there is no information about it, the reference material can be spiked with known amounts of it and then subjected to the analysis. In the latter case, it is necessary to ensure that the amount used for spiking is sufficiently high to mask the presence of possible traces of the radionuclide in the reference sample. Sometimes it is not possible to use the same radionuclide to spike the reference sample. For these cases, a radionuclide of similar chemical behavior can be used. Alternately, we would need a stable isotope of the radionuclide of interest or even a stable element with similar chemical behavior. The determination of $Y$ must now be done using gravimetric or mass spectrometry methods [51]. Of course, the assumption that $Y$ will be the same in real applications is just a working hypothesis that has many criticizable aspects. This approach has several disadvantages. The main drawback is the availability of identical reference matrices. Not always is the appropriate reference matrix available. This is a severe limitation. Furthermore, it is necessary to assume that $Y$ will be the same when analyzing real samples. This means assuming that the procedure will be identical and that the chemical transformation along it will be identical as well. This method is not advisable considering the extremely low concentrations involved in the real separations and the fact that many chemical transformations are concentration-dependent [22].

In environmental radioactivity, the approach is usually different. It is preferred to know $Y$ for each specific extraction. For that, the sample is spiked with a known activity of a radioisotope of the radionuclide of interest. After homogenization, radiochemical procedures are applied, and the recovered spike is measured. It is supposed that the $Y$ values for the spike and the radionuclide of interest are the same. This technique is known as isotope dilution [51, 52] and can be applied to many radionuclides. The radiochemical yield is measured by determining the recovered amount of the spike used. In the case where both the spike and the problem radionuclide can be measured with the same detector, it is obvious that

$$\frac{\varepsilon_p A_p}{\varepsilon_s A_s} = \frac{N_p}{N_s} \tag{13.8}$$

$\varepsilon$, $A$, and $N$ are the counting efficiency, the activity, and the net counts produced by the spike, s, and the problem radionuclide, p. In some cases, (13.9) becomes even simpler, as the counting efficiencies are the same. Indeed, for $\alpha$-spectrometry, this holds within the usual $\alpha$-energy range. Thus, (13.8) becomes

$$\frac{A_p}{A_s} = \frac{N_p}{N_s} \tag{13.9}$$

Now N stands for the net area under the $\alpha$-peaks. Figure 13.9 shows a typical $\alpha$-spectrum including the peaks of the problem radionuclide and the spike. A simple integration of the $\alpha$-peaks gives the activity of the radionuclide problem, Ap, once the added spike activity is known.

The spike must be selected following several criteria. First, it must be an isotope of the radionuclide of interest and, therefore, follow the same chemical behavior. It must emit readily measurable radiation. When both the spike and the radionuclide emit the same radiation type, the detector used must have a sufficient energy resolution to distinguish the radiation. Its half-life should be long enough to obtain enough activity after the extraction process. The spike must not be present in the sample. Of course, the spike should be reasonably available. In some occasions,

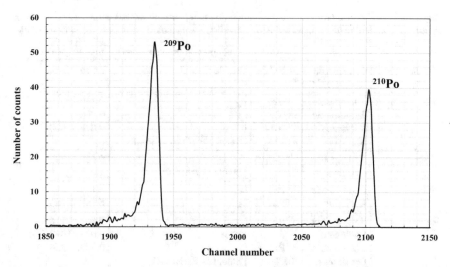

**Fig. 13.9** Alpha spectrum of an environmental $^{210}$Po sample extracted using $^{209}$Po as a spike

the spike is produced ad hoc for the procedure via accelerator nuclear reactions or neutron activation. Should this be the case, very often a previous purification must be performed to avoid interferences. Finally, it is very convenient for the added spike activity to be similar to that expected for the radionuclide problem or within the same order of magnitude. In Table 13.7, a list of spikes used for relevant environmental radionuclides is presented together with emitted radiation [40–42, 44, 53–55]. There are cases where it is very difficult to find the appropriate spike. For these cases, elements of similar chemical behavior can be used. Some examples also appear in Table 13.7. This is the case, for example, for $^{99}$Tc, where some Re isotopes are used to estimate $Y$ [56]. Additionally, when the spike is not easily available, a different approach is used. Now, the average $Y$ can be utilized. It is found that by repeating several times, the analysis of the matrices of interest spiked with known amounts of the tracer. It is assumed that the average $Y$ is representative and can be applied to real problem samples.

The isotope dilution approach is used because the spike and the radionuclide of interest follow the same chemistry and behave identically during the extraction process. This means that the spike must be equilibrated and homogenized in the sample before starting the chemical extraction. In liquid samples, the spike is dissolved, and this is usually enough to achieve that goal. In solids, it is not easy. The sediment or soil sample is impregnated with a solution containing the spike and allowed to dry overnight or during a reasonable time period. In some cases, for instance, in rock samples, it is preferred to carry out a fusion technique (see previous sections) before proceeding with the separation. In this way, the chemical form for spike and problem radionuclide becomes the same.

There are also gravimetry procedures to measure Y. In this case, the role of the spike is played by the carriers, which are usually stable isotopes of the radionuclide

**Table 13.7** Spikes used for radionuclide separation together with the radiation emitted. On some occasions, the yield is determined by using gravimetric methods. In this case, the spike is stable

| Radionuclide | Method | Spike | Radiation emitted (product) | Energy (MeV) |
|---|---|---|---|---|
| $^{89,90}$Sr | Gravimetric | Sr(NO$_3$)$_2$ | Stable (SrCO$_3$) | 0.514 |
| | Radiometric | $^{85}$Sr | $\Gamma$ | |
| $^{99}$Tc | Radiometric | $^{99m}$Tc | $\gamma$ | 0.140 |
| | Radiometric | $^{95m}$Tc | $\gamma$ | 0.766 |
| | Radiometric | $^{97m}$Tc | X, $\gamma$ | 0.018, 0.090 |
| | ICP–MS | Re | Stable, $^{185}$Re by ICP–MS | |
| $^{137}$Cs | Radiometric | $^{134}$Cs | $\gamma$ | 0.605, 0.796 |
| | Gravimetric | CsCl | Stable as AMP/Cs | |
| $^{210}$Po | Radiometric | $^{208}$Po | $\alpha$ | 5.114 |
| | Radiometric | $^{209}$Po | $\alpha$ | 4.883, 4.885 |
| Ra isotopes | Radiometric | $^{223}$Ra | $\alpha$ | 5.607, 5.716 |
| | Radiometric | $^{225}$Ra | X, $\gamma$ | 0.013, 0,040 |
| | Radiometric | $^{133}$Ba | $\gamma$ | 0.303, 0.356 |
| | Gravimetric | BaCl$_2$ | Stable as Ba(Ra)SO$_4$ | |
| U isotopes | Radiometric | $^{232}$U | $\alpha$ | 5.263, 5.320 |
| | | $^{233}$U | $\alpha$ | 4.783, 4.824 |
| | | $^{236}$U | $\alpha$ | 4.445, 4.494 |
| Th isotopes | Radiometric | $^{229}$Th | A | 4.814, 4.845, 4.901 |
| Pu isotopes | | $^{236}$Pu | $\alpha$ | 5.721, 5.767 |
| | | $^{242}$Pu | $\alpha$ | 4.858, 4.902 |
| Am isotopes | | $^{243}$Am | $\alpha$ | 5.233, 5.275 |

of interest. In other cases, they are elements of very similar chemical behavior. This is the case for some determination methods of $^{90}$Sr in environmental samples in the form of carbonates [53]. The sample is spiked with a macroscopic and well-known mass of natural Sr, and the final sample is supposed to be in the SrCO$_3$ form. After drying, the final sample was weighed, and the recovered amount of Sr was deduced. The ratio between the recovered and the added Sr is $Y$. This method has also been used for the determination of $^{226}$Ra and other Ra isotopes, mainly $^{224}$Ra, in environmental samples. Now, the carrier is Ba, and the final form of the prepared sample is BaSO$_4$ [40]. It is assumed that the chemical behaviors of Ba and Ra are very similar and that the calculated $Y$ from the recovered Ba is applicable to Ra. It is very clear that the application of a gravimetric method requires a perfect understanding of the final product and especially of its stoichiometry. On the other hand, the use of macroscopic masses optimizes the quantitative recovery of the radionuclide problem throughout the different steps in the procedure.

## 13.5  Efficiency Calibration of Radiation Counters and Spectrometers

Activity determination requires the calibration of the efficiency of the detectors. As explained in Chap. 8, the counting efficiency depends on several parameters. Equation (8.6) shows that $\varepsilon_{abs}$ is affected by absorption and geometrical effects, as well as by the intrinsic efficiency of the detector. Absorption effects are closely related to the type of particle and its energy. Charged particles are more affected by absorption effects than photons, and, within them, $\alpha$-particles are more affected than $\beta$-particles (see Chap. 7).

In short, once the geometry is fixed, the efficiency depends on the particle energy through absorption and self-absorption effects. The intrinsic efficiency (see Chap. 8) is related to the Detector Physics. In general terms, $\varepsilon_{int} \approx 1$ for charged particles and <1 for electromagnetic radiation. Nevertheless, the expected $\varepsilon_{int}$ can be modified by internal effects such as dead-time losses or color or chemical quenching in liquid scintillation counters. Dead-time losses in Geiger–Müller, for instance, are negligible since we are dealing with very low activities. However, quenching effects must be taken into account since they are independent of the radionuclide concentration.

### 13.5.1  Calibration Curves for Charged Particles

The necessary calibration depends on the counting problem. In the case of the determination of the gross $\alpha$- or $\beta$-radioactivity by gas proportional or GM counting, efficiency curves as a function of the thickness of the sample mass must be constructed. Standard samples are prepared as real samples. For environmental water samples, the method is a simple evaporation of sample aliquots on stainless-steel planchets [38] within a well-defined geometry, usually a circle of a given diameter. The mass of the residue was determined by weighing. Then, it is divided by the circle area to obtain the mass thickness of the sample. Since the counting geometry is well determined, the efficiency will depend on the mass thickness because of absorption effects. The calibration curve for $\beta$-radiation is usually constructed by evaporating well-known amounts of KCl on the planchets. Since the $^{40}$K ($T_{1/2} = 1.248 \times 10^9$ y) isotopic abundance in K is known to be 0.0117%, the activity added to the calibration sample can be easily calculated. In the same way, $\alpha$-radiation calibration sources can be constructed by adding well-known $^{241}$Am ($T_{1/2} = 432.6$ y) or $^{226}$Ra ($T_{1/2} = 1600$ y) activities to water samples that are then evaporated [38]. An example of both calibration curves is presented in Fig. 13.10. In real applications, once the sample mass thickness has been determined, the efficiency is obtained by interpolation in the calibration curve.

For specific $\beta$-emitters, it is necessary to obtain specific calibration sources in the geometry and thickness of interest. The measurement of $^{99}$Tc by gas proportional (or GM) counting can illustrate this case. $^{99}$Tc is a pure $\beta$-emitter, $E_{max} = 297.5$ keV, with a $T_{1/2} = 2.111 \times 10^5$ y. It has some environmental relevance and

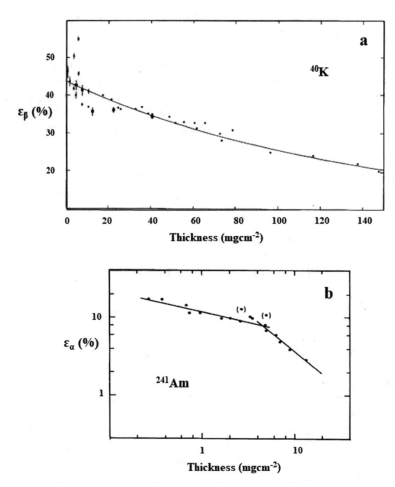

**Fig. 13.10   a** Counting efficiency curve for gross $\beta$ radioactivity measurements. **b** Counting efficiency curve for gross $\alpha$ radioactivity measurements

can be measured with gas ionization detectors [57]. Since its $\beta$-energy is relatively low, counting sources are prepared by electrodeposition after radiochemical extraction [44]. The efficiency calibration is performed by preparing standard samples of different known activities. In Fig. 13.11, the calibration curve is plotted. There, the known disintegrations per minute (dpm) are plotted against the net counts per minute (cpm) obtained with the detector. The data can obviously be fitted to a straight line that serves to obtain the activity of real samples by interpolation. For specific $\alpha$-emitters, for example, the determination of Ra isotopes in environmental water samples by gas counting, the approach is different. Now, the standard samples are prepared following the radiochemical procedure for Ra separation. In this case, Ra is coprecipitated with Ba as sulfate. For that, the water is acidified with $H_2SO_4$. Then, $BaCl_2$ is dissolved in the sample, and the Ba-Ra sulfate precipitates because it is very insoluble under these conditions [40]. The precipitate was filtered, and the filter was dried and counted. The calibration samples are prepared in the same way by spiking the water samples with a well-known $^{226}Ra$ activity. The activity is, in turn, sufficiently high compared to those expected in real samples. Different amounts of $BaCl_2$ are used to cause precipitation, and for that, it is possible to obtain different sample thicknesses. The efficiency calibration curve obtained in this way is presented in Fig. 13.12. Thus, once the real sample has been prepared, the mass thickness is determined, and the efficiency is obtained by interpolation.

In $\alpha$-spectrometry with Si detectors, the counting efficiency is virtually the geometric factor. In fact, the samples are very thin since they are prepared by electroplating techniques. However, the measurements are performed under vacuum conditions. Absorption effects can be neglected, and it can be accepted that the efficiency is constant within a wide energy range [58]. Furthermore, since isotope dilution methods are used (see 13.9), there is no need for efficiency calibration.

As already described in Chap. 10, the effects of color or chemical quenching affect the counting on liquid scintillation counters [59]. The counting efficiency must be determined taking into account these effects. In practice, calibration curves are constructed in which the efficiency is plotted as a function of quenching. For that, the scintillation cocktail is spiked with a known activity of the interesting

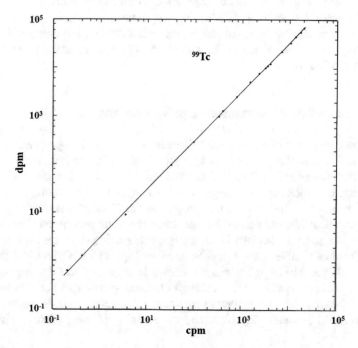

**Fig. 13.11** $^{99}Tc$ counting efficiency calibration curve for environmental applications

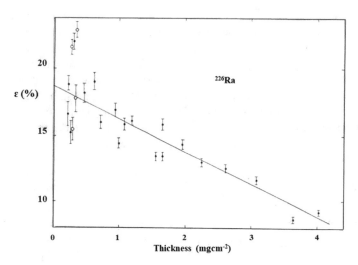

**Fig. 13.12** $^{226}$Ra counting efficiency as a function of sample thickness

radionuclide together with variable concentrations of color or chemical quenching effects, usually $Na_2Cr_2O_7$ and $CCl_4$, respectively [59]. The effects of the colorant or the chemical agent can be observed using the SQP parameter [60]. Thus, the measurement of $\varepsilon$ as a function of SQP gives an efficiency calibration curve that takes into account quenching effects. Two examples are given in Fig. 13.13. The SQP parameter can be obtained during the real sample measurement. For that, it can be used to interpolate in the calibration curve to obtain a more realistic counting efficiency.

## 13.5.2 Calibration Curves for Gamma Radiation

In the case of $\gamma$-radiation, the relevant parameter is the photopeak efficiency as defined in Chap. 8, (8.10). The use of photopeak efficiencies allows the conversion of the net areas under the photopeak into the activity of the radionuclide of interest. The photopeak efficiency (see Chaps. 8 and 10) depends on the counting geometry and the absorption effects due to the sample matrix. Of course, it is closely related to the photoelectric cross section but includes other total absorption phenomena.

There are several standard counting geometries in $\gamma$-spectrometry used for liquid or solid samples. They can be seen in Fig. 13.14. Cylindrical beakers, Petri dishes, and Marinelli beakers are conventionally used in these experiments. Marinelli beakers provide a very efficient counting geometry since a hole in the center of the cylinder permits one to introduce the detector into the beaker. Thus, the solid angle becomes very high, and the counting efficiency is optimized. Of

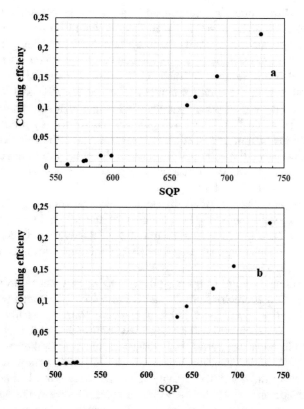

**Fig. 13.13** Quenching curves in LSC, counting efficiency versus SQP for **a** $^3$H in plastic vial and **b** in glass vial. Courtesy of J.-L- García-León

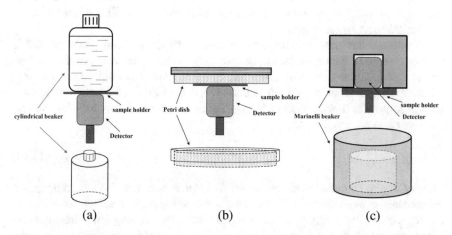

**Fig. 13.14** Standard geometries for $\gamma$-spectrometry: **a** cylinder beaker, **b** Petri dishes, and *c* Marinelli beakers

course, many other geometries are utilized in $\gamma$-spectrometry, whose main advantage is to avoid chemical separations. In fact, it would be possible to measure virtually all the geometries by using mock-up samples for efficiency calibration.

In the case of liquid samples, a calibration cocktail that covers a wide $\gamma$-energy range is diluted in the sample. In Table 11.3, a list of usual calibration samples is given, together with the energies emitted. The use of $^{133}$Ba and $^{152}$Eu is very convenient, but they emit cascade $\gamma$-rays. This means that sum-coincidence corrections must be made (see Chap. 11). We will return to this problem later. In any case, once the sample is spiked, it is measured under the same conditions as the real samples. Since the activities are known, the photopeak efficiencies can be calculated by applying (8.10). A plot of photopeak efficiency curves for different geometries is given in Fig. 11.13.

The curves are usually fitted by using least squares algorithms or the Levenberg–Marquardt method. The goodness of the fitting is measured by the chi-square function, $\chi_R^2$, which must be minimized in the fitting. Other algorithms, such as stimulated annealing, are also used [61]. Some functions utilized for fitting are given in Table 11.4.

The calibrations for solid samples, soil, sediments, and others are very similar. In fact, the standard geometries are the same. From an experimental point of view, some problems arise. First, the calibration standard sources, usually available as solutions, must be well mixed and homogenized within the solid sample. For that, the solid sample must be ground and powdered and then spiked with a known activity. Once the added solution has dried, careful mixing must be performed to homogeneously distribute the activity. The efficiency curves obtained in this way have a shape similar to that of water samples. However, the application to real samples is not always straightforward. The reason is that the compositions of the standard and real samples are not the same and, in some cases, certainly different. Solving this problem requires one to prepare calibration standards for each real sample. This is not a realistic approach, especially when a large number of samples must be measured.

To address this problem, the transmission method has been proposed and is widely used [62]. Since the counting geometry is the same, both for the calibration and real samples, the only difference is the sample composition; i.e., the only difference arises from absorption effects. If the photopeak efficiency of the calibration sample at a given energy is $\varepsilon_{cal}$, the corresponding efficiency for the real sample, $\varepsilon_s$, is expected to be

$$\varepsilon_s = f\varepsilon_{cal} \qquad (13.10)$$

where $f$ is the so-called self-absorption factor that accounts for the composition differences between the calibration and the real samples. The self-absorption factor depends on the $\gamma$-energy and can be calculated by carrying out $\gamma$-transmission experiments such as those depicted in Fig. 13.15. A point-like $\gamma$-source of interesting energy is used to study the transmission of $\gamma$-radiation through the real sample and through the calibration sample (Cutshall 83). If the net count rate under the

**Fig. 13.15** $\gamma$-transmission experiment setup

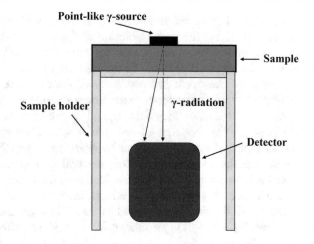

photopeak of interest is S and C for the real and calibration samples, respectively, it can be found [62] that

$$f = \frac{1 - \frac{S}{C}}{\ln \frac{C}{S}} \tag{13.11}$$

Then, once $\varepsilon_{cal}$ is known, $\varepsilon_s$ can be easily found from (13.10).

The photopeak efficiency can be calculated by applying Monte Carlo simulation methods [20]. The Monte Carlo method is a very powerful simulation tool that can be used to study radiation transport in matter. Its application is not easy since it requires one to know the composition of the sample and the geometry of counting very precisely. Regarding the counting geometry, the main problem lies with the detector dimensions, which in some cases are poorly known. It has not been until recently that the manufacturer has provided laboratories with precise detector dimensions. In the first days of Monte Carlo applications, the internal detector dimensions were determined by X-ray radiography. The bulk composition of the sample must also be determined. Additionally, it is necessary to have a good dataset that describes the transport of $\gamma$-radiation across the sample. This is usually contained in the Mote Carlo software but is a source of uncertainty that should be carefully checked. A very important problem in Monte Carlo simulation is related to the charge collection in the detector. This is not a well-understood process related to low-electric field zones within the detector that can produce a significant deviation in the results [63].

As explained in Chap. 10, the $\gamma$-emission in the cascade gives place to the so-called sum peak. This is a consequence of the simultaneous detection of two or more photons emitted in a cascade. In Fig. 13.16, the decay schemes of $^{133}$Ba and $^{152}$Eu are shown, and several photons can potentially produce sum peaks since they are emitted in cascade. The existence of coincidence effects is a consequence of the detector time resolution. If two or more $\gamma$-photons experiment photoelectric effects within the time resolution of the Ge crystal, it will be interpreted as a unique photoelectric effect with an energy that will be the sum of the energies of the involved photons. In the case of very high-activity samples, random coincidences can occur independently of the existence of photon cascades. For low-activity samples, random coincidences are very unlikely, but true coincidences will be possible since they only depend on the decay scheme [64]. As explained in Chap. 10, the existence of sum peaks is a very important error source in $\gamma$-spectrometry. The sum peaks can be interpreted as real peaks, and the counts in the sum peaks are subtracted from the real photopeaks. The situation becomes worse if the sum peak energy coincides with the energy of other photons emitted by the source. Sum-coincidence effects, therefore, must be taken into account for efficiency calibration since several of the calibration standards emit cascade photons (see Fig. 13.16). On the other hand, coincidence effects will be more likely for highly efficient geometries, such as the Marinelli geometry, where the probability of simultaneous photon detection is higher. In Fig. 13.18, we give an example of a calibration curve using $^{133}$Ba and $^{152}$Eu without coincidence corrections. The behavior of the efficiency curve is not smooth within some energy ranges, which in turn are those affected by possible coincidence due to cascade emission.

Coincidence corrections are not trivial. There are several theoretical and empirical methods. Pioneering works from Debertin and Helmers [65] proposed theoretical expressions that take into account emission radiation intensities, $I$, individual photopeak efficiencies, $\varepsilon$, and total efficiencies, $\varepsilon_{tot}$. For a decay scheme such as that presented in Fig. 13.17, involving three emissions, the following

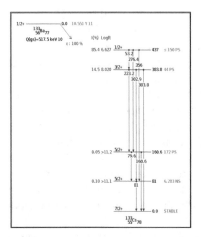
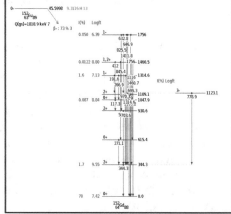

**Fig. 13.16** $^{133}$Ba and $^{152}$Eu decay scheme showing the potential sum cascade. Data have been taken from https://www.nndc.bnl.gov/nudat2/

theoretical corrections are found for point sources [65].

$$C_1 = \frac{n_{10}}{n_1} = \frac{1}{1 - \varepsilon_{tot2}}$$

$$C_2 = \frac{n_{20}}{n_2} = \frac{1}{1 - \frac{I_1}{I_2}\varepsilon_{tot1}}$$

$$C_3 = \frac{n_{30}}{n_3} = \frac{1}{1 + \frac{I_1}{I_3}\frac{\varepsilon_1\varepsilon_2}{\varepsilon_3}} \tag{13.12}$$

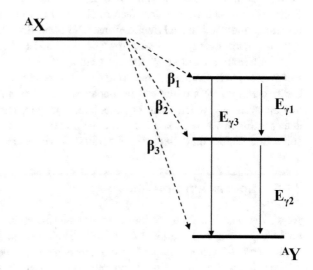

**Fig. 13.17** Three-photon decay scheme

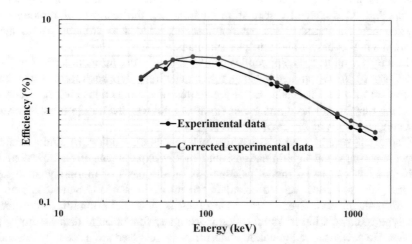

**Fig. 13.18** Coincidence corrections on efficiency calibration curves

$C_i$ are the correction factors, $n_{i0}$ is the count rate for photon $i$ in the absence of coincidence effects, and $n_i$ is the count rate for the photon, but now in the case it is affected by the coincidence effect. These equations are valid for point sources, and in the absence of Bremsstrahlung effects, the angular correlation between the photons and others. This approach is correct but very limited and, of course, of application only to point sources.

There are also empirical methods [66] that involve measuring radionuclides of interest ($^{133}$Ba or $^{154}$Eu) and some monochromatic $\gamma$-emitters at two different geometries. One is that used in real measurements, and the other must be coincidence-effect free. This geometry is usually a point-like geometry at a large distance from the detector, more than 15 cm, which helps to minimize the coincidence effects. The geometrical factors obtained in this way are fitted to analytical functions and used in real applications. The main drawback of this method is the lack of a wide variety of monochromatic $\gamma$-emitters, which limits the fitting.

An alternative method for the calculation of correction factors is Monte Carlo simulation [64]. Coincidence correction factors can be calculated using Monte Carlo methods by simulating the decay schemes of multienergy emitters. Real and apparent efficiencies can be calculated for the geometry of interest once the detector dimensions and sample composition are known. The effect of coincidence correction factors on the efficiency calibration curves can be seen in Fig. 13.18.

## 13.6    Speciation Studies

As we discussed in Chap. 5, the environmental behavior of radionuclides depends on their physical and chemical forms [67]. Their environmental mobility, bioavailability, and radiological impact on humans depend on such forms. From this point of view, the assessment of environmental risks based on total concentrations lacks this information that, in many cases, is crucial and, for that result, insufficient [2, 68, 69]. Consequently, speciation studies, i.e., the analysis of radionuclide physicochemical forms in the environment, are needed to properly assess their environmental behavior and impact on humans.

In the environment, radionuclides can be found in the forms described in [67] (see Chap. 5). As we studied in Chap. 5, radionuclides adopt such forms depending on several factors. The main one is the source of radionuclides and the scenario in which they are released into the environment but also the transport mechanisms and characteristics of the environment [68].

There are several methods to carry out speciation studies: in silico studies, direct environmental determinations, and model experiments [67]. So-called in silico studies consist of model calculations on the basis of thermodynamic and kinetic data that allow one to calculate the distribution of radionuclides among the different phases. Even in the case of using powerful computing resources, this type of calculation lacks precision since many fundamental data are not well known. However, the radionuclide concentrations involved are extremely low, and many effects, such as glass-wall sorption, which become significant in this case, are difficult to reproduce in model calculations. The same can be said for colloids that are difficult to include in models. On the other hand, the environmental kinetic processes are poorly known. This makes this procedure just a first approximation of the problem. Direct analysis of real samples is the preferred method for studying radionuclide speciation [68]. Of course, the method poses several problems, starting from sampling and sample storage. Both procedures must be carefully selected to avoid changes in chemical species during sampling or after prolonged or uncontrolled sample storage. In fact, it is very convenient that some separations are carried out on site, as in the case of particle-borne radionuclides in waters. For that, filtrations are (or should be) carried out on site. The analysis is performed by applying sequential extraction procedures in search of the separation of the species described in [67]. Or at least a relevant number of them. It is true that the use of some chemical procedures can alter the relative concentration of radionuclide species. For that, the sequential methods are designed to avoid such changes. Additionally, as previously noted, extremely low concentrations of the studied radionuclides alter the chemical behavior of the corresponding element. In addition, at such low concentrations, the sorption process in glassware and laboratory instrumentation becomes relevant [67].

For atmospheric aerosols, the use of cascade impactors and activated charcoal cartridges helps to separate radionuclides associated with different particle sizes and gaseous forms, respectively [2].

Radionuclides can be present in water as ions and molecules associated with colloids or pseudocolloids [2]. This situation is not static, as exchange among the different forms can occur. Radionuclides associated with low molecular mass are known to be more mobile than those associated with high molecular mass, such as colloids and macroscopic particles. Low molecular mass forms are, although, also more bioavailable [68]. Physicochemical speciation in environmental waters includes filtration and ultrafiltration, as well as centrifugation and ultracentrifugation procedures [2]. Once size separation is performed, the corresponding radiochemical analysis is applied.

In the case of soils or sediments, the different radionuclide species are extracted using sequential extraction methods. There are several sequential extraction methods in the scientific literature [68]. All of these methods are based on the selective desorption or extraction of exchangeable radionuclide forms or the selective dissolution of solid phases where the radionuclide is associated. The complexity of the procedures varies, and in most cases, it is convenient to combine them to obtain as much complete information as possible. In Table 13.8, some sequential extraction methods are presented [67, 68]. In general, they all try to distinguish the following radionuclide forms: free-bound or exchangeable, bound in carbonates, associated with Fe and Mn (hydr)oxidic coatings, and associated with organic matter and residual forms in crystal lattices of minerals. Each extracted phase was then analyzed by the corresponding radiochemical procedure.

In model experiments, radionuclide speciation is studied in laboratory-controlled experiments with simulations of natural conditions. The results can be interpreted in an easier way than direct determination methods. On the other hand,

**Table 13.8** Sequential extraction procedures for speciation studies. $Ac$ = acetic, $Cit$ = citrate. Generally, the phases considered are free-bound or exchangeable, bound in carbonates, associated with Fe and Mn (hydr)oxidic coatings, and associated with organic matter and residual forms in crystal lattices of minerals. Some authors [67–69] included isotope exchange with stable isotopes as an extraction step for radionuclides to determine exchangeable forms

| Step | Reagents | Reagents | Reagents | Reagents | Reagents | Reagents |
|---|---|---|---|---|---|---|
| F1 | 1 M $MgCl_2$ | 0.1 M NaCl | 0.05 M $CaCl_2$ | $H_2O$ | 1 M $MgCl_2$ | $H_2O$ |
| F2 | 1MNaAc pH = 5 (+HAc) | 1 M HCl | 0.5 M HAc | 1 M $NH_4Ac$ | 0.1 M NaOH | 1MNH$_4$Ac in 25% HAc, pH soil |
| F3 | 0.04 M $NH_2OH.HCl$ + 25%HAc | 0.11 M EDTA + 1.7 M $NH_4OH$ | 0.1 M $Na_4P_2O_7$ | 6N HCl | 0.2 M $(NH_4)_2Ox$ + 0.2 M $H_2Ox$ | 1MNH$_4$Ac in 25% HAc, pH 4 |
| F4 | 0.02 M $HNO_3$ + 30% $H_2O_2$ | Acid digestion | 0.175 M $(NH_4)_2Ox$ + 0.1 M $H_2O_X$ | Acid digestion | 1MNaAc pH 5 (+HAc) | 0.04 $NH_2OH$ · HCl in 25% HAc pH 2 |
| F5 | Acid digestion | – | – | – | 0.175 M NaCit + 5% $Na_2S_2O_3$ | 30% $H_2O_2$ in 0.02 M $HNO_3$, pH 2 |
| F6 | | | | | | 7 M $HNO_3$ |

the different parameters involved can be changed with precision according to the experimental needs. Thus, speciation can be studied under very precisely known experimental conditions [67]. Laboratory experiments depend on the environmental compartment of interest. Thus, specific experiments have been developed for atmospheric aerosols, waters, soils, sediments, and others [67]. Of course, the results obtained in these studies are significant and, as said before, produced under well-controlled conditions. However, they are radiotracer experiments or, in other words, experiments carried out by adding radionuclides of interest to natural or synthetic samples. They are expected to behave identically to the radionuclides present in real samples. From this viewpoint, the results can provide valuable information but not necessarily the exact information that is needed to understand in depth the behavior of radionuclides in the environment.

## 13.7 Quality Assurance

Laboratory managers must implement quality control practices to ensure the reliability of the results. These quality control practices encompass all measures that show the reproducibility and precision of the results obtained [70].

Quality control activities begin with sampling procedures and continue through sample pretreatment, separation procedures, including chemical recovery estimation, and finally the activity measurement and its uncertainty. Quality control procedures must also consider the maintenance of laboratory instrumentation, glassware, and the purity of reagents. From this point of view, it is very convenient to have in written the different protocols that affect all these activities. These 'Good Practice' protocols must be followed as a guarantee of the reproducibility and accuracy of the results. More specifically, reproducibility has to do with internal practices in the laboratory, including sampling, while accuracy has to do with intercomparison exercises and external controls of the results.

Internal laboratory practices are several and, as mentioned above, include all aspects of the radioanalytical procedure [7, 71]. Thus, protocols for sampling, chemical procedures, and counting must exist and serve as a guide for laboratory staff. With regard to the chemical procedure, it is important to carry out blank analysis in which the extraction is done without sample, but using all the reagents and following all the steps in the procedure. Blank samples inform the contribution of the reagents, glassware, and laboratory material to the activity being measured. It also served to control the quality of the reagents used. The measurement of blank samples served to correct the analysis for systematic errors, contamination of the reagents, or laboratory material. Blank samples must be prepared periodically to keep the procedure under control. This must also be done when a new reagent bottle is received in the laboratory. It is very convenient to prepare a blank sample after the measurement of a high-activity sample, which could give place to memory effects in the laboratory material. When possible, replicating the analysis of the same sample is a good control practice of the method. This is not always possible, as the number of samples available frequently does not allow for it. It is also compulsory to periodically check the radiochemical yield obtained by the different procedures implemented in the laboratory. This helps to discover systematic deviations from the protocols, bad practices, reagent failures, etc.

Regarding the counting or spectrometry procedure, it is necessary to periodically carry out detector background determinations. It is important to ensure that the background value used in the determination of the activity is adequate. Usually, the background does not change significantly in well-conditioned laboratories, i.e., those where temperature, humidity, and electrical current are stable. However, this must be checked periodically. There is no other option to ensure that this is true. A periodic check of the gain stability of the amplifiers and of the potential provided by high-voltage sources is also advisable. In any case, the efficiency curves must be remeasured periodically to establish the correct $\varepsilon$ for the specific measurement. Needless to say, the calibration standards used to construct such curves must have certified activity, including the uncertainty of the offered value and date of calibration.

To check the accuracy of the obtained results, it is necessary to apply the procedures to standard samples, Standard Reference Materials (SRM), which have well-known activities of the radionuclides of interest. The periodic analysis of

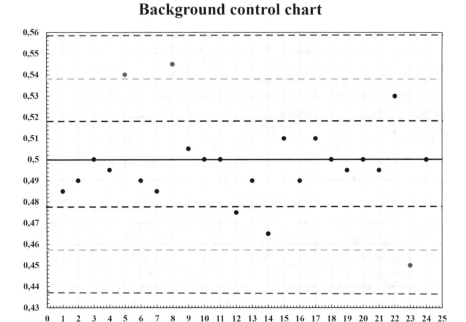

**Fig. 13.19** An example of a control chart for quality assurance of laboratory practices. In this case, the detector background is the controlled parameter

SRM helps to control the quality of the applied procedures. There is an extensive catalog of SRM materials. The International Atomic Energy Agency (IAEA) is able to produce very varied SRMs with different environmental matrices and radionuclides; thus, soils, sediments, waters, seaweed, etc., with well-known activities of many interesting radionuclides can be found in the IAEA catalog [72]. Other national laboratories or even private companies can provide these reference materials [73]. It is interesting to note that the publication of new radiochemical procedures is not practically possible unless the method has been tested with SRM. It is also necessary to participate in blind intercomparison exercises in which different laboratories analyze a reference sample where the activities are known only by the organizing institution. This is a very common practice in national or international laboratory networks. Intercalibration exercises are launched to determine radionuclide concentrations in the new SRM. Only veteran and well-recognized laboratories participate in these exercises, which in turn serve to check the quality of the procedures.

A fairly good practice in laboratories is the construction of control charts in which a determined procedure is periodically checked, and the obtained value is plotted against time. For instance, the radiochemical yield of a procedure, the blank sample activity, the radionuclide concentrations obtained for an SRM, the detector background or its counting efficiency, etc. An example is given in Fig. 13.19. It is expected that the value obtained lies within a given range that can be determined by statistical considerations, i.e., 1, 2, or $3\sigma$ from the average. The systematic appearance of outliers shows that the quality of the procedure must be checked.

**Exercises**

1. $^{99m}$Tc serves as a tracer in experiments to determine $^{99}$Tc in natural samples. We spike 10 g of sediments with a certain $^{99m}$Tc activity and extract, by means of a radiochemical procedure, the Tc present in the sample. The final prepared source was measured with a NaI (Tl) detector, obtaining 10,000 counts during 2 min. Thirty minutes later, and with the same detector, a source of identical geometry was measured in which we added the same tracer activity, obtaining 10,000 counts in 1 min. Using the NaI (Tl) background, we calculated the chemical yield of the process.

2. A few days later, we measured the sample with a GM detector, obtaining 1200 counts in 1000 min. If the detector background was found to be 1600 counts in 2000 min and its counting efficiency 40%, we calculated the $^{99}$Tc activity concentration in mBq g$^{-1}$ in the sediment.

3. To determine the Pb content of an aerosol sample, 10 mg is dissolved in concentrated HNO$_3$. One milliliter of a standard $^{210}$Pb solution was added to the solution, and Pb precipitated as PbCl$_2$. One milligram of the precipitate is recovered, dried, and measured with a NaI(Tl) giving 100,000 counts in 5 min. Find the Pb concentration in the aerosol if 10 ml of the standard solution, measured under the same conditions, gives 20,000 cpm.

4. A BaSO$_4$ sample containing some radioactive $^{35}$S has a specific activity of $7 \times 10^{10}$ dpsg$^{-1}$. A saturated aqueous solution was prepared and diluted to 1/100. Then, 1 ml is taken and evaporated. The residue was counted with a 1% efficiency counter. How many cps will we get if the solubility of BaSO$_4$ in water is $19.8 \times 10^{-6}$ equivL$^{-1}$?

5. What proportion of S is $^{35}$S provided its half-life is 87 d?

6. $^{238, 239, 240, and 241}$Pu isotopes are of environmental interest. Their determination in water samples is planned, and a yield tracer is necessary. Select the tracer using https://www.nndc.bnl.gov/nudat2/ and choose the yield determination method.

7. $^{137}$Cs can be measured by direct $\gamma$-spectrometry or radiochemical methods. When would you need a yield tracer? Select the tracer and choose the yield determination method.

8. Select the yield tracer for the case of natural U. It should be noted that several U isotopes are present in a natural sample.

9. $^{241}$Am in marine sediments is measured by $\alpha$-spectrometry. $^{243}$Am is used as a yield tracer. Sketch the expected $\alpha$-spectrum obtained for the final sample.

10. Suppose that the standard sample $^{243}$Am has a radioactivity concentration of 10 mBqL$^{-1}$ and that 10 g of the marine sediments has been spiked with 1 ml of the $^{243}$Am solution. Find the $^{241}$Am radioactive concentration of the sediment if the count numbers were 200 and 400 under the $^{241}$Am and $^{243}$Am peaks, respectively, obtained over 3 days.

11. Analyze the following chemical separation scheme and comment on the need for each of the method steps.

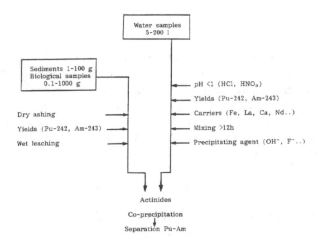

12. $^{90}$Sr will be determined in marine waters by using a chemical method. The decontamination factors for $^{106}$Ru-$^{106}$Rh are found to be $10^{-5}$, for $^{137}$Cs $7 \times 10^{-7}$, and for $^{152}$Eu $10^{-3}$. If $^{90}$Sr is to be measured by $\beta$-counting, evaluate the possible interferences of the nuclides mentioned above in the counting.

13. A standard $^{14}$C sample was counted with a GM detector, giving 5000 counts in 10 min on March 10, 2021. The activity of the standard is known to be 0.175 $\mu$Ci on March 10, 1980 with a relative uncertainty of 5% at the $3\sigma$ confidence level. Calculate the detector efficiency and its error with a $1\sigma$ confidence level. The background of the counter was found to be 40 cpm.

14. he counting efficiency was measured at different $\beta$ energies with a GM detector. For that, some $\beta$-emitters of well-known energies and activities were used.

| Radionuclide | Efficiency (%) |
|---|---|
| $^{14}$C | 1 |
| $^{147}$Pm | 2 |
| $^{99}$Tc | 2.8 |
| $^{90}$Sr | 4 |
| $^{36}$Cl | 6 |
| $^{210}$Bi | 6 |

The results are shown in the table.

(a) Plot the efficiency against the maximum $\beta$-energy
(b) It is known that GM detectors do not provide a spectrometry response. So explain why the counting efficiency depends on the energy.

(c) Discuss the shape of the curve. Explain why there is a limit on the efficiency.

15. Discuss the behavior of $s$ in Fig. 13.18.

# References

1. M. Isaksson, Sampling methods: a survey of methods in use in the Nordic countries. NKS (Nordic nuclear safety research)-17 (2000)
2. B. Salbu, O.C. Lind, Analytical techniques for charactering radioactive particles deposited in the environment. J. Environ. Radioact. **211**, 106078 (2020)
3. *Sediment Sampling Guide and Methodologies*, 2nd edn. (State of Ohio Environmental Protection Agency, USA, 2001)
4. N. Cavanagh, R.N. Nordin, L.G Swain, L.W. Pommen, *Lake and Stream Bottom Sediment Sampling Manual.* (BC Ministry of Environment, Lands, and Parks, Water Quality Branch, Canada, 2001)
5. E. Ceballos, Ph.D. D. Thesis, Universidad de Sevilla (2019)
6. https://icecores.org/about-ice-cores
7. A Guidebook, Measurement of radionuclides in food and the environment, in *IAEA Technical Report Series.* STI/DOC/10/295 (1989)
8. J.M. Gómez-Guzmán, E. Holm, N. Niagolova, J.M. López-Gutiérrez, A.R. Pinto-Gómez, J.M. Abril, M. García-León, Influence of releases of $^{129}$I and $^{137}$Cs from European reprocessing facilities in Fucus vesiculosus and seawater from the Kattegat and Skagerrak areas. Chemosphere **108**, 76–84 (2014)
9. V.K. Dhargalkar, D. Kavlekar, *Seaweeds- A Field Manual.* (National Institute of Oceanography, Dona Paula, Goa, India, 2004)
10. Sampling, storage, and sample preparation procedures for X-ray fluorescence analysis of environmental materials. (IAEA, 1997), IAEA-TECDOC-950
11. A.E. Masresha, L. Skipperud, B.O. Rosseland, G.M. Zinabu, S. Meland, H.-C. Teien, B. Salbu, Speciation of selected trace elements in three Ethiopian Rift Valley Lakes (Koka, Ziway, and Awassa) and their major inflows. Sci. Total Environ. **409**, 3955–3970 (2011)
12. https://www.vliz.be/en/Niskinbottle
13. HASL-300, *EML Procedures Manual V*, Chapter 2, 28th edn. (1997)
14. J.M. Gomez-Guzmán, J.M. López-Gutiérrez, R. García-Tenorio, L. Agullo, J.I. Peruchena, G. Manjón, M. García-León, Estimating the impact from Fukushima in Southern Spain by 131I and Accelerator Mass Spectrometry detection of $^{129}$I. J. Environ. Radioact. **166**, 36–44 (2017)
15. M. Mohery, A.M. Abdallah, Z.M. Al-Amoudi, S.S. Baz, Activity size distribution of some natural radionuclides. Radiat. Protect. Dosim. **158**, 435–441 (2014); M. Grundel, J. Porstendorfer, Differences between the activity size distributions of different natural radionuclide aerosols in outdoor air. Atmos. Environ. **38**, 3723–3728 (2004)
16. R. Leifer, Z. Russell Juzdan, The high altitude sampling program: radioactivity in the stratosphere—final report. EML-458 (1986)
17. Noble gas sampling system, EMSL-LV-539-7. (US Environmental Protection Agency, 1977)
18. Multi-Agency Radiological Laboratory Analytical Protocols Manual (MARLAP), U.S. Environmental Protection Agency (EPA), Department of Energy (DOE), Department of Defense (DOD), Department of Homeland Security (DHS), Nuclear Regulatory Commission (NRC), National Institute of Standards and Technology (NIST), U.S. Geological Survey (USGS) and Food and Drug Administration (FDA) and from the Commonwealth of Kentucky and the State of California, vol. 2, chapter 10 (2004)
19. Generic procedures for monitoring in a nuclear or radiological emergency, IAEA 1999, IAEA-TECDOC-1092

20. S. Hurtado, M. García-León, R. García-Tenorio, GEANT4 code for simulation of a germanium gamma-ray detector and its application to efficiency calibration. Nucl. Instrum. Method. Phys. Res. A **518**, 764–774 (2004)

21. K. Lieser, 'Nuclear and Radiochemistry: Fundamentals and Applications', Second Edition (Wiley-VCH, 2001).

22. W.D. Loveland, D.J. Morrissey, G.T. Seaborg, *Modern Nuclear Chemistry*, 2nd edn, chapter 19. (Wiley, 2017)

23. E. Holm, J. Rioseco, R.B.R. Persson, Surface barrier detectors for the determination of $^{99}$Tc by beta spectrometry, in *Methods of Low-Level Counting and Spectrometry*. (IAEA Proceedings Series, 1981), STI/PUB/592, pp. 277–281

24. *Multi-Agency Radiological Laboratory Analytical Protocols Manual (MARLAP)*, vol. 2, chapter 14 (2004)

25. Ibid. Chapter 12

26. B. Smodis, Radiochemistry for neutron activation analysis, *IAEA Workshop on Nuclear Data for Activation Analysis*. (Abdus Salam International Centre for Theoretical Physics, Trieste, 2005)

27. X. Hou, Activation analysis for the determination of long-lived radionuclides. Radioact. Environ. **11**, Elsevier (2008)

28. *Multi-Agency Radiological Laboratory Analytical Protocols Manual (MARLAP)*, vol. 2, chapter 13 (2004)

29. T.W. May, R.H. Wiedmeyer, A Table of polyatomic interferences in ICP-MS. Atom. Spectrosc. **19**, 150–155 (1998)

30. M. López-Lora, E. Chamizo, M. Villa-Alfageme, S. Hurtado, N. Casacuberta, M. García-León, Isolation of $^{236}$U and $^{239,240}$Pu from seawater samples and its determination by accelerator mass spectrometry. Talanta **178**, 202–210 (2018)

31. X.-L. Zhao, A.E. Litherland, J.P. Doupe, W.E. Kieser, The potential for AMS analysis of $^{10}$Be using BeF$^-$. Nucl. Instrum. Method. Phys. Res. B **223–224**, 199–204 (2004)

32. F. El-Daoushy, The determination of Pb-210 and Ra-226 in lake sediments and dating applications. Uppsala University Institute of Physics report, UUIP-979 (1978)

33. K. Tagami, S. Uchida, Fundamental studies using ICP-MS for the measurement of technetium-99 in a dried-up deposition sample. J. Radioanal. Nucl. Chem. **190**, 31–36 (2005)

34. E. Englund, A. Aldahan, G. Possnert, V. Alfimov, A routine preparation method for AMS measurement of $^{129}$I in solid material. Nucl. Instrum. Method. Phys. Res. B **259**, 365–369 (2007)

35. Villa, G. Manjón, Low-level measurements of tritium in water. Appl. Radiat. Isot. **61**, 319–323 (2004)

36. G. *Choppin*, J. Rydberg, J.-O. Liljenzin, C. Ekberg, *Radiochemistry and Nuclear Chemistry*, 4th edn, chapter 21. (Elsevier, 2013)

37. B.A. Adams, L. Holmes,Adsorptive properties of synthetic resins. J. Chem. Soc. **54**, 1–6 (1935)

38. M. García-León, E. García-Montaño, Simultaneous total $\alpha$ and $\beta$ assay of environmental samples, in *Society for Radiological Protection, 3rd International Symposium, Inverness*, vol. 2 (1982), pp. 637–642

39. A.E. Lally, K.M. Glover, Source preparation in alpha spectrometry. Nucl. Instrum. Method. Phys. Res. **223**, 259–265 (1984)

40. M.C. Morón, R. García-Tenorio, E. García-Montaño, M. García-León, G. Madurga, An easy method for the determination of Ra isotopes and actinide alpha emitters from the same water sample. Appl. Radiat. Isot. **37**, 383–389 (1986)

41. M. Garcia-León, G. Manjón, C.I. Sánchez-Angulo, $^{99}$Tc/$^{137}$Cs activity ratios in rainwater samples collected in the South of Spain. J. Environ. Radioact. **20**, 49–56 (1993)

42. A. Martinez-Aguirre, M. Garcia-León, M. Ivanovich, The distribution of U, Th and 226Ra derived from the phosphate FERTILISER industries on an estuarine system in Southwest Spain. J. Environ. Radioact. **22**, 155–177 (1994)

43. R. García-Tenorio, M. García-León, Electrodeposition of Ra from a HCl + CH$_3$-COONH$_4$ aqueous solution. Appl. Radiat. Isot. **37**, 441–442 (1986)
44. E. Holm, J. Rioseco, M. García-León, Determination of $^{99}$Tc in environmental samples. Nucl. Instrum. Method. Phys. Res. **223**, 204–207 (1984)
45. J. Tian, W.Yang, M. Chen, M. Zheng, D. Chen Y. Qiu, Autodeposition of $^{210}$Bi and $^{210}$Po on nickel discs and their application in the analysis of natural waters. J Radioanal. Nucl. Chem. **331**, 1039–1049 (2022)
46. H. Schoch, K.O. Münnich, Routine performance of a new multicounter system for high-precision $^{14}$C dating, in *Methods of Low-Level Counting and Spectrometry, IAEA Proceedings Series.* (1981), pp. 361–370, STI/PUB/592
47. W.W. Bowman, M.B. Hughes, Proportional counting techniques for routine tritium analysis, in *Methods of Low-Level Counting and Spectrometry, IAEA Proceedings Series.* (1981), pp. 353–360, STI/PUB/592
48. F. El-Daoushy, R. García-Tenorio, Low-level measurements of Ra-226/Rn-222by pulse ionization chambers. Nucl. Instrum. Method. Phys. Res. B **34**, 512–517 (1988)
49. http://www.ntsincorg.com/marinelli_beakers.htm; http://www.ga-maassociates.com/products/liquid-and-solid-beakers/500-ml-marinelli-beakers/
50. D. Thorburn, K. Danzer, A. Townshend, Recommendations for the use of the term 'recovery' in analytical procedures. (IUPAC Recommendations, 2001)
51. M. Thompson, S.L.R. Ellison, A. Fajgelj, P. Willetts, R. Wood,Harmonised guidelines for the use of recovery information in analytical measurement, in Symposium on Harmonisation of Quality Assurance Systems for Analytical Laboratories. (Orlando, USA, 4–5 September 1996), held under the sponsorship of IUPAC, ISO, and AOAC International
52. J.M.B. Moreno, M. Betti, G. Nicolaou, Determination of caesium and its isotope composition in nuclear samples using isotope dilution–ion chromatography–inductively coupled plasma-mass spectrometry. J. Anal. Atomic. Spectrom. **14**, 875–879 (1999)
53. K. Južnič, S. Fedina, Radiochemical determination of $^{90}$Sr and $^{86}$Sr in soil. Z. Anal. Chem. **323**, 261–263 (1986)
54. I. Vioque, G. Manjón, R. García-Tenorio, F. El-Daoushy, Determination of alpha-emitting Pu isotopes in environmental samples. Analyst **127**, 530–535 (2002)
55. N. Vajda, C.K. Kim, Determination of $^{241}$Am isotope: a review of analytical methodology. J. Radioanal. Nucl. Chem. **284**, 341–366 (2010)
56. J.L. Mas, K. Tagami, S. Uchida, Method for the detection of Tc in seaweed samples coupling the use of Re as a chemical tracer and isotope dilution inductively coupled plasma-mass spectrometry. Anal. Chim. Acta **509**, 83–88 (2004)
57. M. García-León, C. Piazza, G. Madurga, $^{99}$Tc in surface air samples during the years 1965–1967. Int. J. Appl. Radiat. Isot. **35**, 961–963 (1984)
58. G.F. Knoll, *Radiation Detection and Measurements,* chapter 11, 4th edn. (Wiley, 2010)
59. M. Villa, G. Manjón, M. García-León, Study of color quenching effects in the calibration of liquid scintillation counters: the case of $^{210}$Pb. Nucl. Instrum. Method. Phy. Res. A **496**, 413–424 (2003)
60. J.L. García-León, Master Thesis, Universidad de Sevilla (2021)
61. S. Hurtado, M. García-León, R. García-Tenorio, A fitting algorithm based on simulated annealing techniques for efficiency calibration of HPGe detectors using different mathematical functions. Nucl. Instrum. Method. Phys. Res. A **594**, 362–367 (2008)
62. J. Mantero, M.J. Gázquez, S. Hurtado, J.P. Bolívar, R. García-Tenorio, Application of gamma-ray spectrometry in a NORM industry for its radiometrical characterisation. Radiat. Phys. Chem. **116**, 78–81 (2015)
63. F. Hernandez, F. El-Daoushy, Accounting for incomplete charge collection in Monte Carlo simulations of the efficiency of well-type Ge-detectors. Nucl. Instrum. Method. Phys. Res. A **498**, 340–351 (2003)
64. S. Hurtado, R. García-Tenorio, M. García-León, Coincidence summing corrections in gamma-ray spectrometry using GEANT4 code. IEEE Trans. Nucl. Sci. **56**, 1531–1536 (2009)

65. K. Debertin, R.G. Helmer, Gamma and X-ray spectrometry with semiconductor detectors. (North-Holland, 1988)

66. B. Quintana, F. Fernández, An empirical method to determine coincidence-summing corrections in gamma spectrometry. Appl. Radiat. Isot. **46**, 961–964 (1995)

67. H.R. von Gunten, P. Benes, Speciation of radionuclides in the environment. Radiochim. Acta **69**, 1–29 (1995)

68. L. Skipperud, B. Salbu, Sequential extraction as a tool for mobility studies of radionuclides and metals in soils and sediments. Radiochim. Acta **103**, 187–197 (2015)

69. C. Bresson, E. Ansoborlo, C. Vidaud, Radionuclide speciation: a key point in the field of nuclear toxicology studies. J. Anal. At. Spectrom. **26**, 593–601 (2011)

70. D, Harvey, *Analytical Chemistry*, chapter 15. (2002), Open Education Resource (OER) Libre-Texts Project (https://LibreTexts.org)

71. *Multi-Agency Radiological Laboratory Analytical Protocols Manual (MARLAP)*, vol. 2, chapter 9 (2004)

72. https://nucleus.iaea.org/sites/ReferenceMaterials/SitePages/Home.aspx.

73. https://www.nist.gov/srm

# Principles of Low-Level Counting and Spectrometry

# 14

**ABSTRACT**

The measurement of environmental radioactivity consists of the determination of a very-low-activity concentration in the studied sample. This requires the use of specific counting or spectrometry techniques, different from those conventionally used. The principles of these new Low-Level Counting (LLC) Techniques are given in this chapter. The evaluation of LLC systems is carried out by studying the concept called Figure of Merit (FOM), which is presented in the chapter. To understand its meaning, some statistical concepts are used, as radioactivity and radioactivity determination are statistical in nature. A generalized FOM is introduced, which helps to design a whole experiment of environmental radioactivity, since it gathers all the relevant aspects, including sampling.

## 14.1 Need of Low-Level Counting Techniques (LLC)

Measurement of radionuclide concentrations in the environment at levels such as those presented in Chaps. 4 and 5 requires special counting or spectrometry techniques. Low-Level Counting Techniques (LLC) is the traditional term used to describe them. LLCs are needed when low or very low radionuclide concentrations are determined through direct observation of the radiation they emit. It is not possible to define precisely when LLC must be used instead of conventional techniques. It is very clear, however, that LLC are needed to give reasonable uncertainties for low-activity determination. We could say that it occurs when dealing with net counting rates of roughly $\leq 1$ cpm (1 count per minute). The use of conventional techniques under these conditions would require counting times that are too long to obtain acceptable uncertainties in the measurement. The reasons are given in Sect. 14.2.

© The Author(s), under exclusive license to Springer Nature Switzerland AG 2022
M. García-León, *Detecting Environmental Radioactivity*, Graduate Texts
in Physics, https://doi.org/10.1007/978-3-031-09970-0_14

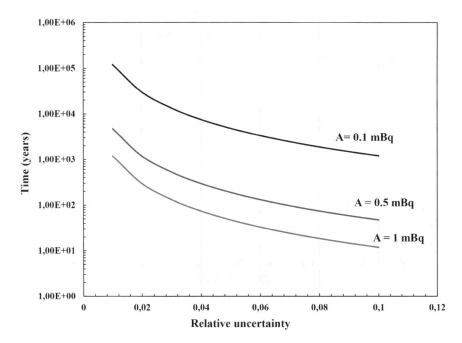

**Fig. 14.1**   Counting time needed for the determination of activities from 0.1 to 1 mBq with relative uncertainties ranging from 1 to 10%

## 14.1.1 Levels of Radioactivity in the Environment

As described in Chaps. 4 and 5, the concentrations of radionuclides in the environment are low or very low. Indeed, levels of some mBq/g or mBq/l or even less are very common, as seen in the mentioned chapters.

The time needed to determine these activities (close to or below mBq) with relative uncertainty from 1 to 10% is plotted in Fig. 14.1. The calculations were carried out using (8.7) and (14.7). A counting efficiency of 30% and a background of approximately 10 cpm were assumed. They are representative of real conventional detector systems.

The obtained counting times are not reasonable at all, and the need for specific counting techniques is apparent. For these cases, LLC must be used.

## 14.1.2 Problems Requiring LLC

The spectrum of problems that require LLC is large, although they are mainly used for the determination of radioactivity in nature. In Table 14.1, inspired by that in [1], an account is given of the fields that benefit from LLC. In addition, LLC is also used for the study of rare processes in Nuclear, Particle, and Astroparticle Physics [2]. Problems such as spontaneous fission, unusually long half-life determination, dark matter identification, Neutrino Physics, and double $\beta$ decay find in LLC an important investigation tool. In Fig. 14.2, a plot of the impact of LLC in different disciplines is presented, where we can see how useful it is for many fields.

As mentioned before, the main fields of application are related to the determination of radioactivity in nature for many purposes. Consequently, LLC are essential for the measurement of radionuclides in the environment. As we will see in the next chapter, LLC evolved from conventional radiometric techniques, with the goal of detecting $^{14}$C for dating purposes. It was W. F. Libby [3] who proposed a first LLC system capable of detecting $^{14}$C in very low-activity samples. He was able to reduce the background of the detector and increase its efficiency in such a way that $^{14}$C dating of ancient samples was possible. This concept originated a generation of LLC systems that were (and are being) used to address different scientific problems, such as those presented in Table 14.1. A good quality LLC system must meet several requirements that can be defined mathematically, according to some statistical criteria. For that, the concept of Figure of Merit (FOM) can be used. Before we define it, we should recall some ideas about the statistical nature of radioactivity.

## 14.2 Counting Statistics

### 14.2.1 The Random Nature of Radioactivity

As we saw in Chap. 2, the disintegration constant $\lambda$ gives us the probability per unit of time that a nucleus disintegrates. The only information we have about the disintegration process is $\lambda$, which connects the initial and final state of the nuclei involved in the disintegration. We are not sure whether a nucleus will disintegrate in a given time. The only thing we know is the probability that it will happen. In other words, the radioactivity is random in nature.

Therefore, the measurement of radioactivity is subject to statistical fluctuations. Thus, if we measure the activity of a sample several times (or any other related magnitude) during a given time period, we will obtain a distribution of data instead of a unique value. Therefore, as is done in many other fields, to determine the activity of a sample, several measurements must be carried out under the same conditions. The data group thus obtained is thought to represent the measurement we are performing. To characterize this data group, usually a mean value is calculated as well as a standard deviation of the data, which is a measure of the spread of the data around the mean value, i.e., of the precision of the mean. However, the theoretical distribution of data is infinite, and formally, the mean value and the standard deviation calculated this way only give an estimation of the parameters. In fact, they are called 'estimators' of the real parameters, known as 'statistics moments' describing the theoretical distribution. While the estimators describe the experimental data group, the statistical moments describe the theoretical distribution.

**Table 14.1**  Spectrum of problems that require LLC. General field, specific topics, and some of the radionuclides used. This table is inspired by Table 1 of [1]

| Field | Topics | Radionuclides |
|---|---|---|
| Environment monitoring | Nuclear installations impact<br>NORM materials<br>Radioactive worldwide control<br>Radionuclide fallout | Fission products mainly $^{137}$Cs<br>Activation products<br>Nuclides of the $^{238}$U, $^{235}$U, and $^{232}$Th series |
| Radiological effects | Radiation dose estimation<br>Radionuclide pathways in living organisms | Fission and activation products<br>$^{14}$C, $^{26}$Al |
| Oceanography | Vertical and horizontal transport of seawater<br>Air-sea exchanges<br>Global ocean circulation<br>Ocean sedimentation processes | $^{137}$Cs, $^{99}$Tc, $^{129}$I, Pu isotopes<br>U isotopes<br>$^{14}$C<br>$^3$H |
| Atmospheric processes | Circulation of air masses<br>Troposphere–stratosphere exchanges<br>Greenhouse effects<br>Behavior of aerosols | Radioactive noble gases, $^{85}$Kr and $^{222}$Rn<br>Cosmogenic radionuclides, $^{14}$C, $^{36}$Cl, $^{10}$Be, etc<br>$^{137}$Cs, $^{90}$Sr, Pu isotopes |
| Hydrology | Surface water transport<br>Dating of Groundwater | $^3$H, $^{14}$C, U isotopes,<br>$^{36}$Cl. $^{39}$Ar |
| Glaciology | Dating of glaciers and polar ice<br>Accumulation rates | $^3$H, $^{14}$C, $^{210}$Pb, $^{55}$Fe<br>Fission products, $^{137}$Cs |
| Limnology | Lake mixings<br>Lake history<br>Sedimentation in Lakes | $^{14}$C, $^{137}$Cs, and $^{210}$Pb |
| Archaeometry | Dating of Archaeological Artifacts<br>Paleobotany<br>Paleoenvironment | $^{14}$C<br>$^{10}$Be, $^{36}$Cl |
| Geological dating | Earth's crust<br>Erosion processes | $^{10}$Be, $^{87}$Rb/$^{87}$Sr, $^{40}$K/$^{40}$Ar<br>U isotopes, Th isotopes |
| Cosmic rays | Cosmic-ray flux<br>Cosmic irradiation dating<br>Galactic and solar cosmic ray flux<br>Earth's magnetic field | Cosmogenic radionuclides |
| Nuclear physics | Long half-lives<br>Spontaneous fission<br>Rare nuclear processes, in general | |
| Particle physics | Double $\beta$ decay<br>Dark matter<br>Solar neutrinos | |

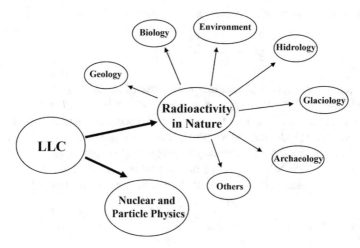

**Fig. 14.2** The impact of LLC in different scientific fields. The main applications of LLC relate to the measurement of radionuclides in nature

Table 14.2 gives a list of statistics moments and their corresponding estimators [4].

Under some conditions, it is possible to foresee $P(x)$. the probability distribution function of the data obtained when a measurement is repeated. This function gives the probable value of each of the individual data obtained in the measurement. Probability distributions are defined by two parameters: the mean value ($\mu$), which represents the distribution, and the standard deviation ($\sigma$), which gives a measure of the width of the distribution (see Table 14.2).

The binomial distribution of probability [5] describes random processes and is known to accurately represent radioactivity. According to its formulation $P(x)$, the

**Table 14.2** Statistics moments and estimators. Moments are representative of the infinite data population or distribution, whereas estimators give an 'estimation' of the moments from the obtained experimental data group obtained. The summation in column 3 extends to the total number of experimental data $n$

| Statistics Moment | Name | Estimator |
|---|---|---|
| $\mu_x$ | Mean first moment | $\bar{x} = \frac{1}{n}\sum_{i=1}^{n} x_i$ |
| $\sigma_x^2$ | Variance-second central moment | $s_x^2 = \frac{1}{n-1}\sum_{i=1}^{n} (x_i - \bar{x})^2$ |
| $\sigma_x$ | Standard deviation | $s_x = \sqrt{s_x^2}$ |
| $\sigma_{\bar{x}}$ | Standard deviation of the mean | $s_{\bar{x}} = \frac{s_x}{\sqrt{n}}$ |
| $\sigma_{xy} = \sigma_{yx}$ | Covariance | $s_{xy} = s_{yx} =$ $\frac{1}{n-1}\sum_{i=1}^{n}(x_i - \bar{x})(y_i - \bar{y})$ |
| $\frac{\sigma}{\mu}$ | Relative standard deviation | $u_x = \frac{s_x}{\bar{x}}$ |

probability of obtaining $x$ successes in a measurement is given by

$$P(x) = \frac{n!}{(n-x)!x!} p^x (1-p)^{(n-x)} \tag{14.1}$$

$n$ is the number of trials, and p is the probability of success in a trial.

In the case of radioactivity, $x$ would be the number of counts, $N$, obtained during a time interval $\Delta t$, i.e., the number of successes, and $n$ the number of nuclei in the sample, the number of trials. It is $n \sim N_A$, where $N_A$ is the Avogadro number. The probability of success (the probability of detecting a count) would be $p \sim (1 - e^{-\lambda \Delta t})$, where $\lambda$ is the decay constant. In the case of the binomial distribution, the mean value and standard deviation are, respectively [5].

$$\overline{N} = np \tag{14.2}$$

and

$$\sigma_N = \sqrt{\overline{N}(1-p)} \tag{14.3}$$

In many cases, $p \ll 1$. This means, in real measurements, that $\Delta t \ll T_{1/2}$ and $\varepsilon$ < 1 or both. Since $n \sim N_A$, , (14.1) approximates

$$P(N) \approx \frac{\overline{N}^N}{N!} e^{-\overline{N}} \tag{14.4}$$

Equation 14.4 is the Poisson distribution of probability [6]. Now, the standard deviation is $\sigma = \sqrt{\overline{N}}$, and therefore, the Poisson distribution is totally defined with just one parameter, $\overline{N}$. Moreover, when $p \ll 1$ and $\overline{N} \gg$, (14.4) becomes [7]

$$P(N) \approx \frac{1}{\sqrt{2\pi \overline{N}}} e^{-\lceil \frac{(N-\overline{N})^2}{2\overline{N}} \rceil} \tag{14.5}$$

which is the Gaussian or Normal distribution of probability. Since it derives from the Poisson distribution, it preserves its properties and, therefore, is also totally defined with only one parameter, $\overline{N}$. In addition, according to the properties of the Gaussian distribution, we know that

$$\overline{N} - \sqrt{\overline{N}} \leq N \leq \overline{N} + \sqrt{\overline{N}} \tag{14.6}$$

in 68% of the cases.

This is a very interesting result since in a reasonable approximation, a single measurement of the number of counts is well representative of the distribution of data. Indeed, provided the conditions leading to (14.5) applies, which is the common case, $N \approx \overline{N}$ as (14.6) shows. In other words, the probability distribution

**Table 14.3** Characteristics of the probability distributions followed by radioactivity. $N$ is the number of counts obtained during a time period $\Delta t$, and $p \sim (1 - e^{-\lambda \Delta t})$ is the probability of detecting a count during such a period of time. The number of nuclei in the sample is $n \sim N_A$, with $N_{A\,being}$ the Avogadro number and $\overline{N}$ being the count number mean value representing the distribution. See the text for an easier comprehension

| Distribution | Mean | Standard deviation |
|---|---|---|
| Binomial | $\overline{N} = np$ | $\sigma_N = \sqrt{N(1-p)}$ |
| Poisson | $\overline{N} = np$ | $\sigma = \sqrt{N}$ |
| Gauss or normal | $\overline{N} = np$ | $\sigma = \sqrt{N}$ with $\overline{N} - \sqrt{\overline{N}} \leq N \leq \overline{N} + \sqrt{\overline{N}}$ in 68% of the cases |

of the number of counts obtained during a time interval, $\Delta t$, is well defined with a single parameter, $N$, now being $\sigma \approx \sqrt{N}$. Consequently, the uncertainty of a single measurement, in a very good approximation, is $\sqrt{N}$ and the relative uncertainty is $\frac{\sqrt{N}}{N}$. Both parameters adequately represent the whole distribution of the number of counts associated with the measurement. In other words, for radioactivity, a single measurement is enough to describe the data distribution. The important thing is that the conditions that lead to (14.5) must be met, and this is more likely to occur as N increases.

For the sake of clarity, we present the characteristics of the different probability distributions in Table 14.3. In Fig. 14.3, the Poisson and Gaussian functions are compared.

## 14.2.2 Uncertainty Calculations in Radioactivity Measurements

According to the statistical nature of radioactivity, the well-known error propagation formula applies. Indeed, let us assume that $z = f(x_1, x_2, x_3, \ldots, x_n)$, $x_i$ being variables that are not statistically correlated. The uncertainty of $z$ can be calculated according to the well-known formula [4]

$$\sigma_z^2 = \sum_{i=1}^{n} \left(\frac{\partial f}{\partial x_i}\right)^2 \sigma_{x_i}^2 \tag{14.7}$$

where $\sigma_{x_i}$ is the uncertainty of each of the variables.

The relative uncertainty of $z$ would be $\frac{\sigma_z}{z}$. In Table 14.4, the uncertainties of some common one or two variable functions are presented. Finding them would be an interesting exercise. Although this is not a matter of this book, it is important to make comments regarding the uncertainty calculations. Of course, (14.7)

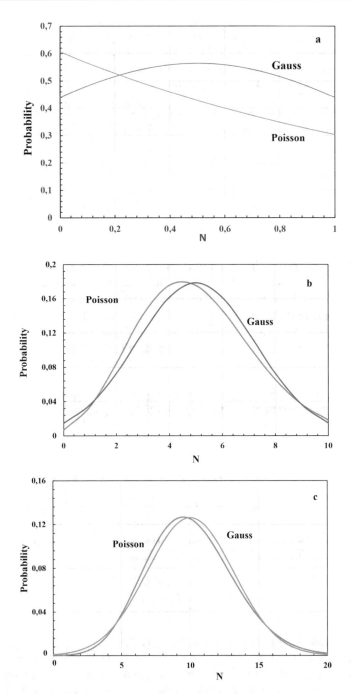

**Fig. 14.3** Poisson and Gaussian probability distributions of probability for three cases: **a** for $\overline{N} = 0.5$; **b** $\overline{N} = 5$; and **c** $\overline{N} = 10$. As $\overline{N}$ increases, the Gaussian distribution approximates the poisson distribution

**Table 14.4** Uncertainties and relative uncertainties calculated for one or two variable common functions. The variables $x$ and $y$ are assumed not to be correlated. The relative uncertainty can be calculated as $\frac{s_x}{z}$ in all cases

| Function | Approximate variance |
|---|---|
| $z = x + y$ | $s_z^2 = s_x^2 + s_y^2$ |
| $z = xy$ | $\left(\frac{s_z}{z}\right)^2 = \left(\frac{s_x}{x}\right)^2 + \left(\frac{s_y}{y}\right)^2$ |
| $z = \frac{x}{y}$ | $\left(\frac{s_z}{z}\right)^2 = \left(\frac{s_x}{x}\right)^2 + \left(\frac{s_y}{y}\right)^2$ |
| $z = x^2$ | $s_z^2 = 4x^2 s_x^2$ |
| $z = \sqrt{x}$ | $s_z^2 = \frac{1}{4}\frac{s_x^2}{x}$ |
| $z = ax^{\pm b}$ | $\left(\frac{s_z}{z}\right)^2 = b\left(\frac{s_x}{x}\right)^2$ |
| $z = ae^{\pm bx}$ | $\left(\frac{s_z}{z}\right)^2 = b^2 s_x^2$ |
| $z = a\ln(\pm bx)$ | $s_z^2 = a^2\left(\frac{s_x}{x}\right)^2$ |

applies to random nature variables. However, it is well known that so-called systematic uncertainties also exist. This kind of uncertainty normally refers to biased sources of inaccuracy, not subject to random fluctuations, or sources of inaccuracy of random nature that cannot be described by statistical methods. Systematic uncertainties must be included in the calculation of the total measurement error. Unfortunately, there are no fixed rules to estimate them, but a series of steps to follow, some of them depending on the experience of the scientist who makes the measurements [4].

## 14.3 Figure of Merit (FOM)

### 14.3.1 Definition and FOM Equation

The Figure of Merit (FOM) helps to define the requirements an LLC system has to meet and its quality compared to others. The definition of FOM must include the relevant parameters of a counting (or spectrometry) system for the problem of detecting the radiation emitted from a sample containing radionuclides. These parameters are the background and counting efficiency for the specific radiation one tries to detect.

A very common definition of FOM [1] is

$$FOM = \frac{1}{T} \tag{14.8}$$

where T is the combined time, $T = T_{S+B} + T_B$, needed to measure the source, $T_{S+B}$, and the background, $T_B$, of a detector that minimizes the relative uncertainty of the net source counting rate. Assuming that the counting process can be statistically described by the Poisson probability distribution, the calculation of the FOM is very simple.

Let us suppose that S is the net source counting rate that has been measured during a time $T_{S+B}$. If B is the background counting rate measured in a time $T_B$, the standard deviation of S, its uncertainty, $\sigma_S$, is given by

$$\sigma_S = \left( \frac{S+B}{T_{S+B}} + \frac{B}{T_B} \right)^{\frac{1}{2}} \tag{14.9}$$

just by applying (14.7). On the other hand, the relative standard deviation or relative uncertainty, $d = \frac{\sigma_S}{S}$, would be

$$d = \frac{1}{S} \left( \frac{S+B}{T_{S+B}} + \frac{B}{T_B} \right)^{\frac{1}{2}} \tag{14.10}$$

Under these conditions, that is, given S and B, the combined time $T = T_{S+B} + T_B$ that minimizes d can be calculated by setting its derivative to zero (14.10) with respect to T. Under these conditions, it can be found that

$$\frac{T_{S+B}}{T_B} = \left( \frac{S+B}{B} \right)^{\frac{1}{2}} \tag{14.11}$$

This relationship can be used to calculate both $T_{S+B}$ and $T_B$ as functions of S and B using (14.10). After some mathematics, it can be found that

$$T = \frac{1}{d^2 S^2} \left[ (S+B)^{\frac{1}{2}} + B^{\frac{1}{2}} \right]^2 \tag{14.12}$$

Hence,

$$FOM = d^2 \frac{S^2}{4B} \left[ \frac{2}{\left(1 + \frac{S}{B}\right)^{\frac{1}{2}} + 1} \right]^2 \tag{14.13}$$

In the case of LLC, usually $S \ll B$ and (14.13) becomes

$$FOM \approx d^2 \frac{\varepsilon^2 a^2}{4B} \tag{14.14}$$

where we have substituted S by $\varepsilon a$, $\varepsilon$ being the counting efficiency and a, the radioactivity in the radioactive source.

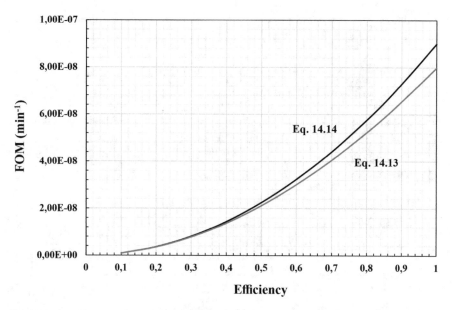

**Fig. 14.4** Equations (14.13) and (14.4) are presented for different $\varepsilon$ values when measuring a source with $a = 1\,mBq$. The background count rate is taken as 1 cpm, and the desired relative uncertainty is 1%

A plot of (14.13) and (14.4) is presented in Fig. 14.4. It is very clear that both equations are equivalent for a low net sample count rate, that is, low $\varepsilon a$. When the sample count rate increases, the FOM is slightly overestimated by the approximation taken in (14.4).

In Fig. 14.5, the variation of FOM (14.4) with the background is presented for different counting efficiencies. Again, $a = 1\,mBq$ and $d = 0.01$. Low backgrounds and high efficiencies improve the Figure of Merit.

### 14.3.2 Analysis of the FOM Equation

Equation 14.14 is normally used in LLC, and the conclusions one can derive from it are clear. FOM improves for low background and high counting efficiency systems. Both conditions cannot be met simultaneously without modifying conventional counting or spectrometry systems. Indeed, increasing the efficiency just by maximizing the relative size source-detector means simultaneously increasing the background. In contrast, the decrease in the background just by diminishing the detector size to avoid the influence of the different background sources means a simultaneous decrease in the counting efficiency. The main goal of LLC is just to obtain lower backgrounds without lowering the counting efficiency. More precisely, the objective is to maximize the FOM, which can be achieved by decreasing

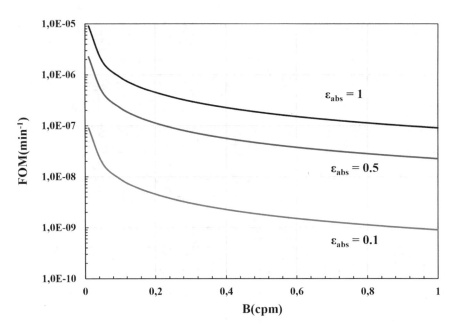

**Fig. 14.5** FOM against background count rate, calculated as in (14.4). $a = 1m\,Bq$, and the desired relative uncertainty is 1%

and increasing the background and efficiency as much as possible, respectively, by using the different techniques presented in Chap. 15.

It is interesting to note the presence of $a$ in (14.14). FOM can also be improved by increasing the activity at the source. This fact reveals that the FOM is not only a matter of the counting or spectrometry system but also of the whole LLC experiment, including radiochemistry and sampling. Indeed, in what follows, we define a generalized FOM that helps to better plan an experiment on the determination of environmental radioactivity.

## 14.4    Generalized Figure of Merit

### 14.4.1 Definition and Equation

In real experiments, the activity in the source we are measuring has been extracted from a sample collected in the environment. If the activity present in the whole sample is A, then we can reformulate the FOM as

$$FOM_g \approx d^2 \frac{\varepsilon^2 Y^2 A^2}{4B} \qquad (14.15)$$

where $Y$ is the radiochemical yield (see Chap. 13) obtained in the extraction of radionuclides chemically from the collected sample. $FOM_g$ is a generalized Figure of Merit since it includes additional parameters related not only to the counting or spectrometry systems but also to the whole experiment.

### 14.4.2 Analysis of the Equation

Indeed, it is apparent that $FOM_g$ can be improved by increasing $Y$ or $A$ or both at the same time. In other words, the chemical extraction method can be improved or the sample size can be increased. Unlike the case of background and efficiency, $Y$ can increase independently of $A$ or, better said, almost independently. As we saw in the previous chapter, the sample size could have some influence on the recovery yield, since the size makes the sample handling more complicated. However, in principle, both could be considered independent. However, (14.15) contains all the interesting variables of the experiment and can be used to compare the behavior of LLC systems with regard to a given problem but also to design the whole experiment, as we will see in the next paragraph.

## 14.5 Designing an LLC Experiment

Conventionally, the Figure of Merit, specifically that of (14.14), serves as a criterion to compare how powerful the counting or spectrometry systems are to provide a given measurement. Based on statistical arguments, it is found that the background and the counting efficiency must decrease or increase, respectively, to obtain a better FOM, i.e., to obtain the best precision in the measurement with the shortest counting time possible. The measurement of radioactivity in the environment, in which we are interested in this book, is not only a matter of using the best counter or spectrometer. In contrast, one must consider the process of sample preparation or even the sampling itself. Equation (14.15) shows that both processes are directly related to the quality of the experiment and, therefore, must be included when designing the LLC experiment. Equation (14.15) contains some hints to improve the results from the point of view of sampling and sample preparation. Indeed, since $Y$ and $A$ should be as high as possible, a careful selection of the radiochemical procedure must be carried out, and the sample must contain as many radionuclides of interest as possible.

### 14.5.1 Sampling Strategy

In most cases, the application of LLC to environmental problems arises from the interest in learning about the behavior of nature. In Table 14.1, we can see several examples that illustrate that idea. One given radionuclide is selected to investigate a given problem since it is supposed to perfectly trace the transformations taking place during the natural process of interest. In other words, the radionuclide must be conservative with respect to the environmental process. For instance, if one is interested in the study of oceanographic currents or the dynamics of mass waters and exchanges among them, the selected radionuclide should be as soluble as possible in seawater. In this way, the radionuclide will follow the dynamics of the water and participate in the exchanges it is supposed to trace. If we want to investigate the exchange of matter between water and sediments in a lake, we have to select a radionuclide with an important tendency to associate with solid matter.

Many other examples can be given to illustrate this idea. However, in general, it is important to make an adequate choice of radionuclide or radionuclides that could help us to understand the natural transformation. Chaps. 4 and 5 contain enough information on the behavior of radionuclides in the environment.

In the case of environmental monitoring applications, radionuclide selection is usually performed following the needs of radiological safety of the population and the control of the environmental impact of some nuclear installations. In any case, the samples must be representative and, of course, of a sufficient size to meet the requirement of a large A in (14.15). The collection of large samples in environmental applications is very common. This is due to the very low concentration expected for the radionuclides of interest. Hundreds or thousands of liters of seawater or rainwater are normally needed to determine interesting radionuclides with acceptable uncertainties. In Fig. 14.6, some pictures of the sampling work are given. We will see in Chaps. 17 and 18 that the use of mass spectrometric techniques reduces the required sample size significantly, in some cases dramatically, for instance, when using Accelerator Mass Spectrometry. In some cases, it is possible to take advantage of the natural or artificial concentration process to collect the appropriate sample. An example is the use of the water cleaning process at water treatment plants, in which the resulting muds contain not only conventional contaminants but also radionuclides [9, 10]. Indeed, in many water filtration plants, alumina is used as a coagulant to perform a physical cleaning of water. The floc formed by precipitation or coagulation causes the removal of suspended or colloidal matter, as well as some ionic species. Radionuclides are associated with these forms and therefore appear in the flocs. The important thing is that hundreds of grams of flocs are equivalent to thousands of liters of raw water. Consequently, it is a large concentration process that can be used advantageously as a sensitive method for the determination of radionuclides in the environment. Furthermore, the collection of floc or sludge is considerably easier than that of water. However, handling grams of sludge is also easier than handling a thousand liters of water. In Table 14.5, some data are given on the use of this strategy for the determination of several radionuclides in surface waters.

The preservation of samples is also an important issue. As described in Chap. 13, sample storage is normally necessary between the collection and the preparation of the radioactive source for measurement. Representativeness therefore includes adequate sample preservation to maintain the information it has on the environment.

The choice of the radiochemical procedure for source preparation depends on the sample and its size. Several procedures are described in Chap. 13. Evaporation, coprecipitations, ionic exchanges, solvent extractions, etc., can be used. The key

**Fig. 14.6** Some pictures of sampling work in different environments. From left to right and from up to down we can see, respectively, some members of the Applied Nuclear Physics Research Group of the University of Sevilla sampling bottom sediments in an estuarine system; a sediment core sampler used for $^{210}$Pb dating of lake sediments; some deferent designs of marine sediment traps, and finally some other members of the University of Sevilla group sampling soils in cultivated areas. Pictures courtesy of Profs. M. Villa and G. Manjón. In [8], a complete description of the methodology for marine sediment sampling is given

**Table 14.5** Activity concentrations for some radionuclides in flocs and in the corresponding waters as presented in [10]. The authors were able to detect fallout radionuclides attributed to the Chinese thermonuclear test of November 17, 1976. However, it was possible in flocs collected from Canadian water treatment planes, not directly in freshwater. Details on the conversion of floc activity concentrations into water activity concentrations can be found in [9], where the retention of radionuclides in flocs and the correspondence between floc mass and water volume are given. The case of $^{99}$Tc [13] is different. The authors only gave the concentration in flocs taken from Spanish water treatment plants to show the possibility of $^{99}$Tc detection in freshwater

| Radionuclide | Concentration in floc (Bq/kg) | Concentration in water (Bq/m$^3$) |
|---|---|---|
| $^{7}$Be | 580 | 9 |
| $^{54}$Mn | 2 | 0.032 |
| $^{58}$Co | 6 | 0.09 |
| $^{60}$Co | 0.7 | 0.01 |
| $^{95}$Zr | 315 | 4.9 |
| $^{103}$Ru | 235 | 3.7 |
| $^{125}$Sb | 23 | 0,36 |
| $^{137}$Cs | 59 | 0.91 |
| $^{141}$Ce | 160 | 2.5 |
| $^{226}$Ra | 10 | 0.16 |
| $^{228}$Th | 68 | 1 |
| $^{235}$U | 1 | 0.02 |
| $^{99}$Tc | 1.3 | |

question in this step is to obtain a yield Y as high as possible and, again, to meet the requirement of (14.15).

## 14.5.2 Counting or Spectrometry, or Both

The selection of the detector is very important. This seems to be obvious, but it is not superfluous to pay attention to this problem. Indeed, depending on the type of radiation emitted by the radionuclide, it will be convenient to use a given detector, or it will be better to perform a counting or spectrometry experiment. The analysis of the radionuclide decay scheme greatly helps in making an adequate selection. Specifically, the intensity of the radiation in the scheme is a relevant parameter to select the most prominent radiation and consequently the detector. This normally applies but not always. In some cases, the most likely radiation is not the easiest to measure. See Chap. 2 for a recap of all these concepts.

For $\gamma$- or $\alpha$-emitters, the trivial choice is to perform a spectrometry experiment by using Ge or Si semiconductor detectors. In the case of $\beta$-emitters, the selection is not as evident, since the continuum nature of the $\beta$ spectrum makes its measurement very complicated. In most cases, $\beta$-emitters are determined by counting experiments, although some applications of Si spectrometers for LLC $\beta$-emitter determination can be documented in the literature [11]. When performing a $\gamma$- or $\alpha$-spectrometric measurement, the peaks clearly determine the measured radionuclide. This does not happen in counting experiments, where it is not possible to unambiguously identify the radionuclide. This is a serious handicap in the case of $\beta$-emitters. In fact, sometimes additional measurements are needed to verify the identity of the radionuclide. Absorption measurements, half-life determination, and other radiation emitted are some of them. In [12], the variation with the energy of the shape of the plateau in a gas-flow proportional counter is used to distinguish $\beta$-emitters. We can see this effect in Fig. 14.7 taken from this reference. The method is applied to the determination and identification of $^{99}$Tc in environmental samples [13].

Although this is not usual, there are several examples in the literature [14–17] where coincidence experiments have been carried out for the determination of radionuclides in environmental samples. In Fig. 14.8, an $\alpha$–$\gamma$ coincidence setup is depicted together with a typical $\gamma$ spectrum from the $^{241}$Am and Pu isotopes obtained in coincidence with the $\alpha$ particles emitted by the radionuclides. It is very apparent that $\varepsilon$ decreases greatly with this setup, and this is not convenient according to (14.15). However, $B$ also decreases greatly, which can compensate for the decrease in $\varepsilon$.

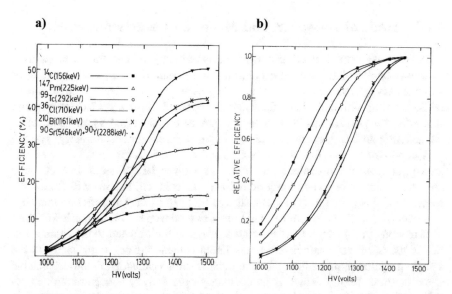

**Fig. 14.7 a** Plateau for several $\beta$-emitters in a gas-flow proportional counter. **b** Normalized plateaus to the efficiency value at 1450 V. The ratios at a given HV are characteristic of the $\beta$-emitter. For high energies, this ratio cannot be distinguished unambiguously. Figures reprinted from [12] with permission of Elsevier

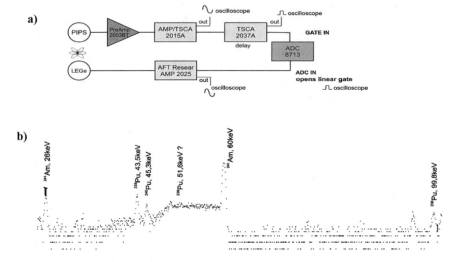

**Fig. 14.8  a** Experimental setup of an $\alpha$-$\gamma$ coincidence system based on a Si PIPS detector and a low-energy Ge detector. **b** $\alpha$-$\gamma$ coincidence spectrum for a Pu isotopes and [241]Am sample. The 60 keV peak from [241]Am is clearly observed, as well as the 43.5 keV and 45.3 keV peaks of [238]Pu and [240]Pu, respectively. The figure is reprinted from [14], with permission by Elsevier, in which the system is described and applied to safeguard measurements

## 14.6   Limit of Detection and Minimum Detectable Activity

Also based on statistical criteria, the concept of limit of detection can be used as well to analyze LLC system capabilities or compare different LLC counting or spectrometry systems. This concept is mostly used in works related to environmental monitoring, although theoretically, it is applicable to every LLC experiment. According to [18], the limit of detection has to be conceived as an estimate of the activity that can or cannot be detected by a counting (or spectrometry) system. It is important to note that this is just an estimate and that it must be applied as an a priori concept to evaluate the utility of a given LLC system. In fact, there are several definitions and expressions for its calculation, but we will follow the formulation developed in [18] and analyzed by [19], as they are of common use.

According to them, we must distinguish two concepts. One is the lower limit of detection (LLD) which is an estimated detection limit related to the characteristics of the counting system itself, i.e., without considering any other factor as the sample preparation procedure, etc. The other is the so-called Minimum Detectable Concentration (MDC), which is an estimate of the activity concentration achievable with an LLC system, including the method used for sample preparation in general. MDC is often called Minimum Detectable Activity (MDA) although the use of Minimum Detectable Concentration is more representative of the real situation since it includes the size of the sample handled in the real experiment.

MDC is very useful in regulatory programs and includes characteristics of the instrumentation and the methodology used for the activity determination.

LLD is defined as the 'smallest amount of activity in a sample that will produce a net count rate from which there is a confidence at a predetermined level that the activity is present' [20]. LLD will inform the detection capabilities of the system.

Assuming that Poisson statistics can be applied to the problem, L. A. Currie derived several expressions for LLD in 1968 [18]. In the case of very low background systems, LLD can be given by

$$LLD = \frac{1}{\varepsilon}(2.71 + 4.65\sigma_B) \qquad (14.16)$$

where $\sigma_B$ is the standard deviation of the detector background measured during time t and $\varepsilon$ is the counting efficiency. By definition, LLD has to be used as an a priori criterion to evaluate the quality or capability of a counting or spectrometry system.

MDC is a level of activity that can be achieved in real experiments by using a given counting or spectrometry system and a method for sample preparation. For that, it can be formulated as follows.

$$MDC = K(2.71 + 4.65\sigma_B) \qquad (14.17)$$

with

$$K = \frac{1}{YVt\varepsilon_{abs}e^{-\lambda\Delta t}} \qquad (14.18)$$

where $Y$ is the radiochemical yield, $V$ is the sample size, t is the counting time and $\varepsilon_{abs}$ is the absolute efficiency, which includes the counting efficiency as well as the self-absorption and absorption effects. The exponential term is the correction factor for radioactive decay during the time elapsed between sample collection and counting, $\Delta t$. In most cases, exponential correction is not necessary since $\Delta t \ll T_{1/2}$.

It is apparent that MDC decreases as $Y$, $V$, $\varepsilon_{abs}$, and $t$ increase. This is expected and compatible with the results previously described for FOM. Figure 14.9 helps to appreciate the dependence of MDC on the sample size. A counting time of 1500 min is assumed, while the recovery yield is supposed to be 80% and the absolute efficiency 50%. The exponential correction and the background count rate are assumed to be 1 and 1 cpm, respectively. Consequently, the total background will be 1500 counts throughout the counting period, and $\sigma_B = \sqrt{1500}$.

Equation 14.17 is very useful when preparing environmental monitoring programs. They usually refer to the study of the environmental radioactive impact of nuclear installations or activities, where it is mandatory to be able to detect activity concentrations that, according to radiological regulations, could be dangerous to the general population. MDC is an a priori criterion for the planning of sampling works, the selection of chemical procedures, detectors, etc., to accomplish the conditions foreseen in the environmental control program.

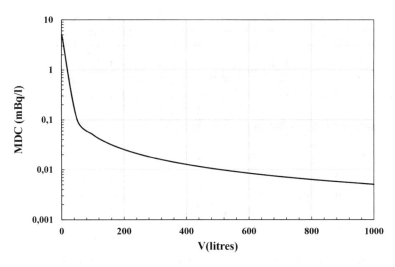

**Fig. 14.9**   MDC versus sample size (volume) in liters

## Exercises

1. The experimental values of two physical quantities, in the appropriate units, are
   $A = 300 \pm 6$ and $B = 15.0 \pm 0.6$. What is the relative error of each of them in
   %?
2. Calculate the value of the following variables obtained from the above-
   mentioned experimental values along with the absolute and relative uncertainties
   (%).

$$a = A + B; \quad b = A + 5B; \quad c = AB; \quad d = AB^2; \quad g = \ln(A/2B).$$

3. The background of a detector reached 845 counts in 30 min. A radioactive
   source increases this value to 80 cpm. Estimate the counting time to determine
   the source net counts with a relative uncertainty of 3%.
4. A detector has a background of 50 cpm, measured over 10 min. A radioactive
   source containing a radionuclide was counted in the same detector, giving 1683
   counts in 10 min and 914 counts 24 h later in 10 min. Find the radionuclide half-
   life and its uncertainty. Consider the statistic as the only source for the half-life
   error.
5. A counter has a background of 90 cpm determined for 1 h. A small sample,
   likely nonradioactive, is counted for 5 min, and 475 counts are found. Is the
   sample radioactive? If during 20 min the sample produces 1900 counts, could
   we think that the sample is radioactive?
6. A sample is counted for 10 min, and 4500 counts are recorded. The background
   of the counter was measured, giving 4000 counts over 10 min. Is the sample
   radioactive? Assume confidence limits of 68 and 95%.

7. The data given below have been obtained with two sources, $A$ and $B$, of the same isotope in the same detector that has a background of 51 counts in 10 min. Source $A$ produced 251 counts in 5 min, and source $B$ produced 717 counts in 2 min. What is the ratio of the activity from source $B$ to source $A$ and its relative uncertainty?

8. A radioactive sample produces 256 counts on a detector in 100 min. What is the relative uncertainty of the count number? For how long should we measure the sample to give the count number with 1% relative uncertainty?

9. Another sample of the same isotope and in the same geometry produces 60,275 counts in 3 min. The sample activity is 0.021 µCi ($\pm 1.2\%$). What activity (with its error) does the sample in the previous exercise have in Bq, if the background at the time of the measurement was 1370 counts in 853 min.

10. What parameters can we play with to improve the uncertainty of the measurement? What would be the limit of the relative uncertainty in the absence of systematic errors?

11. The radioactivity of a long half-life $\beta$ source is measured with a Geiger–Müller detector. The detector background was found to be 3200 counts in $t_1$ min. With the source, we obtained 3200 counts in $t_2$ minutes. Show that the relative uncertainty (%) of the net count number is:

$$\frac{100}{\sqrt{3200}} \frac{\sqrt{t_1^2 + t_2^2}}{(t_1 - t_2)}$$

12. The count rate of the background pulses is equal to $n_b = 15$ pulses per minute, and the count rate of a sample studied in the presence of the background is $n_{pb} = 60$ pulses per minute. Let $t_b$ and $t_{pb}$ be the measurements of the background and those of the sample in the presence of the background. Find the optimal ratio $t_b/t_{pb}$ at which the sample count rate is determined with the highest accuracy for the given total measurement time ($t_b + t_{pb}$).

13. Using the data of the foregoing problem, find the minimum values of $t_b$ and $t_{pb}$ at which the preparation count rate can be determined with accuracy $\eta = 0.050$.

14. The background of the detector was found to be 900 counts during the 30 min counting time. A source increases the counting rate to 80 cpm. Find the time at which the source must be counted to give the net count rate with a relative uncertainty of 3%.

15. The dead time of a GM detector is 100 s. A sample is counted, giving 10,000 cpm with a background of 10 cpm. If the counting efficiency is 30%, find the activity of the sample in $Bq$ and its uncertainty.

16. Make the same calculation as those presented in Fig. 14.1 but for $A = 0.01$ mBq and $A = 10$ mBq. Discuss the results.

17. Use (14.7) to find the uncertainties and relative uncertainties of Table 14.4.

18. Demonstrate (14.13).

19. With the help of the corresponding reference of Currie 1968, find (14.16).

20. Find the counting time needed to obtain the same MDC values in Fig. 14.9 if the counting efficiency is 100%. Calculate for 200, 400, and 600 L. Discuss the results.

# References

1. H. Oeschger, M. Wahlen, Low-level counting techniques. Ann. Rev. Nucl. Sci. **25**, 423–463 (1975)
2. P. Povinec (ed.), Rare nuclear processes, in *Proceedings of the 14th Europhysics Conference on Nuclear Physics*. (Bratislava, World Sci. Pub. Co., 1992)
3. W.F. Libby, *History of Radiocarbon Dating*, in *Radioactive Dating and Methods of Low Level Counting*, IAEA STI/PUB/152, (1967), pp. 3–26
4. W. Bambynek, Uncertainty assignment in radionuclide metrology, in *Low-Level Measurements and their Applications to Environmental Radioactivity*, ed. by M. García-León, G. Madurga (World Sci. Pub. Co., 1988), pp. 3–36
5. G.F. Knoll, *Radiation Detection and Measurement*, 4th edn, chapter 3. (Wiley, 2010)
6. N. Tsoulfanidis, S. Landsberger, *Measurement and Detection of Radiation*, 5th edn, chapter 2. (CRC Press, 2021)
7. W.R. Leo, *Techniques for Nuclear and Particle Physics Experiments: A How-To Approach*, 2nd edn, chapter 4. (Springer-Verlag, 1994)
8. E. Ceballos, Study of the key processes in carbon export and storage in the oceans through radioactive pairs: $^{234}$Th-$^{238}$U and $^{210}$Po-$^{210}$Pb', Ph. D. Thesis, University of Sevilla (2019)
9. J.-C. Roy, C. Barbeau, J.-E. Côté, J. Turcotte, A very sensitive sampling method for the measurement of radioactivity in waters from various sources. Nucl. Instrum. Method. **160**, 187–191 (1979)
10. J.-C. Roy, J. Turcotte, J.-E. Côté, P. Beaulieu, Fallout of $^{58}$Co, $^{60}$Co and $^{54}$Mn in Eastern Canada from nuclear weapons testing'. Int. J. Appl. Radiat. Isot. **32**, 423–428 (1981)
11. E. Holm, J. Rioseco, R.B.R. Persson, Surface barrier detectors for the determination of $^{99}$Tc by beta spectrometry, in *Methods of Low-Level Counting an Spectrometry, Proceedings of a symposium*, IAEA, STI/PUB/592. (Berlin, 1981), 277–281
12. M. García-León, E. García-Montaño, G. Madurga, Characterization of $^{99}$Tc by the shape if its plateau with a gas-flow proportional counter. Int. J. Appl. Radiat. Isot. **35**, 195–200 (1984)
13. M. García-León, C. Piazza, G. Madurga, On the determination of $^{99}$Tc in environmental waters. Int. J. Appl. Radiat. Isot. **35**, 957–960 (1984)
14. K. Breitenecker, D. Donohue, H. Eisenwagner, A.P. Maddison, H. Siegmund, Configuration of an alpha-gamma coincidence spectrometer for utilization of safeguards measurements. Appl. Rad. Isot. **67**, 2088–2091 (2009)
15. J.R. Cadieux, G.A. Fugate, G.S. King III., An alpha–gamma coincidence spectrometer based on the photon–electron rejecting alpha liquid scintillation (PERALS$^R$) system. Nucl. Instrum Method. Phys. Res. A **783**, 22–27 (2015)
16. M. Yazdanpanah-Kejani, F. Abbasi, M. Lamehi-Rachti, V. Doost-Mohammadi, Measurement of beta-gamma coincidence with a multiparameter analyzer system. Radiat. Phys. Chem. **130**, 243–246 (2017)
17. X. Feng, J. Wengang, L. Xuesong, H. Xiaobing, Z. Jiamei, Y. Gongshuo, Single channel beta–gamma coincidence system for radioxenon measurement using well-type HPGe and plastic scintillator detectors. Nucl. Instrum. Method. Phys. Res. Sect. A **729**, 900–904 (2013)
18. L.A. Currie, Limits for qualitative detection and quantitative determination. Anal. Chem. **40**, 586–593 (1968)

19. J.E. Watson (ed.), Upgrading Environmental Radiation Data, Health Physics Society Committee Report HPSR-1'.US Environmental Protection Agency 1980
20. B.S. Pasternack, N.H. Harley, Detection limits for radionuclides in the analysis of multicomponent gamma spectrometer data. Nucl. Instrum. Method. **91**, 533–540 (1971)

# Low-Level Counting and Spectrometry Techniques

# 15

**ABSTRACT**

The principles of LLC techniques were already introduced in the previous chapter. Here, we present specific techniques to improve FOM. Thus, the different methods for background suppression and counting efficiency increase are described. Passive and active shielding techniques are of paramount importance in Environmental Radioactivity as well as the use of high-counting-efficiency detectors. Both allow one to design the LLC experiment with better FOM parameters.

## 15.1 Techniques for Detector Background Suppression

In the previous chapter, the need for background suppression in a counting or spectrometry system has been discussed to reach better FOM, LLD or MDC. From a technical point of view, lowering the background without decreasing the counting efficiency is not trivial.

It was [1] who made the first proposal to lower the background of his detectors to address the problem of $^{14}C$ dating of old samples. In Fig. 15.1, we have the original Libby's system. Using the different shieldings, he was able to reduce the background of the detector up to 1% of the original value. This design has originated a group of techniques that are currently used in many fields, such as those presented in Fig. 14.2. It is interesting to recall the information already given in Chap. 8 on the sources of background in radiation detectors. The techniques deployed tend to suppress them with specific methods for each of the background components. Of course, regarding the instrumental sources of background, there has been great progress since the first time of LLC. Today, the stability of electronics, the conditioning of the detector environment, the temperature and humidity of the detector room, the isolation of the laboratory and the detector against vibrations, and mechanical sources of background have improved considerably.

© The Author(s), under exclusive license to Springer Nature Switzerland AG 2022
M. García-León, *Detecting Environmental Radioactivity*, Graduate Texts
in Physics, https://doi.org/10.1007/978-3-031-09970-0_15

**Fig. 15.1** Original Libby's scheme for the detection of low levels of $^{14}$C [1]. Picture reprinted from W. F. Libby, 'History of radiocarbon dating.' In: Radioactive Dating and Methods of Low-Level Counting. Proceedings of the Symposium Vienna, STI/PUB/152, IAEA 1967, 3–26. With permission by IAEA

For the fundamental sources of background, i.e., those related to physical phenomena affecting the counting, there are several techniques, shielding techniques, that will be explained in what follows. According to the original Libby scheme, roughly speaking, there are two classes of shielding techniques: passive shielding and active shielding. Both are designed to protect the detector from specific background sources.

### 15.1.1 Passive Shielding

Passive shielding is especially effective in reducing the background that comes from the gamma radiation field around the detector. In Fig. 15.2, the shielding of a modern LLC system is presented. The scheme is basically that of Fig. 15.1 but in a modern realization of the idea. The material selected for passive shielding is Pb, which has a high atomic number ($Z = 82$). This favors the photoelectric effect of $\gamma$ rays (see Chap. 7), which produces a significant attenuation of the $\gamma$ radiation flux over the detector. As seen in Table 15.1, some 10 cm thick Pb shielding is enough to reduce the γ flux over the detector to 0.2%. Systematically, the current Pb shielding is 10 to 15 cm thick. Thicker Pb layers can increase the contribution to the background of cosmic-ray μ-induced Bremsstrahlung [2]. In some cases, Pb shielding contains trace amounts of $^{210}$Pb. This is an important drawback, as it gives rise to secondary radiation inside the shielding that affects the detector background. In fact, $^{210}$Pb emits low-energy β radiation ($E_{max}$ = 17 keV and $E_{max}$ = 63.5 keV, with intensities of 84% and 16%, respectively) to decay into $^{210}$Bi and $^{210}$Po by β decay. The more intense β emission feeds the excited $^{210}$Bi state at 46.5 keV. Thus, γ radiation of this energy is also observed with an intensity of 4.25%. All of this radiation can produce X-ray fluorescence inside the shielding. Furthermore, $^{210}$Bi is also a β emitter with $E_{max}$ = 1162 keV capable of producing Bremsstrahlung in shielding. In addition, α particles, 5.407 meV, can be observed in $^{210}$Po [3]. In other words, the purity of Pb shielding must be checked and is compulsory [4, 5]. Many laboratories use old, or very old, Pb from seventeenth- or eighteenth-century roof buildings and even from Roman times. Lead anchors recovered from Roman shipwrecks have also been used [6].

The interaction of cosmic rays with passive shielding leads to the appearance of secondary radiation. This is the case for X-ray fluorescence induced by the interaction of μ with Pb. The X-rays produced can be detected by gas detectors and, of course, can modify the response of γ spectrometers at low energies.

**Fig. 15.2** The passive shielding of a low-level Ge γ-spectrometer is presented. The material selected for shielding is Pb, as it has a high atomic number (Z = 82), which makes it very effective for γ-ray attenuation. The internal lining with Cu attenuates the X-rays produced by fluorescence in Pb by cosmic μ. Picture courtesy of Prof. J. Mantero

**Table 15.1**  Pb, Hg, and Fe thicknesses in cm are needed to reduce the γ radiation to some percentages for the case of $^{40}$ K γ emission of 1461 keV. For instance, $X_{0.5}$ means the required thickness in cm to reduce the γ beam to 50% of the original value. The rest of the thickness can be interpreted in the same way. The reduction has been calculated by applying the well-known exponential equation $I(x) = I(0)e^{-\mu x}$, where $x$ is the thickness of the shield and $\mu$ is the attenuation coefficient

| Element | $X_{0.5}$ | $X_{0.2}$ | $X_{0.1}$ | $X_{0.01}$ | $X_{0.005}$ | $X_{0.002}$ | $X_{0.001}$ | $X_{0.0001}$ |
|---------|-----------|-----------|-----------|------------|-------------|-------------|-------------|--------------|
| Fe      | 1.80      | 4.19      | 6.0       | 11.99      | 13.80       | 16.18       | 17.99       | 23.98        |
| Hg      | 0.99      | 2.29      | 3.281     | 6.56       | 7.55        | 8.85        | 9.84        | 13.12        |
| Pb      | 1.17      | 2.72      | 3.89      | 7.78       | 8.95        | 10.50       | 11.67       | 15.56        |

In Fig. 15.3, Pb X-rays are observed in a background γ-spectrum obtained with a Ge spectrometer. To reduce the influence of X-rays, the Pb shielding is internally lined with a Cu layer (see Fig. 15.2), some a few millimeters thick, which is enough to attenuate the X-ray beam. This layer of obviously very pure Cu, precisely oxygen-free Cu, is also used for gas detector shielding and has been shown to be very efficient in reducing this component of the detector background [3].

In Fig. 15.3, the effect of this conventional passive shielding compound of Pb and the internal Cu layer is shown for a Ge γ-spectrometer, and it can be seen that this very simple setup is sufficient to reduce the background of the spectrometer by two orders of magnitude or three in some energy regions [7].

The nucleonic component of the background can induce nuclear reactions and then the production of neutrons outside and in the shielding. Neutrons induce secondary nuclear reactions, especially of the activation type, that give place to the appearance of radioactive nuclei in the surroundings of the detector. Neutrons can also produce reactions or collisions with the nuclei that form part of the detector material. The energy deposited by the recoil nuclei can produce an electric signal that contributes to the background of the detector. Of course, radioactive nuclei produced in the eventual reaction contribute to the background as well. As we shall see in the next paragraph, the most effective way to solve that problem is by using active shielding techniques. However, in many cases, some specific passive shielding for neutrons is used. In fact, Cd, borated paraffin, or borated polyethylene is used to absorb or thermalize and then absorb neutrons, avoiding in this way activation of the materials surrounding or forming part of the detector [3]. In some cases, the neutron shielding material coats the lead castle externally and, in other cases, internally. The last option is preferred [2, 3]. Another way to avoid the contribution of the neutron reaction products to the background is to select the materials used for the detectors' construction, including the electrical contacts [8]. In modern detectors, the selected materials have a low or very low neutron activation cross section, which greatly diminishes the appearance of the consequent background. The influence of specific neutron-produced radionuclides can be seen in Fig. 15.4 [12].

In Fig. 15.5, a schematic view of an LLC system including neutron shielding is depicted. The influence on the background spectrum is not easy to see. It depends

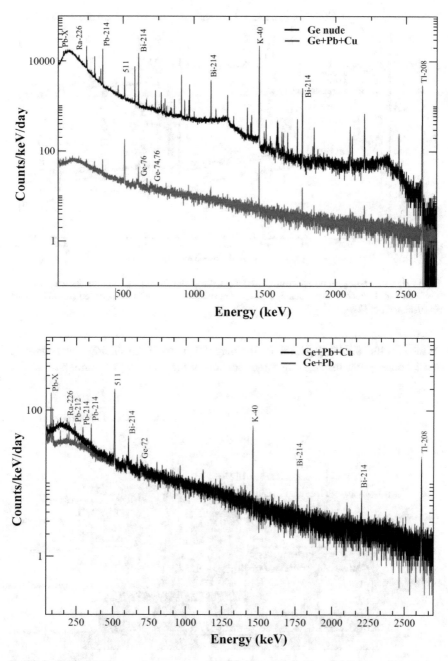

**Fig. 15.3** Typical background γ spectrum of a Ge spectrometer. The contribution of Pb X-rays due to fluorescence induced by μ is observed at photon energies of 74.97 and 84.80 keV. Figure courtesy of Prof. S. Hurtado

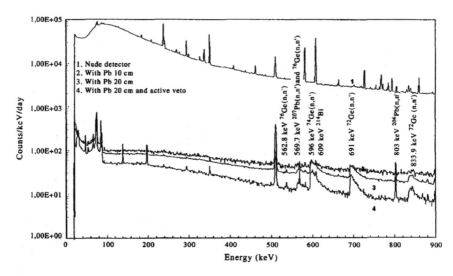

**Fig. 15.4** γ-ray background spectrum of a Ge spectrometer including some peaks from neutron reactions with materials in the detector producing radionuclides. Figure reprinted from [12] with permission from Elsevier

on the specific shielding geometry, the materials used in the shielding arrangement and laboratory construction, and the location of the counting laboratory.

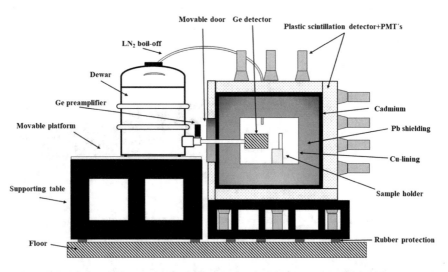

**Fig. 15.5** Gamma-spectrometer shielding includes some elements for neutron absorption. A plastic scintillator detector surrounds the Pb shielding as an anticoincidence protection. However, as a plastic, it also serves to slow down the external neutrons that can then be absorbed by the Cd layer

Although not very much used today, it is interesting to mention that Fe was considered as an alternative Pb in the construction of passive shielding [9]. Because its atomic number is small compared to that of Pb, the volume of the shielding must be higher than that of Pb in the case where the same $\gamma$-ray attenuation is expected. In Table 15.1, some calculations are given on the thickness of Pb and Fe needed to reduce the $\gamma$ flux by some determined percentage. It is applied for the case of the $^{40}K$ $\gamma$ emission of 1461 keV, one of the main contributors to the $\gamma$ background in detectors. Fe is cheaper than Pb and is more easily obtained and machined. However, it was found to contain traces of $^{60}Co$ and $^{137}Cs$ [2, 3] that dissuaded its use as a shield against $\gamma$ radiation. Hg [9] has long been considered another alternative to Pb. Hg ($Z = 80$) has a very similar behavior to Pb as a shield against $\gamma$ radiation due to its atomic number (see Table 15.1). Furthermore, it can be obtained in forms that are purer than those of Pb. However, it is more expensive, toxic, and difficult to manage because it is a liquid at ambient temperatures.

Rn and its progeny contribute very importantly to the counter background. It is present everywhere, and since it is a gas, it diffuses through the passive shielding and reaches the counting chamber. In addition, its contribution to the background can vary unexpectedly. The concentration in the laboratory depends on the ventilation conditions and the components of the wall [10]. It is convenient to adequately ventilate the laboratory room to mitigate emanation from walls, floors, and ceilings by covering them with a suitable paint layer. It is also convenient to fill the counting chamber with radioactive-free material. This is very easy in the case of low-background Ge spectrometers, since it is possible to inject the boiling gas from the Dewar into the counting chamber. Since the air has been stored for a long time before filling the Dewar, it is supposed to be Rn-free. The injected boiling $LN_2$ is heavier than the air and therefore removes it from the counting chamber. Continuous introduction of $LN_2$ keeps the ambient environment around the detector reasonably free of Rn. However, this very easy solution has some drawbacks. The main reason is that the $LN_2$ flux over the detector is not constant, since it depends on the Dewar internal gas pressure [11]. Another option is to isolate the LLC system from the ambient air by introducing it into a plastic chamber, for instance, maintaining an inert air environment around it.

In [12], a very complete account of the passive shielding effects in the background of a Ge spectrometer, including the effects of using active shielding techniques, can be found, as described in what follows.

## 15.1.2 Active Shielding

The cosmic-ray component of the detector background, essentially $\mu$ (see Chap. 8), cannot be suppressed by the Pb shield. On the other hand, the passage of $\mu$ through matter is similar to that of electrons. Thus, a charged-particle counter could detect the $\mu$ flux. Based on that, W. Libby (see Fig. 15.1) [1] proposed to surround the main detector by an array of guard (or veto) detectors connected in anticoincidence with it. The operating principle is very simple. Once a particle is detected in one of the guard detectors, the counting in the main detector is stopped for some time, avoiding interference in the measurement. This basic principle is very effective in suppressing the μ contribution to the background and that of the nucleonic component of cosmic rays. Even secondary radiation produced by the interaction of cosmic rays with materials around the detector, including Pb shielding, is also suppressed [11]. The veto detectors produced a background reduction of up to 1% of the unshielded value in the original Libby design [1]. More recent versions of this proposal make it possible to reduce the background up to 0.1% of the unshielded value [9]. This type of shielding requires appropriate electronics for signal treatment and analysis. For that, they are called active shielding techniques. In general, we call active shielding those techniques that need the use of electronic components.

The original Libby design basically remains today. However, changes can be seen in the literature regarding the type of detector utilized for the anticoincidence (or veto) system and its geometry. Gas ionization detectors, proportional or Geiger–Müller, are used when a gas ionization detector is the main component of the LLC system. Nevertheless, examples exist in the literature where scintillation veto detectors are used. LLC liquid scintillation counters normally incorporate liquid scintillation or plastic scintillation counters as an anticoincidence system [13]. Ge γ-spectrometers are generally covered by scintillation detectors (organic or inorganic) as anticoincidence systems [11]. We describe the case of LLC γ-spectrometry in detail below. An interesting debate is still open on the geometry of the veto system. In general terms, the anticoincidence detector should cover the whole main detector volume, but for many applications, this is not necessary. In Fig. 15.6, an example is given of the geometries used for the anticoincidence system.

For conventional α- and β-activity environmental monitoring with gas detectors, a typical geometry of the veto system is presented in Fig. 15.7. The main detector (or detectors in the case of multidetector systems) is covered by a parallelepiped shape anticoincidence detector. Parts of the detector are not covered, but the main part is, and this setup is sufficient to reduce the background to levels that allow the measurement of radioactivity at environmental levels [14].

According to the Libby scheme, the anticoincidence detector is located inside the Pb shield. This is the original proposal, and many modern LLC systems still follow that design [15]. However, there exists a debate on the opportunity to place the veto inside or outside the Pb shield. This debate is stronger for LLC γ-spectrometers. Both concepts have advantages and drawbacks. Internal veto systems are usually made of NaI(Tl) or BGO detectors, both of the scintillation type. One of them is presented in Fig. 15.8. The main advantage of this configuration is the suppression of the Compton continuum in the measured γ-spectrum, which can be essential for low-level activity concentrations. Indeed, the annular geometry of this system allows the detection of the Compton dispersed γ rays in the main detector with high probability. Once the detection is produced, the coincident pulse in the main detector is rejected. The result is a γ-spectrum in which part of the continuum has disappeared. Of course, the anticoincidence system also works

**Fig. 15.6** An example of an anticoincidence system placed outside the passive shielding. Compare this geometry to that presented in Fig. 15.8. Picture courtesy of Prof. J. Mantero

as a veto detector for cosmic radiation [16]. In Fig. 15.9, a comparison is made between γ-spectra collected with and without the anti-Compton system.

Very clearly the (partial) absence of the Compton continuum, especially that produced by the 1460 keV photon from $^{40}$K, helps very much to identify the photopeaks of interest and to discover new photopeaks that under normal counting conditions could not be detected. Unfortunately, the use of anti-Compton systems greatly reduces the sample size that can be measured since the counting chamber is small in general. On the other hand, conventional NaI(Tl) or BGO detectors and photomultipliers contain radioactive impurities that affect the measurement

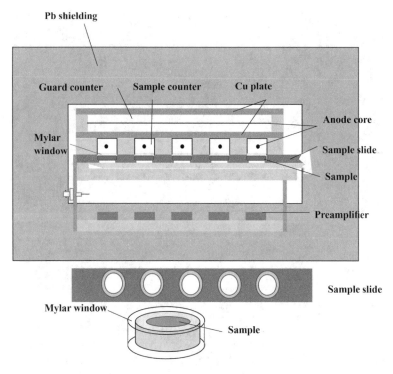

**Fig. 15.7** LLC gas (proportional or GM) was counter with an anticoincidence system. Courtesy by Prof. G. Manjón

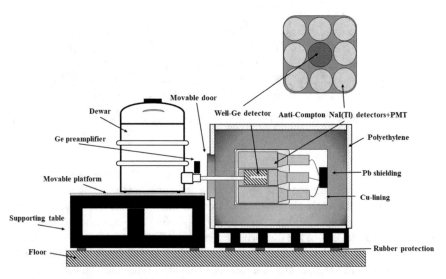

**Fig. 15.8** Anti-Compton LLC γ-spectrometer. The Ge γ-spectrometer is of well-type in this specific case

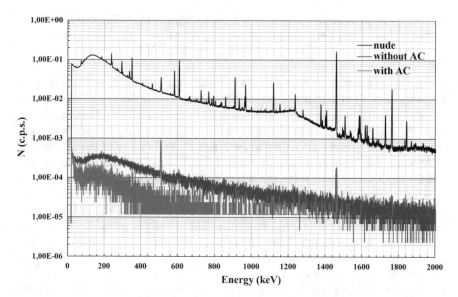

**Fig. 15.9** Comparison of background γ-spectrum with anti-Compton and without anti-Compton detector. Data courtesy of Prof. S. Hurtado

background. Furthermore, since such scintillation detectors are very efficient for γ detection, this configuration is not advisable when γ-cascade emitters are present in the sample. Nevertheless, anti-Compton systems are very often used for some special applications; one of them is $^{210}$Pb dating, in which the anti-Compton ring surrounds a well Ge detector [17, 18].

The most extended concept is the use of external anticoincidence systems in LLC γ-spectrometry [2, 11]. An example is presented in Fig. 15.5. This type of veto configuration suppresses the muonic cosmic background and the neutrons that could produce nuclear reactions in the surroundings of the detector. A very economical option is the use of gas proportional counters as anticoincidence detectors. They are effective for μ but not for neutrons. The most popular choice, however, is the use of plastic scintillation detectors that cover the whole system or most of it (see Fig. 15.5). They can detect both μ and neutrons. Furthermore, the time response of the scintillation detectors is faster than that of the gas detectors (see Chaps. 9 and 10), which makes them more suitable for an anticoincidence electronic setup.

Some other active shielding methods can be found in the literature. The principle of internal anticoincidence is used in [9, 15] for the measurement of $^{3}$H with gas proportional counters. This method is useful for low-energy β emitters that are introduced in the detector with the filling gas (see Sect. 15.2). The authors use a multianode proportional detector, and it is supposed that, being of low energy, the interesting emission will produce a signal only in one anode. Since the anodes are connected in anticoincidence, all coincident signals in several anodes are suppressed. This rejects the contribution of the external background, especially that produced by cosmic $\mu$. This concept has been used for the detection of $^3$H and $^{14}$C.

Another application of the internal anticoincidence method is discrimination by the range of charged particles in gases. This is very similar to the method described above. Cosmic $\mu$ will produce electric signals in several anodes. These events are rejected, keeping only those that give an electric signal in a given number of anodes that are assigned according to the range of the $\beta$ energy of the radionuclide of interest. This method has been applied for $^{14}$C dating [15]. In Fig. 9.10, a schematic view of one of these systems is depicted.

A very refined technique for background suppression is the analysis of the electric pulse shape, specifically, the pulse rise time in proportional detectors (and Ionization Chambers). The pulse rise time depends on the extension of the ionization track in the detector. Low-energy $\beta^-$, X-ray, and Auger electron emitters are expected to produce short ionization tracks close to the central anode and consequently short pulse rise times [9]. Pulse shape analysis is also possible in scintillation detectors. In this case, the pulse decay time is the tool for discriminating the type of particles. As is known in organic scintillators, there are two types of fluorescence processes that govern the emission of light after the excitation of the molecules (see Chap. 10). Most of the de-excitation follows a prompt decay process, with a 3–5 ns decay time. However, some molecules are excited up to a more long-lived excited state, with a decay time of 200–400 ns. Both types of states are always populated, but the ratio of short- to long-lived excited states depends on the energy density deposited in the scintillator, i.e., on the particle type. In this way, the particles can be distinguished by analyzing the pulse decay time. This technique is very widely used in liquid scintillation systems [19].

The use of coincidence techniques for the determination of radionuclides at very low concentration levels can also be considered an active shielding technique, since electronic components are used to activate the shielding. As we saw in Chap. 14, this is a method that causes a lowering of the background, since only the specific particle-$\gamma$ or $\gamma$-$\gamma$ coincidence is counted. Some examples were given in the previous chapter, but additional applications of this concept can be found in [20–22].

### 15.1.3 Underground Laboratories

A further decrease in the background requires the location of the laboratory in underground places, since the overburden greatly helps to decrease the contribution of cosmic radiation to the background of counters [3]. In Fig. 15.10, we can observe the decrease in the cosmic-ray flux with depth measured in meters of water equivalent (This unit is used to adequately compare the depth of the different underground laboratories that are constructed under different geological

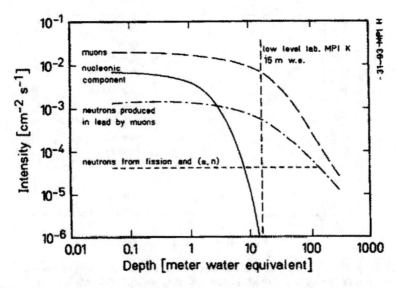

**Fig. 15.10**  Flux of the different components of the cosmic-ray radiation as a function of the depth in meters water equivalent. Figure reprinted from [61] with permission by Elsevier

substrates.). It is very clear that at 10 mwe, the nucleonic component practically disappears, while the rest of the contributions are significantly reduced. In Fig. 15.11, the reduction of the $\mu$ flux with depth is presented [23]. It is apparent that the location of counting laboratories underground is very advantageous in largely suppressing the background of counters. Figure 15.12 shows the change in the background of a gas counter and that of a Ge spectrometer in underground conditions [24, 25].

Underground laboratories are necessary for many applications related to the detection of rare events in Nuclear, Particle or Astroparticle Physics. Dark-matter studies, neutrinoless double $\beta$ decay, Neutrino Physics in general, etc., need backgrounds of a few counts per month or year [26]. A list of underground laboratories and their comparative depth can be found in [27].

Although not strictly necessary, placing conventional environmental measurement laboratories under underground conditions is very convenient. The decrease in background opens the door to new applications and to improve the uncertainty in measurement. In fact, several underground laboratories devoted to fundamental physics have environmental measurement programs [28].

The background in underground laboratories is very low. To fully exploit this fact, an analysis of the purity of the materials used in the experiments is needed. Passive shielding materials, electronics, detector materials, and even construction materials must be carefully selected. For that, they have to be monitored for radioactivity content.

Primordial and cosmogenic radionuclides (see Chap. 3) play a major role in the contamination of the materials handled during the laboratory construction and

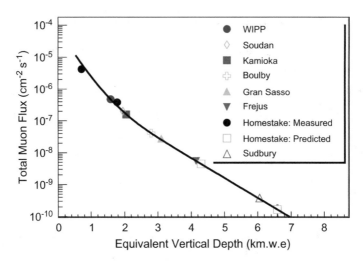

**Fig. 15.11** μ flux variation with depth as measured in different underground laboratories as a function of km water equivalent depth. Figure reprinted from [62] with permission by Elsevier

the experimental setup. However, $^{60}$Co or $^{137}$Cs should also be taken into account. This is a key concern for Underground Laboratories. Fortunately, there is today extensive experience in the presence of these radionuclides in specific materials and the ways in which they can be rejected during the fabrication of the experimental component or the metallurgy of shielding. In Table 15.2, we provide an account of the potential radioactive impurities present in several materials used in the construction of laboratories or detectors, as well as in shielding and electronic components [4, 5, 29].

## 15.2  Techniques for Increasing Counting Efficiency

Equation 14.14 shows the need to increase the counting efficiency to improve the FOM associated with an experiment. A general definition of the counting efficiency can be found in Chap. 8. The factors that affect the counting efficiency are related to different experimental aspects. Absorption and self-absorption factors relate to sample preparation, the environment of the counting chamber, and the thickness of the detector window. The intrinsic efficiency is connected to the Physics of the detector we are using and the solid angle to the counting geometry. All these factors can be modified to obtain good counting efficiency and should be considered to improve the FOM. It is also important to recall that in real experiments, the counters must be calibrated in efficiency since they depend on several parameters, such as the radiation energy and the thickness and shape of the sample. Some calibration curves can be seen in previous chapters.

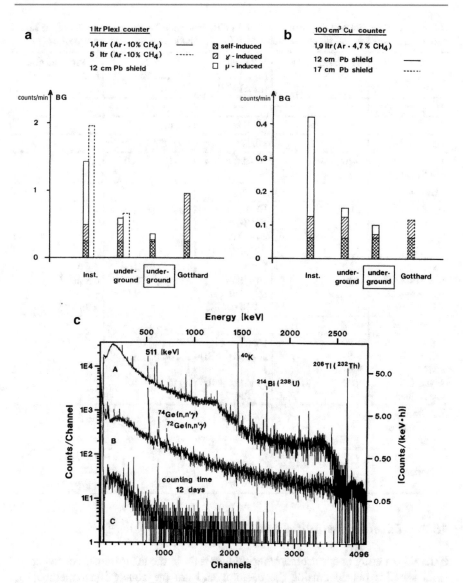

**Fig. 15.12** Background changes when located at underground conditions of **a** a gas proportional counter, reprinted from [24] with permission of IAEA and of **b** a Ge spectrometer at A (nude), B (passive, Pb + Cu, shielding) and C (passive shielding +underground), reprinted from [25] with permission of World Scientific Pub. Co

**Table 15.2** Potential radioimpurities present in different materials and components of the laboratories and experimental setups. See Chapter 4 for information on construction materials

| Material | $^{40}$K (mBqkg$^{-1}$) | $^{238}$U and (daughters) (mBqkg$^{-1}$) | $^{232}$Th and (daughters) (mBqkg$^{-1}$) | $^{222}$Rn exhalation rate |
|---|---|---|---|---|
| *Detector* | | | | |
| Si | | <0.012 | <0.003 | |
| Ge | | <0.012 | <0.003 | |
| Plastic scintillator | 1–10 | 50–300 ($^{214}$Bi) | 20–100 ($^{208}$Tl) | |
| *Shielding* | | | | |
| Pb | 0.3–10$^4$ | 10–5 × 10$^4$ ($^{214}$Bi) 20 ($^{210}$Pb) 0.25 ($^{226}$Ra) | 10–10$^4$ ($^{208}$Tl) | |
| Fe Stainless steel Electrolytic Cu | 1–10$^4$ 1–10$^2$ | 1–2 × 10$^4$($^{214}$Bi) 1($^{226}$Ra) 1–200 ($^{214}$Bi) | 1–10$^4$ ($^{208}$Tl) 1–100 ($^{208}$Tl) | 5–10 μBqm$^{-2}$ s$^{-1}$ 1.2–1.7 μBqm$^{-2}$ s$^{-1}$ |
| *Accessories* | | | | |
| Cable Wires PMT Mylar foil Glass fiber | 1–50 1–50 10$^2$–10$^3$ up to 20 × 10$^3$ | 1–20 ($^{214}$Bi) 50–500 ($^{214}$Bi) 1 up to 20 × 10$^3$ ($^{226}$Ra) | 1–10 ($^{208}$Tl) 5–100 ($^{208}$Tl) 1 up to 20 × 10$^3$ ($^{228}$Th) | |
| *Laboratory construction material* | | | | |
| Shotcrete Concrete | | | | 2–4 mBqm$^{-2}$ s$^{-1}$ 1–70 mBqkg$^{-1}$ h$^{-1}$ |

## 15.2.1 External Counting and Spectrometry

External counting or spectrometry techniques refer to the measurement of radiation from samples placed outside the detector and are the conventional method for determining radioactivity. Under these conditions, the improvement of the counting efficiency comes essentially from maximizing the sample-detector solid angle and minimizing the absorption effects. In gas counters, this can be accomplished by increasing the detector size, and an example can be seen in [30], where a cylinder-shaped detector is presented. In this case, the diameter of the cylinder is greater than its height, facilitating the measurement of large area samples and allowing samples to be prepared with small thicknesses. In this way, the solid angle is increased, and the self-absorption effects are diminished. Generally, it is easy to

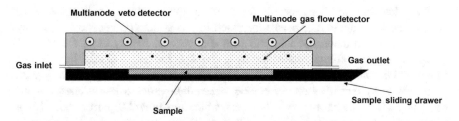

**Fig. 15.13** Gas-flow proportional counter in a $2\pi$ geometry

obtain $2\pi$ or even $4\pi$ geometries in which the efficiency can reach 50% or 100% (see Fig. 15.13) [31, 32].

The increase in the active area in Si semiconductor detectors for $\alpha$-spectrometry is the adopted strategy for this detector family. Efficiency has increased greatly in recent years as technology for the construction of Si detectors has developed. It has allowed the construction of larger detectors while maintaining exceptional energy resolutions, which is the main contribution of this detector family to the field of nuclear radiation measurement. Table 11.5 shows an account of the features of several Si detectors with increasing active areas. It is highlighted that the energy resolution provided by the manufacturer practically does not change with the detector size. A similar case is that of Ge detectors. In recent years, high purity Ge detector construction technology has greatly evolved, and currently, very high relative efficiency (see Chap. 11) Ge detectors can be found in the market. This increase facilitates the measurement of low-level radioactivity samples with extremely good energy resolution. Some features of this type of detector can be found in Table 15.3.

**Table 15.3** Example of Ge detector features, including relative efficiency (see Chap. 11) and energy resolution

| Relative efficiency (%) | FWHM (keV) at 122 keV | FWHM (keV) at 1332 keV |
|---|---|---|
| 10 | 1 | 2 |
| 20 | 1.1 | 2 |
| 30 | 1.2 | 2 |
| 40 | 1.2 | 2 |
| 50 | 1.2 | 2.1 |
| 60 | 1.25 | 2.2 |
| 70 | 1.25 | 2.2 |
| 80 | 1.3 | 2.3 |
| 90 | 1.3 | 2.3 |
| 100 | 1.4 | 2.3 |
| 110 | 1.4 | 2.3 |
| 120 | 1.5 | 2.3 |

Well detectors are very useful for specific applications (see Fig. 15.8). Of course, the solid angle increases very much in this geometry, but the sample size and its shape are determined by the well dimensions. In any case, the Ge well γ-spectrometer has found application in several specific cases. One of them is sediment [210]Pb dating [16, 18]. In this case, [210]Pb is determined by measuring the 46.5 keV γ-ray that it emits. Of course, determining such a low γ energy is not easy, and a careful photopeak efficiency calibration must be performed. This is a drawback of the method. However, one of its advantages is that [226]Ra can be measured simultaneously through the γ rays emitted by its daughters. Therefore, supported and unsupported [210]Pb can be quantified with one single measurement.

Additionally, in γ-spectrometry, special counting geometries are used to increase the efficiency. The case of Marinelli geometries is very instructive (see Fig. 13.14). In this case, the detector is covered by the Marinelli beaker in such a way that the solid angle increases greatly compared to conventional beakers. When using Marinelli beakers in γ-spectrometry, the size of the counting chamber around the Ge detector is a relevant parameter. Indeed, as can be observed in Fig. 15.2, for example, if the sample is too close to the internal shielding walls, the Compton effects of the radiation emitted by the sample could contribute to the continuum of the collected γ-spectrum. This affects the background and hinders the determination of the photopeaks in the low-energy region of the spectrum.

Absorption effects decrease the counting efficiency. Radiation can be absorbed in the sample itself and by the air from the sample to the detector. The thickness of the detector window is also an important parameter. Avoiding the self-absorption effects means the design of careful sample preparation methods whose main goal is to obtain thin radiation sources (see Chap. 13). In the case of α-spectrometry, the use of electrodeposition methods for the preparation of the radiation source is absolutely compulsory. For the absorption effects on the way to the detector, a trivial rule is to reduce the distance sample detector and place the sample as close as possible. It is true that this can increase backscattering effects. This is not an important problem provided that the effect is taken into account when performing the efficiency calibration. An extreme case is the use of SZn(Ag) disks to detect α-particles. Now, the sample is placed in direct contact with the disk [33, 34]. The absorption effects are clearly minimized, and the solid angle is formally increased to 50%.

As described in Chap. 11, Si α- (or β-) spectrometers should be placed under vacuum conditions to decrease the particle energy loss from the sample to the detector. In this manner, it is possible to take advantage of the very good energy resolution of the Si spectrometers. However, it is convenient to study the adequate vacuum conditions for counting, especially during α measurements. Indeed, the recoil nuclei could take enough kinetic energy to reach the detector window and be implanted in it. A perfect vacuum makes the arrival of the nuclei much easier. In real experiments, the adequate vacuum conditions must be determined, balancing the need for a very good energy resolution with that of avoiding contamination of the detector with recoil nuclei. The thickness of the detector window is also an important parameter to optimize the counting efficiency. Thin windows are necessary to maximize it by diminishing the window absorption effects. Today, very thin windows are possible in Si or Ge semiconductor spectrometers (see Chap. 11), allowing the measurement of very low $\alpha$, $\beta$, and $\gamma$ energies.

For gas counters, the window must be thick enough to contain the gas without losses. The use of very thin windows, which allow the measurement of very low-energy $\beta$ emitters or $\alpha$ particles, is only possible with gas-flow counters [35, 36]. In these systems, the gas flows continuously through the detector; otherwise, it can diffuse through the detector window. Now, the gas flux is an important parameter in the optimization of the counter, since the gas pressure in the detector must be constant to maintain the reproducibility of the counting conditions. This is a key issue in proportional counters, where gas multiplication depends on the gas pressure (see Chap. 9).

## 15.2.2 Internal Counting and Spectrometry

Internal counting or spectrometry is possible when the sample can be introduced into the active volume of the detector, allowing them to be placed in close contact. Under these conditions, it is expected that all the absorption effects diminish or disappear, while the solid angle increases formally up to 100%. This technique has been extensively used with gas ionization detectors when the sample can be prepared in gaseous form [14, 37]. The $^{14}C$ count benefits greatly from this possibility since it can be prepared as $CH_4$ or $CO_2$. Both gases are introduced into a gas detector, normally of proportional type, as gas counting or mixed with another gas to optimize the response of the detector. $^{3}H$ can also be prepared as $CH_4$, acetylene, benzene, or other products and can be counted with a gas detector [37, 38]. Of course, the internal counting of $^{14}C$ and $^{3}H$ is very advantageous, since both radionuclides are low-energy $\beta$ emitters. This technique enables us to greatly avoid absorption effects. Internal gas counting can be combined with the use of small or very small gas proportional detectors, some few $cm^3$ in volume. Under these conditions, the background of the detectors decreases without decreasing the counting efficiency [39, 40]. As explained previously (see Chap. 9), the response of gas detectors depends on the filling gas. The pulse shape, height, and time response of the detector depend on the probability of the charge-transfer reactions in the gas. This depends fairly on the geometry of the counter but mostly on the filling gas. Consequently, this is an important issue when using internal counting. $CO_2$, for example, is very sensitive to electronegative impurities in the gas counter (see Chap. 9). For that reason, it is very convenient to select the gas form in which the sample is prepared and to check how necessary it is to mix the sample gas with a different counting gas. In some cases, it is convenient to purify the gas during the filling process using chemical or physical traps for electronegative impurities [38, 41]. The use of internal gas counting boosted the $^{14}C$ dating method by increasing the sensitivity of the counters and therefore expanding the time period accessible by [14]C. This, together with the construction of small gas detectors discussed above, also increased the throughput of the technique.

Noble gas determination in the environment is also carried out by internal gas counting. A very popular case is that of [85]Kr, which is used as a tracer of atmospheric dynamics [9, 42]. In this case, the filling gas is P-10 (see Chapter 9), which is mixed with the Kr sample. Internal counting has also been applied for the determinations of [37]Ar, [39]Ar, [41]Kr, and [127]Xe [9, 43–45]. The measurement of [222]Rn for [210]Pb dating can be carried out by internal gas counting [46]. In this case, $N_2$ is used as the filling gas of the Ionization Chamber. A very interesting case is that of using $GeH_4$ as a gas detector to measure [71]Ge in the search for solar neutrinos [47, 48].

For the case of [3]H, the use of internal gas counting in combination with the internal anticoincidence technique gives very good results (see previous section).

Internal gas counting is an excellent choice for measuring very low-energy radiation. For that, it has been used even in the case where the sample cannot be prepared as a gas. [55]Fe is an EC (see Chapter 1) radionuclide and emits Auger electrons and soft X-rays. Its determination at environmental levels is very difficult. An internal gas counting technique has been proposed for its measurement in environmental samples [49]. Of course, [55]Fe does not form gaseous compounds at ambient temperatures, but it is possible to electroplate it onto the walls of a cylindrical gas counter (the cathode), as presented in Fig. 15.14. With this arrangement, an almost $2\pi$ solid angle is obtained. In addition, the effects of self-absorption and absorption are minimized.

Liquid scintillation counting (see Chap. 10) is also a very well-suited technique for internal counting. The sample in this case can be dissolved in, or mixed homogeneously with, the scintillation cocktail to have an almost $4\pi$ solid angle counting setup. Under these conditions, the counting efficiency is close to 100%, at least formally. Liquid scintillation is currently applied to a wide variety of problems and radionuclides. Thus, it has been applied to the determination of [55]Fe, [90]Sr, [99]Tc, [210]Pb, [226]Ra, [241]Am, etc. [51]. On the other hand, it has provided a fantastic alternative for counting [14]C and [3]H [52]. In fact, it became the choice for [14]C dating before the arrival of AMS (see Chap. 18) [13]. The combination of liquid scintillation counters with [3]H enrichment in the preparation of the radiation source [55] has allowed its determination in modern environmental samples.

Modern liquid scintillation systems addressing LLC problems have a very low background due to a very selective choice of detector construction materials, including the anticoincidence system and the purity of the lead shield. In addition, the possibility of analyzing the pulse shape, as explained before, opens the door to the adequate rejection of spurious pulses that interfere with the measurement. A typical modern liquid scintillation counter is presented schematically in Fig. 15.15. Quenching effects, originating from either chemical or color effects, are very important in environmental problems because of the origin of the treated samples. Several estimations and calibrations to quantify them have been published in the literature (see Chap. 10). Thus, the necessary corrections can be made safely.

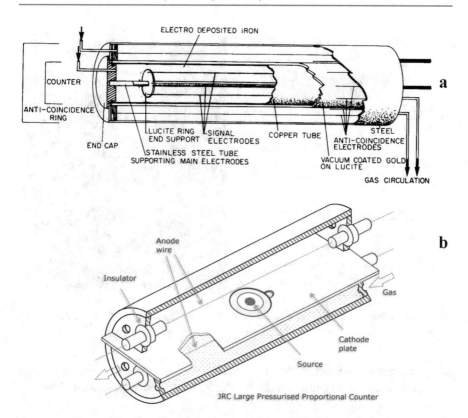

**Fig. 15.14** $^{55}$Fe gas detectors. Fe is electroplated onto the cathode of the gas detector in a). After the electrodeposition, the detector was filled with P-10 counting gas. The whole system is surrounded by another gas detector in anticoincidence. This gives a low background (0.3 cpm on average) and high counting efficiency (60% of the maximum expected with this geometry) system for $^{55}$Fe determination. Figure reprinted from. In case b), the electroplated $^{55}$Fe sample is placed in the center of the detector, as seen in the scheme. Figure reprinted from [50]. Both Figures with permission from Elsevier

Consequently, the applications of liquid scintillation to environmental radioactivity are expanding daily. Cerenkov radiation (see Chap. 7) can be observed in liquid scintillation counters by using the appropriate photomultipliers. This provides a method to determine high-energy β emitters in the environment (see Chap. 10). This is the case for $^{90}$Sr, whose daughter $^{90}$Y emits β radiation with energies well above the Cerenkov threshold, 263 keV, in distilled water. $^{210}$Pb, through its daughter $^{210}$Bi, can also be measured with this technique. In both cases, the scintillation cocktail is not necessary, and the sample is directly diluted in distilled water and placed in the vial.

The use of Ge spectrometers enriched with $^{76}$Ge to measure the neutrinoless double β decay of $^{76}$Ge can be considered another internal counting technique but

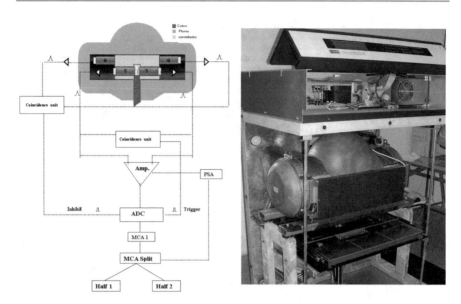

**Fig. 15.15** Schematic drawing of a liquid scintillation counter for LLC. Courtesy of Prof. G. Manjón

with semiconductor detectors [56–58]. This is a very specific LLC application with important fundamental consequences for the Standard Model in Particle Physics.

### 15.2.3 Radiation Coincidence Techniques

As explained above in this chapter and in Chap. 14, the use of radiation emitted in coincidence for the determination of some radionuclides in the environment is an option that can be considered a method for background suppression but also somehow for counting efficiency increase. The choice of a specific disintegration pathway eliminates the interference of other radiation and consequently increases the signal-to-background ratio. This could be considered a way to improve the counting efficiency. Some examples can be seen in the literature [59, 60], where $4\pi\beta$-$\gamma$ systems are used to measure EC- or low-energy $\beta$-emitters.

**Exercises**

1. In Fig. 15.8, we have a scheme of a very low-background intrinsic Ge detector. Comment on the usefulness (what are they for, what effects do they produce? etc....) of each of the shielding elements.
2. In the original Libby design, describe the utility of the different shielding elements.
3. Consider a passive shielding for a gas counter consisting of (from outer to inner part) Pb (10 cm) + Cu (4 cm) + borated polyethylene (20 cm) + Cu (1 cm).

Explain the utility of each shielding layer. Do you consider any of the elements to be superfluous?

4. Suppose now that the previous shielding is used for a NaI(Tl) γ-spectrometer. Make the same discussion.

5. The stainless-steel linear attenuation coefficient for 660 keV γ radiation is $0.433$ cm$^{-1}$, while for concrete, it is $0.144$ cm$^{-1}$. Find the thickness of both materials necessary to reduce the γ flux from a source by half. Discuss the utility of both materials as a shielding option for a γ-spectrometer.

6. We want to choose the passive shielding of a detector for the measurement of very small activities. We have to choose between Fe and Pb. To make the decision, two γ attenuation experiments are carried out using Fe and Pb sheets. Suppose that the thickness of the sheets was 10 cm. Demonstrate that Pb shielding is more effective than Fe shielding; use the data in the table below and make calculations with at least three energies of gamma radiation.

| Linear | Attenuation | Coefficient (cm$^{-1}$) |
|---|---|---|
| Energy (keV) | Pb (Z $=$ 82) | Fe (Z $=$ 26) |
| 50 | 65.0 | 15.2 |
| 100 | 62.0 | 2.93 |
| 150 | 21.8 | 1.54 |
| 200 | 10.7 | 1.15 |
| 300 | 4.29 | 0.866 |
| 400 | 2.49 | 0.740 |
| 500 | 1.72 | 0.662 |
| 1000 | 0.798 | 0.471 |
| 2000 | 0.524 | 0.334 |

7. Use the foregoing table to find the necessary Fe thickness to attenuate 1.0 MeV γ radiation in the same proportion as a 10 cm Pb sheet would.

8. The linear or mass hemireduction thickness of a material is defined as the thickness that reduces the incident γ radiation flux by half.
   (a) Calculate the linear and mass hemireduction thicknesses of Pb and Fe for the photons emitted by $^{40}$K (Eγ $=$ 1460 keV).
   (b) Calculate the mass and linear thickness of Fe that produces the same reduction factor for this γ energy as a 10 cm thick Pb shielding.

9. Calculate the mass of Pb and Fe necessary to build an equivalent shield in a parallelepipedal shape and a 4π geometry with a thickness of 10 cm for the case of Pb and the thickness calculated in part b) of the previous exercise. Suppose that the counting chamber has dimensions of 25 cm × 25 cm × 25 cm.
   Data: $m_{Fe} = 0.38$ cm$^{-1}$, $m_{Pb} = 0.581$ cm$^{-1}$, $r_{Fe} = 7.874$ gcm$^{-3}$, and $r_{Pb} = 11.35$ gcm$^{-3}$.

10. $^{210}$Pb is extremely useful for sediment dating. According to its decay scheme, there are at least three methods to measure $^{210}$Pb activity. Propose one of the possible methods and choose the appropriate detectors. Since its measurement is an LLC problem, we propose shielding methods and their effect to improve the FOM of the experiment.

11. Based on the $^{226}$Ra decay scheme, propose some methods to measure its activity. Choose the appropriate detectors. Assume that the $^{226}$Ra samples have very low activity. Design LLC methods with your chosen detectors and make a qualitative comparison between each of them using the FOM concept.

12. Suppose that you plan to determine $^{239}$Pu in marine waters. Using its decay scheme, design two methods for the determination of $^{239}$Pu. Include radiochemical separations if needed and consider it an LLC problem.

13. Using Eq. 14.15 compare the designed methods with respect to $FOM_g$. Consider that the activity concentration of $^{239}$Pu is approximately 1 $mBqL^{-1}$ and that you need to determine $^{239}$Pu with 1% relative uncertainty. Take the values of the parameters of interest from the literature.

14. Make the same study of exercises 12 and 13 for the case of $^{90}$Sr in seawater. Take 1 $mBqL^{-1}$ and 1% relative uncertainty as well.

15. From the knowledge you have about the background of detectors, analyze comparatively the importance of the different background components for the following detectors: GM, SZn (Ag), and Ge.

16. Based on the knowledge you have about the background of detectors, compare the importance of the different background components for the following detectors: proportional, Ge, and LSC.

17. Using Eq. 8.11 design appropriate passive shielding for Ge, NaI(Tl), gas-flow, and LSC detectors.

18. Select the appropriate detector from the previous exercise to determine $^3$H, $^{90}$Sr, $^{99}$Tc, and $^{137}$Cs in environmental. Add the active shielding elements needed to improve background reduction.

## References

1. W.F. Libby, *Radiocarbon Dating*, 2nd edn. (University of Chicago Press, 1955); W.F. Libby, History of radiocarbon dating, in *Radioactive Dating and Methods of Low-Level Counting. Proceedings of the Symposium Vienna*, STI/PUB/152, IAEA 1967, 3–26; Dating by radiocarbon. Acc. Chem. Res. **5**, 289–295 (1972)

2. G. Heusser, Low-radioactivity background techniques. Ann. Rev. Nucl. Part. Sci. **45**, 543–590 (1995)

3. P. Theodorsson, *Measurement of Weak Radioactivity*, Chapter 11 (World Scientific Publishing Co., 1996)

4. S. Hurtado, R. García-Tenorio, M. García-León, $^{210}$Pb determination in lead shields for low-level γ-spectrometry applying two independent radiometric techniques. Nucl. Instrum. Methods Phys. Res. A **497**, 381–388 (2003)

5. M. Laubenstein, Screening of materials with high purity germanium detectors at the Laboratori Nazionali del Gran Sasso. Int. J. Modern Phys. **32**, 1743002 (2017)

6. A. Alessandrello, C. Cattadori, G. Fiorentini, E. Fiorini, G. Gervasio, G. Heusser, G. Mezzo-rani, E. Pernicka, P. Quarati, D. Salvi, P. Sverzellati, L. Zanotti, Measurements on radioactivity of ancient roman lead to be used as shield in searches for rare events. Nucl. Instrum. Methods Phys. Res. Sect. B **61**, 106–117 (1991)

7. S. Hurtado, Ph.D. Thesis, Universidad de Sevilla

8. J.-C. Roy, J.-E. Côté, R.W. Durham, S.R. Joshi, A study of the indium and germanium photo-peaks in the background spectra of Ge spectrometers with a passive shield. J. Radioanal. Nucl. Chem. **130**, 221–230 (1989)

9. H. Oeschger, M. Wahlen, Low level counting techniques. Ann. Rev. Nucl. Sci. **25**, 423–463 (1975)

10. P. Theodorsson, *Measurement of Weak Radioactivity*, Chapter 9 (World Scientific Publishing Co., 1996)

11. S. Hurtado, M. García-León, R. García-Tenorio, Optimized background reduction in low-level gamma-ray spectrometry at a surface laboratory. Appl. Radiat. Isot. **64**, 1006–1012 (2006)

12. R. Núñez-Lagos, A. Virto, Shielding and background reduction. Appl. Radiat. Isot. **47**, 1011–1021 (1996)

13. P.P. Povinec, A.E. Litherland, K.F. von Reden, Developments in radiocarbon technologies: from the Libby counter to compound-specific AMS analyses. Radiocarbon **51**, 45–78 (2009)

14. I.U. Olsson, Low-level counting using gas filled counters as applied to $^{14}$C dating with emphasis on reliability, in *Low-Level Measurements and Their Applications to Environmental Radioactivity*, ed. M. García-León, G. Madurga (World Scientific Publishing Co., 1988), pp. 171–223

15. P. Povinec, Low level gas counting, in *Low-Level Measurements of Man-Made Radionuclides in the Environment*, ed. M. García-León, G. Madurga (World Scientific Publishing Co. 1991), pp. 38–71

16. F. El-Daoushy, R. García-Tenorio, Well Ge and semi-planar Ge (HP) detectors for low-level gamma-spectrometry. Nucl. Instrum. Methods Phys. Res. A **356**, 376–384 (1995)

17. F. Arnaud, O. Magand, E. Chapron, S. Bertrand, X. Boës, F. Charlet, M.-A. Mélières, Radionu-clide dating ($^{210}$Pb, $^{137}$Cs, $^{241}$Am) of recent lake sediments in a highly active geodynamic setting (Lakes Puyehue and Icalma—Chilean Lake District). Sci. Total Environ. **366**, 837–850 (2006)

18. N. Markovic, P. Roos, S.P. Nielsen, Low-level gamma-ray spectrometry for the determination of $^{210}$Pb. J. Radioanal. Nucl. Chem. **311**, 1473–1478 (2017)

19. L. Pujol, J.A. Sánchez-Cabeza, Optimisation of liquid scintillation counting conditions for rapid tritium determination in aqueous simples. J Radioanal. Nucl. Chem. **242**, 391–398 (1999)

20. M.O. Ranki, The use of coincidence counting techniques for analyzing low level plutonium contamination on filters. Nucl. Instrum. Methods **24**, 221–226 (1963)

21. R. de la Fuente, B. de Celis, V. del Canto, J.M. Lumbreras, B. de Celis Alonso, A. Martín-Martín, J.L. Gutiérrez-Villanueva, Low level radioactivity measurements with phoswich detec-tors using coincident techniques and digital pulse processing análisis. J. Environ. Radioact. **99**, 1553–1557 (2008)

22. N. Markovic, P. Roos, S.P. Nielsen, Digital gamma-gamma coincidence HPGe system for environmental analysis. Appl. Radiat. Isot. **126**, 194–196 (2017)

23. P. Povinec, Underground low-level counting, in *Low-Level Measurements of Radioactivity in the Environment*, ed. M. García-León, R. García-Tenorio (World Scientific Publishing Co., 1994), pp. 113–140

24. H. Oeschger, J. Beer, H.H. Loosli, U. Schotterer, Low-level counting systems in deep under-ground laboratories, in *Methods of Low-Level Counting an Spectrometry, Proceedings of a symposium, IAEA, STI/PUB/592* (Berlin, 1981), pp. 459–474.

25. R. Wordel, D. Mouchel, A. Bonne, P. Meynenedonckx, H. Vanmarcke, Low level Gamma-ray measurements in a 225 m deep underground laboratory, in Low-Level Measurements of Radioactivity in the Environment, ed. M. García-León, R. García-Tenorio (World Scientific Publishing Co., 1994), pp. 141–154

26. M. Laubenstein, I. Lawson, Low background radiation detection techniques and mitigation of radioactive backgrounds. Front. Phys. **8**, 577734 (2020)
27. J. Heise, Worldwide underground facilities, in *XXIX International Conference on Neutrino Physics and Astrophysics*, Jun 22–Jul 2, 2020; that can be found in https://indico.fnal.gov/event/43209/contributions/187839/attachments/129296/159530/Heise-UGLabs_Worldwide-Neutrino2020-2020_07_01.pdf
28. M. Köhler, D. Degering, M. Laubenstein, P. Quirin, M.-O. Lampert, M. Hult, D. Arnold, S. Neumaier, J.-L. Reyss, A new low-level γ-ray spectrometry system for environmental radioactivity at the underground laboratory Felsenkeller. Appl. Radiat. Isot. **67**, 736–740 (2009)
29. F. Marchegiani, F. Ferella, S. Nisi, Material screening with mass spectrometry. Physics **3**, 71–84 (2021)
30. A.G. Burn, X. Lin, D.K. Haines, M.M. Hussain, K.X. Yang, J.C. Marrantino, U.-F. Syed, A. Bari, T.M. Semkow, Improved sample preparation procedure for gross alpha counting on gas-flow proportional counter. J. Radioanal. Nucl. Chem. **325**, 417–423 (2020)
31. A. Ravindra, D.B. Kulkarni, L. Joseph, M.S. Kulkarni, D.A.R. Babu, A primary standard for the measurement of alpha and beta particle surface emission rate from large area reference sources. Appl. Radiat. Isot. **107**, 87–91 (2016)
32. L.E. King, J.M.R. Hutchinson, M.P. Unterweger, A new large-area $2\pi$ proportional counting system at NIST. Appl. Radiat. Isot. **66**, 877–880 (2008)
33. M. Ardid, J.L. Ferrero, A. Herrero, Study of the background on a ZnS(Ag) alpha counter with a plastic veto detector. Nucl. Instrum. Methods Phys. Res. A **557**, 510–515 (2005)
34. S. Yamamoto, I. Aoki, T. Higashi, Optical fiber-based ZnS(Ag) detector for selectively detecting alpha particles. Appl. Radiat. Isot. **169**, 109495 (2021)
35. V. Jobbágy, U. Wätjen, J. Meresova, Current status of gross alpha/beta activity analysis in water samples: a short overview of methods. J. Radioanal. Nucl. Chem. **286**, 393–399 (2010)
36. A.G. Burn, X. Li, D.K. Haines, Improved sample preparation procedure for gross alpha counting on gas-flow proportional counter. J. Radioanal. Nucl. Chem. **325**, 417–423 (2020)
37. P. Povinec, Developments in radio analytics: from Geiger counters to single atom counting. J. Radioanal. Nucl. Chem. **318**, 1573–1585 (2018)
38. E. Mace, C. Aalseth, T. Alexander, H. Back, A. Day, E. Hoppe, M. Keillor, J. Moran, C. Overman, M. Panisko, A. Seifert, Simultaneous measurement of tritium and radiocarbon by ultra-low background proportional counting. Appl. Radiat. Isot. **126**, 171–174 (2017)
39. E.V. Sayre, G. Harbottle, R.W. Stoenner, R.L. Otlet, G.V. Evans, Small gas proportional counters for the $^{14}$C measurement of very small simples, in *Methods of Low-Level Counting an Spectrometry, Proceedings of a Symposium*, IAEA, STI/PUB/592 (Berlin, 1981), pp. 393–408
40. K. Jelen, M. Geyh, A low-cost miniature counter system for radiocarbon dating. Radiocarbon **28-2A**, 578–585 (1986)
41. W. Rau, G Heusser, $^{222}$Rn emanation measurements at extremely low activities. Appl. Radiat. Isot. **53**, 371–375 (2000)
42. C. Chao-Feng Chen, Y. Li-Tao, H. Yan-Jun, S.-G. Zhi-Hong, G. Gui-Ying, Y. Jian-Lin Yao, D.S. Xiang-Dong, A study on $^{85}$Kr measurement with an internal gas proportional counter. Appl. Radiat. Isot. **155**, 108948 (2020)
43. R. Purtschert, R. Yokochi, W. Jiang, Z.-T. Lu, P. Mueller, J. Zappala, E. Van Heerden, E. Cason, M. Lau, T.L. Kieft, C. Gerber, M.S. Brennwald, T.C. Onstott, Underground production of $^{81}$Kr detected in subsurface fluids. Geochim. Cosmochim. Acta **295**, 65–79 (2021)
44. E.K. Mace, C.E. Aalseth, A.R. Day, E.W. Hoppe, J.I. McIntyre, A. Seifert, R.M. Williams, Direct low-energy measurement of $^{37}$Ar and $^{127}$Xe in a radiotracer gas using low-background proportional counters. J. Radioanal. Nucl. Chem. **318**, 125–129 (2018)
45. T.R. Alexander, C.E. Aalseth, H.O. Back, T.W. Bowyer, A.R. Day, E.S. Fuller, J.C. Hayes, E.W. Hoppe, T.W. Hossbach, P.H. Humble, D.T. Keller, M.E. Keillor, L.S. Lidey, E.K. Mace, J.I. McIntyre, J.M. Mendez, H.S. Miley, S.J. Morris, A.W. Myers, J.L. Orrell, C.T. Overman, M.E. Panisko, A. Seifert, G.A. Whyatt, R.M. Williams, Characterization of a low background proportional counter for a high throughput Argon-37 collection and measurement system. Nucl. Instrum. Methods Phys. Res. **A954**, 161794 (2020)

46. F. El-Daoushy, An ionization chamber and a Si-detector for lead-210 chronology. Nucl. Inst. Methods **188**, 647–655 (1981)
47. T. Bowles, V.N. Gavrin, The status of the solar neutrino problem. Annu. Rev. Nucl. Part. Sci. **43**, 117–164 (1993)
48. F. Kaether, W. Hampel, G. Heusser, J. Kiko, T. Kirsten, Reanalysis of the Gallex solar neutrino flux and source experiments. Phys. Lett. B **685**, 47–54 (2010)
49. L.D. Labeyrie, H.D. Livingston, A.G. Gordon, Measurement of [55]Fe from nuclear fallout in marine sediments and seawater. Nucl. Instrum. Methods **128**, 575–580 (1975)
50. S. Pommé, H. Stroh, R. Van Ammel, The [55]Fe half-life measured with a pressurized proportional counter. Appl. Radiat. Isot. **148**, 27–34 (2019)
51. X. Hou, X. Dai, Environmental liquid scintillation analysis, in *Handbook of Radioactivity Analysis*, vol. 2 (Elsevier, 2020), pp. 41–135
52. H.A. Polach, Perspectives in radiocarbon dating by radiometry. Nucl. Instrum. Methods Phys. Res. **B29**, 415–423 (1987)
53. H. Polach, G. Calf, D. Harkness, A.G. Hogg, L. Kaihola, S. Robertson, Performance of new technology liquid scintillation counters for [14]C dating. Nucl. Geophys. **2**, 75–79 (1988)
54. H. Polach, L. Kaihola, S. Robertson, H. Haas, Small sample [14]C dating by liquid scintillation spectrometry. Radiocarbon **30**, 153–155 (1988)
55. B. Kumar, L.-F. Han, L.I. Wassenaar, P.M. Klaus, G.G. Kainz, D. Hillegonds, D. Brummer, M. Ahmad, D.L. Belachew, L. Araguás, P. Aggarwal, A compact tritium enrichment unit for large sample volumes with automated-re-filling and higher enrichment factor. Appl. Radiat. Isot. **118**, 80–86 (2016)
56. A. Morales, Results on neutrinoless double beta decay of [76]Ge to the excited states of [76]Se in the Frejus experiment, in *Low-Level Measurements and Their Applications to Environmental Radioactivity*, ed. M. García-León, G. Madurga (World Scientific Publishing Co., 1988), pp. 292–358
57. M. Agostini, A.M. Bakalyarov, E. Andreotti et al., Characterization of [76]Ge enriched broad energy Ge detectors for GERDA phase II. Eur. Phys. J. C **79**, 978 (2019)
58. A.J. Zsigmond for the LEGEND Collaboration, LEGEND: the future of neutrinoless double-beta decay search with germanium detectors. J. Phys. Conf. Ser. **1468**, 012111 (2020)
59. R. Gunnink, L.J. Colby Jr., J.W. Cobble, Absolute beta standardization using 4 pi beta-gamma coincidence techniques. Anal. Chem. 31, 796–798 (1959)
60. J.I. McIntyre, M. Cooper, J. Ely et al., Further developments of a robust absolute calibration method utilizing beta/gamma coincidence techniques. J. Radioanal. Nucl. Chem. **296**, 693–698 (2013)
61. G. Heusser, Cosmic ray-induced background in Ge-spectrometry. Nucl. Instrum. Methods Phys. Res. B **83**, 223–228 (1993)
62. F.E. Gray, C. Ruybal, J. Totushek, D.-M. Mei, K. Thomas, C. Zhang, Cosmic ray muon flux at the Sanford underground laboratory at Homestake. Nucl. Instrum. Methods Phys. Res. A **638**, 63–66 (2011)

# Principles of Mass Spectrometry

# 16

**ABSTRACT**

The determination of environmental radioactivity by direct radiation counting or spectrometry is not the best option in many cases. On some occasions, when the radionuclide is long lived, its concentration is very low, or both, it is convenient to directly count the number of atoms instead of the emitted radiation. Therefore, mass spectrometry techniques are very useful, and the main goal of this chapter is to introduce their general principles. In addition, different conventional mass spectrometry methods and their applications to environmental radioactivity determination are also presented.

## 16.1 Limitations of Radiometric Methods. Need for Mass Spectrometry Techniques

The determination of radionuclides in the environment is based on the measurement of very low levels of radioactivity. As shown in Fig. 16.1, there are several interesting problems that require the determination of a few disintegrations per minute, hour, day, year, or even century in some cases. This poses a hard challenge for conventional radiometric methods, since the activity concentrations are extremely low, and it is not possible to give reasonably precise results even when using the LLC techniques described in Chaps. 14 and 15.

### 16.1.1 Loss of Information by Counting Emitted Radiation

It is very interesting to analyze (2.8) presented in Chap. 2. The activity, A, is proportional to the number of nuclei, N, through the disintegration constant, $\lambda$, which

© The Author(s), under exclusive license to Springer Nature Switzerland AG 2022
M. García-León, *Detecting Environmental Radioactivity*, Graduate Texts
in Physics, https://doi.org/10.1007/978-3-031-09970-0_16

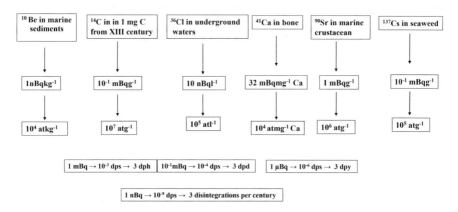

**Fig. 16.1**  Some interesting problems in the determination of radioactivity in the environment for different purposes. The activities are very low and pose an important challenge for radiometric methods

gives the probability per unit time that the nucleus disintegrates. For long half-lives or small N, or both, A will be too low to give accurate results in reasonable counting times with radiometric techniques.

In Table 16.1, we give the number of nuclei involved in the radioactivity levels presented in Fig. 16.1. They can be easily calculated by applying (2.8). It is apparent that the number of nuclei is very large compared to the number of disintegrations per unit time and the activity of the sample. If we were able to access the number of nuclei, we could probably carry out a more precise experiment. It is true, however, that when determining the radioactivity of a sample, we are interested in the measurement of the radioactive nuclei content. The use of the emitted radiation is just a way to find the information we are looking for: N in (2.8). Therefore, somehow, we are not taking advantage of all the experimental information about the sample if we measure the emitted radiation instead of the nuclei number directly.

**Table 16.1**  Corresponding number of nuclei to the activities presented in Fig. 16.1

| Radionuclide | $T_{1/2}$ (y) | Activity (mBq) | Number of atoms |
|---|---|---|---|
| $^{10}$Be | $1.51 \times 10^6$ | $10^{-6}$ | $6.8 \times 10^4$ |
| $^{14}$C | 5700 | $10^{-1}$ | $2.6 \times 10^7$ |
| $^{36}$Cl | $3.01 \times 10^5$ | $10^{-5}$ | $1.36 \times 10^5$ |
| $^{41}$Ca | $9.94 \times 10^4$ | 32 | $1.44 \times 10^{11}$ |
| $^{90}$Sr | 28.90 | 1 | $1.3 \times 10^6$ |
| $^{137}$Cs | 30.08 | $10^{-1}$ | $1.4 \times 10^5$ |

## 16.1.2 Counting Atoms Instead of Emitted Radiation

Counting atoms instead of emitted radiation gives access to all of the experimental information contained in the sample. It seems to be very advantageous, especially in the case of long-lived radionuclides, extremely low-activity samples, or both.

Mass spectrometric techniques help to count N directly. Historically, such techniques were developed for the determination of the mass of nuclei. In our case, we can use it to count the nuclei of a known mass. In what follows, we recall the principles of mass spectrometry and describe different mass spectrometers. In general, these instruments are more complicated and expensive than detectors. This should be considered when planning an experiment, since there are still many problems in which LLC techniques are sufficient to afford the measurements. Nevertheless, mass spectrometry is currently available for most laboratories, and compact systems are available on the market that can be used even by nonexpert staff.

## 16.2 Basics of Mass Spectrometry

It is known that charged particles change their trajectories under an electric or magnetic field. This is the basic principle of mass spectrometry. Thus, a mass spectrometer consists of an ion source followed by a combination of magnets and electrostatic analyzers and finally an ion detector where the mass of interest is measured. The sample is introduced into the ion source in a form that will depend on its working principle. Once the sample is ionized, it is extracted from the ion source at energies of approximately a few or tenths of keV. With the help of some ion optics, the beam is injected into the magnets and the electrostatic analyzers, where it is analyzed for mass and energy. The selected mass is transported toward a detector where the number of ions of the mass of interest is counted. A general scheme is presented in Fig. 16.2. There are different types of mass spectrometers, depending on the ion source used. We will describe some of them in the next sections.

The advantage of using mass spectrometry instead of radiation counting is obvious in many cases. It is easy to find that the counting time or the amount of sample needed to obtain a given uncertainty in the measurement of radionuclide concentrations at environmental levels is smaller in mass spectrometry than in radiometric counting. In Table 16.2, we present the time needed to obtain 1% uncertainty in the measurement of some $10^{-1}$ mBq/g for several radionuclides both by mass spectrometry and by counting the emitted radiation. In Chap. 18, Table 18.6, the sample amount necessary to obtain 1% uncertainty is also given by mass spectrometry and radiation counting. The advantage of using mass spectrometry seems apparent.

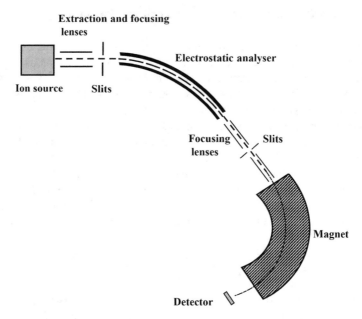

**Fig. 16.2** Schematic presentation of a mass spectrometer as described in the text

**Table 16.2** Counting time needed to determine approximately $10^{-1}$ mBq/g for some radionuclides either by mass spectrometry or radiation counting. The radiation counting efficiency is taken
as 1, the time needed by radiation counting is approximately 3 y, while the efficiency for mass
spectrometry is taken as 0.01%, which is quite close to an average experimental value

| Radionuclide | $T_{1/2}$ (y) | Number of atoms | Counting time Mass spectrometry (s) |
|---|---|---|---|
| $^{10}$Be | $1.51 \times 10^6$ | $6.8 \times 10^9$ | 0.015 |
| $^{14}$C | 5700 | $2.6 \times 10^7$ | 3.85 |
| $^{36}$Cl | $3.01 \times 10^5$ | $1.36 \times 10^9$ | 0.07 |
| $^{41}$Ca | $9.94 \times 10^4$ | $4.5 \times 10^8$ | 0.22 |
| $^{90}$Sr | 28.90 | $1.3 \times 10^5$ | 769 |
| $^{137}$Cs | 30.08 | $1.4 \times 10^5$ | 714 |

## 16.2.1 Electrostatic and Magnetic Rigidity

When a charged particle passes through a region where an electric or magnetic
field is present, the trajectory deviates depending on the intensity of the field as
well as on the mass, charge, and energy of the particle. Figure 16.3 illustrates this
effect.

Specifically, if the electric or magnetic field is applied in a direction perpendicular to the plane of the particle trajectory, it is very easy to find that the radii
of the new trajectories, $\rho_E$ and $\rho_{B,}$ provoked by the application of an electric or

**Fig. 16.3** Charged-particle trajectories under E and B fields applied perpendicular to the plane of the particle movement

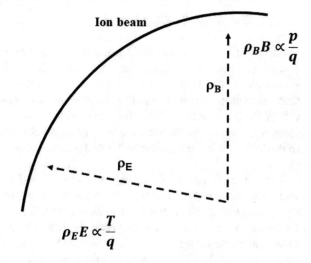

magnetic field, respectively, are

$$\rho_E = \frac{Mv^2}{qE} \tag{16.1}$$

and

$$\rho_B = \frac{Mv}{qB} \tag{16.2}$$

$M$ and $q$ are the mass and charge of the particle, respectively, while $E$ and $B$ are the applied electric and magnetic fields, respectively. The velocity of the particle is $v$.

The quantities $\rho_E E$ and $\rho_B B$ are called electric and magnetic rigidity, respectively, and it is very clear that

$$\rho_E E \propto \frac{T}{q} \tag{16.3}$$

and

$$\rho_B B \propto \frac{p}{q} \propto \frac{\sqrt{MT}}{q} \tag{16.4}$$

In other words, once the electric or magnetic rigidities are set in a mass spectrometer, we select the particle kinetic energy, $T$, or the momentum, $p$. More specifically, we are selecting a value of $\frac{T}{q}$ and $\frac{M}{q}$. Under these conditions, the particles with the selected $T$, $M$, and $q$ will reach the detector, which is placed in a position according to the expected $\rho_E$ or $\rho_B$.

## 16.2.2 The Mass-Energy Plane

However, things are more complicated in the real world. For instance, the kinetic energy of the particles leaving the ion source will follow a distribution rather than a constant value. This means that particles with the same M will not have the same electric or magnetic rigidity. In addition, the charge of the particles at the exit of the ion source will also follow a distribution. For that, it is not possible to construct a mass spectrometer based only on an electric field or a magnetic field. In contrast, it is more convenient to combine the action of both fields on a charged beam.

Let us plot $\frac{M}{q}$ against $\frac{T}{q}$, as shown in Fig. 16.4. If we had a mass spectrometer with only one electrostatic analyzer, it can be seen that many combinations of $M$ and $q$ are compatible with the same electric rigidity, i.e., with $\frac{T}{q}$ constant (16.3). On the other hand, setting the magnetic rigidity to a constant value, it is very easy to see that there are many possible combinations of $M$, $T$, and $q$ that give the same $\rho_B B$ value (see 16.4). This would be the case for a mass spectrometer based only on a magnetic field. However, by setting both $\rho_E E$ and $\rho_B B$ to known values, we define a point in the mass-energy plane, as shown in Fig. 16.4. Mathematically, the interception of both curves, i.e., the combination of an electric field and a magnetic field defines a $(\frac{M}{q}, \frac{T}{q})$ point in the mass-energy plane without ambiguity. This means that we should expect to count the selected mass, M, at the detector of our mass spectrometer.

As commented before, the importance of mass spectrometry has been crucial for the development of Nuclear Physics as it has allowed the measurement of atomic and nuclear masses, as well as the introduction of the concept of isotopes [1]. Pioneering spectrometers were based on the ideas described previously, that is, in the combination of electrostatic and magnetic deflection. The first mass spectrograph, the Aston mass spectrograph (see Fig. 16.5a), was based on the concept of velocity

**Fig. 16.4** $(\frac{M}{q}, \frac{T}{q})$ for a mass spectrometer. Many $\frac{M}{q}$ values are possible for a fixed electric rigidity. In contrast, many different $M$, $T$, and $q$ combinations make the same magnetic rigidity possible. The intersection of both curves fixed a point in the $(\frac{M}{q}, \frac{T}{q})$ plane

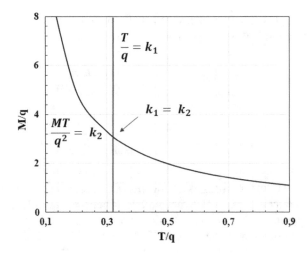

focusing [2], while the next generation of mass spectrographs, represented by the Mattauch-Herzog mass spectrograph (see Fig. 16.5b), was based on the concept of double focusing on mass and direction [2–4]. In both cases, the analyzed masses were recorded on photographic plates. For that, they are called mass spectrographs. Compared to modern mass spectrometers, both instruments gave poor mass resolution, some 1% in the case of the Aston concept and 100 times better for Mattauch instruments. Real mass spectrometers appeared when photographic plates were substituted by current collectors. Dempster and Niers spectrometers are examples of such spectrometers [5, 6]. Currently, other detection systems, such as Faraday cups or electron multipliers, channel plates, channeltrons, photomultipliers, and others, are used [2].

To compare the ability to distinguish masses among mass spectrometers, the so-called mass resolution, $R$, is defined as [7]

$$R = \frac{M}{\Delta M} \tag{16.5}$$

$\Delta M$ is the mass difference between two masses that can be separated, and $M$ is the observed mass. Thus, taking $\Delta M = 1$, $R$ will give the maximum mass that can be distinguished from its neighboring mass. A mass spectrometer with $R = 600$

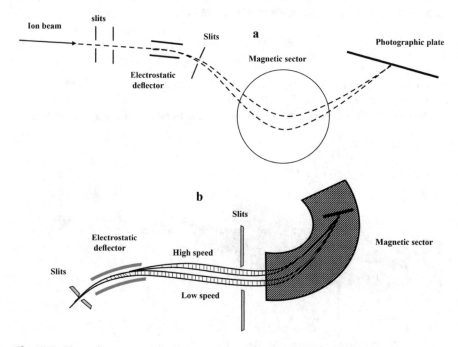

**Fig. 16.5** Pioneering mass spectrometers. **a** Aston mass spectrograph; **b** Mattauch mass spectrograph

will not be able to distinguish masses beyond 600. In contrast, it is expected that the instrument can adequately resolve smaller masses.

## 16.2.3 The Dynamic Approach

The use of static fields provides methods with good or very good mass resolution. However, achieving static fields, especially static magnetic fields, is not easy from a technical point of view. Of course, there are currently solutions for the problem, so it is possible to use this approach. Nevertheless, there exists another approach to the problem that avoids the use of static magnetic fields. We can call it the dynamic approach, and it is based on the use of radiofrequency fields.

In fact, it was at the beginning of the 1950s that the utility of radiofrequency quadrupole fields for mass analysis was recognized [8, 9]. Quadrupole mass filters (QMFs) have many advantages, the main one being the absence of static magnetic fields and the possibility of selecting the desired mass electronically. QMF-based mass spectrometry has a lower mass resolution than the static field approach, although its handling is much simpler and the results are sufficiently good for many applications. In fact, today, QMF is the most common option for mass selection in most of the so-called low-energy mass spectrometers [10].

Basically, a QMF consists of four parallel electrodes, as shown in Fig. 16.6. Two opposite electrodes are connected to a potential

$$\Phi = [U + V \cos (\omega t)] \tag{16.6}$$

$U$ is a direct current voltage, $V$ is the alternate voltage amplitude, $\omega$ is the angular frequency, and $t$ is the time. The other pair of opposite electrodes are connected to a potential $-\Phi$. The selection of the applied voltages and of the angular frequency

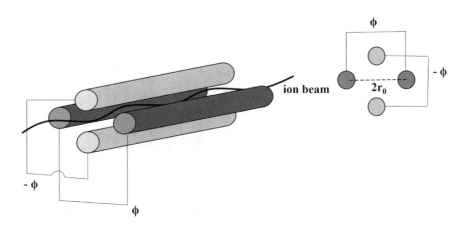

**Fig. 16.6** A quadrupole mass filter

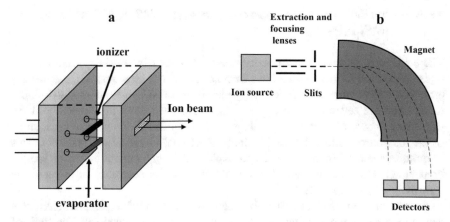

**Fig. 16.7**  **a** Schematic view of an ion source in a TIMS system; **b** a basic view of a complete TIMS system

permits the passage of the ions of interest, i.e., those of a given $\frac{M}{q}$ ratio. The remaining ions are thrown out of the QMF central path. This is a consequence of the solution of the ion motion equations inside the QMF. Let us suppose that the charge state is $q = 1$; then,

$$\frac{d^2x}{dt^2} + \frac{2e}{Mr_0^2}(U + V \cos \omega t)x = 0$$

$$\frac{d^2y}{dt^2} - \frac{2e}{Mr_0^2}(U + V \cos \omega t)y = 0 \tag{16.7}$$

$$\frac{d^2z}{dt^2} = 0$$

$e$ is the electron charge, and $M$ is the mass of the ion of interest. According to Fig. 16.6, $r_0$ is half of the distance between the electrodes, and it is interesting to note that there is no movement along the z-axis. Equation 16.7 are known as Mathieu equations [10], which have stable solutions for the ion trajectories only for determined values of $U$, $V$, and $\omega$. It can also be found that the mass resolution is $\propto \frac{U}{V}$ for QMF.

According to the results obtained, a QMF spectrometer can be operated in two modes at a fixed $R$. Keeping $\omega$ constant, $\frac{M}{q}$ can be selected by changing $U$ and $V$, which allows one to obtain a mass spectrum. Lighter or heavier ions than those selected will have unstable trajectories along the x-axis or y-axis. This is the most common operation. Alternatively, $U$ and $V$ can be fixed and $\omega$ can be varied to tune the QMF for a given $\frac{M}{q}$. The rejection of lighter and heavier species follows the same criteria.

## 16.3   Low-Energy Mass Spectrometers: TIMS, SIMS, GDMS, RIMS, ICP–MS

The sensitivity of mass spectrometry makes it a very interesting technique that finds application in many disciplines such as Chemistry, Geology, Art, Archaeometry, and Environmental Sciences in general. To accomplish so many different tasks, there are several designs. The difference among them lies basically in the ion source, which is selected depending on the application. As we shall see, some of the ion sources allow the sample analysis without chemical pretreatment. This is quite convenient in many cases to avoid cross-contamination during sample preparation. However, in most cases, radiochemistry is necessary to increase the sensitivity of the experiment.

Several of the techniques described here are more suitable for multielemental determinations, whereas others are for monoelemental applications. The limit of detection (LLD, see Chap. 14) varies from one mass spectrometer to another, but it can be estimated to range from ng to pg levels. We will pay attention to those spectrometers most commonly used for environmental radioactivity determination, which is the objective of the book. Needless to say that there exist some other concepts not included here because thus far they have not been applied to our problem.

*Thermal Ionization Mass Spectrometry (TIMS)*
Thermal Ionization Mass Spectrometry is based on the thermionic effect. In this way, ions are obtained by heating the sample. In practical terms, the ion source in a TIMS spectrometer is usually a filament where the sample solution is deposited. On other occasions, there are more filaments, one of them being used for the evaporation of the sample and the other (or others) for the ionization itself. A basic scheme of a TIMS spectrometer is given in Fig. 16.7. Assuming that the energy distribution from the ion source is narrow, many TIMS systems use only a sector magnet to select the mass of interest [11]. In many cases, this is not enough to quantify the less abundant isotopes, and an electrostatic analyzer, or a quadrupole, is used as an additional mass filter [12].

According to the Saha-Langmuir equation [12], the population of positive ions, when heating a sample, follows an exponential function such as

$$\frac{N_+}{N_0} \propto e^{\left[\frac{(W_e - I_p)}{kT}\right]} \tag{16.8}$$

where $N_+$ and $N_0$ are the number of positive ions and neutral atoms or molecules after heating at a temperature $T$, respectively; $W_e$ is the work function for electron emission from the metal surface; and $I_p$ is the ionization potential. $K$ is the Boltzmann constant.

Under adequate vacuum conditions and using Ta, Re, and W filaments, which are refractory elements with high $W_e$, most elements with $I_p$ below 7 eV may be ionized at temperatures from 800 to 2000 °C. For that, this technique can be considered a multielement technique since the variation of T would give the possibility of ionizing most of the elements. It is also possible to obtain a certain distribution of negative ions, $N^-$, when a sample is heated. Now

$$\frac{N_-}{N_0} \propto e^{\left[\frac{(E_e - W_e)}{kT}\right]} \tag{16.9}$$

$E_a$ is the electron affinity of the element. In this way, halogen-negative ions can be produced using low-We filaments. This use of TIMS is not common [11].

*Secondary Ion Mass Spectrometry (SIMS)*
SIMS consists of bombarding a sample with an ionic current, the primary ion beam. It sputters negative or positive ions, or neutral species, the secondary ion beam, from the sample. The detached species are analyzed for masses with a combination of electric and magnetic fields according to the well-known procedure. A scheme of this dispositive is given in Fig. 16.8. The secondary ion beam is counted with a channeltron device, a channel plate scintillator-photomultiplier, or a Faraday cup [13].

SIMS is especially efficient for analyzing solid samples. In fact, it is used for direct sample analysis without the need for a previous chemical treatment. SIMS may also provide information on the distribution of radionuclides along the surface of the sample or even at depth. $Ar^+$, $O_2^+$, $O_2^-$, $Cs^+$, and $Ga^+$ beams are commonly used to irradiate samples at energies up to 20 keV [13]. High primary ion currents

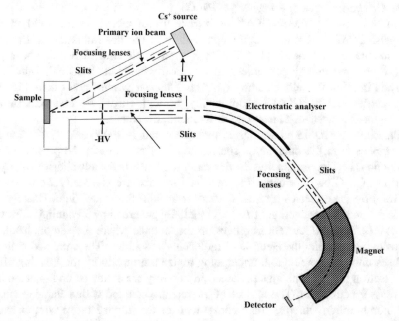

**Fig. 16.8**  Schematic view of a SIMS system

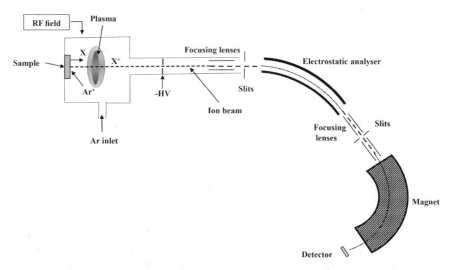

**Fig. 16.9**  Schematic view of a GDMS system

are normally used for elemental depth profiling studies. Under these conditions, the instrument provides 3D information, which acts as a microprobe with a spatial resolution of approximately 20 μm. Low-current primary ion SIMS gives surface information about the sample.

*Glow Discharge Mass Spectrometry (GDMS)*
As in the case of SIMS, GDMS is a very useful technique for solid sample analysis. GDMS has found application mainly in the determination of traces in semiconductors and metals [14]. However, some applications have also been performed on nonconducting samples and specifically in the nuclear field. Thus, trace analysis in nuclear fuels, cladding materials, nuclear wastes, and others [15] has been carried out by GDMS. Additionally, as with SIMS, no chemical sample pretreatment is needed. This is one of its main advantages.

Basically, a GDMS is similar to that presented in Fig. 16.9. The difference in comparison to the other mass spectrometry systems lies in the process of ion production, i.e., the ion source. In this case, a GDMS takes advantage of the production of a Glow Discharge in a gas at low pressure (typically from 0.01 to 10 Torr) by the effect of a direct current or a radio frequency field. Usually, the gas is Ar, and the production of $Ar^+$ is essentially governed by a Penning ionization effect [16]. The $Ar^+$ current impinges on the cathode where the sample has been deposited, generating the ejection of molecules and atoms. The sputtered material diffuses into the glow region, undergoing ionization mostly by the Penning effect and electron impact. The ions produced in this way are extracted and injected into the mass spectrometer. The masses of interest are counted with a Faraday cup or a photomultiplier, which is more convenient for the minor isotope component of the beam.

*Resonance Ionization Mass Spectrometry (RIMS)*

In the case of RIMS, the sample must be thermally evaporated prior to ion production. The evaporated atoms, either in the ground state or in low-lying excited states due to heating, are then irradiated by a laser beam tuned at energies that cause their resonant excitation in one or two steps [12].

The highly excited atoms are ionized by photons either directly or indirectly by exciting the atom to an autoionizing state that decays, emitting one electron. This causes the atom to ionize. In both cases, the ionization is very selective since the laser absorption is resonant, taking place at a characteristic energy of atom one of interest. In Fig. 16.10, we depict a representation of the two-step ionization. It is apparent that RIMS is not a multielement technique. In contrast, it is highly selective since the common bandwidth of pulsed lasers is $10^{-4}$ eV, or even much lower in the case of continuous-wave lasers (cw lasers). Indeed, being the density of atomic excited levels, respectively, 1 eV$^{-1}$ and 100 eV$^{-1}$ at low and high energies, with typical line widths of approximately $10^{-8}$ eV, it is easy to understand that virtually there will not be any interference of unwanted atomic species [17]. Consequently, the suppression of isobars and neighboring masses is almost complete. The sources of background are other atomic species ionized during sample heating or by nonresonant photoionization.

Once the ions are produced, they are mass analyzed by a time-of-flight setup (see Chap. 12), which is sufficient in many cases (see Fig. 16.11). The laser pulse provides the start signal for the time analysis. This system provides enough sensitivity for many radionuclide determinations at environmental levels [18], including the measurement of isotope ratios [19]. However, the use of cw lasers and the coupling of a quadrupole mass selector at the exit of the ion source allows the determination of radionuclides at ultratrace levels.

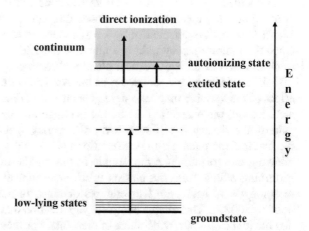

**Fig. 16.10** Excitation process in a RIMS ion source

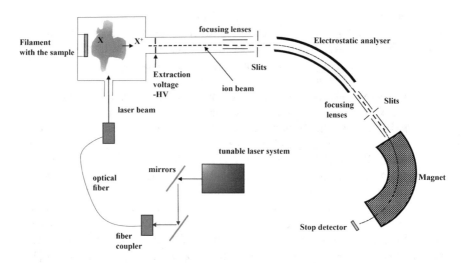

**Fig. 16.11** The RIMS system included a cw laser-ion source and a quadrupole mass selector. The mass analysis is carried out by a TOF system. The start signal is provided by the laser pulse, while the stop signal is provided by the stop detector

## Inductively Coupled Plasma–Mass Spectrometry (ICP–MS)

ICP–MS is likely the most popular mass spectrometry method and by far the most applied to the determination of radionuclides and their isotopic ratios in the environment. The reason is the high ionization efficiency obtained with ICP-based ion sources and the possibility of simultaneous ionization of many elements, which gives a multielemental character to the technique. Ionization is produced by the heat provided by an Ar plasma in the so-called plasma torch. Usually, the torch is made of three concentric quartz tubes through which the Ar is fluxed at 8–20 L/min. The torch is surrounded by a coil connected to a radio frequency generator. Once the radiofrequency field is applied, a variable magnetic field appears in the torch oriented at the coil vertical plane and traversing the Ar gas. A spark from a Tesla coil initiates the ionization of the Ar. The cations and electrons under the effect of the variable magnetic field generate additional ions, and a plasma is created. Should the magnetic field be high enough and the gas flows in a symmetric pattern, the plasma is self-sustained [20]. A very specific feature of this setup is the fact that the main plasma current flows close to the external walls of the torch, allowing one to introduce the sample in the middle of the plasma torch without disturbing effects from the plasma itself. With typical radiofrequency generators of some 40–50 MHz, the temperatures obtained in the plasma torch range from 4500 to 8000 K. This allows complete vaporization and atomization of the sample and the ionization of a wide range of elements. For that, the technique has a multi-element character. An ICP ion source is presented in Fig. 16.12, while a complete ICP–MS is depicted in Fig. 16.13.

The sample is extracted from the ion source in a two-stage process. A low-pressure gradient separates the sampler region from the torch, allowing us to

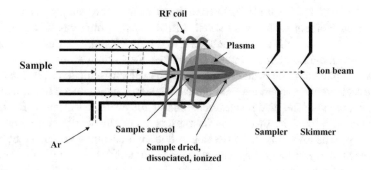

**Fig. 16.12** Schematic view of an ICP ion source

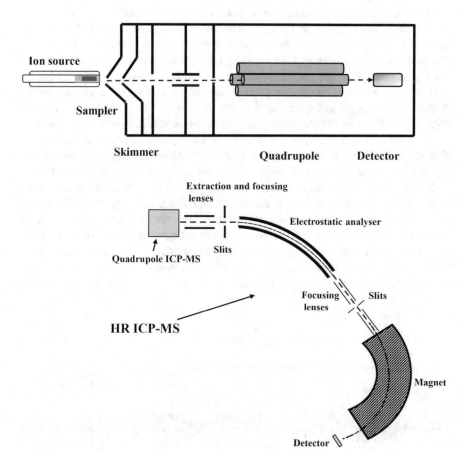

**Fig. 16.13** Schematic view of an ICP–MS

extract the ions by a simple difference in pressure. The sampler is a metallic cone-shaped element with a very small orifice to avoid high gas flux in this first low-pressure zone. This area is connected to a high vacuum area through another cone-shaped metallic element called a skimmer. The skimmer orifice is also very small and is placed in the center of the gas jet. An electrostatic lens helps to extract the cations from the gas, which after a focusing step are introduced in a quadrupole for mass analysis. The electrostatic lens allows for a first selection of the kinetic energy because the ions in the gas jet present a certain distribution of energies. In some systems after this first selection and after focusing, a second electrostatic lens is used to reject neutral species and make an additional selection of energies before the introduction of the ion beam in the quadrupole. The selected mass or masses are usually counted with an electron multiplier [12].

The use of quadrupoles as mass selectors is the conventional option in ICP–MS. Nevertheless, the capacity to reject isobars and the mass resolution of a conventional ICP–MS instrument could be insufficient for some applications in the field of environmental radioactivity. As a solution, a double-focusing mass spectrometer is coupled to an ICP ion source, resulting in high resolution ICP–MS (HR-ICP–MS). In this way, a combination of an electrostatic sector and a magnet serves to determine without ambiguity a $\frac{M}{q}$ point in the $(\frac{M}{q}, \frac{T}{q})$ plane. These systems significantly improve the mass resolution of conventional ICP–MS. Most recently, the concept of a collision/reaction cell (Fig. 16.14) cell has been introduced in ICP–MS [21]. These cells are filled with a gas, usually $O_2$, $N_2$, $NH_3$, and others, and the ion beam is passed through the cell normally after a first mass selection with a quadrupole. The change in ion trajectory caused by collisions with the gas or the production of new compounds by chemical reactions provokes the deviation and rejection of many interferences from the main ion beam. After passing through the cell, the beam is injected into a second quadrupole where a new mass selection is produced. The improvement in mass sensitivity is apparent (see Table 16.3), and the ability to measure isotopic ratios has increased greatly compared to conventional ICP–MS systems [21, 22].

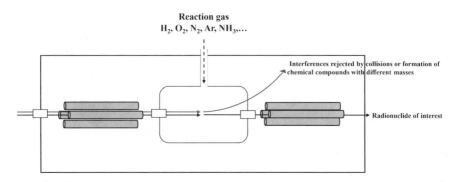

**Fig. 16.14**  A collision/reaction cell in an ICP–MS system

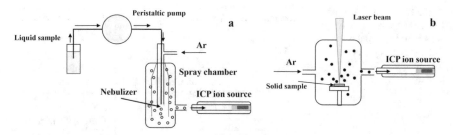

**Fig. 16.15** **a** Pneumatic Nebulizers and Ultrasonic Nebulizers; **b** Laser Ablation system. Both types of sample introduction systems are used for ICP–MS

However, a crucial step in ICP–MS is the introduction of the sample. Gas, liquid, or solid samples are accepted in an ICP–MS, although the common sample form is liquid or solid in environmental radioactivity. Liquids are usually introduced in the ion source as fine aerosols. For that, a basic or acid solution of the sample is introduced into a nebulizer to produce fine particles. They are vaporized and atomized within the plasma torch, provided they are small enough. There are several procedures for producing the nebulization of the sample. In the so-called pneumatic nebulizers, the sample is absorbed by a capillary tube with a peristaltic pump and exposed to a high-speed gas stream. It produces very fine droplets that are carried toward the plasma torch [20]. Another procedure uses ultrasonic nebulizers in which the liquid sample is introduced into a piezoelectric transducer that breaks the liquid air surface under the action of an ultrasonic generator (200 kHz to 10 MHz frequency) that drives it (Fig. 16.15a). Droplets are produced at a higher rate than in the case of pneumatic nebulizers [12]. For pneumatic and ultrasonic systems, additional variables appear that have to be optimized to increase the sensitivity of the ICP–MS and the interference rejection. In fact, the flow of the liquid and carrier gas or the ultrasound frequency could be an important parameter to consider when optimizing the whole system.

Solids are introduced directly without previous chemical dissolution by using Electrothermal Vaporization (ETV-ICP-MS) or Laser Ablation (LA-ICP-MS) techniques. ETV-ICP-MS [23] allows for direct volatilization of the solid sample. The aerosol produced is introduced into the plasma torch. This system increases the sensitivity of the ICP by some two orders of magnitude compared to that of LA-ICP–MS, but it is very expensive.

LA-ICP–MS is the most popular technique for the analysis of solid samples [24]. As shown in Fig. 16.15b, the sample is irradiated by a Nd-YAG laser. The material is vaporized under the action of the laser beam and is introduced into the plasma torch with an Ar stream. In principle, this technique can be used for many types of samples. On the other hand, as the laser beam can be focused on a given area in the sample, LA can provide information on the spatial distribution of radionuclides in the sample [25]. More examples of introduction designs can

**Table 16.3** Comparison of the mass resolution and sensitivity among the different concepts of low-energy mass spectrometry

| Radionuclide | Technique | LLD (mBq) or MDC $(mBqml^{-1}$ or $g^{-1})$ |
|---|---|---|
| $^{41}Ca$ | RIMS | 1–30 mBq |
| $^{90}Sr$ | RIMS<br>Reaction Cell-ICP–MS<br>HR-ICP–MS<br>GDS-MS | 1.5–4 mBq<br>500 mBqml$^{-1}$<br>55 mBqml$^{-1}$ in waters<br>50 mBqg$^{-1}$ in soils |
| $^{99}Tc$ | RIMS<br>HR-ICP–MS<br>ICP–MS | $20 \times 10^{-3}$ mBq<br>0.05–0.3 mBq<br>10 mBq |
| $^{129}I$ | Reaction Cell-ICP–MS<br>Reaction Cell-ICP–MS<br>HR-ICP–MS | 0.037 mBqml$^{-1}$ in waters<br>0.003 mBqg$^{-1}$ in sediments<br>0.015 mBqml$^{-1}$ in waters |
| $^{135}Cs$ | HR-ICP–MS<br>Reaction Cell-ICP–MS<br>TIMS | $5 \times 10^{-4}$ mBqml$^{-1}$ in riverwater<br>mBqml$^{-1}$ in rainwater<br>$6 \times 10^{-4}$ mBqg$^{-1}$ in soils |
| $^{137}Cs$ | HR-ICP–MS<br>Reaction Cell-ICP–MS | 23 mBqml$^{-1}$ in riverwater<br>19 mBqml$^{-1}$ in rainwater |
| $^{226}Ra$ | TIMS<br>ICP–MS<br>HR-ICP–MS | 0.037 mBq<br>0.1–0.5 mBq<br>0.002 mBqml$^{-1}$ in seawater |
| $^{228}Ra$ | TIMS | 12 mBq |
| $^{230}Th$ | ICP–MS | 3.4 mBqml$^{-1}$ in riverwater |
| $^{232}Th$ | ICP–MS<br>HR-ICP–MS<br>Reaction Cell-ICP–MS | 0.003 mBqml$^{-1}$ in riverwater<br>$3 \times 10^{-6}$ mBqml$^{-1}$ in natural water |
| $^{235}U$ | ICP–MS<br>HR-ICP–MS | $2 \times 10^{-6}$ mBqml$^{-1}$ in riverwater |
| $^{236}U$ | ICP–MS | $2 \times 10^{-5}$ mBqml$^{-1}$ in riverwater |

(continued)

**Table 16.3** (continued)

| Radionuclide | Technique | LLD (mBq) or MDC $(mBqml^{-1}$ or $g^{-1})$ |
|---|---|---|
| $^{238}U$ | ICP–MS | $3 \times 10^{-5}$ mBqml$^{-1}$ |
| | HR-ICP–MS | in riverwater |
| | TIMS | $1.3 \times 10^{-5}$ mBqml$^{-1}$ |
| | | in riverwater |
| | | 31 mBqg$^{-1}$ in soil |
| $^{239}Pu$ | TIMS | • mBq |
| | HR-ICP–MS | • mBq |
| | RIMS | • mBq |
| | ICP–MS | 0.78 mBqml$^{-1}$ in |
| | HR-ICP–MS | fresh waters |
| | HR-ICP–MS | 0.03 mBqml-1 in |
| | | seawater |
| | | 0.06 mBqg-1 in |
| | | sediments |
| $^{240}Pu$ | TIMS | • mBq |
| | HR-ICP–MS | • mBq |
| | RIMS | • mBq |
| | ICP–MS | 3.6 mBqml$^{-1}$ in |
| | HR-ICP–MS | fresh water |
| | HR-ICP–MS | $7 \times 10^{-4}$ mBqml$^{-1}$ |
| | | in seawater |
| | | 0.001 mBqg-1 in |
| | | sediments |
| $^{241}Pu$ | TIMS | 0.4 mBq |
| | HR-ICP–MS | 11.9 mBq |
| | RIMS | 20 mBq |
| | HR-ICP–MS | 0.36 mBqg$^{-1}$ in |
| | | sediments |
| $^{241}Am$ | HR-ICP–MS | 0.013 mBqg$^{-1}$ in soil |

be found in the literature. For example, coupling an ICP–MS to a Liquid Chromatographer combines the selectivity of Liquid Chromatography with ICP-MS [26].

As will be seen in the next section, all the different mass spectrometry methods described above have been applied intensively to the determination of radionuclides in the environment. Depending on the Physics behind the method and the instrumentation used, they are more or less sensitive. In Table 16.3, a comparison is given on the sensitivity of the methods described in this section [22, 27].

## 16.4 Applications to Environmental Radioactivity

There are many examples in the literature on the application of mass spectrometry to the determination of radionuclides in the environment. Some good reviews can be found in [12, 21, 22]. ICP–MS is by far the most used in environmental studies, but other mass spectrometry techniques have succeeded in some specific applications.

TIMS has been applied to the determination of Pu isotopes in urine samples from radiation-exposed workers [28]. The characterization of the $\frac{^{237}Np}{^{239}Pu}$ and $\frac{^{240,241,242}Pu}{^{239}Pu}$ global fallout signatures of isotope ratios was carried out in [29]. Regarding the Fukushima accident (see Chap. 4), measurements of U and Pu isotopes, as well as $^{135}$Cs and $^{137}$Cs, have also been determined in soils and plant samples (see Chap. 4) by TIMS [30, 31].

Few examples can be found for SIMS, but there are still some. SIMS has been used in the determination of U isotopes [32–34] in hot particles (see Chap. 5). An interesting application consisted of the determination of U isotopes in particles from the Kosovo area that identified depleted uranium in the composition of the samples [35, 36]. $^{129}$I, $^{137}$Cs, and $^{238}$U have been determined in soils and grass samples [37].

The applications of GDMS to environmental radioactivity determination have been concentrated in the problem of nuclear fuels, waste, cladding materials, etc. An interesting paper is that presented in [38]. More closely related to environmental measurements, we can find results on the determination of U isotopes in soils [39, 40], as well as $^{237}$Np in the Irish Sea [41].

Some specific environmental applications of RIMS can be found. Taking advantage of the isotope shifts at the atomic level, it was possible to measure $^{89}$Sr and $^{90}$Sr in air samples [42] using a continuous-wave laser source. Pulsed laser RIMS has been applied to the determination of actinides, mainly Pu isotopes, in environmental samples [43, 44]. Furthermore, the determination of $^{99}$Tc at environmental levels was carried out in [43]. Here, the laser-ion source was coupled to a double-focusing mass spectrometer.

As described above, ICP–MS is perhaps the most popular mass spectrometric technique for the determination of radionuclides in the environment. Thus, it is possible to find many papers in the scientific literature devoted to the application of ICP–MS under different setups to environmental radioactivity. Some examples follow.

HR-ICP–MS has been applied to determine Pu isotope ratios in Semipalatinsk soil samples [45]. The U isotopes and their ratios have been measured in soils affected by the Chernobyl accident [46]. Conventional quadrupole ICP–MS has found room in the study of many aspects of environmental studies. For example, fission products, $^{137}$Cs and $^{90}$S, as well as actinides have been determined in nuclear waste [47]. In NORM-contaminated soil samples, U isotopes have also been measured [48].

$^{99}$Tc has been determined in soil, atmospheric deposition, seawater, and vegetation samples [49–53].

A very interesting review is published on the application of different mass spectrometry methods to the determination of radionuclides in environmental samples affected by the Fukushima accident (see Chap. 4) [21]. Among the information provided in the paper, the applications of ICP–MS stand out. Specifically, we used

HR-ICP–MS or ICP–MS in combination with collision/reaction cells. Thus, it can be seen how this technique is applied for the determination of $^{90}$Sr, $^{135}$Cs, $^{129}$I, $^{132}$Te, $^{137}$Cs, U isotopes, Pu isotopes, and $^{241}$Am in a variety of environmental samples [54–60].

**Exercises**

1. Calculate the number of nuclei in Table 16.1 corresponding to the activities in Fig. 16.1.
2. Find the counting time needed to determine approximately 1 mBq/g for some radionuclides by mass spectrometry or radiation counting, according to the conditions of Table 16.2.
3. Find (16.1) and (16.2).
4. Justify (16.3) and (16.4).
5. Equation 16.5 gives the mass resolution of a mass spectrometer. Discuss the range of utility of some spectrometers with R = 50, 100, 200, and 400. Suppose that it is necessary to determine the $\frac{^{240}Pu}{^{239}Pu}$ atom ratio in an atmospheric sample. Choose the appropriate spectrometer(s) to perform the measurement and discuss the reasons.
6. Make the selection now to detect $^{99}$Tc in environmental waters. Discuss the choice.
7. A Ca isotope of mass 39.96 amu will be analyzed by a mass spectrometer. In the first stage, it is ionized by an electron beam to Ca$^+$. In the second stage, the ion is accelerated by applying a potential difference of 1500 V. The ion beam is transported toward a magnet where a 2.5 T uniform magnetic field is applied.
   (a) Find the equation to calculate the ion speed and its trajectory radius at the exit of the magnet.
   (b) Calculate the velocity and radius of the trajectory of the ion beam.
8. Explain what happens when we increase the extraction voltage. Does the radius increase or decrease? If the voltage increase is twice, does the radius also increase in the same proportion? Explain what happens to the radius of when the magnetic field increases. If this increase is twice, how does the radius change?
9. A magnetic field-based mass spectrometer is used to determine $^{12}$C in a sample. For that, $^{12}$C+ is extracted from an ion source with an extraction potential V. It is observed that the $^{12}$C+ ions are collected after describing a trajectory with a radius of 9 cm. Unexpectedly, another unknown ion was collected with a radius of 10.4 cm. Find its mass and guess the nuclear species. Suppose that both ions are injected at the same charge state. Of course, the applied magnetic field is the same.
10. The figure shows a mass spectrograph based on the deviation suffered by the nuclei of charge q when passing through a uniform magnetic field B perpendicular to the trajectory of the particle. Each nucleus of mass m follows a path whose radius of curvature is r, which can be found with the attached formula,

where v is the velocity of the nucleus. However, all nuclei with the same v and m/q ratio will reach the same position on the photographic plate. In other words, the method has some abilities when measuring masses. Design a method to unambiguously determine the mass of nuclear species using nuclear detectors instead of photographic plates.

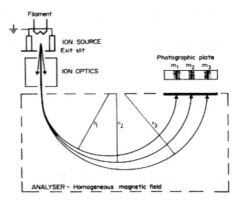

FIG. 2.2. The principle of the mass spectrograph (-meter, if the photographic plate is replaced by an ion current meter).

$$q/m = v/Br$$

11. Sketch the mass spectrum obtained for a natural U sample, a 20% U enriched sample, and a 0.2% depleted U sample. Explain the differences.

# References

1. F. Aston, Isotopes and atomic weights. Nature **105**, 617–619 (1920)
2. G. Münzenberg, Development of mass spectrometers from Thomson and Aston to present. Int. J. Mass Spectrom. **349–350**, 9–18 (2013)
3. J. Mattauch, R. Herzog, Über einen neuen Massenspektrographen. Z. Physik **89**, 786–795 (1934)
4. I. Takeshita, Mattauch-Herzog type mass spectrograph with a two-stage electrostatic field. Z. Naturforschg. **21**, 14–25 (1966)
5. A.J. Dempster, A new method of positive ray analysis. Phys. Rev. **11**, 316 (1918)
6. A.O. Nier, A mass spectrometer for routine isotope abundance measurements. Rev. Sci. Instrum. **11**, 212 (1940)
7. T.R. Ireland, Invited review article: recent developments in isotope-ratio mass spectrometry for geochemistry and cosmochemistry. Rev. Sci. Instrum. **84**, 011101 (2013)
8. W. Paul, H. Steinwedel, Ein neues Massenspektrometer ohne Magnetfeld Z. Naturforsch., Teil A **8**, 448 (1953)
9. R.F. Post, University of California Radiation Laboratory, Report No. UCRL-2209, Berkeley, CA (1953)
10. J.E. Campana, Elementary theory of the quadrupole mass filter. Int. J. Mass Spectrom. **33**, 101–117 (1980)

11. S. Bürger, J. Vogl, U. Kloetzli, L. Nunes, M. Lavelle, Thermal ionisation mass spectrometry, in *New Developments in Mass Spectrometry No. 3: Sector Field Mass Spectrometry for Elemental and Isotopic Analysis*, ed. by T. Prohaska, J. Irrgeher, A. Zitek, N. Jakubowski, Chapter 14 (Royal Society of Chemistry 2015), pp. 385–442

12. C. Walther, K. Wend, Radioisotope mass spectrometry, in *Handbook of Radioactivity Analysis*, vol. 1, 4th edn. Chapter 8 (Academic Press, 2020), pp. 861–898

13. A. Benninghoven, K.T.F. Janssen, J. Tümpner, H.W. Werner, *Secondary Ion Mass Spectrometry* (SIMS VIII, Wiley, 1992)

14. V. Hoffmann, M. Kasik, P.K. Robinson, C. Venzago, Glow discharge mass spectrometry. Anal. Bioanal. Chem. **381**, 173–188 (2005)

15. M. Betti, L.A. de las Heras, Glow discharge spectrometry for the characterization of nuclear and radioactively contaminated environmental simples. Spectrochim. Acta, Part B **59**(9), 1359–1376 (2004)

16. D. Carazzato, M.J. Bertrand, Characterization of a glow discharge ion source for the mass spectrometric analysis of organic compounds. J. Am. Soc. Spectrom. **5**, 305–315 (1994)

17. M.R. Savina, M.J. Pellin, E. Tripa, I.V. Veryovkin, W.F. Calaway, A.M. Davis, Analyzing individual presolar grains with CHARISMA. Geochim. Cosmochim. Acta **67**, 3215–3225 (2003)

18. J.V. Kratz, Ultratrace analysis of long-lived radionuclides by resonance ionization mass spectrometry (RIMS). Radioanal. Nucl. Chem. **303**, 1361–1366 (2015)

19. N. Trautmann, G. Passler, K.D.A. Wendt, Ultratrace analysis and isotope ratio measurements of long-lived radioisotopes by resonance ionization mass spectrometry (RIMS). Anal. Bioanal. Chem. **378**, 348–355 (2004)

20. M. Schönbächler, Inductively coupled plasma–mass spectrometry (ICP–MS), in *Encyclopedia of Geochemistry: A Comprehensive Reference Source on the Chemistry of the Earth*, ed. By W.M. White (Springer, 2016), pp. 723–728

21. W. Bu, Y. Ni, G. Steinhauser, W. Zheng, J. Zheng, N. Furutaf, The role of mass spectrometry in radioactive contamination assessment after the Fukushima nuclear accident. J. Anal. At. Spectrom. **33**, 519–546 (2018)

22. X. Hou, P. Roos, Critical comparison of radiometric and mass spectrometric methods for the determination of radionuclides in environmental, biological and nuclear waste simples. Anal. Chim. Acta **608**, 105–139 (2008)

23. J. Comte, P. Bienvenu, E. Brochard, J.-M. Fernandez, G. Andreoletti, Determination of selenium-79 in solutions of fission products after pre-treatment by ion exchange chromatography and ETV-ICP-MS. J. Anal. At. Spectrom. **18**, 702–707 (2003); P. Grinberg, S. Willie, R.E. Sturgeon, Determination of natural Sr and $^{90}$Sr in environmental samples by ETV-ICP-MS. J. Anal. At. Spectrom. **22**, 1409–1414 (2007)

24. A.L. Gray, Solid sample introduction by laser ablation for inductively coupled plasma source mass spectrometry. Analyst **110**, 551–556 (1985)

25. Krachler, Z. Varga, A. Nicholl, M. Wallenius, K. Mayer, Spatial distribution of uranium isotopes in solid nuclear materials using laser ablation multicollector ICP–MS, Microchem. J. **140**, 24–30 (2018)

26. J.W. Olesik, J.A. Kinzer, S.V. Olesik, Capillary electrophoresis inductively coupled plasma spectrometry for rapid elemental speciation. Anal. Chem. **67**(1995), 1–12 (1995)

27. E.R. Gonzales, S.R. Garcia, C. Mahan, W. Hang, Evaluation of mass spectrometry and radiation detection for the analysis of radionuclides. J. Radioanal. Nucl. Chem. **263**, 457–465 (2005)

28. W.C. Inkret, D.W. Efurd, G. Miller, D.J. Roskop, T.M. Benjamin, Applications of thermal ionization mass spectrometry to the detection of $^{239}$Pu and $^{240}$Pu intakes. Int. J. Mass Spectrom. **178**, 113–120 (1998)

29. J.M. Kelley, L.A. Bond, T.M. Beasley, Global distribution of Pu isotopes and $^{237}$Np. Sci. Total Environ. **1999**(237–238), 483–500 (1999)

30. M.S. Snow, D.C. Snyder, N.R. Mann, B.M. White, Method for ultra-trace caesium isotope ratio measurements from environmental samples using thermal ionization mass spectrometry. Int. J. Mass Spectrom. **381–382**, 17–24 (2015)

31. J.A. Dunne, D.A. Richards, H.W. Chen, Procedures for precise measurements of 135Cs/137Cs atom ratios in environmental samples at extreme dynamic ranges and ultra-trace levels by thermal ionization mass spectrometry. Talanta **174**, 347–356 (2017)

32. M. Betti, A.D. Heras, G. Tamborini, Mass spectrometric determination of trasnuranium elements. Appl. Spectrosc. Rev. **41**, 491–514 (2006)

33. G. Tamborini, M. Betti, Characterization of radioactive particles by SIMS. Mikrochim. Acta **132**, 411–417 (2000)

34. Y. Ranebo, M. Eriksson, G. Tamborini, N. Niagolova, O. Bildstein, M. Betti, The use of SIMS and SEM for the characterization of individual particles with a matrix originating from a nuclear weapon. Microsc. Microanal. **13**, 179–190 (2007)

35. O.C. Lind, B. Salbu, K. Janssens, K. Proost, P.R. Danesi, Characterisation of DU particles from Kosovo and Kuwait, in *Radioactive Particles in the Environment*, ed. by D.H. Oughton, V. Kashparov (2009), pp. 57–67

36. S. Torok, J. Osan, L. Vincze, S. Kurunczi, G. Tamborini, M. Betti, Characterization and speciation of depleted uranium in individual soil particles using microanalytical methods. Spectrochim. Acta, Part B **59–5**, 689–699 (2004)

37. O. Lefèvre, M. Betti, L. Koch, C.T. Walker, EPMA and mass spectrometry of soil and grass containing radioactivity from the nuclear accident at Chernobyl. Mikrochim. Acta **13**, 399–408 (1996)

38. M. Betti, Use of a direct current glow discharge mass spectrometer for the chemical characterization of samples of nuclear concern. J. Anal. At. Spectrom. **11**(1996), 855–860 (1996)

39. D.C. Duckworth, C.M. Barshick, D.H. Smith, Analysis of soils by glow discharge mass spectrometry. J. Anal. At. Spectrom. **8**, 875–879 (1993)

40. M. Betti, L.A. de las Heras, Glow discharge spectrometry for the characterization of nuclear and radioactively contaminated environmental samples. Spectrochim. Acta B **59**, 1359–1376 (2004)

41. L. Aldave de las Heras, E. Hmecek, O. Bildstein, M. Betti, Neptunium determination by dc glow discharge mass spectrometry (dc-GDMS) in Irish Sea sediment simples. Anal. At. Spectrom. **17**, 1011–1014 (2002)

42. K. Wendt, G.K. Bhowmick, B.A. Bushaw, G. Herrmann, J.V. Kratz, J. Lantzsch, P. Müller, W. Nörtershäuser, E.W. Otten, R. Schwalbach, U.-A. Seibert, N. Trautmann, A. Waldek, Rapid trace analysis of 89,90Sr in environmental samples by collinear laser resonance ionization mass spectrometry. Radiochim. Acta **79**, 183–190 (1997)

43. G. Passler, N. Erdmann, H.-U. Hasse, G. Herrmann, G. Huber, S. Köhler, J.V. Kratz, A. Mansel, M. Nunnemann, N. Trautmann, A. Waldek, Application of laser mass spectrometry for trace analysis of plutonium and technetium. Kerntechnik **62**, 85–90 (1997)

44. C. Grüning, G. Huber, J.V. Kratz, G. Passler, N. Trautmann, A. Waldek, K. Wendt, 'Resonance ionization mass spectrometry for ultratrace analysis of plutonium with a new solid-state laser system. Int. J. Mass Spectrom. **235**, 171–178 (2001)

45. M. Yamamoto, A. Tsumura, Y. Katayama, T. Tsukatani, T., Plutonium isotopic composition in soil from the former Semipalatinsk nuclear test site. Radiochim. Acta **72**(1996), 209–215 (1996)

46. S.F. Boulyga, J.S. Becker, Isotopic analysis of uranium and plutonium using ICP-MS and estimation of burn-up of spent uranium in contaminated environmental samples. J. Anal. At. Spectrom. **17**, 1143–1147 (2002)

47. W.F. Kinard, N.E. Bibler, C.J. Coleman, R.A. Dewberry, Radiochemical analyses for the defense waste processing facility startup at the Savannah River Site. J. Radioanal. Nucl. Chem. **219**, 197–201 (1997)

48. S. Uchida, R. García-Tenorio, K. Tagami, M. García-León, Determination of U isotopic ratios in environmental samples by ICP-MS. J. Anal. Atom. Spectrom. **15**, 889–892 (2000)

49. S. Morita, C.K. Kim, Y. Takaku, R. Seki, N. Ikeda, Determination of technetium-99 in environmental simples by inductively coupling plasma mass spectrometry. Appl. Radiat. Isot. **42**, 531–534 (1991)
50. N. Momoshima, M. Sayad, Y. Takashima, Analytical procedure for technetium-99 in seawater by ICP-MS. Radiochim. Acta **63**, 73–78 (1993)
51. K. Tagami, S. Uchida, Analysis of technetium-99 in soil and deposition samples by inductively coupled plasma-mass spectrometry. Appl. Radiat. Isot. **47**, 1057–1060 (1996)
52. S. Uchida, K. Tagami, W. Rühm, E. Wirth, Determination of $^{99}$Tc deposited on the ground within the 30-km zone around the Chernobyl reactor and estimation of $^{99}$Tc released into atmosphere by the accident. Chemosphere **39**, 2757–2766 (1999)
53. J.L. Mas, M. García-León, J.P. Bolívar, $^{99}$Tc detection in water samples by ICP–MS. Radiochimica **92**, 39–46 (2004)
54. W. Bu, J. Zheng, Q. Guo, T. Aono, H. Tazoe, K. Tagami, S. Uchida, M. Yamada, A method of measurement of $^{239}$Pu, $^{240}$Pu, $^{241}$Pu in high U content marine sediments by sector field ICP-MS and its application to Fukushima sediment simples. Environ. Sci. Technol. **48**, 534–541 (2014)
55. T. Ohno, Y. Muramatsu, Y. Shikamori, C. Yoyama, N. Okabea, H. Matsuzakic, Determination of ultratrace $^{129}$I in soil simples by triple quadrupole ICP-MS and its application to Fukushima soil simples. J. Anal. At. Spectrom. **28**, 1283–1287 (2013)
56. G.S. Yang, H. Tazoe, K. Hayano, K. Okayama, M. Yamada, 'Isotopic compositions of U-236, Pu-239, and Pu-240 in soil contaminated by the Fukushima Daiichi nuclear power plant accident. Sci. Rep. **7**, 13619 (2017)
57. A. Shimada, K. Sakatani, Y. Kameo, K. Takahashi, Determination of $^{129}$I in the accumulated radioactive water and processed water of the Fukushima Daiichi nuclear power plant. J. Radioanal. Nucl. Chem. **303**, 1137–1140 (2015)
58. L. Cao, J. Zheng, H. Tsukada, S. Pan, Z. Wang, K. Tagami, S. Uchida, Simultaneous determination of radio cesium ($^{135}$Cs, $^{137}$Cs) and plutonium ($^{239}$Pu, $^{240}$Pu) isotopes in river suspended particles by ICP-MS/MS and SF-ICP-MS. Talanta **159**, 55–63 (2016)
59. Z.T. Wang, J. Zheng, L.G. Cao, K. Tagami, S. Uchida, Method for ultratrace level $^{241}$Am determination in large soil samples by sector field-inductively coupled plasma-mass spectrometry: with emphasis on the removal of spectral interferences and matrix effect. Anal. Chem. **88**, 7387–7394 (2016)
60. J. Zheng, K. Tagami, S. Uchida, Rapid analysis of U isotopes in vegetables using ICP-MS: application to the emergency U monitoring after the nuclear accident at TEPCO's Fukushima Dai-ichi power station. J. Radioanal. Nucl. Chem. **292**, 171–175 (2012)

# Principles of Particle Accelerators

# 17

**ABSTRACT**

Conventional or low-energy mass spectrometry has some limitations. To overcome them, Accelerator Mass Spectrometry (AMS) will be introduced in Chap. 18. Before we describe in this chapter a new device, the accelerator, which coupled to a mass spectrometer, helps to improve the sensitivity of mass spectrometric applications in environmental radioactivity. Thus, electrostatic and radiofrequency field-based accelerators, such as Van de Graaff, Cockcroft–Walton, LINACs, cyclotrons, and synchrotron devices, are presented. In addition, the ion sources and the most important elements of ion optics that make possible the transport of the ion beam along the accelerator are also described. This chapter aims to clarify the accelerator operating principles to facilitate the understanding of AMS. Some applications of accelerators to environmental problems are also included.

## 17.1  Need of Accelerators

Nuclear reactions are very useful for understanding the structure of the nucleus and the forces that act inside. The first nuclear reaction was provoked in 1919 by Rutherford , who irradiated N with $^{214}$Po $\alpha$ particles [1]. The emission of protons led him to deduce that a nuclear transmutation was occurring. Currently, we identify that reaction as

$$^{14}N(\alpha, p)^{17}O$$

This reaction was possible because the energy of the $\alpha$ particles was 7.7 MeV, which is above the $^{14}$N Coulomb barrier [2]. A similar reaction, now with Al, led to F. Joliot  and I. Curie  in 1934 [3] to produce $^{30}$P. In fact, this was the first time

© The Author(s), under exclusive license to Springer Nature Switzerland AG 2022    521
M. García-León, *Detecting Environmental Radioactivity*, Graduate Texts
in Physics, https://doi.org/10.1007/978-3-031-09970-0_17

that artificial radioactivity was generated. In this case, the nuclear reaction was.

$$^{27}\text{Al}(\alpha, n)^{30}\text{P}$$

and $^{30}$P ($T_{1/2} = 2.5$ min) was produced by irradiating Al with a 'strong polonium preparation of 100 millicuries' [3]. Although the Po isotope is not identified in the paper, it appears that it was mainly $^{210}$Po, which emits 5.3 MeV $\alpha$ particles.

However, as soon as the charge of the target nucleus increases, higher projectile energies are needed for nuclear reactions to occur [2]. In addition, depending on the reaction of interest, it is convenient to have different projectiles. This is not always possible with common radioactive sources. These requirements have boosted the investigation of a new instrument called an accelerator that is able to supply sufficient energy to charged particles to overcome the Coulomb barrier and produce nuclear reactions.

One of the first achievements in the race to construct these new instruments was obtained in 1932 by Cockcroft and Walton [4, 5], which induced the reaction.

$$^{7}\text{Li}(p, \alpha)^{4}\text{He}$$

with a direct current accelerator of Cockcroft–Walton type, as we shall see in the next sections.

Today, accelerators are playing an increasingly important role in the determination of radionuclides in the environment. As we will see in Chap. 18, Accelerator Mass Spectrometry (AMS) is the most sensitive technique for many applications. Since AMS consists of the coupling of one accelerator to a mass spectrometer, it is convenient to know the operating principles of accelerators before describing it.

## 17.2  Parts of an Accelerator

An accelerator consists basically of three parts: the ion source, the acceleration step, and the reaction chamber . In addition, some Ion Optics  is necessary for the proper transport of the beam. The ion beam is produced in the ion source, accelerated in the acceleration step, and finally, the nuclear reactions take place in the reaction chamber . A schematic view of an accelerator is depicted in Fig. 17.1. In what follows, the different parts of the instruments are described.

### 17.2.1 The Ion Source

The projectiles are produced at the ion source, where an ionization process of the sample takes place. The simplest way to produce ions is by electron  impact in gases. Thus, electrons  released by the thermionic effect in a cathode at a given T enter a chamber where a gas at low pressure ($10^{-4}$ mbar) is confined. Electron collisions cause ionization of the gas atoms or molecules (see Fig. 17.2a). The

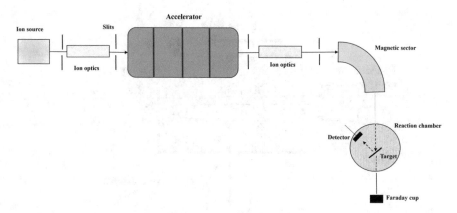

**Fig. 17.1**   A schematic view of an accelerator

positive ions obtained in this way are extracted by a cathode potential (positive potential) that in turn rejects the electrons . The ions are injected into the accelerator. In many cases, the electrons are confined in helical trajectories within the chamber by a magnetic field to increase the ionization efficiency and the ion current. There are many designs for this very simple concept [5]. The electron current can be obtained by an arc discharge, for instance (Fig. 17.2b), that passes through the gas, producing ionization.

As we saw in Chap. 16, the combination of an electron source with the application of RF fields provokes the appearance of plasmas in gases that generate the ionization of the sample of interest. Several of the concepts described there are used in accelerators as ion sources.

All these ion sources produce positive ions, although in some cases, it is possible to use charge exchangers to obtain negative ions, which are most commonly used in tandem accelerators. Nevertheless, negative ions are commonly obtained by using Cs Sputtering Ion Sources . They are described in Chap. 18, and for that, we will not go into detail here (see Fig. 18.3).

## 17.2.2 The Acceleration Step

With the help of optical and mechanical elements described in the following, the ions are injected into the acceleration step. There, static or radiofrequency electric fields are used to accelerate them. The different accelerator types will be presented in the next sections.

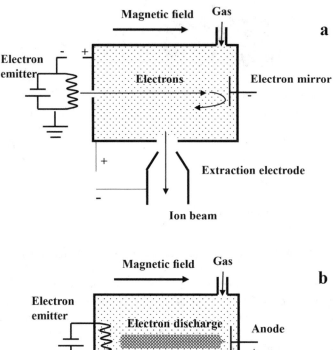

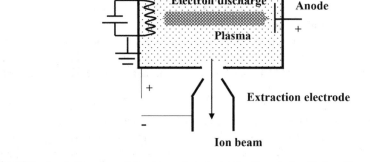

**Fig. 17.2** An electron impact ion source based on **a** the thermionic effect and **b** arc discharge

## 17.2.3 The Reaction Chamber

The accelerated ions are directed toward the reaction chamber where the target is placed. Appropriate detectors are located inside or around the reaction chamber to detect the emitted radiation during the reaction. In Fig. 17.3, a typical reaction chamber is depicted. The selection of the detectors will depend on the expected radiation. In many cases, the setup is more complicated since it is necessary to study the coincidences among the different radiations. On other occasions, the mass or charge of the reaction products must be studied, and $\Delta$E–E telescopes or TOF spectrometers (see Chap. 12) can be used.

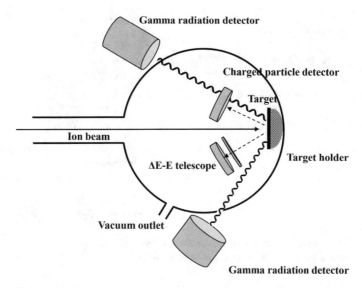

**Fig. 17.3** Typical reaction chamber

As we will see in Chap. 18, the reaction chamber is directly substituted by detectors in Accelerator Mass Spectrometry. In this case, the main goal is the measurement of the mass of the ion beam components, not the production of nuclear reactions.

## 17.2.4 Ion Optics and Other Elements

The ion beam consists of several charged species that must be transported all along the accelerator to reach the reaction chamber. For that, electric and magnetic components are used to focus and deviate the beam according to experimental needs. For instance, the extraction of ions from an ion source is usually carried out by an extraction potential that simultaneously rejects the electrons or the positive ions accompanying the interesting ion species. The beam is injected into the accelerator line, as shown in Fig. 17.4. It is important to keep the accelerator line in good vacuum conditions, typically $10^{-3}$ to $10^{-6}$ mbar, to avoid inacceptable losses of intensity and energy [6]. Vacuum values of $10^{-9}$ to $10^{-12}$ mbar are needed in storage rings but are not of significant use in our problems [7].

As seen in Fig. 17.4, electrostatic or static magnetic fields are used to select the mass and energy of the interesting beam components and transport them along the accelerator line. This is done as described in Chap. 16. Focusing the beam is achieved with Einzel lenses [8]. Usually, it consists of three cylindrical lenses, the first and the third connected to the same potential, while the second is connected to a higher potential (see Fig. 17.5a). The beam is accelerated in the first part of the lens and decelerated when passing from the second step to the third. The final

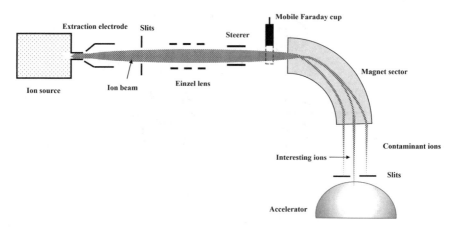

**Fig. 17.4** Low-energy side of an accelerator showing the different optical and mechanical elements used to conduct the beam along the accelerator lines

effect is that the energy of the beam remains unchanged. However, at the same time, the defocusing effect produced by the first acceleration is compensated by the focusing effect the beam undergoes when passing from the second cylinder to the third cylinder. The final effect is a beam focus that depends on the geometry of the lenses and the voltage ratios, $R = \frac{V_{\text{einzel}} - V_0}{V_0}$. This arrangement is known as an accelerating Einzel lens , which is the most common setup. However, it is possible to construct a decelerating Einzel lens [9], $R < 0$, which has a higher focusing power, although the beam energy remains unchanged as well.

Magnetic quadrupoles are also used as a beam-focusing element. In Fig. 17.6, a scheme of a magnetic quadrupole is depicted. The magnet combination has four poles.

Depending on the orientation of the poles, the beam is focused on the x-axis and defocused on the y-axis, with the z-axis being that of the beam direction [10]. To achieve a total focusing effect, two magnetic quadrupoles can be set in tandem but with an appropriate distance among them; otherwise, the focusing effect would cancel out. In this way, the second quadrupole corrects the defocusing effect of the first element.

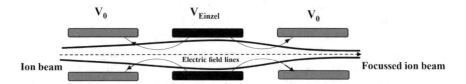

**Fig. 17.5   a** Schematic view of an Einzel lens

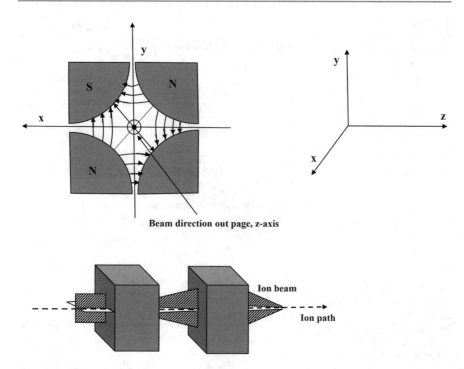

Beam direction out page, z-axis

Ion beam

Ion path

**Fig. 17.6** Schematic view of a magnetic quadrupole

Electrostatic steerers are used to deflect the beam independently of the ion mass and are usually used to optimize the injection of the beam into the accelerator step [11]. However, it can be used throughout the accelerator. The simplest electrostatic steerer design consists of a pair of parallel plates connected to the opposite potential, ±V (see Fig. 17.7a). It can be shown that the steering angle, that is, that formed between the original and the deflected beam trajectories, is $\theta \propto \frac{Vl}{V_b d}$, where l is the length of the plates, $d$ is the distance between the plates, and $V_b$ is the beam accelerating potential.

In the case of light elements such as $H^+$ or $He^+$, it is possible to use a magnetic steerer to provide a small deflection to the beam (Fig. 17.7b) [12].

Faraday cups (Fig. 17.8a) are used along the accelerator line to monitor the intensity, size, and form of the beam. As we will see, for Accelerator Mass Spectrometry , Faraday cups are essential instruments since they provide the measurement of the major element in the beam.

The slits are used as an additional beam line selector (Fig. 17.8b). By opening or closing the slit, it is possible to increase or decrease the beam current. In mass spectrometers (see Chap. 18), they can help to reject unwanted masses that could have similar trajectories to the ion of interest.

Beam Profile Monitors (BPM) are essential in accelerators and supply the so-called beam diagnostic. BPM can measure the energy, intensity, brightness

**Fig. 17.7** Schematic representation of **a** Electrostatic steerer  and **b** magnetic steerer

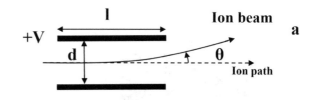

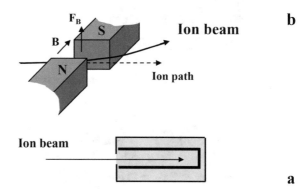

**Fig. 17.8** Schematic representation of **a** Faraday cups , **b** Slits on the low-energy side, and **c** Beam Profile Monitor  on the low-energy side

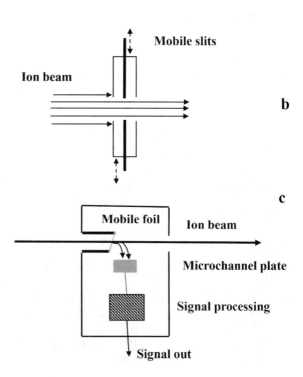

(intensity per unit area and solid angle), and emittance (distribution of the particles in the beam according to their position and momentum) of the beam, among other parameters that help to establish the beam status. This is very important operational information about the ongoing experiment. There are different BPM designs [13]. Some of them are presented in Fig. 17.8c. BPMs are not supposed to disturb the beam when measuring the different parameters, and this is a key point in their design and construction.

## 17.3   Electrostatic Accelerators

In electrostatic accelerators, the ions are injected into a region where a direct current voltage, $V$, is applied. The electric field created in this way supplies the ion with an energy, $E$, given by

$$E = qeV \qquad (17.1)$$

$e$ is the electron charge, and q is the charge state of the ion. There are two designs providing steady high voltages: the Cockcroft and Walton and Van de Graaff concepts presented in 1930 [14] and 1931 [15], respectively.

### 17.3.1 Van de Graaff Accelerators

Van de Graaff accelerators are based on the concept of Van de Graaff electrostatic generators [15]. The Van de Graaf proposal is presented in Fig. 17.9 [16]. Basically, high voltage is achieved by the accumulation of charge at the accelerator terminal. As shown in Fig. 17.9, the accelerator tank contains two pulleys connected to a conveyor belt that mechanically transport the charge from the earthed pulley to the terminal pulley. The belt is made of insulating materials and is positively charged at the bottom by a corona discharge that pulls electrons from the belt. This corona discharge is induced by a voltage source connected to a metallic brush close to the earthed pulley. This metallic component induces a charge dissociation in the gas contained in the accelerator tank. The charge is directed toward the earthed pulley but stops in the belt since it is an insulating element. The belt is positively charged this way. When it reaches the terminal pulley, the belt induces a positive charge at the accelerator terminal by another corona discharge that neutralizes the positive charge of the belt. The belt continues its journey to be charged once again; in addition, so on. In this way, the terminal is charged, and a high voltage appears between the terminal and the ground of the tank. To avoid sparks inside the tank, this is filled with insulating gases. First, high-pressure $N_2$ or $CO_2$ was used. They provide the material for corona discharges at the same time.

In the original design, the ion source was inside the tank, and the ions were created there. They were injected into a tube under high vacuum conditions ($\leq 10^{-6}$ mbar). The energy gained by the ions is given by (17.1).

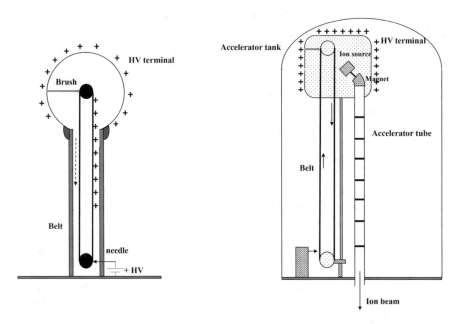

**Fig. 17.9**  The original Van de Graaf proposal and a modern version of the Van de Graaff accelerator

In modern designs, the charge transport is made by pellets that are metallic cylinders (Fig. 17.10). The pellets are joined by insulating material and charged by induction effects. This system is known to provide a more stable voltage inside the terminal, which provides more stable energies to the ions. In addition, currently, the accelerator tank is filled with a dielectric gas, $SF_6$, at very high pressure (5 to 6 bar), which allows one to achieve higher voltages. With this mechanism, it is possible to generate very high potentials, the technical limit being approximately 25 MV. According to (17.1), the maximum energy that can be supplied with this accelerator depends on the terminal tension but also on the charge state. A way to further increase the energy with electrostatic accelerators was presented by R. J. Van de Graaff in 1960 [17], although many laboratories developed the idea in previous works. He introduced the concept of multistage accelerators and specifically the concept of tandem accelerators. Although this was implemented for the first time in Van de Graaff accelerators, it can be adapted as well to Cockcroft–Walton accelerators.

According to that proposal, the ion beam can be accelerated in a two-stage process, as seen in Fig. 17.11, where the design by R. J. Van de Graaff is presented. The accelerator tank is now modified, and the positive terminal voltage is applied in the middle of the tank in a place where a new component, the stripper, is located. The negative-ion beam is accelerated gaining an energy given by (17.1) with $q = 1$. Then, it passes through the stripper, where some of its electrons are removed by a series of charge-exchange reactions. The positively charged ion beam receives

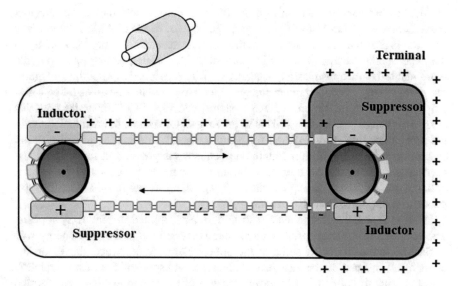

**Fig. 17.10**   Pellet alternative for charge transport

an additional acceleration since it is repelled by the positive terminal tension. The total energy of the beam is

$$E = (1 + qe)V \tag{17.2}$$

where $q$ is the charge state after the stripper.

As we will see in the next chapter, the stripping process plays a key role in Accelerator Mass Spectrometry (AMS) . In fact, during the stripping process, a dramatic reduction in background is produced, which is the main contribution

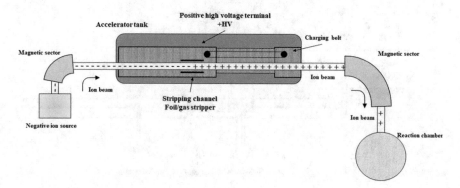

**Fig. 17.11**   The tandem  (two-stage) accelerator

of AMS to the measurement of radionuclides by mass spectrometry. The stripper consists of a very thin foil of C or a gas, $O_2$, $N_2$, Ar, or He in AMS systems, at a pressure higher than the vacuum conditions in the accelerator tube. Depending on the experimental needs, C or gas is more suitable. Usually, gas strippers provide more stable beam transmission, whereas C foils provide the possibility of gaining more energy because higher charge states are populated after stripping . Gas stripping is the usual option in AMS systems, and Fig. 17.12 illustrates how this charge-exchange process takes place in an Ar gas. Of course, the residual gases in the accelerator line are a problem for beam transport. For that, an effective pump system must be installed to maintain an adequate gas pressure in the stripper and prevent gas release to the line. A turbomolecular pump is needed to maintain a nominal gas thickness of approximately 2 $\mu gcm^{-2}$ oxygen in this case. The gas pressure is approximately $10^{-2}$ mbar.

After stripping , a charge-state distribution, as presented in Fig. 18.9, appears. The selection of the appropriate charge state is carried out at the high-energy side of the accelerator by using some of the optical elements described above. On the other hand, the use of the stripper modifies the transport of the beam, which is dispersed due to collisions. Moreover, because of the strong positive ions formed in the stripper, Coulomb repulsion effects appear, which may lead to losses of the beam intensity. In contrast, this effect is very interesting for AMS. In fact, it is responsible for the lowering of the molecular background in AMS, as we will describe in the next chapter.

The use of the tandem  concept means that negative ions must be injected into the accelerator. In the original design [17], a positive ion source was used, and a charge-adding element converted the positive beam into a negative beam. Today, the common option is the use of Cs-sputtering ion sources to directly inject negative ions (see Chap. 18). It is possible to construct accelerators with more

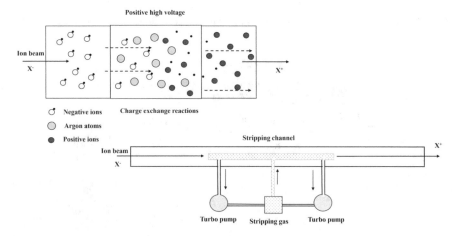

**Fig. 17.12**  Schematic view of the stripping  process and a gas stripper

than two acceleration stages. In fact, Van de Graaff proposed three and even four stages to increase energy [17].

### 17.3.2 Cockcroft–Walton Accelerators

The Cockcroft–Walton concept was proposed in 1930 [14] and, in fact, was the first to achieve a real nuclear reaction as described above [4, 18].

In the original proposal, the voltage was obtained by rectifying a current from a low-frequency transformer. This evolved toward the instrument shown in [4]. The general principle underlying the concept is presented in Fig. 17.13, which is similar to the original figure appearing in the reference. They were able to multiply the original voltage $E$ into $3E$ by mechanically switching the capacitors alternatively. A more evolved instrument is also presented in the same reference, where the mechanical switching is substituted by the use of diodes. A modern version of the Cockcroft–Walton concept is presented in Fig. 17.14. The so-called cascade generator consists of a coupling of rectifying diodes and capacitors applied to an alternating-current generator. A multistage cascade generator (Fig. 17.14) allows one to amplify the original amplitude by a factor that depends on the number of stages. Generally, if we have N diodes and N capacitors, the maximum voltage would be $V_{max} = NV_0$. The maximum voltages obtained in this way are approximately a few MV. As in the case of Van de Graaf, the accelerator tank is filled with $SF_6$ to avoid sparks and maximize the voltage.

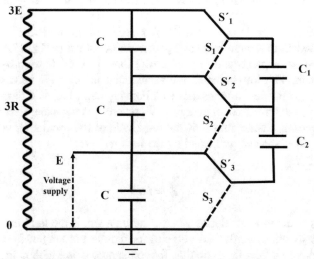

Capacitors $C_1$ and $C_2$ are connected to capacitors C alternatively by doted-, S, and full-line, S′, connections

**Fig. 17.13**  Original proposal by Cockcroft and Walton

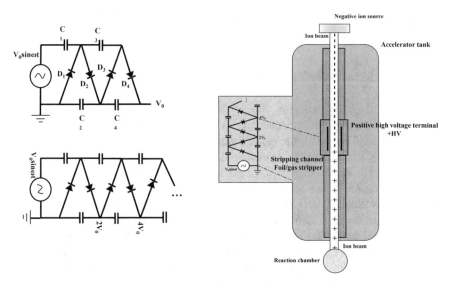

**Fig. 17.14** A modern realization of the Cockcroft–Walton concept and a modern Cockcroft–Walton accelerator

## 17.4   Linear Accelerators (LINACS)

The energy  provided by electrostatic accelerators is limited by the maximum voltages reached with the Van de Graaf and Cockcroft–Walton  concepts. To increase the energy, new methods are needed. One of them, the use of oscillating fields, was first proposed by G. Ising in 1924 [19] and developed by R. Widerøe  in 1928 [20].

The Widerøe  accelerators consist of a series of cylindrical electrodes (drift tubes) connected to a radiofrequency generator, as shown in Fig. 17.15. Each tube is connected to the next tube in such a way that an oscillating voltage is applied between them. The ions injected are accelerated in the gaps between the tubes since the space inside the tubes acts as a Faraday cup , i.e., as a field-free region. In each step, the ion gains an energy $qeV$, where q is the charge state of the ion, e is the electron  charge, and V is the amplitude of the oscillating voltage. Thus, for $n$ drift tubes, the energy gained by the ions will be

$$E = nqV \qquad (17.3)$$

Obviously, the ions accelerate when entering a gap if the field has an adequate sign value. In other words, the outgoing drift tube must repel the ion, and the next tube should attract it. To do this, the ion should exit a tube at the appropriate time, i.e., in phase with the radiofrequency field. This means that the lengths of the drift tubes must be adjusted to the increasing energy of the ions. To achieve

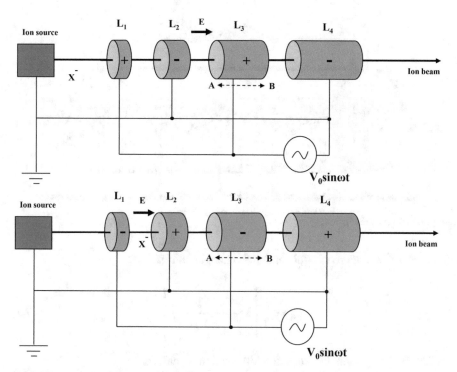

**Fig. 17.15** Original Wideröe LINAC

that condition, the time the ion spends going from A to B in Fig. 17.15 should be $\frac{T}{2}$, with $T$ being the period of the radiofrequency field . It is clear that

$$\frac{L_n}{v_n} = \frac{T}{2} = \frac{1}{2f} = \frac{\lambda}{2c} \tag{17.4}$$

where $L_n$ and $v_n$ are the length of the n drift tube and the ion speed in it, respectively; $f$ and $\lambda$ are the frequency and wavelength of the field, respectively, and $c$ is the speed of light in vacuum.

On the other hand, in a classical approach, it is

$$\frac{1}{2}Mv_n^2 = nqeV + C \tag{17.5}$$

where $C$ is the injection energy and $M$ is the mass of the ion. Then, by combining (17.4) and (17.5), we have

$$L_n = \frac{1}{2f}\sqrt{\frac{2(nqV + C)}{M}} = \lambda\sqrt{\frac{nqV + C}{2Mc^2}} \tag{17.6}$$

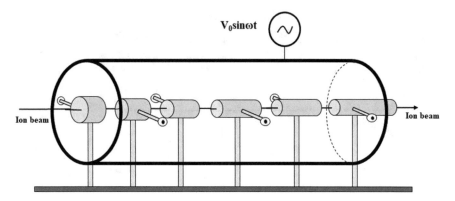

**Fig. 17.16**  Alvarez  LINAC

in a relativistic approach, this equation changes into [21]

$$L_n = \frac{\lambda}{2} \left\{ 1 - \frac{1}{\left[ \frac{nqV+C}{Mc^2} \right]^2} \right\}^{\frac{1}{2}} \tag{17.7}$$

which basically contains the same information as before.

According to that equation, the length of each drift tube, and consequently that of the accelerator, will depend on the frequency of the field. This posed strong limits to the application of this idea to real Nuclear Physics experiments at the beginning, since the radiofrequency fields  by that time were too small $f$ or, in other words, two long $\lambda$. It was after the Second World War when it was possible to develop high-frequency fields useful for the acceleration of e and p, which boosted the construction and improvements of LINACS .

A modified version of this design is the so-called Alvarez  LINAC  [21, 22], in which the ions are accelerated inside a resonant structure in which the field pattern is repeated in every tube (Fig. 17.16). When the field changes the polarity, the ions are shielded inside the drift tubes, gaining energy in the gap as before. Certainly, LINACS  are able to supply energies higher than those of electrostatic accelerators, some few GeV, but on the basis of an increase in the accelerator length poses a limit on the development of that concept. Circular accelerators are an alternative to this problem.

## 17.5  Circular Accelerators

Circular accelerators combine the use of alternating electric fields with the application of a magnetic field normal to the ion motion planes. This confines the ion beam into a circular trajectory that allows one to gain energy without the need for an excessively long accelerator.

### 17.5.1 Cyclotrons

The cyclotron is the first circular accelerator and was invented by E. Lawrence , who patented the cyclotron in 1932 [23]. Together with Livingston, they proved it to be an excellent tool in Nuclear Physics because they were able to accelerate protons up to 1.22 MeV [24].

A cyclotron consists of two semicylindrical hollow metal plates connected to an oscillating field and a magnet that provides a uniform magnetic field normal to the plates and to the ion trajectory (see Fig. 17.17). Because of the magnetic field, $B$, the trajectories of the ions will be circular with radius $r$. Thus,

$$qevB = \frac{Mv^2}{r} \tag{17.8}$$

where $v$ is the speed of the ion, $M$ is its mass, and $q$ is the charge state. Under these conditions, the angular frequency $\omega$ of the ions would be

$$\omega = \frac{v}{r} = \frac{qeB}{M} \tag{17.9}$$

As in the case of LINACS , it is expected that the ions gain energy each time they pass the semicylindrical plates. This occurs when the field frequency, $f$, resonantly coincides with the cyclotron frequency . In other words,

$$f = \frac{\omega}{2\pi} = \frac{qeB}{2\pi M} \tag{17.10}$$

Since $\omega$ should be constant, $r$ must increase each time the ions are accelerated. This means that the trajectories become spirals inside the accelerator. After several

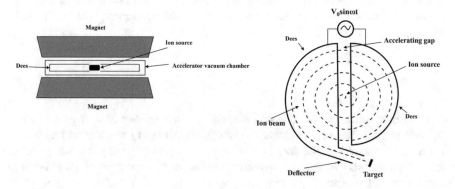

**Fig. 17.17**  A schematic view of a cyclotron accelerator

acceleration steps, it can be found that the energy $E$ of the ions is

$$E = \frac{p^2}{2M} = \frac{(qeBR)^2}{2M} \tag{17.11}$$

$R$ is the radius of the cyclotron . However, when the ions reach velocities close to c, for instance, 0.9c, their mass changes, so (17.10) must include relativistic corrections. Now, the oscillating field frequency must be changed according to the following.

$$f = \frac{qeB}{2\pi M}\sqrt{1 - \frac{v^2}{c^2}} \tag{17.12}$$

Thus, a new generation of cyclotrons  appeared [25] synchrocyclotrons, in which the frequency of the alternating field is changed and modulated according to the energy variations. An alternative to this solution is the so-called isochronous cyclotrons  in which the magnetic field increases as long as the radius increases. This requires an azimuthal variation of the field to keep the ions in the orbit. This strong focusing solution is thought to have been theoretically anticipated by L. H. Thomas in 1938 [26] and carried into real application by E. Courant et al. in 1952 [27].

## 17.5.2 Synchrotrons

The strong focusing principle or the alternating-gradient focusing principle boosted the design of a new concept, the synchrotron , which helps to overcome the limitations in energy of the cyclotrons . A synchrotron  combines the use of alternating fields with that of changing magnetic fields to confine the accelerated particle in a determined circular trajectory. It consists of a ring divided into sectors that play the role of drift tubes. A magnetic field is applied to the whole instrument to keep the ions in a trajectory whose radius should be that of the accelerator (see Fig. 17.18). Thus,

$$R = \frac{vM}{qeB} \tag{17.13}$$

Of course, since the velocity and mass of the ions change as they gain energy, $B$, and the field frequency, $f$, have to change over time to keep the ion trajectory, that is, $R$, constant. For the case of electrons , $v \sim c$, at relatively low energies, so it is only necessary to change B to compensate for the variation of M. However, for the case of protons, the relativistic regime is reached at higher energies, so it is necessary to change f as well.

With synchrotrons , it has been possible to reach energies around TeV. For that, they played, and play, a very important role in Particle Physics. A very important

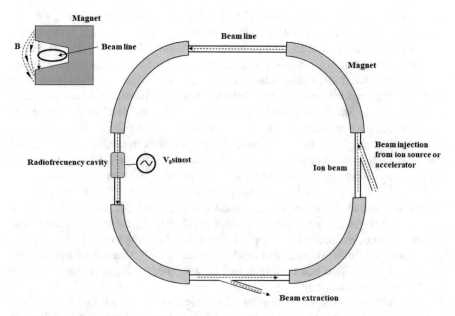

**Fig. 17.18**  Schematic view of a synchrotron

characteristic of synchrotrons  is the appearance of synchrotron radiation . This electromagnetic radiation is emitted by charged particles when they move under the action of a magnetic field normal to its trajectory. The emitted radiation causes a loss of energy per loop $\Delta T$ that approximately

$$\Delta T = 8.85x10^4 \frac{E^4}{R} \tag{17.14}$$

$E$ in GeV, $R$ in m, and $\Delta T$ in eV. It is obvious that the longer $R$ is, the smaller $\Delta T$ is. This being the reason why modern synchrotrons  tend to be large installations.

Key scientists in the development of the synchrotron  were M. Oliphant, who proposed the idea of a magnet pulsed ring [28], as well as [29] and [34], who simultaneously developed the idea of accelerating $e$ and $p$ with that principle.

## 17.6    Applications of Accelerators to Environmental Problems

The development of accelerators in all their versions has always been driven by the search for the constituents of matter and the fundamental interactions between them. However, from the beginning of their history, accelerators have contributed to other scientific areas, and their impact on them has been clear. In Fig. 17.19, we see how Betghe [31] illustrates this idea. The so-called Betghe accelerator tree

shows how accelerators have been applied to Medicine, Material Science, technological problems, etc., since 1940, close to the appearance of such instruments.
Furthermore, at the end of the 1970s, many accelerators that were essential during
the first three decades of Nuclear Physics became obsolete because the energies
they provided were too small to learn more about the nucleus. The experimental
infrastructures created around those facilities began to search for a new source of
scientific life: applications. What had previously been a collateral vocation became
the basic objective of some laboratories. A consequence of this was the production
of intense activity in the field of accelerator applications. In such a short period of
time, the design of accelerators specifically intended for applications became an
autonomous discipline that continues. It is not possible to devote too much space
to describing all the possible applications of accelerators. Only with the problem
of Accelerators in Medicine could we fill a book. We concentrate on possible
applications to the environment. From that point of view, the most relevant contribution of accelerators to these applications arises from the use of two groups of
techniques: the Ion Beam Analysis (IBA) techniques and Accelerator Mass Spectrometry (AMS) . This last technique will be described in the next chapter. Some
details are given about IBA .

IBA (Ion Beam Analysis) techniques basically consist of the irradiation of samples with ion beams to produce reactions at the nuclear or atomic level, or both at
the same time. Figure 17.20 shows the basis of the IBA techniques. The spectrum

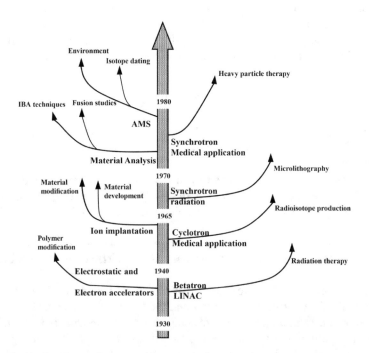

**Fig. 17.19** The Betghe accelerator tree

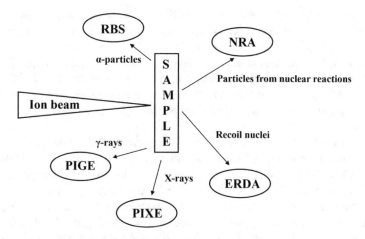

**Fig. 17.20**  Schematic view of IBA techniques

of radiation obtained after the reactions is collected with the appropriate detectors. This allows us to identify and quantify the elemental or isotopic composition of the irradiated sample.

There are several IBA techniques [32]. One of the most widely used methods is *Rutherford Backscattering Spectrometry (RBS)*, which allows the determination of concentration profiles of heavy elements distributed in light element matrices. It is a highly sensitive and nondestructive technique based on the detection of projectiles, normally some few MeV α particles, backscattered from the sample. The energy spectrum after the backscattering identifies the nucleus against which the α particle collided and the depth (up to a few tens of μm) at which the collision occurred. In this way, for example, concentration profiles of implanted impurities can be obtained in different matrices [32]. Thin film thicknesses and interfaces can also be measured, as well as compound stoichiometry. Furthermore, the study of backscattered particles on the directions of maximum symmetry of single crystals can reveal the phenomenon of ion channeling, which permits the location of impurities in the crystalline structure, quantification of the crystalline quality of a single crystal, determination of defect concentration, composition and thickness of surface amorphous layers, etc. RBS has also been applied [33] to determine the percentage composition of U and Pu of hot radioactive particles (see Chap. 5) taken from Palomares in Spain where a nuclear accident occurred (see Chap. 4).

*Analysis by Nuclear Reactions (NRA)* is another technique that has great potential for the detection of traces of light elements in matrices of heavy elements. With the typical energies provided by small- or medium-sized accelerators, nuclear reactions take place primarily through collisions with light nuclei. We can cite some examples: $^{14}$N (d, α)$^{12}$C at 1400 keV, $^{16}$O(d, p γ)$^{17}$O$^{*}$ at 860 keV, or $^{12}$C(d, p)$^{13}$C at 950 keV. When we detect γ radiation, we speak of *PIGE (Particle-Induced γ-Ray Emission)*. This technique has been applied to the determination of U signatures in various materials for nuclear forensic purposes [34].

Another widely used technique, probably the most popular, is the so-called *PIXE (Particle-Induced X-ray Emission)*. PIXE consists of the irradiation of a sample with a charged-particle beam to produce characteristic X-rays. In this way, a very sensitive (nondestructive and multielement) analysis of the sample ($Z \geq 12$, basically) can be performed, from major to trace elements. NRA and PIGE complement PIXE, as they can report on $Z < 12$ elements. In Fig. 17.21, we have a PIXE X-ray spectrum of a hot radioactive particle [33] in which U and Pu can be identified.

The so-called micro-PIXE technique is a variant of the previous technique, which allows the analysis of samples with spatial resolutions on the order of 1 or 2 $\mu$m [35]. This is possible with microprobe lines, in which the proton beam is narrowed to such dimensions by a suitable combination of quadrupoles. This allows characterizing the composition of a sample by measuring directly on it (areas of a rock, radioactive particles , areas of accumulation of contaminants in a plant, etc.) with great spatial resolution. By sweeping the microbeam over the sample, we can even obtain the spatial distribution of the different present elements (multielemental 'mappings') in preselected areas. A U map of a radioactive particle obtained by $\mu$-PIXE is presented in Fig. 17.22.

Finally, in the analysis by ERDA (Elastic Recoil Detection Analysis) also based on elastic dispersion, the particles that come out of the surface of a sample when

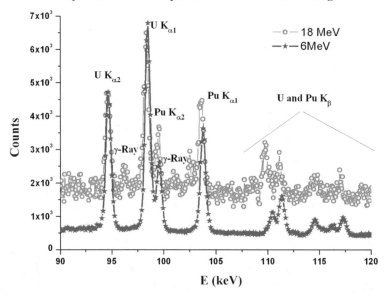

**Fig. 17.21** U-PIXE spectrum of a hot radioactive particle . Figure reprinted from [33] with permission by Elsevier

**Hot particle A1-5: U mapping 100 x 100 mm2**

**Fig. 17.22**  U map of a hot radioactive particle  obtained by μ-PIXE analysis. Figure courtesy of Dr. M. C. Jiménez-Ramos

bombarded are detected. It is generally used to study surface concentration profiles of light elements using ions of higher atomic mass as projectiles.

Globally, IBA  techniques are powerful tools for material characterization. They are sensitive, fast, and nondestructive. Its versatility allows us to tackle very competitive problems in areas as diverse as Materials Sciences, Archaeology, Biomedicine, Environment, etc. An interesting review on the environmental applications of IBA  can be found in [36].

To close this chapter, it is interesting to provide information on the worldwide inventory of accelerators. This can be found in several websites that include the main features and achievements [37]. The reader is invited to make a web tour and see how the accelerator technology has evolved over time and how it has contributed to Science.

**Exercises**

1. We wish to accelerate $^4$He particles using a tandem  accelerator. Assuming that the terminal voltage is 1 MV and that after passing the beam through the stripper all possible charge states (including neutral) can be produced with a certain probability, calculate the final energies obtained as a function of the final particle charge.
2. How can we separate the beams of different energies?
3. What is the possible maximum energy that can be provided to a beam of deuterons and alpha particles with a 2 MV tandem  accelerator?

4. Explain the formula that relates the voltage (V) generated in a tandem accelerator, the state of charge (q) of the ion after passing through the stripper, and the final energy (in MeV) of the particle. What is the maximum energy of a beam of protons and $^3$He using a 3 MV tandem accelerator?

5. Demonstrate (17.6).

6. Consider a LINAC with 30 stages. He$^{2+}$ is accelerated. The RF field features are V $=$ 150 kV and 100 MHz. Find the maximum energy achieved by the projectile and the length of the last drift tube.

7. Justify (17.11).

8. In a 1.5 m diameter and 10 MHz frequency cyclotron , the magnetic field necessary to achieve the maximum energy for protons and deuterons is found.

9. A proton beam is accelerated up to 10 GeV in a synchrotron with a magnetic field of 15 T. Find the radius of the proton orbit.

10. Suppose that there is a synchrotron with the above-calculated mean radius. Find the energy of the synchrotron radiation emitted per loop.

# References

1. E. Rutherford, Collision of a particle with light atoms. IV. An anomalous Effect in Nitrogen. Philos. Mag. **37**(6), 581 (1919)
2. K.S. Krane, *Introductory Nuclear Physics*, Chapter 11 (Wiley, 1988)
3. F. Joliot, I. Curie, Artificial production of a new kind of radio element. Nature **133**, 201–202 (1934)
4. J.D. Cockcroft, E.T.S. Walton, Experiments with high velocity positive ions. (I) Further developments in the method of obtaining high velocity positive ions. Proc. R. Soc. A **136**, 619–630 (1932)
5. J.D. Cockcroft, E.T.S. Walton, Experiments with high velocity positive ions. II. The disintegration of elements by high velocity protons. Proc. R. Soc. A **137**, 229–242 (1932)
6. H. A. Enge, 'Introduction to Nuclear Physics', Chapter 12, Addison-Wesley Publishing Company, 1966.
7. C. Benvenuti, R. Calder, O. Gröbner, Vacuum for particle accelerators and storage rings. Vacuum **37**, 699–707 (1987)
8. F. Hinterberger, Ion optics with electrostatic lenses, in *CAS-CERN Accelerator School: Small Accelerators*, ed. by D. Brandt (2006), pp. 27–44
9. T. Kalvas, Beam extraction and transport, in *CAS-CERN Accelerator School: Ion Sources*, ed. by R. Bailey (2012), arXiv:1411.2445v1 [physics.acc-ph]
10. S. Henderson, J. Holmes, Y. Zhang, Fundamentals of accelerator physics and technology, magnetic fields and magnets designs (Us Particle Accelerator School, 2009), https://uspas.fnal.gov/
11. P. Veltri, M. Cavenago, G. Chitarin, D. Marcuzzi, E. Sartori, G. Serianni, P. Sonato, Electrostatic steering and beam let aiming in large neutral beam injectors. AIP Conf. Proc. **1655**(2015), 050005 (2015)
12. C. Mühle, Magnets and special magnets, in *CAS - CERN Accelerator School: Beam Injection, Extraction and Transfer*, ed. by B. Holzer (2018)
13. J.L. Vignet, A. Delannoy, E. Guéroult, P. Gangant, J.C. Foy, S. Cuzon, C. Houarner, M. Blaizot, The beam profile monitors for SPIRAL 2, in *Proceedings of the 9th European Workshop on Beam Diagnostics and Instrumentation for Particle Accelerators* (2009), pp. 176–178

14. J.D. Cockcroft, E.T.S. Walton, Experiments with high velocity positive ions. Proc. Roy. Soc. A **129**, 477–489 (1930)
15. R.J. Van de Graaff, A 1,5000,000 volt electrostatic generator. Phys. Rev. **38**, 1919 (1931)
16. R.J. Van De Graaff, K.T. Compton, L.C. Van Atta, The electrostatic production of high voltage for nuclear investigations. Phys. Rev. **43**, 149 (1933)
17. R.J. Van De Graaff, Tandem electrostatic accelerators. Nucl. Instrum. Methods **8**, 195–202 (1960)
18. J.D. Cockcroft, E.T.S. Walton, Experiments with high velocity positive ions. (II) The disintegration of elements by high velocity protons. Proc. R. Soc. A **137**, 229–242 (1932)
19. G. Ising, Prinzip einer Methode zur Herstellung von Kanalstrahlen hoher Voltzahl. Arkiv för Matematik, Astronomi, and Fysik. Bands **18–30**, 1–4 (1924)
20. R. Widerøe, Über Ein Neues Prinzip Zur Herstellung Hoher Spannungen. Archiv für Elektronik und Übertragungstechnik **21–4**, 387–406 (1928)
21. D. Alesini, *Proceedings of the CAS–CERN Accelerator School: Free Electron Lasers and Energy Recovery Linacs*, Hamburg, Germany, ed. by R. Bailey, CERN Yellow Reports: School Proceedings, vol. 1/2018, CERN-2018-001-SP
22. M. Conte, W.W. MacKay, *An Introduction to the Physics of Particle Accelerators*, 2nd edn., Chapter 1 (World Scientific Publishing Co., 2008)
23. E.O. Lawrence, Method and apparatus for the acceleration of ions. Application January 26, 1932. Serial No. 589,033, United States Patent Office
24. E.O. Lawrence, M.S. Livingstqn, The production of high-speed light ions without the use of high voltages. Phys. Rev. **40**, 19–35 (1932)
25. J.R. Richardson, B.T. Wright, E.J. Lofgren, B. Peters, Development of the frequency modulated cyclotron. Phys. Rev. **73**, 424 (1948)
26. L.H. Thomas, The paths of ions in the cyclotron I. Orbits in the magnetic field. Phys. Rev. **54**, 580 (1938)
27. E.D. Courant, M.S. Livingston, H.S. Snyder, The strong-focusing synchrotron-a new high energy accelerator. Phys. Rev. **88**, 1190 (1952)
28. M.O. Oliphant, The acceleration of particles to very high energies, unpublished manuscript (University of Birmingham, 1943); M.O. Oliphant, J.S. Gooden, G.S. Hide, The acceleration of charged particles to very high energies. Proc. Phys. Soc. **59**, 666 (1947)
29. V.I. Veksler, A new method of accelerating relativistic particles. Comptes **43–8**, 346–348 (1944)
30. E.M. McMillan, The synchrotron - a proposed high-energy accelerator. Phys. Rev. **68**, 143 (1945)
31. K. Betghe, Industrial applications of accelerators. Nucl. Phys. News **9/1**, 20 (1999)
32. M. Nastasi, J.W. Mayer, Y. Wang, *Ion Beam Analysis: Fundamentals and Applications* (CRC Press, 2015)
33. M.C. Jiménez-Ramos, J. García López, M. Eriksson, J. Jernstrom, R. García-Tenorio, PIXE analysis of U and Pu from hot particles: K-lines vs. L-lines. Nucl. Instrum. Methods Phys. Res. B **273**, 118–121 (2012)
34. A.I. Apóstol, A. Pantelica, I. Ortega, N. Marginean, O. Sima, M. Straticiuc, M.C. Jimenez-Ramos, V. Fugaru, Ion beam analysis of elemental signatures in uranium dioxide samples: importance for nuclear forensics. J. Radioanal. Nucl. Chem. **311**, 1339–1346 (2017)
35. J. García-López, M.C. Jiménez-Ramos, M. García-León, R. García-Tenorio, Characterisation of hot particles remaining in soils from Palomares (Spain) using a nuclear microprobe. Nucl. Instrum. Methods Phys. Res. B **260**, 343–348 (2007)
36. K.G. Malmqvist, Ion beam analysis for the environment. Nucl. Instrum. Methods Phys. Res. B **85**, 84–94 (1994)
37. http://www-elsa.physik.uni-bonn.de/accelerator_list.html; https://nucleus.iaea.org/sites/accelerators/Lists/Accelerator%20List%20full/AllItems.aspx?Paged=TRUE&p_Title=Germany&p_City=Dresden&p_ID=189&PageFirstRow=61&&View=%7B9C7A4A50-21A5-4E7A-B5D2-7224770E0336%7D

# Accelerator Mass Spectrometry (AMS)

# 18

**ABSTRACT**

In this chapter Accelerator Mass Spectrometry (AMS) is introduced as a mass spectrometry technique that enables overcoming the limitations of conventional (or low energy) mass spectrometry. In AMS, the coupling of an accelerator, conventionally of tandem type, to a mass spectrometer generates a dramatic lowering of the background or interferences in the determination of radionuclides of interest. This facilitates the resolution of many interesting problems in environmental radioactivity. The technique is described in depth, and the modern approach of the so-called low-energy AMS systems is studied. The main applications to environmental radioactivity are also presented.

## 18.1 Experimental Challenges in the Determination of Radioactivity in the Environment

Conventional mass spectrometry has greatly helped to determine radionuclides in the environment. However, many problems still exist that pose important challenges. In Fig. 18.1, we give some typical isotopic ratios for selected radionuclides in some natural compartments. It is well known that the measurement of these ratios provides unique information on interesting natural processes. However, the determination of such extremely low levels requires very sensitive measurement techniques.

## 18.2 Limitations of Low-Energy Mass Spectrometry

Conventional (or low energy) mass spectrometry cannot afford those problems because of some technical limitations. They come mostly from the background associated with the measurements, which severely interferes with the determination

© The Author(s), under exclusive license to Springer Nature Switzerland AG 2022     547
M. García-León, *Detecting Environmental Radioactivity*, Graduate Texts
in Physics, https://doi.org/10.1007/978-3-031-09970-0_18

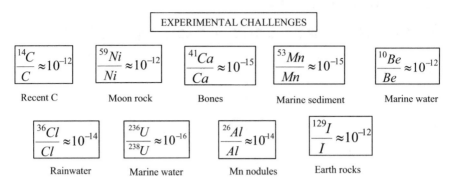

**Fig. 18.1** Isotopic ratios in different environmental compartments. The experimental challenge for their detection is important. Radiometric techniques are not useful in these cases. Conventional mass spectrometry can be used, although more sensitive techniques are needed. Accelerator Mass Spectrometry can help

of very low concentrations of radionuclides in natural samples. There are two main contributions to the background: the isobaric and molecular background and the so-called mass-to-charge ratio background. Both are described below.

## 18.2.1 Isobar and Molecular Background

Isobars and molecules of the same mass as the radionuclide of interest may interfere with its measurement. Some examples are given in Table 18.1. In many cases, isobar and molecular interfering species are more abundant in the sample than the radionuclide of interest. To give a precise and realistic measurement, this background must be suppressed.

## 18.2.2 Mass-To-Charge Ratio Background

As described in Chap. 16, a mass spectrometer truly measures the ratio M/q instead of M, where M is the mass of the interesting species and q its charge. This means that several combinations of M and q could give the same ratio. This ambiguity has to be broken, and in very few cases, it is possible with conventional mass spectrometry. The use of collision cells in ICP–MS helps to do this, but it is not enough in most cases, especially for specific radionuclides or when extremely low concentrations are involved in the measurement. Some examples of this background source are given in Table 18.2. Within this context, a new analytical technique, Accelerator Mass Spectrometry (AMS), appears to be an alternative to cope with such difficult experimental challenges in the presence of such large background sources.

As described below, an AMS consists of the coupling of a mass spectrometer to an accelerator, generally of tandem type. This helps to overcome the most common

**Table 18.1** Examples of isobars and molecules that can interfere with the measurement of different radionuclides by mass spectrometry. In some cases, the radionuclide is injected as an oxide in the spectrometer because its production rate at the ion source is clearly higher than in the elemental form

| Radionuclide | Potential interference molecules |
|---|---|
| $^{10}$Be as $^{10}$BeO | $^{10}$BO, $^{9}$Be$^{16}$OH, $^{9}$Be$^{17}$O, $^{12}$C$^{14}$N, $^{13}$C$_2$ |
| $^{14}$C | $^{14}$N, $^{12}$CH$_2$, $^{13}$CH, $^{14}$NH, $^{7}$Li$_2$ |
| $^{26}$Al | $^{26}$Mg, $^{14}$N$^{12}$C, $^{10}$B$^{16}$O, $^{13}$C$_2$, $^{25}$MgH, $^{26}$MgH |
| $^{36}$Cl | $^{36}$S, $^{36}$Ar |
| $^{41}$Ca | $^{41}$K, |
| $^{129}$I | $^{129}$Xe, $^{97}$Mo$^{16}$O$_2$, $^{127}$IH$_2$, $^{128}$TeH |
| $^{236}$U as $^{236}$UO | $^{235}$U$^{16}$O, $^{235}$UH$^{16}$O, $^{235}$U$^{18}$O |
| $^{237}$Np as $^{237}$NpO | $^{238}$U$^{15}$N, $^{238}$UH$^{14}$N, $^{238}$U$^{13}$CH$_2$, $^{235}$UH$_2$$^{16}$O, $^{235}$U$^{18}$O |
| $^{239}$Pu, $^{240}$Pu as $^{239,240}$PuO | $^{238}$U$^{16}$O, $^{238}$U$^{17}$O, $^{238}$UH$^{16}$O, $^{238}$U$^{17}$O, $^{238}$U$^{18}$O, $^{238}$UH$^{17}$O |

**Table 18.2** Examples of ions of the same M/q ratios that can potentially interfere with the measurements of the radionuclides of interest

| Ion | Same M/q ratio interfering species |
|---|---|
| $^{14}$C$^{+}$ | $^{28}$Si$^{2+}$ |
| $^{14}$C$^{+}$ | $^{12}$C$^{16}$O$^{2+}$ |
| $^{14}$C$^{2+}$ | $^{7}$Li+ |
| $^{36}$Cl$^{3+}$ | $^{12}$C$^{+}$ |
| $^{36}$Cl$^{4+}$ | $^{9}$Be$^{+}$ |
| $^{36}$Cl$^{4+}$ | $^{18}$O$^{2+}$ |

limitations of conventional, or low energy, mass spectrometry. However, first, we will give some data about the history of AMS.

## 18.3 History of AMS

### 18.3.1 Early Measurements. The $^{14}$C Dating Problem

The first application of an accelerator-based mass spectrometry technique was carried out by Alvarez and Cornog in 1939 [1], when they measured $^{3}$He in He with a cyclotron. Nevertheless, the modern form of AMS, i.e., that based on tandem accelerators, appeared in 1977 [2] as a way to overcome the limitations that radiometric techniques had for $^{14}$C dating problems. The discovery at that time of the instability of $^{14}$N$^{-}$ was the starting point of this new adventure, as it is the main isobar for counting $^{14}$C atoms (see Table 18.1). Indeed, since $^{14}$N$^{-}$ was found to disappear as it is formed in the ion source, counting $^{14}$C atoms with a tandem accelerator was possible and realistic. At the same time, this provided excellent sensitivity to $^{14}$C dating problems by reducing the sample size needed and the counting time for precise dating. Therefore, it also increased the throughput of the technique. Once the door was open, many other radionuclides were targets of AMS systems: $^{10}$Be, $^{32}$Si, $^{36}$Cl, and $^{129}$I, among others. On the other hand, many tandem accelerators became obsolete for Nuclear Physics research at these times. Many of them found in AMS a new opportunity to prolong its useful life.

## 18.4    Principles and Contributions of AMS

### 18.4.1 Typical AMS Systems

Basically, an AMS system consists of coupling of a mass spectrometer to a tandem accelerator [3, 4]. The typical layout is depicted in Fig. 18.2a, and some realistic AMS systems are presented in Fig. 18.2b. In general terms, three parts can be distinguished in one AMS system: the low-energy side, the accelerator, and the high-energy side. The low-energy side is generally formed by a combination of magnets and electrostatic analyzers in addition to the ion source. The accelerator is normally of a Van de Graaff or Cockcroft–Walton tandem type, and finally, the high-energy side is also formed by a combination of magnets and electrostatic analyzers. Radionuclides of interest are counted by a nuclear radiation detector. As described in what follows, this simple scheme helps to dramatically decrease the main background sources.

### 18.4.2 The Ion Source and Low-Energy Side. Advantages of Accelerating Negative Ions

To take advantage of the stripping process in tandem accelerators, it is necessary to inject negative ions from the ion source. The best option in AMS is the use of Cs-sputtering ion sources because of their efficiency and stability. In addition, it is possible to obtain an ample family of negative ions with them. In Fig. 18.3, a scheme of these ion sources is presented. Negative ions are produced by bombarding the sample with a current of $Cs^+$ ions. This is obtained by heating a vapor of Cs up to 1200 °C, a temperature at which the atoms are thermally ionized. The $Cs^+$ current is accelerated toward the sample surface, which is connected to a negative potential with regard to the ionizing area. The contact of $Cs^+$ with the atoms or molecules in the sample produces the formation of negative ions through a charge-exchange reaction. Negative ions of interest are extracted by a positive extraction potential, $V_0$.

The use of negative ions produces a rejection of interfering elements, which means a reduction of the expected background. In some cases, this is a crucial step in the process. In fact, it is a key question in the case of $^{14}$C, as said before. Indeed, since $^{14}$N$^-$ is not stable, the main isobaric interference for $^{14}$C vanishes in the ion source. This happens to other radionuclides of interest either in elemental or molecular form. In Table 18.3, an account of several of these cases is given. We

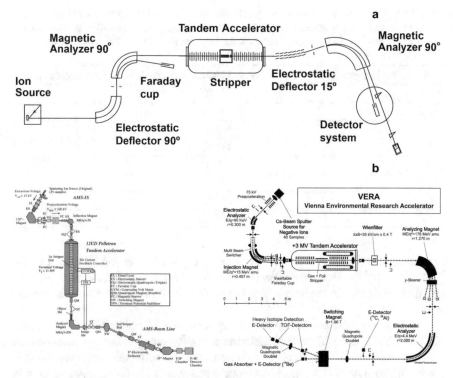

a

b

**Fig. 18.2**   Figure a) is a schematic layout of a typical AMS system. The figure is courtesy by Prof. J. M. López [9] and is based on the ETH, Zürich, 6 MV tandem Van de Graaff AMS system. In addition, two realistic systems based on a 14 MeV tandem Van de Graaff Pelletron at Tsukuba (reprinted from [8] with permission of Elsevier) and the other on a 3 MV tandem Van de Graaff Pelletron at Vienna (reprinted from [69] with permission from Elsevier) were used

**Fig. 18.3**   Schematic view of a Cs-sputtering ion source

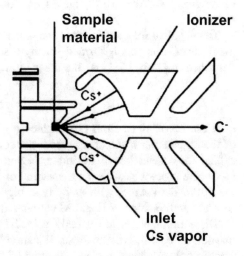

**Table 18.3** Use of negative ions allows for the first selection in AMS, since some isobaric pairs are not stable at the negative charge state. This also occurs with some molecules. In such cases, it is convenient to inject the ion under a given molecular form, since the parallel molecule formed by the isobaric pair is not stable at a negative charge state

| Radionuclide in elemental or molecular form | Isobaric pair that does not form negative ions |
|---|---|
| $^{14}C$ | $^{14}N$ |
| $^{26}Al$ | $^{26}Mg$ |
| $^{36}Cl$ | $^{36}Ar$ |
| $^{129}I$ | $^{129}Xe$ |
| $^{10}BeF$ | $^{10}BF$ |
| $^{32}SiH_3$ | $^{32}SH_3$ |
| $^{41}CaH_3$ | $^{41}KH_3$ |
| $^{41}CaF_3$ | $^{41}KF_3$ |

can see how the use of negative ions is a very good option for $^{14}C$ but also for $^{26}Al$, $^{36}Cl$, or $^{129}I$. Even for $^{10}Be$ or $^{32}Si$ and others, the isobars can be rejected by injecting them into the accelerator in a molecular form under which the isobars are unstable.

Unfortunately, the instability of the negative ions of noble gases makes conventional AMS unuseful for their determination. Recently, the use of positive ions in AMS experiments has received attention again to solve this problem [5]. It is interesting to note that in the pioneering AMS works [1, 6, 7], positive ions in cyclotrons were used. Studies are in progress on that problem by using Electron Cyclotron Resonance (ECR) ion sources coupled to a superconducting LINAC accelerator. According to the results presented in that paper, $^{39}Ar$ can be measured with this system, which is a very relevant result, since the low abundance of $^{39}Ar$ in Nature ($^{39}Ar/Ar \sim 10^{-16}$) makes its measurement impossible by conventional methods.

The negative ions are injected into the low-energy side of the spectrometer. In many cases, this part is formed just by a sector magnet that selects the mass as described in Chap. 16. Now, the magnetic rigidity is

$$\rho_B B = \sqrt{2MV_0} \qquad (18.1)$$

as it can be obtained from (16.2) since $q = 1$ and $V_0$ is the extraction potential of the ion source. If the energy from the source was constant, this magnet would suffice for a selection of the mass. Nevertheless, this is not the case since the sputtering process produces a dispersion of the energies of the outgoing negative ions from the source, with a $\frac{1}{E^2}$ type high-energy tail. Under these conditions, several combinations of E and M can give the same magnetic rigidity.

These ambiguities are usually solved by adding, before or after the sector magnet, an electrostatic analyzer. Equation 16.1 shows that the application of an electric field produces a clear selection of the ion energy, E, since $q = 1$. Consequently, as shown in Chap. 16, the adequate application of a magnetic and an electric field to the beam produces a sharp determination of a point in the $(\frac{M}{q}, \frac{T}{q})$ plane. However, after this first selection, several species can still survive. Isobars and molecules of the same mass as well as particles undergoing random collisions with the residual gases in the accelerator tubes or in the tubes themselves. Some of them could follow the trajectory of radionuclides of interest, contributing to the background of the measurement.

The low-energy side also contains several ion optic components that help to focus and transport the charged beam before injection into the accelerator. They were described in Chap. 17.

### 18.4.3 The Tandem. Charge Stripping and Tandem Beams. Molecular Background Suppression

As described in the previous chapter, the ion beam is accelerated in two steps. The total energy after the acceleration is that of (17.2), where q is one of the positive charge states produced after the stripping process. The selection of one of the charge states by an electrostatic analyzer at high energy further reduces the background of the measurement. During the stripping process, the molecules traveling together with the radionuclide of interest lose electrons as well and become unstable. Due to Coulomb repulsion effects, the molecules break into fragments. It is very easy to see that the energy, $T$, of the molecular fragment, for instance, of mass $m_1$, is

$$T = eV(\frac{m_1}{M} + q) \qquad (18.2)$$

$M$ is the total mass of the molecule before breaking, which is that of the radionuclide of interest, and $q$ is the charge state after stripping. It is very clear that the fragment can be easily eliminated by the following electrostatic filters. Thus, a major part of the molecular background is rejected. This is a unique feature of AMS and an important contribution to the field, as it increases the sensitivity of the measurement by orders of magnitude. The lowering of the background is at this point very important, but some possible interfering elements are still present in the beam. Of course, isobars and molecular fragments can simulate the trajectory of the radionuclide of interest. Thus, additional filters are still needed on the high-energy side.

### 18.4.4 High-Energy Side and Ion Detectors. Isobar Rejection

On the high-energy side, a combination of a sector magnet and an electrostatic analyzer allows the selection of a specific point in the $(\frac{M}{q}, \frac{T}{q})$ plane as previously described. However, now

$$\frac{T}{q} = eV\frac{(1+q)}{q} \qquad (18.3)$$

**Fig. 18.4** $(\frac{M}{q}, \frac{T}{q})$ plane with tandem beams, i.e., beams of known charge that can be selected in the high-energy side. The red circle highlights the selected $(\frac{M}{q}, \frac{T}{q})$ point at these conditions

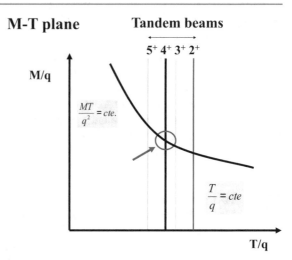

This means that the electrostatic analyzer is in practical terms selecting one of the charge states from the distribution leaving the tandem. This imposes a restriction in the $(\frac{M}{q}, \frac{T}{q})$ plane, as seen in Fig. 18.4, and the pair $(\frac{M}{q}, \frac{T}{q})$ is more precisely selected. In other words, most of the remaining background is removed with this combination of electric and magnetic fields. Nevertheless, isobars are still present in the beam, as are other molecular fragments that can reach the detector after random collisions with the residual gases or with the accelerator tube.

Since the beam energies are $\sim \frac{MeV}{amu}$, it is possible to use nuclear radiation detectors. Apart from counting the arriving atom number, the detector can inform about some new characteristics of the ions and thus identify the radionuclide of interest unambiguously.

In most cases, a $\Delta$E-E telescope is used as a final detector. Multianode gas ionization chamber telescopes are the most popular choice, although those based on Si detectors are also used (see Chap. 12). In both cases, the mass resolution of these instruments is good enough to distinguish the radionuclide of interest from its isobars. Some $\Delta$E–E plots are presented in Fig. 18.5 for $^{14}$C, $^{10}$Be, $^{26}$Al, and $^{36}$Cl.

Time-of-flight spectrometers (TOFs) are also used. Now, the mass of the ion can be known by the measurement of the time one ion spends traveling a given distance, since the kinetic energy is also known. Details can be found in Chap. 12, but a TOF spectrum is depicted in Fig. 18.6 for $^{129}$I measurements.

There are additional isobar suppression methods. One of them is based on the use of laser light for electron detachment of elements and molecules [10]. Selective electron detachment can be carried out by using laser radiation of the appropriate wavelength. For that, a laser beam is sent collinearly to the ion beam before acceleration. The beam–laser interaction can produce a selective removal of elemental or molecular isobar species provided that there exists a difference between

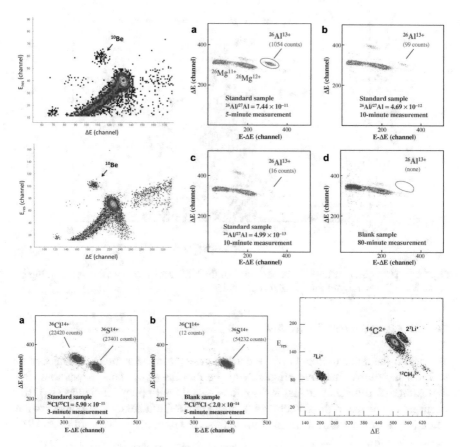

**Fig. 18.5** ΔE-E plots for $^{14}$C, $^{10}$Be, $^{26}$Al, and $^{36}$Cl measurements. Together with the spots of the interesting radionuclides, the spots of the main interfering ions are also included. The $^{26}$Al and $^{36}$Cl figures are reprinted from [8] with permission from Elsevier. The $^{10}$Be Figure from [27] with permission of Elsevier and the $^{14}$C Figure is courtesy of Dr. F. J. Santos

the electron affinity of the isobaric interference and that of the radionuclide of interest.

Originally, this technique was studied for the separation of $^{36}$S from $^{36}$Cl. Recently, new technical developments in laser light sources have permitted the application of this technique to other radionuclides. The so-called ILIAMS method, after ion laser interaction mass spectrometry, is currently a promising field of study [11].

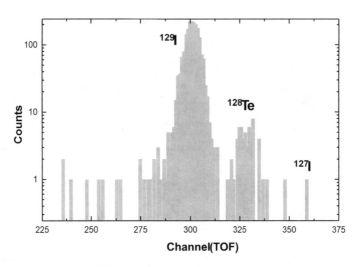

**Fig. 18.6** TOF spectrum for $^{129}$I measurement in the ETH Zürich 6 MV AMS system. Figure taken from [9]. Courtesy of Prof. J. M. López Gutiérrez

## 18.4.5 Measurements in AMS

To convert the detector counts into the atoms present in the sample, it is necessary to know the global counting efficiency of the process ($\varepsilon_{tot}$). This means knowing the extraction efficiency at the ion source ($\varepsilon_{is}$), the stripping efficiency ($\varepsilon_{st}$), the counting efficiency at the detector ($\varepsilon_d$), and a general efficiency that considers the beam losses of the atoms of interest throughout the system ($\varepsilon_t$). According to [12], $\varepsilon_{tot}$ can be expressed as

$$\varepsilon_{tot} = \varepsilon_{is}\varepsilon_{st}\varepsilon_d\varepsilon_t \tag{18.4}$$

In many cases, the product $\varepsilon_{st}\varepsilon_t$ is called the overall transmission. Generally, $\varepsilon_{is}$ ranges from 0.5 to 30%, while the overall transmission reaches values from 10 to 50%. Considering that the counting efficiency $\varepsilon_d$ is practically 100%, the total efficiency $\varepsilon_{tot}$ can be a few%. Of course, each partial efficiency depends on the previous efficiency. Thus, (18.4) is not an exact representation of the $\varepsilon$ value. However, it helps to give an idea of how complicated it is to obtain a precise $\varepsilon_{tot}$ value.

However, conventionally, the AMS measurements are given in terms of isotopic ratios, i.e., as the ratio between the number of atoms of interest and the number of atoms of, normally, the more abundant stable isotope of the isotopic series. Thus, the obtained counts in the detector can be, in fact, normalized to those found for the stable isotope. This normalization allows us to cancel all the efficiencies in (18.4) provided that the stable isotope is measured under the same conditions as the radionuclide of interest. In Table 18.4, the most important AMS radionuclides are given together with the isotopic ratios used in their measurements. For the sake

**Table 18.4** List of the most important AMS radionuclides together with the isotopic ratio used for their measurement. As a reference, the typical ratios found in environmental samples are also given. For the case of $^{237}$Np, $^{239}$Pu, and $^{240}$Pu, no stable isotope exists in nature, and as explained in the text, $^{242}$Pu is used as a reference isotope. $^{36}$Cl is compared to stable Cl, which is formed by $^{35}$Cl (75.77%) and $^{37}$Cl (24.23%). Both stable isotopes must be measured with Faraday cups placed in the appropriate position

| AMS Radionuclides | Isotopic ratio | Environmental isotopic ratio |
|---|---|---|
| $^{10}$Be | $^{10}$Be/$^{9}$Be | $10^{-11}$–$10^{-5}$ |
| $^{14}$C | $^{14}$C/$^{12}$C | $10^{-14}$–$10^{-11}$ |
| $^{26}$Al | $^{26}$Al/$^{27}$Al | $10^{-13}$–$10^{-12}$ |
| $^{36}$Cl | $^{36}$Cl/Cl | $10^{-15}$–$10^{-12}$ |
| $^{41}$Ca | $^{41}$Ca/$^{40}$Ca | $10^{-14}$–$10^{-11}$ |
| $^{129}$I | $^{129}$I/$^{127}$I | $10^{-12}$–$10^{-7}$ |
| $^{236}$U | $^{236}$U/$^{238}$U | $10^{-12}$–$10^{-6}$ |
| $^{237}$Np | – | – |
| $^{239}$Pu, $^{240}$Pu | – | – |

of comparison, their typical environmental values are also given. Of course, if the concentration of the stable isotope is known, that of the radionuclide of interest immediately follows.

The measurement of stable isotopes is only possible by using Faraday cups. Their abundance would overload the radiation detectors, making their use unrealistic. In principle, Faraday cups can be placed on the lower- or high-energy side of the spectrometer. However, the measurement must be performed under the same conditions as for the radionuclide of interest. For that, the best option is to locate the Faraday cup on the high-energy side (Fig. 18.7). This configuration better reproduces the whole process. In a real AMS system, however, there also exist fixed or mobile Faraday cups at the low-energy side, normally used for beam diagnostics and testing of the ion current from the ion source.

The processes of production, transmission, and counting of isotopes are very complex in an AMS system. Thus, it is not always ensured that all the efficiencies considered in (18.4) are the same for the different isotopes involved in the measurement. There are fractionation events that could affect the isotopes in a different way. The experimental isotopic ratio obtained, $R_{exp}$, might not be the real ratio, $R$. For that, it is very convenient to use a standard of known composition and refer to it. This is the normal practice in AMS and so

$$R = R_{exp} \frac{S}{S_{exp}} \tag{18.5}$$

$S$ is the standard isotopic ratio, which is well known, and $S_{exp}$ is the standard measured ratio. In addition, background effects must be considered. In fact, the background produced during the operation of the spectrometer or during the preparation of the sample could affect the measurement. For that, a more exact

## High Energy side

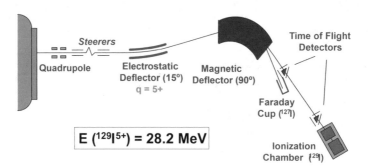

**Fig. 18.7** Location of the Faraday cups on the high-energy side of the ETH Zürich 6 MV AMS system. Figure taken from [9]. Courtesy of Prof. J. M. López-Gutiérrez

expression for $R$ is

$$R = (R_{exp} - R_b)\frac{S}{S_{exp}} \tag{18.6}$$

$R_b$ is the background isotopic ratio.

Some radionuclides do not have stable isotopes, and the ratios must be measured by selecting another reference isotope. This is the case for $^{239}$Pu, $^{240}$Pu, and $^{237}$Np, where $^{242}$Pu is used as a reference isotope (see Table 18.4). In real measurements, the samples are spiked with known amounts of $^{242}$Pu. $^{242}$Pu is also radioactive, and therefore, very low masses have to be used for spiking. In real terms, some picograms are the typical mass added to the sample [12]. It is apparent that now the quantification of $^{242}$Pu must be done with the radiation detectors and not with the Faraday cup.

All these conditions make AMS a very powerful analytical technique that finds applications in many fields. Nuclear Physics, Astrophysics, Particle Physics, Geology, Archaeology, Medicine, Industry, etc., benefit from its analytical features; and, of course, environmental or paleoenvironmental studies, as we will see in the sections that follow. AMS has found more applications within this field, as it is the most sensitive technique for determining very long-lived radionuclides in natural samples.

In Table 18.5, we give a list of traditional radionuclides determined by AMS in natural samples and the isotopic ratios that some important AMS installations can measure. This list is increasing today, as we will see later, but for is now enough to give us an idea on the sensitivity of the technique. AMS is faster and less destructive, and the throughput is clearly higher than that of other analytical techniques. More specifically, compared to LLC techniques. This is because the required counting times and sample sizes are orders of magnitude shorter and lower, respectively, to provide precise measurements. As a consequence, the number of analyses per unit of time increases. In Table 18.6, a comparison is made between AMS and counting techniques. Indeed, it is very simple to see from (2.8) that the number of counts, $N_{count}$, recorded during a time interval $\Delta t$, when counting a sample containing N radioactive nuclei is

$$N_{count} = \frac{\Delta t}{T_{\frac{1}{2}}} \varepsilon N \ln 2 \qquad (18.7)$$

where $T_{1/2}$ is the half-life of the radionuclide and $\varepsilon$ is the counting efficiency. This equation is valid for $\Delta t \ll T_{1/2}$. Let us assume $\Delta t = 6 \times 10^5$ s, one week of

**Table 18.5** Minimum detectable isotopic ratios (sensitivity) for some AMS radionuclides detectable by several facilities worldwide, as identified in column 1. This table tries to give some examples, and more information can be found in the following references: [13–17]. The Web page of the installations gives this information in many cases, such as for Lawrence Livermore National Laboratory (LLNL). Table 18.7 gathers this information for the Spanish SARA installation at the National Center for Accelerators in Seville, Spain

| Facility | $^{10}$Be | $^{14}$C | $^{26}$Al | $^{36}$Cl | $^{41}$Ca | $^{129}$I | $^{236}$U |
|---|---|---|---|---|---|---|---|
| VERA Austria | $5 \times 10^{-16}$ | $3 \times 10^{-16}$ | $6 \times 10^{-16}$ | $10^{-13}$ | $10^{-13}$ | $2 \times 10^{-14}$ | $6 \times 10^{-13}$ |
| ETH/PSI Switzerland | $3 \times 10^{-15}$ | $4 \times 10^{-15}$ | $5 \times 10^{-15}$ | $2 \times 10^{-15}$ | $2 \times 10^{-12}$ | $10^{-13}$ | $10^{-14}$ |
| LLNL/CAMS USA | $5 \times 10^{-17}$ | $2 \times 10^{-15}$ | $2 \times 10^{-15}$ | $3 \times 10^{-15}$ | $10^{-13}$ | $2 \times 10^{-14}$ | $7 \times 10^{-10}$ |
| JAE-AMS Japan | $5 \times 10^{-15}$ | $2 \times 10^{-16}$ | $10^{-14}$ | | | | |
| Sirius Australia | | | $10^{-11}$ | $10^{-15}$ | | | |
| DREAMS, Germany | $5 \times 10^{-16}$ | | $8 \times 10^{-16}$ | $4 \times 10^{-16}$ | $2 \times 10^{-15}$ | $2 \times 10^{-14}$ | |

**Table 18.6** Comparison between counting techniques and AMS. In column 4, the number of atoms needed for a 1% uncertainty measurement of samples compounded by radionuclides in column 1 is given. In column 5, the number of atoms needed for an AMS experiment under the same conditions is shown. Between 2 or 6 orders of magnitude fewer atoms are needed in AMS. See the text for details of the calculations. Efficiency data can be found in [18–21]

| Radionuclide | Half-life (y) | AMS efficiency | N (counting) | N (AMS) |
|---|---|---|---|---|
| $^{10}$Be | $1.40 \times 10^6$ | $2.00 \times 10^{-3}$ | $1.05 \times 10^{12}$ | $5.00 \times 10^6$ |
| $^{14}$C | $5.73 \times 10^3$ | $4.00 \times 10^{-2}$ | $4.30 \times 10^9$ | $2.50 \times 10^5$ |
| $^{26}$Al | $7.00 \times 10^5$ | $1.00 \times 10^{-4}$ | $5.25 \times 10^{11}$ | $1.00 \times 10^8$ |
| $^{36}$Cl | $3.01 \times 10^5$ | $2.00 \times 10^{-2}$ | $2.26 \times 10^{11}$ | $5.00 \times 10^5$ |
| $^{41}$Ca | $9.94 \times 10^4$ | $1.00 \times 10^{-4}$ | $7.46 \times 10^{10}$ | $1.00 \times 10^8$ |
| $^{129}$I | $1.60 \times 10^7$ | $1.00 \times 10^{-2}$ | $1.20 \times 10^{13}$ | $1.00 \times 10^6$ |

**Table 18.7** Features of the SARA AMS system of the National Center for Accelerators in Seville, Spain, based on a 1 MV tandem Cockcroft–Walton accelerator (see Fig. 18.8). With the charge state presented in column 2, the sensitivity reached for the corresponding isotopic ratio is displayed in column 3. In the case of $^{237}$Np, $^{239,240}$Pu, $^{241,\,243}$Am, there are no reference isotopes in nature. $^{242}$Pu is used instead. For that, the sensitivity in terms of the minimum detectable atom number or mass is given

| Radionuclide | Charge State | Sensitivity | Environmental isotopic ratio |
|---|---|---|---|
| $^{10}$Be | $1^+$ | $2 \times 10^{-14}$ ($^{10}$Be/$^9$Be) | $10^{-11}$–$10^{-5}$ |
| $^{14}$C | $2^+$ | $2 \times 10^{-15}$ ($^{14}$C/$^{12}$C) | $10^{-14}$–$10^{-11}$ |
| $^{26}$Al | $1^+$ | $2 \times 10^{-14}$ ($^{26}$Al/$^{27}$Al) | $10^{-13}$–$10^{-12}$ |
| $^{41}$Ca | $3^+$ | $10^{-12}$ ($^{41}$Ca/$^{40}$Ca) | $10^{-14}$–$10^{-11}$ |
| $^{129}$I | $3^+$ | $2 \times 10^{-14}$ ($^{129}$I/$^{127}$I) | $10^{-12}$–$10^{-7}$ |
| $^{236}$U | $3^+$ | $8 \times 10^{-11}$ ($^{236}$U/$^{238}$U) | $10^{-12}$–$10^{-6}$ ($^{236}$U/$^{238}$U) |
| $^{237}$Np | $3^+$ | $10^6$ atoms (1 fg) | – |
| $^{239,240}$Pu, $^{241,\,243}$Am | $3^+$ | $10^6$ atoms (1 fg) | – |

counting time, which is quite common in normal experiments. On the other hand, let us take $\varepsilon = 1$, the most favorable case. Under these conditions, the number of atoms, N, in the sample needed to give a count number with 1% uncertainty, that is, $N_{count} = 10^4$, is presented in column 4 of Table 18.6. On the other hand, the number of atoms counted in an AMS system, $N_{AMS}$, for the same sample will be

$$N_{AMS} = \varepsilon_{tot} N \qquad (18.8)$$

where $\varepsilon_{tot}$ is the global efficiency described in (18.4). A group of $\varepsilon_{tot}$ values found in the current literature is included in column 3 of Table 18.6. For $N_{AMS} = 10^4$, that is, for an AMS measurement with 1% uncertainty, the number of atoms needed in the sample is presented in column 5 of *Table 18.6*. The AMS atom number is within 2 to 6 orders of magnitude lower than those for counting. This is a clear indication of the advantages of using AMS.

## 18.5   Low-Energy AMS (LEAMS)

### 18.5.1 The Use of Low-Terminal Voltages

The main drawback of AMS is the need for large and expensive facilities. This has prevented its use in many laboratories. Nevertheless, at the beginning of the 2000s, the scenario changed, since it was realized [22, 23] that the use of low-terminal voltages, i.e., 1 MV, was feasible for competitive $^{14}$C dating works. Since that time, the so-called low-energy AMS (LEAMS) has evolved in a relevant way, and it is currently applicable to the measurement of almost all AMS radionuclides. A first consequence of the use of low-terminal voltage machines is the significant reduction in the dimensions of the spectrometer. In fact, it is possible to find compact LEAMS machines that need laboratories of less than 10 m$^2$ surface (see Fig. 18.8). Additional consequences are the ease of operation and reduced costs of these systems, which have made AMS available to many laboratories, and its diffusion worldwide has been very fast.

However, the technical challenges posed by the use of low-terminal voltage are not trivial. Thus, it has been necessary to revise some scientific knowledge on beam transport and charge-state changes in strippers at low energies.

## 18.5.2 Charge Stripping at Low-Terminal Voltages

The charge states primarily produced during the stripping process at low energies are 1$^+$ or 2$^+$ [22]. At these charge states, molecules do not suffer the 'Coulomb explosion' effect, which helps to reject the molecular background in AMS systems, as explained above. Figure 18.9 gives a typical charge-state distribution obtained at low energies. For the case of $^{14}$C measurements, where the molecules $^{13}$CH$^+$ or $^{12}$CH$_2$$^{2+}$ are important components of the background, this situation makes the use of AMS for dating impossible.

However, it was found that such molecules could be destroyed, increasing the density of the stripper gas. This means increasing the gas pressure, which poses several technical problems in the design of LEAMS systems. First, keeping the whole system vacuum at good values is more difficult. Thus, a high pumping capacity must be implemented to avoid residual gases in the accelerator tube that could cause a background increase. On the other hand, higher stripper pressures mean an increase in the angular divergence of the beam after stripping. To avoid beam losses, the dimensions of the housing and exit of the stripper must be enlarged compared to conventional systems. This must be compatible with the need to maintain adequate vacuum in the lines. Additionally, angular divergence introduces new conditions in the design of optics for beam transport along the high-energy side spectrometer. Specifically, the acceptance of the sector magnet must be increased. The energy dispersion is also important when designing the sector magnet and the electrostatic analyzer on the high-energy side. Furthermore, since the ion scattering cross section is $\sim \frac{1}{E^2}$, it is expected that more dispersive processes take place during beam transport. It must be taken into account when designing the system. Of course, the use of low energies makes the selection of the detector window where significant amounts of energy can be deposited crucial, thereby making radionuclide identification difficult. Finally, from a practical viewpoint, the energy of the ions at the ion source exit is not negligible compared to the total energy acquired during the acceleration. This must be considered when

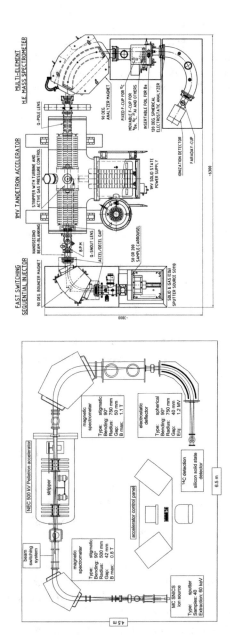

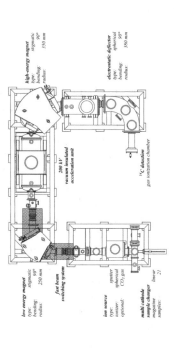

**Fig. 18.8** Some LEAMS compact systems. The Tandy 500 kV AMS system at ETH Zürich (reprinted from [70] with permission of Elsevier) and the 1 MV AMS system at CNA, Sevilla (reprinted from [71] with permission of Elsevier). Finally, the 200 kV Van de Graaff MICADAS (reprinted from [28] with permission from Elsevier)

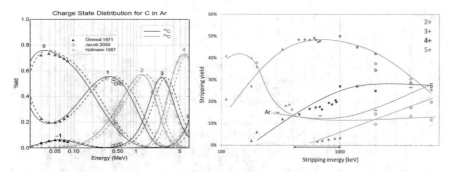

**Fig. 18.9** Charge-state distribution for $^{14}$C and $^{12}$C using Ar as the stripper gas (reprinted from [24] with permission of Elsevier) and for U using He as the stripper gas (reprinted from [72] with permission of Elsevier)

tuning the system for a given mass counting. All these problems have been overcome in recent years, and it is currently possible to find LEAMS compact systems that can competitively measure all AMS radionuclides. In Table 18.7, the features of one such LEAMS system for different interesting radionuclides are given [25].

## 18.5.3  Some Examples of LEAMS Systems

Some examples of modern LEAMS systems can be seen in Fig. 18.8. In Fig. 18.10, taken from [20], the increase in laboratories that use such a compact LEAMS system is given over time.

One pioneering spectrometer was the Tandy AMS system at the ETH Zürich. It is based on a 600 kV terminal voltage van de Graaff tandem accelerator. Together with the 1 MV Cockcroft–Walton Tandetron system of the National Center for Accelerators (CNA) in Spain, the so-called SARA belongs to the first generation of compact LEAMS systems. The dimensions of such spectrometers are relatively small, approximately $4 \times 6$ m$^2$. Originally, Tandy was oriented for $^{14}$C dating purposes. Today, it has evolved toward a similar system, MILEA [26], which is a multielemental spectrometer capable of measuring masses from $^{14}$C to Pu isotopes and other actinides. The CNA system was oriented to multielemental purposes from the beginning, and today, it is possible to measure $^{14}$C, $^{10}$Be, $^{26}$Al, $^{41}$Ca, $^{129}$I, Pu isotopes, $^{236}$U, $^{237}$Np, and other actinides [27]. Even smaller systems than those described can be competitive in AMS. An example is the MICADAS spectrometer, $3 \times 2.5$ m$^2$ in size, based on a 200 kV Van de Graaff accelerator [28]. MICADAS is a $^{14}$C dedicated system that is very competitive for dating purposes (Fig. 18.8).

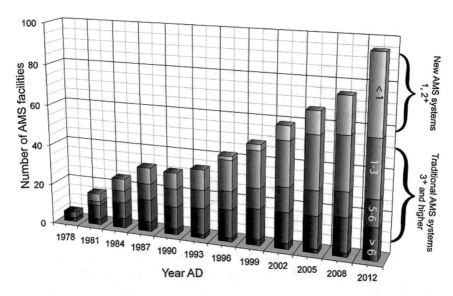

**Fig. 18.10** Temporal evolution of AMS facilities all over the world. The numbers within the bars indicate the terminal voltages. The number of installations that use terminal voltages <1 MV has increased tremendously in the last 10 years. The darker gray at the bottom shows the number of cyclotron AMS systems during the time from 1978 to 2008, when they disappeared. Figure reprinted from [20] with permission of Elsevier

## 18.6    AMS Applications to Environmental Radioactivity

### 18.6.1 AMS Sample Preparation

The sensitivity offered by AMS makes it an unvalued tool for environmental radioactivity studies. The main contributions compared to radiometric methods are the reduction of the counting times and the sample size (see Table 18.6). Counting times of 15 or 20 min for the determination of Pu isotopes or $^{129}$I in samples of no more than a few milligrams or liters, for solid or liquid samples, are sufficient in most measurements. This is simply impossible to do with conventional radiometric techniques. In fact, as presented in Table 18.5, AMS is the only option in many cases for the measurement of some long-lived radionuclides.

The mass resolution of AMS also allows the measurement of interesting isotopic ratios that cannot be measured by conventional detectors. This is, for example, the case for the $^{240}$Pu/$^{239}$Pu isotopic ratio, which is almost impossible to determine by α-spectrometry. It is well known that the information provided by these ratios is crucial to solving the origin and dynamics of radionuclides in the environment. This is another important contribution of AMS to the field of environmental radioactivity.

Of course, the preparation of samples requires a specific radiochemical procedure. As described in Chap. 13, for AMS, care must be taken to avoid the presence

**Table 18.8**  Materials used to prepare the so-called cathode, which is introduced in the ion source. This material increases the electrical and heat conductivity during irradiation with $Cs^+$. In the case of $^{14}C$, the graphite obtained in the sample preparation is directly pressed into the cathode. This graphite contains Co, which is used as a catalyst for the graphitization process, and $^{36}Cl$ is prepared in the form of AgCl. AgBr is added when the amount of AgCl is not enough to prepare the $^{36}Cl$ cathode

| Radionuclide | Material |
|---|---|
| $^{10}Be$ | Nb |
| $^{14}C$ | Graphite |
| $^{26}Al$ | Cu |
| $^{36}Cl$ | Ag Br |
| $^{41}Ca$ | Ag |
| $^{129}I$ | Nb or Ag |
| $^{236}U$ | $Fe_2O_3$ matrix mixed with Nb |
| $^{237}Np$ | $Fe_2O_3$ matrix mixed with Nb |
| Pu isotopes | $Fe_2O_3$ matrix mixed with Nb |

of mass-interfering elements in the final sample, which is not the main goal for radiochemistry oriented to counting radiation. This means that some modifications of the methodology must be performed, and examples can be found in the literature [29, 30]. However, the chemical form of the final sample must be compatible with the capacities of the ion source. It is also interesting to note that in many cases, it is convenient to inject specific molecules into the accelerator instead of pure elements, as presented in Table 18.3. This must be considered when designing a radiochemical method for a given radionuclide. Moreover, the final form of the sample must also be compatible with the sputtering ion source. The common case is that the final chemical compound is in a solid form. In this case, after drying the sample, it is powdered and mixed with a compound that facilitates heat and electricity conduction during irradiation with $Cs^+$. They are selected by optimizing several parameters during the AMS measurement. One of them is the current obtained from the ion source. Normally, the selection is carried out empirically, and a list of compounds used for the preparation of several sources is given in Table 18.8.

## 18.6.2  AMS in Environmental Radioactivity

The impact of AMS on environmental radioactivity studies is shown in Fig. 18.11. However, it is also true that the number of environmental problems addressed by AMS continuously increases as the number of possible AMS radionuclides increases. In what follows, some examples are described.

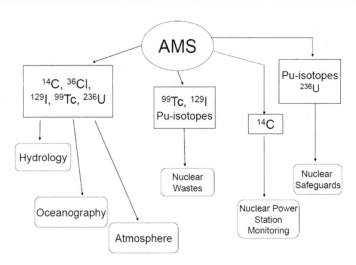

**Fig. 18.11** Impact of AMS in different topics of interest for environmental radioactivity studies by measuring some radionuclides in natural samples. Figure reprinted from [31] with permission of Elsevier

## $^{14}C$

Although the main role of $^{14}C$ in AMS is radiocarbon dating, it has been determined in many types of environmental samples.

$^{14}C$ is used to trace the movements of seawater masses. Seawater profiles can be produced by measuring approximately 0.5 L of water instead of hundreds, which is needed in conventional radiometric counting procedures. AMS counting times of 20 min give uncertainties close to $6°/_{oo}$ or better with some 20 min counting time per sample.

Radiocarbon has also been determined in samples taken from the surroundings of environments affected by the operation of nuclear installations. This is necessary since it is considered to be the main contributor to the collective dose of nuclear fuel processes [32]. $^{14}C$ has been measured in sediments, mussels, and biota collected in the Irish Sea [33]. Furthermore, $^{14}C$ has been measured in tree rings to estimate the impact of nuclear facilities on their close environment [34]. It has also been determined in atmospheric samples, rivers, and drinking waters near nuclear power plants [35].

## $^{129}I$

Before the arrival of the AMS, it was not possible to measure $^{129}I$ in the environment. Some previous trials using Neutron Activation Analysis were performed (see Chap. 19), but a realistic assessment of $^{129}I$ at environmental levels was only possible with AMS sensitivity. The environmental $^{129}I$ data file is large, as it has been measured since the early years of AMS. As $^{129}I$ is very conservative with regard to seawater, it is very much used in oceanographic studies [36]. $^{129}$I can be determined with a relative uncertainty of 10% in only 1 L of water.

The presence of $^{129}$I in the atmosphere also makes it an interesting air mass tracer [37]. Even sites that are not directly affected by nuclear installations. An example is that presented in [38], where the authors showed the influence of Sellafield and La Hague emissions in southern Spain by measuring atmospheric aerosol samples or just 0.5 L of rainwater samples. Using similar techniques, it was possible to detect the arrival of the Fukushima contamination plume in the same geographical area [39]. $^{129}$I can be used to trace the radiological impact of $^{131}$I [40], which is very useful, provided the very short half-life of $^{131}$I (8.02 days) and its paramount radiological importance. Such a short half-life makes it very difficult to perform a realistic measurement of its environmental distribution. Fortunately, $^{129}$I can provide a retrospective estimate of the radiological impact of $^{131}$I.

Chemical speciation studies are very relevant to learn about the dynamics of radionuclides in the environment. Indeed, as described in previous chapters, the chemical form of the radionuclide determines its mobility in the environment. Speciation studies for long-lived radionuclides are simply impossible with conventional radiometric techniques. The sensitivity of AMS facilitates the carrying out of $^{129}$I speciation analysis. Very illustrative studies for sea and freshwater systems can be found in [41, 42].

### $^{36}Cl$

The determination of $^{36}$Cl by AMS started at the beginning of the AMS technique [43]. It was possible to analyze $^{36}$Cl in 1 to 5 L of natural water samples with uncertainties less than 10% in 1 h of counting time. $^{36}$Cl has been applied to the dating of groundwater samples [44]. In addition to underground waters, it has been used as a tracer of accident contamination [45]. $^{36}$Cl measurements in the atmosphere [46] revealed a seasonal time pattern variation with peaks during late spring–early summer. An excess of $^{36}$Cl atmospheric deposition was found compared to results foreseen by cosmogenic models in [47]. They attributed this effect to the retention of the $^{36}$Cl bomb produced in the biosphere and its delayed reintroduction into the lower parts of the atmosphere.

### Pu Isotopes

The determination of Pu isotopes and Pu isotope ratios, mainly $^{240}$Pu/$^{239}$Pu, is of outstanding importance in the field of environmental radioactivity. Only in very special cases can such a ratio be estimated by using $\alpha$-spectrometry deconvolution codes. AMS can do so in shorter time frames and using smaller sample amounts. In fact, Pu isotopes in a variety of environmental samples can be determined at levels down to a few $\mu$Bq, i.e., some 3 orders of magnitude lower than the typical detection limits obtained in conventional $\alpha$-spectrometry [48]. The same authors measured the $^{240}$Pu/$^{239}$Pu isotope ratio in soil samples taken from Palomares in southeastern Spain, where an accident occurred in 1969. During such an accident, some amounts of Pu were dispersed in the environment. When measuring such an isotope ratio, it was possible to distinguish the soils affected by the accident from those affected by the fallout. In fact, the characteristic ratios of highly contaminated samples were found to be 6.57 ± 0.06%, quite far from the expected ratio associated with fallout, typically approximately 17%. The utility of measuring the ratio to unambiguously assign the origin of Pu in the study area was shown once again in [49], where $^{240}$Pu/$^{239}$Pu was determined in a marine sediment core taken offshore from Palomares. Ratios close to 6% in the first cm of the core demonstrated the influence of the accident on the marine area. The rest of the core showing ratios close to 11% is clearly more compatible with fallout ratios.

Due to the sensitivity of AMS, the Pu isotopes and $^{240}$Pu/$^{239}$Pu ratio in soil samples from the Southern Hemisphere were measured [50]. It is well known that contamination by man-made radionuclides in such a hemisphere is considerably lower than that in the Northern Hemisphere because most nuclear bomb tests occur in the Northern Hemisphere. In addition, the nuclear reprocessing activities in the Southern Hemisphere are also very small compared to those maintained in the North. Nevertheless, the authors were able to characterize the influence of French tests in some of the samples by measuring isotope ratios close to 4%, far from the ratios of the fallout in the area, which is 18%. Typical counting times of 20–30 min are needed to obtain results with a good uncertainty of a few percent. Using sample masses of 2–3 g.

A very interesting application of AMS is the measurement of Pu isotopes and isotopic ratios in human urine samples [51]. Routine measurement of Pu in urine samples is a mandatory practice for exposed workers. Conventionally, it is carried out by analyzing large volumes of urine samples by α-spectrometry. This work takes one week, including the radiochemical separation of Pu. The AMS method uses only 1–2 L of sample volume, and the analysis takes 10 h, including chemical processing. The advantage is very clear.

As in the case of $^{129}$I, Pu speciation studies are only possible with the help of AMS. An interesting study is that presented in [52], where Pu speciation is investigated in freshwater systems from the Ob and Yenisey rivers and estuaries in Russia.

## $^{236}$U

The measurement of $^{236}$U is impossible using conventional radiometric methods. Nevertheless, it is a very good oceanographic tracer, and its determination by AMS, although a challenging problem, is certainly feasible. The number of works on the determination of $^{236}$U in the literature has recently increased. Measurements in global fallout [53] show that the isotopic ratio $^{236}$U/$^{238}$U ranges from $10^{-7}$ to $10^{-9}$.

Measurements of just 2 L of river and seawater samples have been published [54]. Or in soils [55], where the inventory of $^{236}$U in the Southern Hemisphere has been found to be one order of magnitude lower than in the Northern Hemisphere. Its uptake by plants has also been studied [56], where the authors showed a preferential uptake of $^{236}$U compared to that of $^{239}$Pu.

Its presence and dynamics in oceans have been investigated by [57] with a sample volume of only 1 L. $^{236}U$ concentrations are higher in surface samples than in deep waters. The possibility of measuring $^{236}U$ by LEAMS was proposed by [58]. It was measured in the northwestern Mediterranean area [59]. Counting times of approximately 25 min and sample volumes of 5 L were enough for very precise determinations.

### $^{237}Np$

$^{237}Np$ is also a potential oceanographic tracer, and its measurement by AMS is receiving much attention. However, its measurement is not easy at all. Furthermore, the lack of a long-lived isotopic tracer makes its determination a very challenging task. Despite this, methods for the sequential separation of Pu isotopes $^{236}U$ and $^{237}Np$ in seawater have been published using $^{242}Pu$ as a tracer, which have made its determination possible in ocean waters [61].

### Others

Some other radionuclides have been determined in environmental samples, although studies in the regular literature are scarce. $^{99}Tc$ is one of them. The first work [62] focused on the technical effort to distinguish it from $^{99}Ru$, its main interference isobar. Refinements were made to the technique in [63]. An application to real samples can be found in [64], where seawater samples, some 200 ml sample volume, or arid soils, 60 to 100 g, were analyzed for $^{99}Tc$.

AMS has also tried $^{90}Sr$ or $^{135}Cs$ in [65] and [66], respectively. The case of $^{135}Cs$ is very special, since Cs does not easily form negative ions. In fact, it has to be injected into the accelerator as $CsF_2{}^-$, and an Isobar Separator for Anions has to be used to reject the Ba contamination of the sample. The paper provides promising first results.

Measurements of Am and Cm isotopes have been carried out by LEAMS [67]. Sensitivities of ~fg or even ~$10^{-1}$ fg for Am and Cm isotopes, respectively, can be reached, and the radionuclides are injected into the accelerator under an oxide form.

Recently, the potential use of the $^{233}U/^{236}U$ ratio has been proposed as a water mass tracer [68]. This is only possible by using AMS to measure $^{233}U$.

### Exercises

1. Demonstrate (18.2).
2. A $^{14}C$ beam containing some $^{12}C^-$ is extracted from an ion source with a voltage V. The beam is deflected by a magnetic field B that normalizes the beam. Find the new trajectory radii of the ratio between $^{14}$ and $^{12}C$.
3. Make the same calculation in the case of an electric deflection.
4. Some $\Delta E$-E plots are given in Fig. 18.5. Describe and interpret the presented results and specifically comment on the spot distribution in the plots.
5. A comparison between radiometric counting techniques and AMS is presented in Table 18.6. Make the calculations in columns 4 and 5.

6. In Tables 18.4 and 18.5 AMS isotopic ratios in the environment and their sensitivity in several facilities are given. Take the $^{14}$C data in Table 18.4 as the actual $^{14}$C content in Nature. Under these conditions, estimate the maximum $^{14}$C date that the laboratories in Table 18.5 can give.

7. Make the same calculation for $^{41}$Ca in the case where a full dating method could be developed based on that radionuclide.

8. Explain how isobars can be distinguished in an AMS experiment.
   (a) with a TOF spectrometer.
   (b) with a $\Delta$E-E telescope.

9. Describe the effect of the stripping process on background lowering in AMS.

10. Give a brief description on the origin of $\frac{M}{q}$ interferences in mass spectrometry and how they are suppressed in AMS.

11. Describe the differences between conventional mass spectrometry and AMS. What are the main advantages of using AMS instead of conventional mass spectrometry?

12. Explain the advantages of using AMS in addition to conventional radiometric techniques for $^{14}$C dating of archaeological samples.

13. The maximum magnetic rigidity at the high-energy side magnet of a tandem Cockcroft–Walton AMS system is 74 meV amu q$^{-2}$. If the radius of the magnet is 85 cm, find the energy that a $^{240}$Pu $^{3+}$ beam can reach to deviate accordingly to the system beamline.

14. Make the same calculation for $^{240}$Pu $^{2+}$, as well as $^{14}$C$^{+}$ and $^{14}$C$^{2+}$. Consider that the maximum terminal voltage is 1 MV.

15. It is expected that during the measurement of a $^{14}$C sample by a tandem 1 MV AMS system, some interfering species are injected into the tandem accelerator along with $^{14}$C$^{-}$. Examples are $^{12}$CH$_2$$^{-}$, $^{13}$CH$^{-}$, and $^{7}$Li$_2$$^{-}$, among others.
   (a) Explain why such species are not rejected on the low-energy side of the spectrometer.
   (b) Explain where you expect (and why) such interfering species could be rejected.

16. Usually, $^{10}$Be is injected as $^{10}$BeO into AMS systems. As in the previous exercise, along with $^{10}$BeO$^{-}$, some interfering species are also injected into the tandem accelerator: $^{10}$BO$^{-}$, $^{9}$BeO$^{-}$, and $^{9}$BeHO$^{-}$.
   (a) Explain why such species are not rejected on the low-energy side of the spectrometer.
   (b) Explain where you expect (and why) such interfering species could be rejected.

17. However, $^{10}$B$^{x+}$, together with $^{10}$Be$^{x+}$, survives after the magnetic and electrostatic filters on the high-energy side. Since their masses are very close, they cannot be distinguished with a conventional $\Delta$E-E telescope. Thus, an absorber foil is usually interposed between the magnet and the electrostatic analyzer on the high-energy side to facilitate their discrimination. Explain why is it possible.

# References

1. L.W. Alvarez, R. Cornog, $^3$He in helium. Phys. Rev. **56**, 379 (1939)
2. H.E. Gove, K.H. Purser, A. Litherland, Accelerator mass spectrometry (AMS) 1977–1987. Nuc. Instrum. Meth. Phys. Res. B **268**, xvii-xxii (2010)
3. C. Tuniz, W. Kutschera, D. Fink, G.F. Herzog, J.R. Bird, *Accelerator Mass Spectrometry: Ultrasensitive Analysis for Global Science* (CRC Press, 1998)
4. R. Finkel, M. Suter, AMS in the earth sciences: technique and applications. Adv. Anal. Geochem. **1**, 1–114 (1993)
5. M. Paul, R.C. Pardo, P. Collonc, W. Kutschera, K.E. Rehm, R. Scott, R.C. Vondrasek, Positive-ion accelerator mass spectrometry at ATLAS: peaks and pits. Nuc. Instrum. Meth. Phys. Res. B **456**, 222–229 (2019)
6. R.A. Muller, Radioisotope dating with a cyclotron. Science **196**, 489 (1977)
7. G. Raisbeck, F. Yiou, Possible use of $^{41}$Ca for radioactive dating. Nature **277**, 42 (1979)
8. K. Sasa, Y. Nagashima, T. Takahashi, R. Seki, Y. Tosaki, K. Sueki, K. Bessho, H. Matsumura, T. Miura, M. He, $^{26}$Al and $^{36}$Cl AMS system at the University of Tsukuba: a progress report. Nuc. Instrum. Meth. Phys. Res. B **259**, 41–46 (2007)
9. J.M. López-Gutiérrez, Aplicaciones de la Espectrometría de Masas con Aceleradores a la determinación de radionúclidos de semivida grande en la Naturaleza. Ph.D. Thesis, University of Sevilla, 1999
10. D. Berkovits, E. Boaretto, G. Hollos, W. Kutschera, R. Naaman, M. Paul, Z. Vager, Selective suppression of negative ions by lasers. Nucl. Instrum. Methods A **281**, 663–666 (1989)
11. M. Martschini, D. Hanstorp, J. Lachner, C. Marek, A. Priller, P. Steier, P. Wasserburger, R. Golser, The ILIAMS project – an RFQ ion beam cooler for selective laser photodetachment at VERA. Nuc. Instrum. Meth. Phys. Res. B **456**, 213–217 (2019)
12. E. Chamizo, Medida de Isótopos de Pu, $^{239}$Pu y $^{240}$Pu, mediante Espectrometría de Masas con Aceleradores de Baja Energía. Ph.D. Thesis, Universidad de Sevilla, 2009
13. M. Christl, C. Vockenhuber, P. W. Kubik, L. Wacker, J. Lachner, V. Alfimov, H.-A. Synal, The ETH Zurich AMS facilities: performance parameters and reference materials. Nuc. Instrum. Meth. Phys. Res. B **294**, 29–38 (2013)
14. Y. Saito-Kokubu, N. Fujita, M. Miyake, T. Watanabe, C. Ishizaka, N. Okabe, T. Ishimaru, A. Matsubara, A. Nishizawa, T. Nishio, M. Kato, H. Torazawa, N. Isozaki, Current status of JAEA-AMS-TONO in the 20th year. Nuc. Instrum. Meth. Phys. Res. B **456**, 271–275 (2019)
15. K.M. Wilcken, T. Fujioka, D. Fink, R.H. Fülöp, A.T. Codilean, K. Simon, C. Mifsud, S. Kotevski, SIRIUS performance: $^{10}$Be, $^{26}$Al and $^{36}$Cl measurements at ANSTO. Nuc. Instrum. Meth. Phys. Res. B **455**, 300–304 (2019)
16. G. Rugel, S. Pavetich, S. Akhmadaliev, S.M. Enamorado Baez, A. Scharf, R. Ziegenrücker, S. Merchel, The first four years of the AMS-facility DREAMS: Status and developments for more accurate radionuclide data. Nuc. Instrum. Meth. Phys. Res. B **370**, 94–100 (2019)
17. M. López-Lora, Low energy accelerator mass spectrometry of actinides ($^{236}$U, $^{237}$Np, $^{239,240}$Pu) at the Centro Nacional de Aceleradores and its applications in oceanography. Ph.D. Thesis, University of Sevilla, 2019
18. M. Suter, R. Balzer, G. Bonani, H.J. Hofmann, E. Morenzoni, M. Nessi, W. Wölfli, Efficiency parameters of the ETH AMS facility. Nucl. Instrum. Methods Phys. Res. B **10/11**, 877–880 (1985)
19. W. Kutschera, Progress in isotope analysis at ultra-trace levels by AMS. Int. J. Mass Spectrom. **242**, 145–160 (2005)
20. H.-A. Synal, Developments in accelerator mass spectrometry. Int. J. Mass Spectrom. **349–350**, 192–202 (2013)
21. W. Kutschera, Applications of accelerator mass spectrometry. Int. J. Mass Spectrom. **349–350**, 203–218 (2013)
22. M. Suter, A new generation of small facilities for accelerator mass spectrometry. Nucl. Instrum. Methods Phys. Res. B **139**, 50–157 (1998)

23. M. Suter, S.W.A. Jacob, H.-A. Synal, Tandem AMS at sub-MeV energies status and prospects. Nucl. Instrum. Methods Phys. Res. B **172**, 144–151 (2000)
24. M. Suter, S. Maxeiner, H.-A. Synal, C. Vockenhuber, Charge-state distributions and charge-changing cross sections and their impact on the performance of AMS facilities. Nucl. Instrum. Methods Phys. Res. B **437**, 116–122 (2018)
25. E. Chamizo, F.J. Santos, J.M. López-Gutiérrez, S. Padilla, M. García-León, J. Heinemeier, Ch. Schnabel, G. Scognamiglio, Status report of the 1 MV AMS facility at the Centro Nacional de Aceleradores. Nucl. Instrum. Methods Phys. Res. Sect. B **361**, 13–19 (2015)
26. https://www.ionplus.ch/milea
27. G. Scognamiglio, E. Chamizo, J. M. López-Gutiérrez, A. M. Müller, S. Padilla, F.-J. Santos, M. López-Lora, C. Vivo-Vilches, M. García-León, Recent developments of the 1 MV AMS facility at the Centro Nacional de Aceleradores. Nucl. Instr. Meth. Phys. Res. Sect. B **375**, 17–25 (2016)
28. H.-A. Synal, M. Stocker, M. Suter, MICADAS: a new compact radiocarbon AMS system', Nucl. Instrum. and Methods in Physics Research Section B259 (2007) 7–13.
29. E. Chamizo, M.C. Jiménez-Ramos, L. Wacker, I. Vioque, A. Calleja, M. García-León, R. García-Tenorio, Isolation of Pu-isotopes from environmental samples using ion chromatography for accelerator mass spectrometry and alpha spectrometry. Anal. Chim. **606**, 239–245 (2008)
30. J.M. Gómez-Guzmán, J.M. López-Gutiérrez, A.R. Pinto, E. Holm, M. García-León, Analysis of $^{129}I$ in lichens by accelerator mass spectrometry through a microwave-based sample preparation method. Nuc. Instrum. Meth. Phys. Res. B **268**, 1171–1174 (2010)
31. M. García-León, Accelerator mass spectrometry (AMS) in radioecology. J. Environ. Radioact. **186**, 116–123 (2018)
32. United Nations Scientific Committee on the Effects of Atomic Radiation UNSCEAR, Report to the General Assembly. Annex B: Exposure of the Public and Workers of Various Sources of Radiation **1**, 247 (2008)
33. K.M. Tierney, G.K.P. Muir, G.T. Cook, G. MacKinnon, J.A. Howe, J.J. Heymans, S. Xu, Accumulation of Sellafield-derived radiocarbon ($^{14}C$) in Irish Sea and West of Scotland intertidal shells and sediments. J. Environ. Radioact. **151**, 321–327 (2016)
34. R. Janovics, D.I. Kelemen, Z. Kern, S. Kapitany, M. Veres, A.J.T. Jull, M. Molnar, Radiocarbon signal of a low and intermediate level radioactive waste disposal facility in nearby trees. J. Environ. Radioact. **153**, 10–14 (2016)
35. Z. Wang, D. Hu, H. Xu, Q. Guo, $^{14}C$ distribution in atmospheric and aquatic environments around Qinshan nuclear power plant, China. Radiocarbon **56**, 1107–1114 (2014)
36. G.M. Raisbeck, F. Yiou, $^{129}I$ in the oceans: origins and applications. Sci. Total Environ. **237**(238), 31–41 (1999)
37. T. Jabbar, G. Wallner, P. Steier, A review on $^{129}I$ analysis in air. J. Environ. Radioact. **126**, 45–54 (2013)
38. F.J. Santos, J.M. López-Gutiérrez, E. Chamizo, M. García-León, H.-A. Synal, Advances on the determination of atmospheric $^{129}I$ by accelerator mass spectrometry (AMS). Nuc. Instrum. Meth. Phys. Res. B **249**, 772–775 (2006)
39. J.M. Gómez-Guzmán, J.M. López-Gutiérrez, R. García-Tenorio, L. Agulló, J.I. Peruchena, G. Manjón, M. García-León, Estimating the impact from Fukushima in Southern Spain by $^{131}I$ and accelerator mass spectrometry detection of $^{129}I$. J. Environ. Radioact. **166**, 36–44 (2017)
40. Y. Muramatsu, H. Matsuzaki, C. Toyama, T. Ohno, Analysis of $^{129}I$ in Fukushima prefecture soils: preliminary reconstruction of $^{131}I$ deposition related to the accident at Fukushima Daiichi nuclear power plant (FDNPP). J. Environ. Radioact. **139**, 344–350 (2015)
41. V. Hansen, P. Yi, X. Hou, A. Aldahan, P. Roos, G. Possnert, Iodide and iodate ($^{129}I$ and $^{127}I$) in surface water of the Baltic Sea, Kattegat and Skagerrak. Sci. Total Environ. **412–413**, 296–303 (2011)
42. K.A. Schwehr, S. Otosaka, S. Merchel, D.I. Kapland, S. Zhang, C. Xua, H.-P. Li, Y.-F. Ho, C.M. Yeager, P.H. Santschi, ASTER Team, Speciation of iodine isotopes inside and outside of a contaminant plume at the Savannah River Site. Sci. Total Environ. **497**(498), 671–678 (2014)

43. D. Elmore, B.R. Fulton, M.R. Clover, J.R. Marsden, H.E. Gove, H. Naylor, K.H. Purser, L. Kilius, R.P. Beukens, A.E. Litherland, Analysis of $^{36}$Cl in environmental water samples using electrostatic accelerators. Nature **277**, 22–25 (1979)

44. W. Balderer, H.-A. Synal, Use of chlorine-36 as tracer for the evolution of waters in geothermal and tectonically active areas in Western Turkey. Nuc. Instrum. Meth. Phys. Res. B **123**, 387–393 (1977)

45. C. Roux, C. Le Gal La Salle, C. Simonucci, N. Van Meir, K. Fifield, ASTER Team, O. Diez, S. Bassot, R. Simler, D. Bugai, V. Kashparov, J. Lancelot, High $^{36}$Cl/Cl ratios in Chernobyl groundwater. J. Environ. Radioact. **138**, 19–32 (2014)

46. F.J. Santos, J.M. López-Gutiérrez, M. García-León, Ch. Schnabel, H.-A. Synal, M. Suter, Analysis of $^{36}$Cl in atmospheric samples from Seville (Spain) by AMS. Nucl. Instrum. Meth. Phys. Res. B **223/224**, 501-506 (2004)

47. A. Blinov, S. Massonet, H. Sachsenhauser, C. Stan-Sion, V. Lazarev, J. Beer, H.-A. Synal, M. Kaba, J. Masarik, E. Nolte, An excess of $^{36}$Cl in modern atmospheric precipitation. Nucl. Instrum. Meth. Phys. Res. B **172**, 537–544 (2000)

48. E. Chamizo, M. García-León, H.-A. Synal, M. Suter, L. Wacker, Determination of the $^{240}$Pu/$^{239}$Pu atomic ratio in soils from Palomares (Spain) by low-energy accelerator mass spectrometry. Nucl. Instrum. Meth. Phys. Res. B **249**, 768–771 (2006).

49. E. Chamizo, M.C. Jiménez-Ramos, S.M. Enamorado, M. García-León, R. García-Tenorio, J.L. Mas, P. Masqué, J. Merino, J.A. Sanchez-Cabeza, Characterization of the plutonium isotopic composition of a sediment core from Palomares, Spain, by low-energy AMS and alpha-spectrometry. Nucl. Instrum. Meth. Phys. Res. B **268**, 1273–1276 (2010)

50. E. Chamizo, M. García-León, J.I. Peruchena, F. Cereceda, V. Vidal, E. Pinilla, C. Miró, Presence of plutonium isotopes, $^{239}$Pu and $^{240}$Pu, in soils from Chile. Nucl. Instrum. Meth. Phys. Res. B **24**, 3163–3166 (2011)

51. H. Hernández-Mendoza, E. Chamizo, A. Yllera, M. García-León, A. Delgado, Measurement of $^{239}$Pu in urine samples at ultra-trace levels using a 1 MV compact AMS system. Nucl. Instrum. Meth. Phys. Res. B **268**, 1331–1333 (2010)

52. L. Skipperud, J. Brown, L.K. Fifield, D.H. Oughton, B. Salbu, Association of plutonium with sediments from the Ob and Yenisey Rivers and Estuaries. J. Environ. Radioact. **100**, 290–300 (2009)

53. M. Srncik, P. Steier, G. Wallner, Depth profile of $^{236}$U/$^{238}$U in soil samples in La Palma, Canary Islands. J. Environ. Radioact. **102**, 614–619 (2011)

54. R. Eigl, M. Srncik, P. Steier, G. Wallner, $^{236}$U/$^{238}$U and $^{240}$Pu/$^{239}$Pu isotopic ratios in small (2 L) sea and river water samples. J. Environ. Radioact. **116**, 54–58 (2013)

55. M. Srncik, S.G. Tims, M. De Cesare, L.K. Fifield, First measurements of $^{236}$U concentrations and $^{236}$U/$^{239}$Pu isotopic ratios in a Southern Hemisphere soil far from nuclear test or reactor sites. J. Environ. Radioact. **132**, 108–114 (2014)

56. M.B. Froehlich, M.M.A. Dietze, S.G. Tims, L.K. Fifield, A comparison of fallout $^{236}$U and $^{239}$Pu uptake by Australian Vegetation. J. Environ. Radioact. **151**, 558–562 (2016)

57. N. Casacuberta, M. Christl, J. Lachner, M. Rutgers van der Loeff, P. Masqué, H.-A. Synal, A first transect of 236U in the North Atlantic Ocean. Geochim. Cosmochim. Acta **133**, 34–46 (2014)

58. C. Vockenhuber, M. Christl, C. Hofmann, J. Lachner, A.M. Müller, H.-A. Synal, Accelerator mass spectrometry of $^{236}$U at low energies. Nucl. Instrum. Meth. Phys. Res. B **269**, 3199–3203 (2011)

59. E. Chamizo, M. López-Lora, M. Bressac, I. Levy, M.K. Pham, Excess of $^{236}$U in the northwest Mediterranean Sea. Sci. Total Environ. **565**(2016), 767–776 (2016)

60. M. López-Lora, I. Levy, E. Chamizo, Simple and fast method for the analysis of $^{236}$U, $^{237}$Np and $^{239}$Pu and $^{240}$Pu from seawater samples by accelerator mass spectrometry. Talanta **200**, 22–30 (2019)

61. M. López-Lora, E. Chamizo, I. Levy, M. Christl, N. Casacuberta, T.C. Kenna, $^{236}$U, $^{237}$Np and $^{239,240}$Pu as complementary fingerprints of radioactive effluents in the western Mediterranean Sea and in the Canada Basin (Arctic Ocean). Sci. Total Environ. **765**, 142741 (2021)

62. L.K. Fifield, R.S. Carling, R.G. Cresswell, P.A. Hausladen, M.L. di Tada, J.P. Day, Accelerator mass spectrometry of $^{99}$Tc. Nucl. Instrum. Meth. Phys. Res. B **168**, 427–436 (2000)
63. L. Wacker, L.K. Fifield, S.G. Tims, Developments in AMS of $^{99}$Tc. Nucl. Instrum. Meth. Phys. Res. **B223**(224), 185–189 (2004)
64. B.A. Bergquist, A. Marchetti, R.E. Martinelli, J.E. McAninch, G.J. Nimz, L.D. Proctor, J.R. Southon, J.S. Vogel, Technetium measurements by accelerator mass spectrometry at LLNL. Nucl. Instrum. Meth. Phys. Res. B **172**, 328–332 (2000)
65. S.J. Tumey, T.A. Brown, T.E. Hamilton, D.J. Hillegonds, Accelerator mass spectrometry of strontium-90 for homeland security, environmental monitoring and human health. Nucl. Instrum. Meth. Phys. Res. B **266**, 2242–2245 (2008)
66. C.M. MacDonald, C.R.J. Charles, X.L. Zhao, W.E. Kieser, R.J. Cornett, A.E. Litherland, Determination of $^{135}$Cs by accelerator mass spectrometry. Nucl. Instrum. Meth. Phys. Res. B **361**, 554–558 (2015)
67. M. Christl, X. Dai, J. Lachner, S. Kramer-Tremblay, H.-A. Synal, Low energy AMS of americium and curium. Nucl. Instrum. Methods Phys. Res. **B331**, 225–232 (2014)
68. E. Chamizo, M. Christl, M. López-Lora, N. Casacuberta, A.-M. Wefing, T.C. Kenna, The potential of $^{233}$U/$^{236}$U as a water mass tracer in the Arctic Ocean. J. Geophys. Res. Oceans **127**, e2021JC017790 (2022)
69. P. Steier, R. Golser, W. Kutschera, A. Priller, C. Vockenhuber, S. Winkler, VERA, an AMS facility for "all" isotopes. Nucl. Instrum. Methods Phys. Res. B **223–224**, 67–71 (2004)
70. H.-A Synal, S. Jacob, M. Suter, The PSI/ETH small radiocarbon dating system. Nucl. Instrum. Methods Phys. Res. B **172**, 1–7 (2000)
71. E. Chamizo, J.M. López-Gutiérrez, A. Ruiz-Gómez, F.J. Santos, M. García-León, C. Maden, V. Alfimov, Status of the compact 1MV AMS facility at the Centro Nacional de Aceleradores (Spain). Nucl. Instrum. Methods Phys. Res. B **266**, 2217–2220 (2008)
72. E. Chamizo, M. López-Lora, Accelerator mass spectrometry of 236U with He stripping at the Centro Nacional de Aceleradores. Nucl. Instrum. Methods Phys. Res. B **438**, 198–206 (2019)

# Neutron Activation Analysis

# 19

**ABSTRACT**

Neutron Activation Analysis is a powerful tool for determining isotopes at trace levels in environmental samples. In some cases, it has been used for environmental radioactivity. The main objective of the chapter is the description of the technique. Thus, the most commonly used neutron reactions are described, and the general equation that allows the determination of the mass of the radionuclide of interest is presented. Usually, radiochemical separation is not necessary, although in some cases neutron activation requires it. Both practical approaches are described along the most important interferences that could affect the measurement. Neutron sources needed for activation are also described. Some specific applications to environmental radioactivity are presented.

## 19.1 Principles of Neutron Activation Analysis (NAA)

### 19.1.1 Neutron Activation of Materials

Neutron Activation Analysis (NAA) is an analytical technique that allows one to quantify the elemental and isotopic composition of a sample. The sensitivity is close to parts per billion (ppb) for most of the elements, and it can be applied virtually to any type of sample. In fact, the list of disciplines that benefit from NAA is large, as seen in Table 19.1 [1]. Hevesy and Levi discovered NAA in 1936 [2]. They used a Ra-Be neutron source (see next Sections) to activate Dy atoms and suggested the possibility of using neutrons to induce radioactivity in samples. The appearance of nuclear reactors during the 1940s boosted NAA as the available neutron fluxes increased by more than six orders of magnitude. In addition, the advances in Detector Physics made possible the analysis of virtually all periodic systems by NAA. Indeed, at the very beginning, only GM or Ionization Chamber detectors were available. This severely limited the analysis capabilities

© The Author(s), under exclusive license to Springer Nature Switzerland AG 2022
M. García-León, *Detecting Environmental Radioactivity*, Graduate Texts
in Physics, https://doi.org/10.1007/978-3-031-09970-0_19

**Table 19.1** List of disciplines that benefit from NAA

| Field | Applications |
|---|---|
| Archaeology | Archaeological objects components |
| Biomedicine | In vivo determination of essential elements (Ca, K, O, P, N, …) |
| Environment | Pollutants in environmental samples, Neutron radiography of materials |
| Forensic | Bomb debris, bullet lead, glass fragments, paint, hair, gunshot residue, biological residues, shotgun pellets |
| Geology | Online analysis of geological substrates, geological exploration, cosmic dust, lunar samples, meteorites, Neutron radiography of materials … |
| Industry | Online analysis of materials<br>Material cleanness, neutron radiography of materials |
| Material science | Material composition (major and minor components), Material purity, oxygen in materials, neutron radiography of materials |
| Nutrition | Protein content of food. Determination of essential elements (Ca, K, O, P, N, …) |
| Nuclear physics | Cross sections determinations, nuclear structure, purity of detector shielding |
| Security | Explosive, chemical weapon, and contraband detection |
| Safeguards | Nuclear material detection |

of the technique. However, scintillation detectors, mostly NaI(Tl), and Ge semiconductor detectors were available during the 1950s and 1960s, respectively. This, together with the development of new electronics, with special mention to the multichannel analyzer, converted NAA into a very powerful analytical technique and a benchmark technique for those developed later.

Basically, NAA consists of irradiating a sample with neutrons to produce nuclear reactions [3]. Indeed, as already described in Chap. 7

$$n + {}^A X \rightarrow \left[ {}^{A+1}_Z X \right]^* \rightarrow b + {}^{A'}_Z Y$$

The compound nucleus formed in this way decays through different channels depending on the neutron energy and the mass number $A$. NAA is based on the study of the radiation emitted after neutron irradiation, which leads to identification ${}^A X$. The most commonly used reaction in NAA is the neutron capture reaction, in which

$$n + {}^A X \rightarrow \left[ {}^{A+1}_Z X \right]^* \rightarrow \gamma + {}^{A+1} X$$

The $\gamma$ radiation is emitted immediately after neutron capture as a product of compound nucleus deexcitation. Usually, ${}^{A+1} X$ is radioactive and emits radiation according to its $T_{1/2}$. Both can be used to identify ${}^A X$. Depending on the radiation used for the determination, two techniques can be distinguished [3, 4]: Prompt Gamma NAA (PGNAA) or Delayed NAA (DNAA). It is the problem that has to be solved or, in many cases, the available experimental setup that advises the use of PGNAA or DNAA.

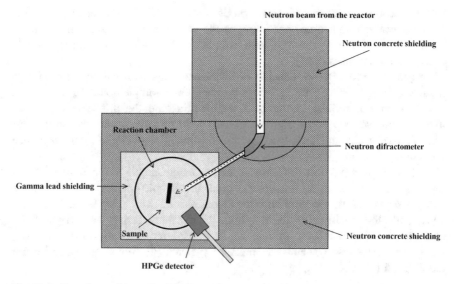

**Fig. 19.1** Experimental setup for PGNAA

## 19.1.2 Prompt Gamma NAA (PGNAA)

PGNAA is based on the detection of $\gamma$ radiation emitted immediately after n capture, i.e., radiation emitted directly from the compound nucleus [5]. This usually happens within the $10^{-14}$ s that follow the neutron capture. The use of PGNAA is recommended when elements of very high neutron capture cross sections have to be analyzed, when $^{A+1}X$ is stable or extremely short-lived, or when very low $\gamma$-radiation intensities are involved.

PGNAA requires special experimental setups since the radiation from the sample must be measured during neutron irradiation. In other words, the irradiated sample cannot be transported manually or automatically to a counting room. In the case of using a nuclear reactor as a neutron source, the most common option (see next sections), the neutron beam has to be extracted and conducted toward the irradiation chamber, which contains a Ge semiconductor detector to record the $\gamma$-spectrum from the sample. A picture describing this situation is presented in Fig. 19.1. Since neutrons do not have an electric charge, transporting them to the irradiation chamber is not easy. In the case presented in Fig. 19.1, this is carried out by using a diffraction crystal that has to be applied to thermal or cold neutrons.

## 19.1.3 Delayed NAA (DNAA)

This is the most common technique and is applied to a wide variety of samples and problems [4]. It consists of measuring the radiation emitted by $^{A+1}X$ or its daughters. It is a very flexible technique since the counting times involved can be selected according to the radioactive properties of $^{A+1}X$. Moreover, it is possible to select the counting times according to the expected interferences in the measurements. In this way, the irradiated sample can be cooled until the short-lived potential interferences decay.

In general terms, DNAA consists of three steps: sample irradiation during time $t$, sample decay during time $t_d$ before counting, and sample counting during $\Delta t$. The equations described in the next section illustrate the differences between them. Of course, depending on the problem, these three time periods can be chosen to speed up the analysis, increase the sensitivity, or decrease the interference influence.

In DNAA, it is possible to transport the irradiated sample to a counting laboratory [3]. Thus, when using a nuclear reactor, the sample can be inserted in areas where the neutron flux is very high inside the reactor itself. This is very advantageous since the sensitivity of the technique greatly increases, as we will see in the next sections. The irradiated sample can be transported manually to the counting laboratory or by using pneumatic transport systems [6], which are safer from a radiological point of view.

### 19.1.4 Radiochemical and Instrumental NAA

DNAA techniques can, in turn, be classified into two categories depending on the need for chemical separation of the radioactive species to be measured.

In radiochemical NAA (RNAA), a previous chemical separation of the radioactive species is needed before counting. This technique is applied for those cases where the radionuclide of interest does not emit $\gamma$-radiation or the interferences are two high. The main advantage is the possibility of isolating the radionuclide, almost definitively discarding the interferences. Separation can be performed before and after irradiation. In the latter case, radiological safety considerations must be followed to avoid too intense operational exposure. This technique is destructive and therefore cannot be applied to valuable samples, such as those of archaeological interest. The main disadvantage is the monoelemental character of the technique. The most widely used NAA technique today is the so-called Instrumental NAA (INAA).

In INAA, the $\gamma$-radiation emitted by the irradiated sample is measured using a Ge semiconductor spectrometer (see Chap. 11). When INAA is used, chemical separations are not necessary since the extremely good energy resolution of Ge detectors allows the simultaneous determination of many radionuclides during the same measurement. This is the main advantage of the technique that makes it very popular. Of course, the technique is not destructive and thus can be applied to many problems.

## 19.2  General Equation

### 19.2.1 The Mass Equation

Let us suppose the following neutron capture reaction.

$$n + X \rightarrow Y + \gamma$$

in which nucleus $X$ produces nucleus $Y$, which in turn is radioactive with a decay constant $\lambda$. Then, the production of $Y$ nuclei over time will be

$$\frac{dN_Y}{dt} = \sigma\phi N_X - \lambda N_Y \tag{19.1}$$

$N_x$ is the number of parent nuclei at time $t$, $\sigma$ is the neutron capture cross section, and $\phi$ is the neutron flux. It is important to note that $t$ is the irradiation time. Integration of (19.1) gives

$$N_Y(t) = \frac{\sigma\phi N_X(0)}{\lambda - \sigma\phi}\left(e^{-\sigma\phi t} - e^{-\lambda t}\right) \tag{19.2}$$

with an obvious meaning for the notation. Usually, $\lambda \gg \sigma\phi$ and $\sigma\phi t \ll 1$. Under these conditions, the activity of the $Y$ nuclei produced at time $t$ will be

$$A_Y(t) = \sigma\phi N_X(t)\left(1 - e^{-\lambda t}\right) \tag{19.3}$$

or

$$A_Y(t) = \sigma\phi N_X(t)F_S(t) \tag{19.4}$$

with

$$F_S(t) = \left(1 - e^{-\lambda t}\right) \tag{19.5}$$

being the saturation factor. When $t \ll T_{1/2}$, the Y nucleus half-life, then $F_S \approx 1$, and

$$A_Y(t) \approx \sigma\phi N_X(t) \tag{19.6}$$

In other words, the $Y$ nuclei are produced at the same rate as they decay. This is the saturation condition, which is similar to the secular equilibrium concept, already explained in Chap. 2. Equation 19.3 can alternatively be expressed as

$$A_Y(t) = \frac{w N_A m\sigma\phi\left(1 - e^{-\lambda t}\right)}{P_M} \tag{19.7}$$

$w$ and $m$ are the isotopic abundance of the nuclei $Y$ and its mass in the sample, respectively. $P_M$ is the atomic weight of the element, of which $Y$ is one isotope. Finally, $N_A$ is the Avogadro number.

As explained above in DNAA, the activity is measured a time $t_d$ after irradiation. In this case,

$$A_Y(t_d) = \frac{w N_A m \sigma \phi \left(1 - e^{-\lambda t}\right)}{P_M} e^{-\lambda t_d} \tag{19.8}$$

and it turns out that

$$m = \frac{P_M A_Y(t_d)}{w N_A \sigma \phi \left(1 - e^{-\lambda t}\right) e^{-\lambda t_d}} \tag{19.9}$$

The mass equation permits the calculation of the mass of the nuclei $Y$ in the irradiated sample. However, in real terms, we do not know $A_Y(t_d)$ but the number of counts provided by the detector during a counting time $\Delta t$. The net number of counts, $n_Y$, obtained during such a period would be

$$n_Y = \varepsilon I \int_{t_d}^{t_d + \Delta t} A_Y(t_d) e^{-\lambda t} dt \tag{19.10}$$

$\varepsilon$ is the counting efficiency, and $I$ is the intensity of the radiation selected for the measurement.

After some mathematics

$$A_Y(t_d) = \frac{n_Y \lambda e^{\lambda t_d}}{\varepsilon I \left(1 - e^{-\lambda \Delta t}\right)} \tag{19.11}$$

and,

$$m = \frac{n_Y \lambda e^{\lambda t_d} P_M}{\varepsilon I w N_A \sigma \phi \left(1 - e^{-\lambda t}\right)\left(1 - e^{-\lambda \Delta t}\right) e^{-\lambda t_d}} \tag{19.12}$$

which is a more realistic equation for the mass.

Usually, under real conditions, the neutron flux has a continuum energy spectrum with a shape similar to that in Fig. 19.2. Since $\sigma$ depends on the neutron energy, it is better to use $\sigma_{\text{eff}}$, and effective cross section, and $\phi_{\text{tot}}$, a total flux, in the above equations. They are defined as

$$\sigma_{\text{eff}} = \frac{\int \sigma(E) \phi(E) dE}{\int \sigma(E) dE} \tag{19.13}$$

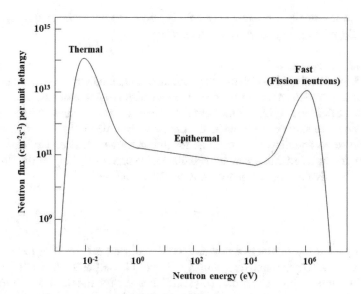

**Fig. 19.2**   Typical neutron flux in a nuclear reactor

$\sigma_{eff}$ represents an average cross section weighted for the neutron energy distribution, and

$$\phi_{tot} = \int \phi(E) dE \tag{19.14}$$

The total neutron flux involved in the experiment. Both $\sigma_{eff}$ and $\phi_{tot}$ depend on the neutron source used for activation and, therefore, on the irradiation facility.

## 19.2.2 Mass Determination Methods

There are different methods for mass determination [7]. Some details about them follow.

*The Absolute Method*
In the absolute method, (19.12) is solved directly by applying the different parameters involved to the nucleus under measurement. Of course, this means perfect knowledge of $\sigma_{eff}$, $\phi_{tot}$, and $\varepsilon$. This method is not widely used due to the uncertainties in the determination of these parameters, especially $\sigma_{eff}$ and $\phi_{tot}$. Indeed, in real samples, depending on several conditions, they can vary along the experiment and even along the sample volume. This certainly makes it difficult to apply this procedure. Today, the use of Monte Carlo simulation to describe the path of neutrons along the sample greatly helps to solve the problem. Nevertheless, this requires knowledge to obtain a very good approximation of the exact composition of the sample. This, on many occasions, especially for environmental samples, is

difficult. Additionally, on some occasions $\lambda$ and $I$ are not well known. This adds uncertainty when using the absolute method.

*The Comparator Method*

To avoid the mentioned problems, the comparator method can be used, and today, it is the most popular method in NAA laboratories. Now, a comparator sample, a compound of known masses of the element or elements we are interested in, is irradiated under the same conditions as the unknown sample. Then, it is also counted under the same counting conditions as for the unknown sample. The mass of the comparator (standard), $m_c$, can be calculated by using (19.12), as the unknown mass, $m$. By dividing both Equations, it can be found that

$$m = \frac{n_Y}{n_{Y,c}} m_c \qquad (19.15)$$

$n_Y$ and $n_{Y,c}$ are the number of counts recorded during $\Delta t$ for both the unknown sample and the comparator, respectively. Since all the parameters are the same for both samples, they cancel out, so (19.15) has such a simple form. In the case where the counting parameters are different, i.e., $\Delta t$, $t_d$, or the irradiation time, $t$, (19.15) becomes slightly more complex, as different time corrections must be made for the unknown sample and the comparator. In any case, $\varepsilon$, $\sigma_{\text{eff}}$, and $\phi_{\text{tot}}$ are the same.

Equation 19.15 can be applied only in the case where the irradiation conditions and counting conditions (essentially counting geometry) are preserved. Of course, this also means that the comparator must be distributed in a sample matrix identical to that of the unknown sample. This is a drawback of the method because it is not easy to achieve. In addition, for multielemental determinations, it is necessary to use different comparators. The number of nuclei will be determined.

*The $k_0$-Comparator or Single Comparator Method*

For the last case, it is more suitable to use the single comparator method [8]. This method has the advantages of the comparator method but also of the absolute method. In the single comparator method, only one reference element is used. Normally it is selected to have well-known nuclear and atomic data, although in the relative version this is not strictly necessary. The basis of the method is exactly the same as that of the comparator method described above. The single comparator and the unknown sample are irradiated and counted, and the results are compared. Of course, to make the comparison possible, the counting and irradiation conditions should be the same. Thus, the irradiation and counting geometry, as well as the unknown sample composition, must be the same. To ensure that, in some cases, a known amount of the single comparator is dissolved or homogenously mixed in the unknown sample. In any case, the composition of the reference and problem sample should at least be very similar.

After irradiation and counting, the single comparator mass, $m_c$, which is known, is determined by (19.12). In the same way, the unknown sample is irradiated, and the mass of element i, $m_i$, can also be deduced from (19.12). The mass ratio is

$$\frac{m_i}{m_c} = \frac{n_{Y,i}}{n_{Y,c}} \frac{\lambda_i e^{\lambda_i t_d} P_{M,i} \varepsilon_c I_c w_c \sigma_{eff,c} \left(1 - e^{-\lambda_c t}\right)\left(1 - e^{-\lambda_c \Delta t}\right)e^{-\lambda_c t_d}}{\lambda_c e^{\lambda_c t_d} P_{M,c} \varepsilon_i I_i w_i \sigma_{eff,i} \left(1 - e^{-\lambda_i t}\right)\left(1 - e^{-\lambda_i \Delta t}\right)e^{-\lambda_i t_d}} \qquad (19.16)$$

with an obvious meaning for the notation. Apart from the Avogadro number, the only parameter that can be canceled out in this case is $\phi_{tot}$, since $\sigma_{eff}$ is different for the single comparator and the unknown element. The decay constants are also different, as are the isotopic abundances, counting efficiencies, and radiation intensities. In any case, canceling out the neutron flux is very important, since it is a significant source of uncertainty in the determination of the elements. The comparator factor $k_0$ is then defined as

$$k_0 = \frac{\lambda_i e^{\lambda_i t_d} P_{M,i} \varepsilon_c I_c w_c \sigma_{eff,c} \left(1 - e^{-\lambda_c t}\right)\left(1 - e^{-\lambda_c \Delta t}\right)e^{-\lambda_c t_d}}{\lambda_c e^{\lambda_c t_d} P_{M,c} \varepsilon_i I_i w_i \sigma_{eff,i} \left(1 - e^{-\lambda_i t}\right)\left(1 - e^{-\lambda_i \Delta t}\right)e^{-\lambda_i t_d}} \qquad (19.17)$$

Under well-defined irradiation and counting conditions, $k_0$ can be readily found provided $\sigma_{\text{eff}}$ for both the unknown and comparator elements are well known. Thus, $m_i$ can be calculated by applying (19.16). The main drawback of this method lies in the fact that the nuclear data must be very well known for the unknown element and the comparator. From this point of view, this procedure has the same drawback as the absolute method. To avoid this problem, an empirical version of this method is used for most applications. In this case, $k_0$ can be experimentally determined simply by irradiating the well-known masses of the elements of interest and of the single comparator. In fact, under these conditions, $k_0$ for element $i$ can be determined from (19.16).

$k_{0,i}$ obtained in this way is only valid for the irradiation and counting conditions used in the experiment. Thus, it can be applied to the determination of unknown samples that are irradiated and then counted under the same conditions. NAA laboratories determine $k_{0,i}$ for a list of elements by performing several measurements with well-known masses of the elements of interest. The $k_{0,i}$ list obtained in this way can be easily applied to unknown samples, since the mass ratio is simply the ratio of the number of net counts in the detector (usually the net area under the photopeak in a Ge detector) multiplied by $k_{0,i}$ and the added mass of the single comparator. That is,

$$m_i = \frac{n_{Y,i}}{n_{Y,c}} m_c k_{0,i} \qquad (19.18)$$

provided that the counting and irradiation conditions are the same as those used in the determination of the $k_0$ coefficients [7].

## 19.3  Sensitivity, Interferences, and Limitations

### 19.3.1 Analysis of the Mass Equation. Design of an NAA Experiment

The analysis of (19.12) provides some hints about the possibility of improving the sensitivity of the method.

The number of counts, $n_Y$, is the key parameter for improving the determination of mass. It has to be increased on the basis of the counting conditions. This means selecting the appropriate detector according to the criteria already described in Chap. 14 for the Figure of Merit (FOM). The counting efficiency must be high, and the detector background must be low. Furthermore, in the case of using chemical separation to isolate the radionuclide or radionuclides to be measured, the yield must be as high as possible. Irradiation, decay, and counting times are also important to improve the sensitivity of the measurement.

From this point of view, the selection of the radionuclide that will be used to determine the element of interest is very important. Conventionally, the activation product (see the previous Section) $Y$ is the choice for the determination of $X$. However, it could be convenient to analyze the possibility of using another radionuclide produced during irradiation or as a decay product. For example, if

$$X \xrightarrow{\sigma\emptyset} Y \xrightarrow{\lambda_Y} Z \xrightarrow{\lambda_Z} W$$

and $T_{1/2}^Z > T_{1/2}^Y$, it is likely that it is better to use $Z$ instead of $Y$ for the determination of $X$. In that case, (19.12) changes, since now we have to solve the following equation:

$$\frac{dN_Z}{dt} = (\sigma\phi N_X + \lambda_Y N_Y) - \lambda_Z N_Z \qquad (19.19)$$

Let us have the following activation sequence.

$$X \xrightarrow{\sigma_X\phi} Y \xrightarrow{\lambda_Y} Z \xrightarrow{\lambda_Z} W$$
$$\sigma_Y\phi \searrow$$
$$N \xrightarrow{\lambda_N} M$$

should $\sigma_Y \phi > \lambda_Y$, it is likely better to use $N$ to determine $X$. The interesting equation is now

$$\frac{dN_N}{dt} = (\sigma_X\phi N_X + \sigma_Y\phi N_Y) - (\lambda_Y N_Y + \lambda_N N_N) \qquad (19.20)$$

**Table 19.2** NAA sensitivity

| Elements | Sensitivity (ng) |
|---|---|
| Dy, Eu | $10^{-3}$ |
| In, Lu, Mn | $10^{-3}$ to $10^{-2}$ |
| Au, Ho, Ir, Re, Sm, W | $10^{-2}$ to $10^{-1}$ |
| Ag, Ar, As, Br, Cl, Co, Cs, Cu, Er, Ga, Hf, I, La, Sb, Sc, Se, Ta, Tb, Th, Tm, U, V, Yb | $10^{-1}$ to 1 |
| Al, Ba, Cd, Ce, Cr, Hg, Kr, Gd, Ge, Mo, Na, Nd, Ni, Os, Pd, Rb, Rh, Ru, Sr, Te, Zn, Zr | 1 to 10 |
| Bi, Ca, K, Mg, P, Pt, Si, Sn, Ti, Tl, Xe, Y | 10 to $10^{2}$ |
| F, Fe, Nb, Ne | $10^{2}$ to $10^{3}$ |
| Pb, S | $10^{4}$ |

There are many possibilities that can be used to obtain information about the content of the element of interest in the sample. All of them must be taken into account when planning the NAA experiment.

The number of counts also increases when high activity is obtained during irradiation. Thus, the radionuclide production rate, $\sigma \phi$, should also be high. The selection of the specific neutron reaction and flux is a key issue for the experiment.

Conventionally, thermal or epithermal neutrons are mostly used in NAA and, more specifically, in INAA. However, on some occasions, it is more convenient to use the resonance range of the neutron excitation function since the cross sections are higher. Selection of this energy range is carried out by absorbing the thermal and epithermal components of neutron flux by covering the sample with Cd or B, which have a very high neutron capture cross section in the thermal and epithermal regions [3]. The use of fast neutrons is less conventional and is more oriented to very specific applications since the cross sections are lower than those for thermal or epithermal neutrons. Additionally, as we shall see below, more specialized neutron sources are needed to obtain fast neutrons. In commonly used nuclear reactors, the fast neutron part of the total spectrum is not greater than 2% [4].

These considerations must be taken into account when planning NAA experiments. Of course, for multielement applications, the selection of the different counting and irradiation parameters must be done by looking for the best general sensitivity possible for the elements under analysis. This is not easy and could mean worsening some specific sensitivities. In Table 19.2, some detection limits as defined in Chap. 14 are presented for different NAA laboratories [7, 9]. Levels of ppb make NAA a very competitive analytical technique.

## 19.3.2 Types of Interferences

The planning of an NAA experiment also includes the study of the potential interferences that can be found during the measurements.

**Table 19.3** A classification of interferences in NAA

| Main reaction | Nuclear interference | Type | Example |
|---|---|---|---|
| $^A X(n, \gamma)^{A+1} X$ | Primary | *First order*<br>$^{A+1}X$ production by other reactions with $n$ | $B(n, x)^{A+1}X$ |
| | | *Second order*<br>$^{A+1}X$ removal by other reactions | $^{A+1}X(n, x)E$ |
| $^A X(n, \gamma)^{A+1} X$ | Secondary | *Direct*<br>$^{A+1}X$ is already present in the sample | |
| | | *Indirect*<br>$^{A+1}X$ production by particles released by n interactions | $A(n, x')C$<br>$D(x', x'')^{A+1}X$ |
| | | *Radioactive interferences*<br>Radionuclides in the sample could interfere the measurement of $^{A+1}X$ radiations | $^{238}$U-, $^{235}$U-, $^{232}$Th-series nuclides, and $^{40}$K |

The main interferences come from the possible nuclear reactions that can take place in addition to the main neutron reaction producing the nucleus of interest. In Table 19.3, the classification of the so-called nuclear interferences is given. In principle, we have primary and secondary interferences.

*Primary Nuclear Interferences*
In this case, the neutrons can directly or indirectly produce the same radionuclide that serves to identify the nucleus of interest, $X$.

Let us suppose that the main reaction is as follows

$$^A X(n, \gamma)^{A+1} X$$

a first-order primary nuclear interference for this reaction would be

$$B(n, x)^{A+1} X$$

$x$ being protons, neutrons, or other particles. These interfering reactions are possible when using a neutron spectrum containing a relevant fast fraction capable of picking up particles from the compound nucleus.

Another possible first-order interference would be

$$C(n, x)D \xrightarrow{\sigma_D} {}^{A+1} X$$

Second-order primary interference would consist of the removal of $^{A+1}X$ nuclides by neutron capture or other neutron reactions in general. Now, the interference reaction would be

$$^{A+1}X(n, x)E$$

$^{A+1}X$ is lost as a consequence of the neutron reaction. It is interesting to note that when thermal neutrons are used, the cross sections for neutron capture are high, especially for high $A$. In other words, the previous reaction could be very likely.

A more complex situation, although possible, would be the following second-order primary interference.

$$F(n, x)G \xrightarrow{\sigma_G} {}^{A}X(n, \gamma)^{A+1}X$$

In this case, an excess of $^{A+1}X$ is observed that of course does not correspond to the presence of $^{A}X$ in the sample.

*Secondary Nuclear Interferences*
Secondary nuclear interferences can also occur. Now $^{A+1}X$ appears as a product of nuclear reactions caused by particles produced in the sample by the neutron reactions. For example,

$$A(n, x')C$$
$$D(x', x'')^{A+1}X$$

In addition to these nuclear interferences, it is necessary to check that the radionuclide used for the analysis is already present in the sample. The presence of natural radioactivity in the sample is always a problem and must be checked before the measurement. $^{238}U$-, $^{235}U$-, $^{232}Th$-series nuclides, and others, mostly $^{40}K$, can seriously interfere with the measurement of the activated radionuclides, either directly or indirectly by overlapping photopeaks in the Ge spectrometer.

Together with the interferences, some other limitations must be taken into account when planning an NAA experiment. Physical limitations are mainly related to changes in $\phi$ during irradiation. These changes can be produced by self-absorption, dispersion, or flux 'hardening'. The reduction in the sample size helps to overcome these effects, but this also leads to a decrease in the sensitivity of the measurement. The use of comparator methods solves the problem, provided that they can empirically simulate the behavior of the neutron flux within the sample matrix.

Of course, limitations of the counters used for radioactivity determination are also important. The background and counting efficiencies limit the sensitivity of counting, as shown in Chap. 14. For INAA, Ge semiconductor detectors are used because of their extremely good energy resolution. In this case, the continuum Compton can greatly modify the spectrum shape, contributing to enhancing the background and, in this way, decreasing the sensitivity. The use of anti-Compton systems is a very good option. In any case, it is highly recommended to use LLC techniques (see Chap. 15) to decrease the background and improve the FOM of the experiment.

Radiochemical separations pose some additional limitations by introducing undesirable interferences or reducing the number of radionuclides produced available for counting. As shown in Chap. 13, all chemical processes have a yield that is rarely 100%.

All the above-mentioned effects must be taken into account together with the interferences resulting from nuclear reactions when an NAA experiment is going to be carried out.

## 19.4   Experimental Systems

In conventional INAA, the setup consists of a neutron source, an irradiated sample transport device, and a Ge semiconductor detector. A pictorial scheme is given in Fig. 19.3 [1]. INAA does not require radiochemistry. The technique is essentially nondestructive, although some sample adequacy to the irradiation geometry is needed [10, 11]. In the case of solid environmental samples, crushing, pulverization, homogenization, and pelletization are usually needed. For liquid samples, a previous evaporation and powdering of the residue might be needed. In some cases, a preconcentration step by coprecipitation or purification by ionic exchange prior to irradiation is convenient. In this case, the sample, once pretreated, must be dried and processed as a solid sample. Details of these steps can be found in Chap. 13.

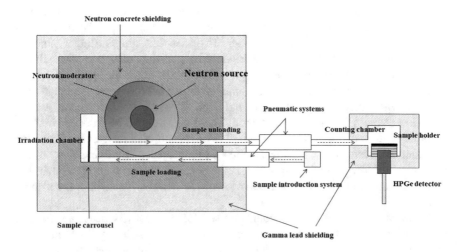

**Fig. 19.3**   General NAA experimental setup

The irradiation process begins with the selection of the sample container. It is true that the composition of the container must be carefully selected to avoid interferences that can affect the count by directly overlapping the peak or increasing the Compton continuum of the $\gamma$-spectrum. Both effects will decrease the MDA of the measurements (see Chap. 14). The container material should be different for long-term irradiation and short-term irradiation [10]. Metals are preferred in the first case, and polyethylene is preferred in the latter case. Some experiments require suppression of the thermal or epithermal neutron spectrum. In this case, the sample to be irradiated is surrounded by a blanket of Cd, which greatly helps to diminish this part of the energy spectrum, since the Cd neutron capture cross section is very high (see Chap. 7). Transport of the irradiated sample to the counting laboratory must be performed following all radiological safety regulations, as it is expected to manipulate a highly radioactive sample [6]. For many applications, it is preferred to cool the sample until the radioactive level is low enough to make handling safer. This could even benefit the experiment since some radioactive species sufficiently decay and do not interfere with counting. In many cases, automatic transport devices are used. After irradiation, the samples are counted with a Ge semiconductor $\gamma$-spectrometer as described in Chap. 11. The multielement character of the analysis makes it convenient to use extended energy-range Ge detectors to be able to detect from the X- to $\gamma$-ray photon range. This expands the analytical possibilities to virtually all periodic tables. Since low-level radioactivity concentrations are expected in many cases, LLC techniques must be used (see Chap. 15).

There are different neutron sources for NAA applications. Although the most commonly used source of neutrons is the nuclear reactor, all possibilities are described here since they can be applied to specific problems.

## 19.4.1 Radioisotope Sources

They are based on $(\alpha, n)$ or $(\gamma, n)$ reactions already described in Chap. 7. For that, a radioactive isotope is surrounded by a blanket of a material containing the nucleus in which the reaction is produced. Table 19.4 provides a compilation of the neutron reactions involved. In Fig. 19.4a, $a$ scheme of this neutron source is given [1].

The most popular $(\alpha, n)$ sources are $^{226}$Ra(Be), $^{241}$Am(Be), $^{210}$Po(Be), and $^{239}$Pu(Be), in which the $\alpha$ radiation emitted by $^{226}$Ra, $^{241}$Am $^{210}$Po, or $^{239}$Pu produces neutrons through $^{9}$Be$(\alpha, n)^{12}$C. The Ra-Be sources live longer than Am-Be and Po-Be, since the half-life of $^{226}$Ra, 1600 y, is higher than that of $^{241}$Am, 432.7 y, and $^{210}$Po, 138.376 d. Despite this, they are less used today, as they emit a significant amount of $\gamma$-radiation from the $^{226}$Ra daughter nuclides. This limits its utility in many cases. Of course, the $^{239}$Pu half-life, 24,110 y, makes it more advantageous, but high-activity $^{239}$Pu sources are less available. The neutron energy spectrum obtained with these sources is continuous and generally

**Table 19.4** Neutron reaction of interest for radioisotope neutron sources and neutron generators

| Neutron source | Type of reaction | Neutron reaction |
|---|---|---|
| Radioisotope sources | $(\alpha, n)$ | $\alpha + {}^9Be \rightarrow {}^{12}C + n$ |
| | $(\gamma, n)$ | $\gamma + {}^9Be \rightarrow 2\,{}^4He + n$ |
| Neutron generator | $(x, n)$ | ${}^2H + {}^2H \rightarrow {}^3He + n$ |
| | | ${}^3H + {}^2H \rightarrow {}^4He + n$ |
| | | ${}^3H + {}^1H \rightarrow {}^3He + n$ |
| | | ${}^7Li + {}^1H \rightarrow {}^7Be + n$ |
| | | ${}^9Be + {}^2H \rightarrow {}^{10}B + n$ |
| | | ${}^7Li + {}^2H \rightarrow {}^8Be + n$ |
| | | ${}^{51}V + {}^1H \rightarrow {}^{51}Cr + n$ |

ranges from 10 to 12 MeV. Neutron fluxes of up to $10^2$–$10^4$ ns$^{-1}$ per MBq of the radioactive source are obtained.

The $(\gamma, n)$ sources are based on reactions of the type ${}^9Be(\gamma, n)^8Be$ and ${}^2H(\gamma, n)^1H$. They are constructed by mixing Be compounds or volumes of heavy water, $D_2O$, with $\gamma$-emitting radionuclides with sufficient $T_{1/2}$. Of course, the energy of the emitted $\gamma$-radiation must be higher than the neutron binding energy in ${}^2H$ or in ${}^9B$. Radionuclides such as ${}^{24}Na$, ${}^{56}Mn$, ${}^{116}In$, ${}^{124}Sb$ are good candidates as the source radioisotopes, with ${}^{124}Sb$ being the most popular due to its longer half-life, $T_{1/2} = 60.20$ d. ${}^{124}Sb$ emits a $\gamma$ photon 1.692 MeV energy, high enough to remove the less bound neutron from the ${}^9B$ nucleus. The usefulness time period is clearly shorter than that of $(\alpha, n)$ sources. However, the neutron energy spectrum is discrete, which makes them very useful for specific applications. Neutron fluxes of up to 10 ns$^{-1}$ per MBq of the radioactive source can be produced.

Fission neutrons from spontaneous fission radionuclides such as ${}^{252}Cf$ ($T_{1/2}$ = 2.645 y) are also used. The neutron spectrum is continuous and fluxes up to $10^5$ ns$^{-1}$ per MBq or $10^{12}$ ns$^{-1}$ per g can be obtained.

Radioisotope neutron sources provide low neutron sources but are portable, which is an interesting advantage for some in situ applications.

### 19.4.2  Nuclear Reactors

Nuclear reactors are the most commonly used neutron source [3, 4]. Thermal ${}^{235}U$ nuclear fission-based reactors working at power levels up to 100 MWt provide neutron fluxes close to $10^{15}$ ncm$^{-2}$ s$^{-1}$, mostly in the thermal range, where fluxes close to $10^{14}$ ncm$^{-2}$ s$^{-1}$ can be found. This is an important advantage with respect to radioisotope sources. Of course, the neutron energy spectrum is a continuum (see Fig. 19.2), which is a drawback that must be overcome when carrying out the calculations (see (19.12)). However, as seen in Fig. 19.2, thermal neutrons are the

**Table 19.5** List of operational research nuclear reactors. Except for one in Russia, the rest provide thermal neutrons

| Country | Number |
|---|---|
| Russia | 52 |
| USA | 50 |
| China | 16 |
| Argentina | 5 |
| Canada | 5 |
| Germany | 5 |
| India | 5 |
| Italy | 5 |
| Brazil | 4 |
| Iran | 4 |
| Kazakhstan | 4 |
| Belarus | 3 |
| Belgium | 3 |
| Czech Republic | 3 |
| France | 3 |
| Indonesia | 3 |
| Japan | 3 |
| Ukraine | 3 |
| Others | 44 |
| Total | 220 |

most abundant, with fluxes representing approximately 93% of the total flux. The rest are shared between fast neutrons, approximately 5%, and epithermal neutrons, approximately 2%. (For a reminder of the classification of neutron energies, see Chap. 7.) This obviously facilitates the analysis since the excitation function for neutrons shows that the neutron capture cross sections are generally higher in the thermal range, as seen in Fig. 19.5. It can be said that the use of thermal or epithermal neutrons is the generalized option in NAA. In Table 19.5, we give a list of operational nuclear research reactors worldwide [12]. With very few exceptions, all of them operate with thermal neutrons. A schematic representation of a nuclear reactor setup for NAA is given in Fig. 19.4b.

### 19.4.3 Neutron Generators. Accelerator-Based Neutron Sources

Neutron generators are today an alternative to nuclear reactors [1]. They are mainly based on $^2$H$(d$, n$)^3$He and $^3$H$(d$, $n)^4$He (see Table 19.4). These reactions take place at relatively low-energy particle beams since they are exothermic. The cross sections have a resonance at 100 keV. Thus, particle beam energies from 100 to

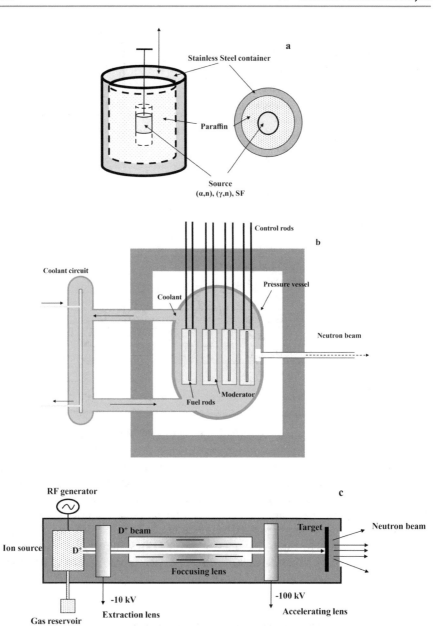

**Fig. 19.4**  Neutron sources: **a** radioisotope source, **b** nuclear reactor, and **c** neutron generator

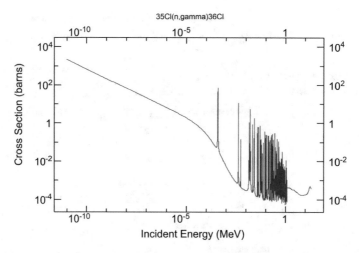

**Fig. 19.5** Typical neutron excitation function. Data retrieved from the IAEA ENDF Database Retrieval System, 2004–2022: https://www-nds.iaea.org/endf/

500 keV are enough to produce neutrons of 2.4 MeV in the first case and 14.1 MeV in the second case.

Under these conditions, it is not necessary to use accelerators to produce the reactions. There are several designs of neutron generators based on the production of a D plasma with radiofrequency fields. The most popular design is based on the Penning ion source [1]. The charged $^2$H nuclei are extracted from the source and accelerated by a high voltage of no more than 110 kV and directed toward a target containing $^2$H, $^3$H, or a mixture (see Fig. 19.4c). An additional advantage of these sources is the possibility of working in pulsed operation mode providing neutron bunches at frequencies from 10 Hz to 20 kHz. Usually, neutron fluxes ranging from $10^5$ to $10^8$ fast ncm$^{-2}$ s$^{-1}$ are obtained with these sources. The neutron spectrum can be thermalized by using filters at the exit of the source. However, this produces a high $\gamma$ background, which is not desirable in many applications and, of course, decreases the flux. Neutron generators based on these reactions find application in problems where fast neutrons provide more sensitivity.

These reactions and others, such as $^2$H($d$, $n$)$^3$He, $^7$Li($p$, n)$^7$Be, $^9$Be($d$, $n$)$^{10}$B, $^7$Li($d$, $n$)$^8$Be, and $^{51}$V($p$, $n$)$^{51}$Cr, can be easily produced in accelerators. The neutron yield in this case is certainly low compared to the fluxes obtained with the sources described above. However, the cinematic features of the neutron beam can be precisely controlled. Thus, the neutron energy is well known, and the uncertainties associated with the neutron capture cross section are greatly reduced.

Neutron sources based on spallation reactions are a very promising alternative to nuclear reactors. Now, a heavy nuclei target is bombarded with a GeV or tenths of a GeV proton beam, producing spallation reactions [13]. A spallation reaction (see Fig. 19.6) can be considered a highly asymmetric fission in which neutrons are also emitted. The target can be solid or liquid and formed by Hg, Ta, W, Pb,

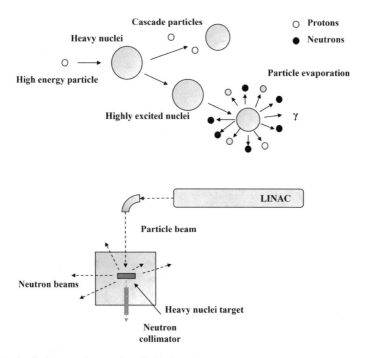

**Fig. 19.6** Spallation reactions and spallation-based neutron sources

Bi, or U. On the other hand, high beam intensities of some mA are used. The source is also pulsed with frequencies ranging from 25 to 60 Hz.

In spallation reactions, the neutron energy spectrum is similar to that obtained in fission, i.e., shifted to the fast side. Neutron fluxes can reach values up to $10^{14}$ ncm$^{-2}$ s$^{-1}$ in the continuous mode and up to $10^{15}$ in the pulse mode. The neutron flux can be thermalized as in the previous cases by using moderator materials within the target.

## 19.5    Other Neutron-Based Analytical Techniques

There are some other alternatives to NAA neutron-based analytical techniques. In what follows, two of the more interesting are described.

### 19.5.1 Neutron Resonance Capture Analysis

As already described in Chap. 7, resonances are present in the neutron nucleus-excitation function (see Fig. 19.5). The resonances take place at well-defined energies that clearly identify the compound nucleus and therefore the target

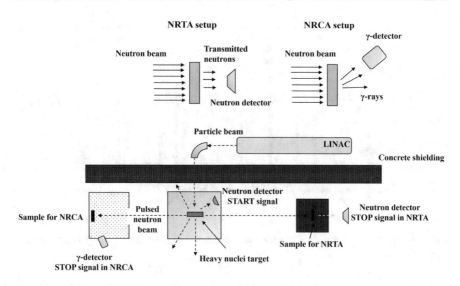

**Fig. 19.7**   NRTA and NRCA experimental setup. Time-of-flight spectrometer scheme for NRCA

nucleus. Neutron Resonance Capture Analysis (NRCA) is based on the identification of resonances occurring within a sample that is bombarded with neutron flux [14]. In NRCA, the neutrons are provided by a spallation source. The neutron flux is pulsed, and the energy of the resulting neutrons is measured by using a time-of-flight system. In Fig. 19.7, a scheme of the experimental setup is given. The start signal for the TOF system is provided by the pulse neutron beam, and the stop signal is provided by the $\gamma$-ray cascade produced during neutron capture by the compound nucleus. The neutron energy in this way measured gives the resonance energy, which specifically corresponds to an absorbing nucleus, immediately identified.

An NRCA spectrum of an archaeological artifact is presented in Fig. 19.8. NRCA is similar to PGNAA, where the prompt $\gamma$-energy identifies the compound nucleus.

### 19.5.2 Neutron Resonance Transmission Analysis

Neutron Resonance Transmission Analysis (NRTA) is an alternative technique to NRCA. Now, the surviving neutrons in a transmission experiment are observed. Those neutron energies absent in the outgoing beam reflect the resonant captures occurring in the sample and therefore the absorbing nucleus present. The experimental setup is similar to that of NRCA, but the stop signal is provided by a neutron detector, generally scintillators based on the $^6\text{Li}(n, \alpha)^3\text{H}$ reaction or organic scintillators, which are better suited for fast neutrons. Although NRCA

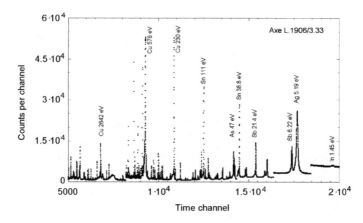

**Fig. 19.8** NRCA spectrum of an archaeological artifact. Figure reprinted from [14] with permission by Springer

finds applications mainly in Archaeometry, NRTA has been used to characterize nuclear fuel [15].

## 19.6    Applications to Environmental Radioactivity

As presented in Table 19.1 NAA finds application in many scientific disciplines. In addition, it helps to answer many problems related to social needs. Fields such as Archaeology, Biomedicine, Geology, Art, Physics, Chemistry, Environment, Environmental pollution, and Industry benefit from NAA [1]. Furthermore, the application of NAA to forensic analysis or nutrition studies or in the field of security with the detection of hidden explosives or chemicals in packages or the control of nuclear safeguard is well known. In general, the multielemental character of NAA as well as the typical sensitivities (see Table 19.2) that can be reached make NAA the technique of choice for many applications. In addition, we cannot forget that NAA provides isotopic information on the sample since it is based on the nuclear interaction of neutrons and target nuclei. From this viewpoint, NAA can be used to study specific nuclear reactions that lead to quantifying the presence of a given nucleus in the analyzed sample. This approach is very useful for environmental radioactivity problems [2, 3, 10]. The main application in this field is the determination of long-lived (or very long-lived) radionuclides in environmental samples for which conventional radiometric techniques are unable to make precise measurements [10]. NAA competes with AMS and ICP–MS (see previous chapters).

In Table 19.6, a list of radionuclides studied by NAA in the environment is presented together with their half-lives. The neutron reactions used for the analysis and the resulting radionuclides are also given. It is interesting to note that NAA is useful insofar as the radionuclide produced can be measured using conventional radiometric techniques. In other words, the half-life of the resulting nuclides and the emitted radiation can be readily managed. In fact, these two conditions reduce the possibilities of NAA application to the radionuclide listed in the Table. Normally, the resulting nuclides emit $\gamma$-rays, allowing the use of Ge detectors that, on the basis of their extremely good energy resolution, permit the determination of the nuclides virtually without interferences. Usually, the so-called advantage factor [10] summarizes how favorable the use of the activation product is to determine the radionuclide of interest. It is defined as the relationship between the activation product activity, $A_Y$, produced during a given irradiation time, $t$, and measured after a decay time, $t_d$, and the radionuclide activity of interest, $A_X = \lambda_X N_X$. In the case of long-lived radionuclides, $A_X$ is supposed to be virtually constant. The advantage factor can be calculated using (19.8). Thus,

$$AF = \frac{A_Y(t_d)}{A_X} = \frac{\sigma\phi\left(1 - e^{-\lambda t}\right)}{\lambda_X}e^{-\lambda t_d} \tag{19.21}$$

$AF > 1$ or $\gg 1$ are necessary for favorable NAA conditions. On the contrary case, it is not convenient to use an NAA method for radionuclide determination.

RNAA must be used in some cases, as radiochemical separations of the produced nuclides are needed. However, INAA is sufficient for some applications, which is very advantageous since it does not require the manipulation of difficult or very radioactive samples. Additionally, for many applications, the radionuclide of interest must be preconcentrated and chemically extracted prior to neutron irradiation. The radiochemical procedure must be designed in such a way that the presence of potential interferences must be avoided. However, the use of tracers to measure the radiochemical yield is also a source of potential interference. This must be taken into account when planning the experiment.

Once the radionuclide has been extracted, it has to be transferred to an irradiation container. Because the radionuclide concentration is expected to be very

**Table 19.6** List of relevant radionuclides for the environment that can be analyzed by NAA

| Radionuclide | Half-life (y) | Neutron reaction | Used radiation |
|---|---|---|---|
| $^{99}$Tc | $2.1 \times 10^5$ | $^{99}$Tc$(n, \gamma)^{100}$Tc | $\beta$ |
| | | $^{99}$Tc$(n, n')^{99m}$Tc | $\gamma$ |
| $^{129}$I | $1.57 \times 10^7$ | $^{129}$I$(n, \gamma)^{130}$I | $\beta, \gamma$ |
| $^{135}$Cs | $2.05 \times 10^6$ | $^{135}$Cs$(n, \gamma)^{136}$Cs | $\beta, \gamma$ |
| $^{230}$Th | $7.54 \times 10^4$ | $^{230}$Th$(n, \gamma)^{231}$Th | $\beta, \gamma$ |
| $^{232}$Th | $1.4 \times 10^{10}$ | $^{232}$Th$(n, \gamma)^{233}$Th | $\gamma$ |
| $^{235}$U | $7.04 \times 10^8$ | $^{235}$U$(n, f)\dots {}^{140}$La, $^{141}$Ce, $^{133}$I or $^{134}$I | $\beta, \gamma$ |
| $^{238}$U | $4.46 \times 10^9$ | $^{238}$U$(n, \gamma)^{239}$U | $\gamma$ |
| $^{237}$Np | $2.14 \times 10^6$ | $^{237}$Np$(n, \gamma)^{238}$Np | $\gamma$ |
| $^{231}$Pa | $8.27 \times 10^4$ | $^{231}$Pa$(n, \gamma)^{232}$Pa | $\gamma$ |
| $^{242}$Pu | $3.75 \times 10^5$ | $^{242}$Pu$(n, \gamma)^{243}$Pu | X-, $\gamma$-rays |

low, the container must be clean and not contain the radionuclide studied or any other trace nuclides that could produce potential interferences during irradiation. Of course, the irradiation has to be planned also taking into account the possibility of consecutive neutron reactions with the produced nuclides. In summary, a careful study of the potential interferences as explained in the previous sections is compulsory, taking into account the very low levels of radioactivity we are dealing with.

Specific details follow on the determination of some long-lived radionuclides by NAA.

### $^{99}Tc$

$^{99}$Tc is a long-lived ($T_{1/2} = 2.1 \times 10^5$ y) low-energy β-emitter. There are two reactions used for its determination [10, 16]. The main reaction is $^{99}$Tc$(n, \gamma)^{100}$Tc, which is also a β-emitter decaying to $^{100}$Ru with $T_{1/2} = 15.8$ s. The cross section for thermal neutrons is 20 b and 30 b for higher energy neutrons. Alternatively, $^{99}$Tc$(n, n')^{99m}$Tc can be used. Now $^{99m}$Tc is a γ-emitter that decays to $^{99}$Tc with $T_{1/2} = 6.0$ h. However, this reaction is not as convenient since the cross section is considerably lower, 0.24 b.

To avoid interference, Tc must be extracted before and after irradiation. For that, rapid methods are necessary because of the very low half-life of $^{100}$Tc. It can be determined by γ-spectrometry with Ge spectrometers measuring 539.52 keV (6.6%) or 590.77% (5.73%) γ-emissions. Alternatively, it can be determined by direct β-counting. In this case, the radiochemical extraction procedure must provide a very pure $^{100}$Tc sample to avoid the presence of interferences that could affect the counting. A limit of detection of approximately 2.8 mBq has been reported for $^{99}$Tc using NAA.

### $^{129}I$

$^{129}$I ($T_{1/2} = 1.57 \times 10^7$ y) is a β-emitter that decays into $^{129}$Xe and feeds an excited level at 39.6 keV. Therefore, a γ-ray is also observed, in addition to X-rays and conversion electrons. The reaction used for the NAA analysis is $^{129}$I$(n, \gamma)^{130}$I [10] with a cross section of 30 b. $^{130}$I is a β-emitter with a $T_{1/2} = 12.3$ h decaying into $^{130}$Xe. In the process, some γ-rays are emitted from the excited states of $^{130}$Xe that can be used to determine $^{130}$I. Several radionuclides, presumably present in environmental samples, can suffer neutron reactions that interfere with the determination of $^{129}$I. This is the case for $^{235}$U, $^{128}$Te, $^{133}$Cs, and others [10]. In other words, $^{129}$I must be chemically extracted before irradiation. A postirradiation chemical procedure is convenient to remove possible interferences during counting [17]. Some examples can be found in the literature on the determination of $^{129}$I in the environment by NAA. $^{129}$I has been analyzed in seawater [17] with detection limits close to $10^{-13}$ g of $^{129}$I using 4 to 10 l of water samples. Data can be found for seaweed [18] with detection limits of approximately 1 mBq.

*$^{135}Cs$*

$^{135}$Cs ($T_{1/2} = 2.05 \times 10^6$ y) is a pure β-emitter that can be analyzed using $^{135}$Cs($n$, $\gamma$)$^{136}$Cs with a cross section close to 8 b. $^{136}$Cs is a β-emitter with a $T_{1/2} = 13.04$ d. Its daughter, $^{136}$Ba, appears in several excited states after $^{136}$Cs decay. Thus, several γ-rays of relatively high intensity can be used to determine $^{136}$Cs and therefore $^{135}$Cs. The most intense are 818.514 keV (99.7%) and 1048.073 (80%).

$^{135}$Cs is present at very low levels in the environment and is virtually impossible to measure by radiometric methods. Mass spectrometry, including AMS and NAA, are the methods of choice [10]. Regarding NAA, the main interferences come from the other Cs isotopes present in the sample, $^{133}$Cs, $^{137}$Cs, $^{134}$Cs, and $^{85}$Rb [19]. In fact, its concentration clearly affects the NAA detection limit for $^{135}$Cs. However, in general terms, a value for LID close to $10^{-1}$ mBq has been reported [10, 19].

*$^{230}Th$*

It is an α-emitter ($T_{1/2} = 7.54 \times 10^4$ y) that can be easily measured by conventional α-spectrometry. However, the cross section for $^{230}$Th($n$, $\gamma$)$^{231}$Th is high, 23.08 b, and therefore the determination by NAA is favorable since AF is also >1 [10]. $^{231}$Th ($T_{1/2} = 25.52$ h) is a β-emitter feeding several excited states of $^{231}$ Pa and thus emitting γ-rays [20].

*$^{232}Th$*

$^{232}$Th is an α-emitter with $T_{1/2} = 1.4 \times 10^{10}$ y. It is very often measured by simple α-spectrometry with Si semiconductor detectors. Nevertheless, in some cases, it is convenient to use it.

$$^{232}Th(n, \gamma)^{233}Th \xrightarrow{\beta, T_{\frac{1}{2}} = 22.3 min} {}^{233}Pa \xrightarrow{\beta, T_{\frac{1}{2}} = 27.0 d} {}^{233}U$$

with a 7.4 b cross section $^{233}$U is an α-emitter with $T_{1/2} = 1.59 \times 10^5$ y; that is, it can be considered stable from an experimental point of view. Usually, $^{232}$Th is determined through the γ-radiation emitted by $^{233}$Pa, mainly 311.91 keV photons with an intensity of 38.2% [20]. In many cases, it is convenient to take advantage of the epithermal neutron cross section of this reaction, which is approximately 72 b. The use of these neutrons allows us to largely reduce some interfering nuclei in the counting as $^{24}$Na, $^{38}$Cl, and $^{32}$P. Detection limits of approximately $10^{-8}$ g have been reported for INAA. They can be improved by using a previous chemical separation that reached values close to $10^{-11}$ g [10, 20].

*$^{235}U$*

$^{235}$U is an α-emitter. Its half-life is $7.04 \times 10^8$ y. It is a fissile radionuclide with a fission cross section for thermal neutrons of 585 b. As $^{238}$U and $^{232}$Th, it can be determined by α-spectrometry with Si detectors. However, its isotopic abundance in nature is very low (0.72%). This makes it difficult to detect for some samples. As an alternative, $^{235}$U can be determined by measuring some fission products that can appear after neutron irradiation of the sample. For that, $^{140}$La, $^{141}$Ce, $^{133}$I, or $^{134}$I have been used in the literature [10].

### $^{238}U$

$^{238}$U ($T_{1/2} = 4.46 \times 10^9$ y) is an α-emitter that can be easily determined by Si detector α-spectrometry. Alternatively,

$$^{238}U(n, \gamma)^{239}U \xrightarrow{\beta, T_{\frac{1}{2}}=23.5min} {}^{239}Np \xrightarrow{\beta, T_{\frac{1}{2}}=2.6d} {}^{239}Pu$$

can be used. The cross section of this reaction is 2.7 b, and the final product $^{239}$Pu is also α-emitter with $T_{1/2} = 24.110 \times 10^3$ y. $^{239}$Np is the nucleus selected for the determination of $^{238}$U. In particular, the 106.123 keV (25.342%), 277.599 keV (14.51%), and 228.183 keV (10.73%) photons it emits are relatively intense. As in the case of $^{232}$Th, the epithermal neutron cross section is higher, some 277 b, making it more advantageous to discriminate from potential interferences [20]. The use of RNAA reduces the LID to $10^{-9}$ g, which is one order of magnitude smaller than that obtained by INAA. There are several examples in the current literature on the application of NAA to environmental samples, as well as urine and human hair [21, 22]. All of them reveal a high potential applicability. However, the use of conventional α-spectrometry by Si detectors is preferred and, of course, available to many laboratories.

### $^{237}Np$

$^{237}$Np is an α-emitter with $T_{1/2} = 2.14 \times 10^6$ y. The neutron reaction is $^{237}$Np($n$, $\gamma$)$^{238}$Np with $\sigma = 169$ b. $^{238}$Np is a β-emitter ($T_{1/2} = 2.1$ d) that decays into $^{238}$Pu, which is an α-emitter. Several γ-rays emitted by $^{238}$Np can be used to determine $^{237}$Np. Specifically, those of high intensity were 984.45 keV (25.2%), 1025.87 (8.75%), and 1028.53 keV (18.23%). Due to the intense interference suffered by the presence of $^{238U}$, a previous chemical separation is needed. Indeed,

$$^{238}U(n, 2n)^{237}U \xrightarrow{\beta, T_{\frac{1}{2}}=6.75d} {}^{237}Np(n, \gamma)^{238}Np$$

$$^{235}U(n, \gamma)^{236}U(n, \gamma)^{237}U \xrightarrow{\beta, T_{\frac{1}{2}}=6.75d} {}^{237}Np(n, \gamma)^{238}Np$$

can produce $^{238}$Np. Because the abundance of U in the environment is very high compared to that of Np, it is compulsory to make a chemical separation prior to irradiation. The epithermal neutron cross section for $^{237}$Np($n$, $\gamma$)$^{238}$Np is approximately 600 b. The use of these neutrons significantly reduces the interference from other nuclides present in the sample. Nevertheless, chemical separation after irradiation is also necessary [10]. NAA has been used for the determination of $^{237}$Np in various environmental matrices, such as soils, sediments, seaweed, and seawater [23].

## $^{231}Pa$

$^{231}$Pa is an α-emitter with $T_{1/2} = 8.27 \times 10^4$ y. After neutron irradiation, $^{232}$Pa ($T_{1/2} = 1.31$ d) is formed via $^{231}$Pa($n, \gamma$)$^{232}$ Pa with $\sigma = 200.6$ b. This isotope is also an α-emitter feeding several excitation states of its descendant $^{232}$U. In this way, it emits γ-rays, which can be easily measured. The selected γ-ray is 969.13 keV with an intensity of 42.3%.

$^{232}$ Pa(n,γ)$^{233}$ Pa ($T_{1/2} = 26.975$ d) with $\sigma = 651.3$ b is an interference reaction that must be taken into account for the determination of $^{231}$ Pa [20].

## $^{242}Pu$

It is an α-emitter ($T_{1/2} = 3.75 \times 10^5$ y), and the thermal neutron AF is $\gg 1$ for the reaction $^{242}$Pu($n, \gamma$)$^{243}$Pu [10]. $^{243}$Pu is a β-emitter with a half-life of 4.956 h that emits a few, not very intense, X- and γ-rays that could be used for $^{242}$Pu determination. NAA can potentially be used in this case, although to date, there is no documentation in the current literature on its feasibility [20, 24].

**Exercises**

1. Derive the general equations of neutron activation when using the first descendant. Assume that all data are known and that irradiation is carried out during a time t, the activity measured at a time $t_1$ after discontinuation of irradiation, and that the measurement is carried out during a time interval $\Delta t$ similar to the half-lives of the recoil nucleus and its descendant.

2. Derive the general equations of neutron activation when using the second descendant. Assume that all data are known and that irradiation is carried out during a time t, the activity measured at a time $t_1$ after discontinuation of irradiation, and that measurement is carried out during a time interval $\Delta t$ similar to the half-lives of the recoil nucleus and its descendant.

3. A neutron activation reaction $^{A}X(n, \gamma)^{A+1}X$ is produced in a nuclear reactor. Find an expression for the mass, m, of the nucleus $^{A}X$ present in the sample from the determination of the activity of $^{A+1}X$ during an irradiation time t. Suppose that the measurement of the activity of the $^{A+1}X$ nucleus is carried out at time $t_1$ after the irradiation has stopped and during time interval $\Delta t > T_{1/2}$ ($^{A+1}X$).

4. A neutron activation experiment is carried out to quantify the amount of $^{23}$Na in a sample. The useful reaction is $^{23}$Na($n, \gamma$)$^{24}$Na. However, it is also observed that the interfering reaction $^{27}$Al($n, \alpha$)$^{24}$Na takes place. Demonstrate that the equation that gives the content of $^{23}$Na in the material (atomic density) as a function of the activity of $^{24}$Na at the end of the activation experiment is.

$$N_{23}(0) = \frac{\lambda_{24} N_{24}(t) - \sigma_\alpha \phi N_{Al}(0)}{\sigma_\gamma \phi}$$

where $\phi$ is the neutron flux, $N_{Al}$ is the number of $^{27}$Al atoms per unit volume, the irradiation time is t, and the cross sections for the main (useful) and interfering reactions are $\sigma_y$ and $\sigma_\alpha$ respectively. The decay constant of $^{24}$Na is $\lambda_{24}$. Assume that the amounts of $^{23}$Na and $^{27}$Al do not vary throughout the irradiation and that $\lambda_{24}$ is much greater than the activation rates of $^{23}$Na and $^{27}$Al.

5. A 1 g $^{59}$Co sample is placed in a reactor in a zone where a thermal neutron flux of $10^{12}$ n/cm$^2$s is present. For these neutrons, the $^{59}$Co cross section for the $^{59}$Co$(n, \gamma)^{60}$Co reaction is 36.3 barn. Calculate the activity of the sample after 2 days and after 4 weeks of irradiation.

6. Two identical silver foils are irradiated with a neutron flux from a nuclear reactor. The $^{108}$Ag-induced activity is measured with a detector for 10 min after a cooling time of 2 min. Find the count ratio expected from the two foils if one has been irradiated for 10 min and the other for 20 min. Suppose that the neutron flux and the counting geometry are identical for both foils.

7. Natural indium contains 95.7% 115In, which has a cross section $\sigma = 160$ b for the reaction $^{115}$In $(n, \gamma)^{116m}$In. A 0.8 g In foil was introduced into a neutron reactor, and after 162 min, it was removed, and its activity was measured for 54 min with an 8% efficiency detector. A total of 25,000 counts were recorded. Find the flux of thermal neutrons in the area where the foil was exposed.

8. The $^{113}$Cd cross section for thermal neutron capture is a $2 \times 10^4$ barn. What natural Cd thickness, in g/cm$^2$, is required to attenuate the intensity of a thermal neutron beam to 0.05% of its initial value?

9. A gold foil (100% $^{197}$Au) is irradiated with a slow neutron flux of $5 \times 10^{14}$ n/cm$^2$s undergoing radiative capture $^{197}$Au$(n, \gamma)^{198}$Au, $\sigma_7 = 98.8$ barn. The $^{198}$Au isotope formed this way decays by $\beta$ with a half-life of 2.7 d and has a large cross section, $2.51 \times 10^4$ barn for $^{198}$Au $(n, \gamma)^{199}$Au. How long after the irradiation has started will we obtain the maximum number of $^{198}$Au atoms?

10. At this moment, what is the relationship between the number of atoms of $^{198}$Au and $^{197}$Au?

11. The $^{197}$Au thermal neutron activation cross section is $\sigma_7 = 98.8$ barn. The $^{198}$Au obtained is a $\beta$-emitter ($T_{1/2} = 2.70$ d) and, in turn, has a large neutron capture cross section $\sigma_8 = 2.5 \times 10^4$ barn. Find the neutron flux at which $^{198}$Au is produced by neutron capture and decays at the same rate.

12. With that flux, how long does it take to get maximum activity? At that moment, what fraction of the initial atoms is still $^{197}$Au and $^{198}$Au, as well as $^{199}$Au and $^{198}$Hg?

13. A mass, $m_x$, of the nuclei of mass number $A$ is irradiated with a neutron flux $\phi$, and the reaction $X(n, b)Y$ occurs with a cross section $\sigma_x$. The nucleus $Y$ is radioactive, with a decay constant $\lambda_y$, and undergoes the reaction $Y(n, c)Z$ with $\sigma_y$. $Z$, in turn, is also radioactive, $\lambda_z$, and undergoes the reaction $Z(n, d)T$ with $\sigma_z$. Assuming that it is constant throughout the irradiation process, find the equation for the evolution of the activity of $Z$ during irradiation time $t$ as a function of known parameters and assume that, at $t = 0$, $Ny = Nz = 0$. Explain the conditions needed for saturation in $Z$ production.

14. Natural Cu consists of 69% $^{63}$Cu and 31% $^{65}$Cu. 64Cu and 66Cu are formed by neutron irradiation of Cu with half-lives of 2.7 h and 5.1 min, respectively.

What is the activity in Bq of each of these isotopes if 1 g of Cu is irradiated with a thermal neutron flux of $10^9$ neutrons/cm$^2$s for 15 min?

Data: $^{63}$Cu$(n, \gamma)^{64}$Cu, $\sigma = 4.4$ barns.
$^{65}$Cu$(n, \gamma)^{66}$Cu, $\sigma = 2.2$ barns.

15. Find the irradiation time to obtain the maximum activity of each of the isotopes? Calculate $A_{max}$ for $^{64}$Cu and $^{66}$Cu.

16. $^{36}$Cl is generated in rocks by neutron activation of the naturally occurring $^{35}$Cl by $^{35}$Cl$(n, \gamma)^{36}$Cl. This can be used to determine how long the rock has been exposed to cosmic radiation. Find the relationship between the number of atoms $^{36}$Cl $/^{35}$Cl that allows this dating to be carried out and obtain the final equilibrium value of this ratio.

17. In a real sample with a chlorine content of 2 mg/g, it was found that $^{36}$Cl/$^{35}$Cl $= (3.05 \pm 0.25) \times 10^{-12}$. How long has the sample been exposed to neutron radiation? How much argon gas (see $^{36}$Cl decay scheme) has 10 t of this sample been released into the atmosphere?

Cl isotopic composition: 75.77% $^{35}$Cl, 24.23% $^{37}$Cl.
Neutron flux: $\Phi = 100$ m$^{-2}$ s$^{-1}$
$^{35}$Cl$(n, \gamma)^{36}$Cl cross section: $\sigma = 43.7 \times 10^{-28}$ m$^2$

18. An experiment is carried out in which 1 g of $^{130}$Te is irradiated with a neutron flux $\phi$. As a consequence, the reaction $^{130}$Te$(n, 2n)^{129}$Te takes place. The resulting $^{129}$Te can be produced in its ground state or in a metastable state, following the decay scheme in Figure:

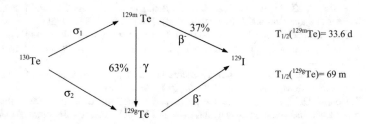

$\sigma_1$ and $\sigma_2$ are the neutron cross sections for $^{129m}$Te and $^{129}$Te, respectively. Write (without solving) the equations of time evolution of the population of each nucleus.

19. Solve the equation for $^{129m}$Te as a function of irradiation time. If some approximations were needed, justify them.

20. The $^{130}$Te foil is irradiated with neutrons for a time $t = 1$ h. Immediately after irradiation, the emitted $\gamma$ radiation is measured for 10 s, obtaining a counting rate of 14,667 cps. What is the efficiency value of the detector used if the background obtained was 600 counts during the same counting time?

Data: $\sigma_1 = 1.100$ barn; $\sigma_2 = 0.670$ barn; $\phi = 10^{12}$ cm$^{-2}$ s$^{-1}$

# References

1. Neutron Generators for Analytical Purposes. STI/PUB/1535 (IAEA, 2012)
2. G. Hevessy, H. Levi, Action of slow neutrons on rare earth elements. Nature **137**, 185 (1936)
3. Z. Alfassi (ed.), *Activation Analysis*, vol. 1 (CRC Press, 1990)
4. Z. Alfassi (ed.), *Activation Analysis*, vol. 2 (CRC Press, 1990)
5. Zs. Révay, T. Belgyain, Principles of PGAA method, in *Handbook of Prompt Gamma Activation Analysis with Neutron Beams*, ed. by G.L. Molnár (Kluwer Academic Publishers, 2004), pp. 1–30
6. Development of an Integrated Approach to Routine Automation of Neutron Activation Analysis (IAEA-TECDOC-1839, IAEA, 2018)
7. S.I. Kafala, T.D. MacMahon, Comparison of neutron activation analysis methods. J. Radioanal. Nucl. Chem. **271**, 507–516 (2007)
8. F. Girardi, G. Guzzi, J. Pauly, Reactor neutron activation analysis by the single comparator method. Anal. Chem. **37**, 1085–1092 (1965)
9. M. Mohammadzadeh, M. Ajami, A. Shadeghipanah, M. Rezvanifard, 'A study of neutron activation analysis compared to inductively coupled plasma atomic emission spectrometry for geological simples in Iran, Nuclear. Eng. Technol. **50**, 1349–1354 (2018)
10. X. Hou, Activation analysis for the determination of long-lived radionuclides, in *Radioactivity in the Environment*, vol. 11 (Elsevier, 2008), pp. 371–405
11. B. Smodis, Radiochemistry for neutron activation analysis, in *IAEA Workshop on Nuclear Data for Activation Analysis*, Abdus Salam International Centre for Theoretical Physics, Trieste, 2005
12. https://www.statista.com/statistics/1203545/global-number-of-operational-nuclear-research-reactors/
13. M.N.H. Comsan, Spallation neutron sources for science and technology, in *Proceedings of the 8th Conference on Nuclear and Particle Physics*, Hurghada, Egypt, 20–24 November 2011.
14. H. Postma, R.C. Perego, P. Schillebeeckx, P. Siegler, A. Borella, Neutron resonance capture analysis and applications. J. Radioanal. Nucl. Chem. **271**(1), 95–99 (2007)
15. H. Postma, P. Schillebeeckx, Neutron resonance capture and transmission analysis, in *Encyclopedia of Analytical Chemistry: Applications, Theory and Instrumentation*, ed. by R.A. Meyers (Wiley, 2009)
16. S. Foti, E. Delucchi, V. Akamian, Determination of picogram amounts of $^{99}$Tc by neutron activation análisis. Anal. Chim. Acta. **60**, 261–268 (1972)
17. X. Hou, H. Dahlgaard, B. Rietz, U. Jacobsen, S.P. Nielsen, A. Aarkrog, Determination of $^{129}$I in seawater and some environmental materials by neutron activation analysis. Analyst **124**, 1109–1114 (1999)
18. C.-L. Tseng, J.H. Chao, Low-level determination of $^{129}$I in environmental simples by neutron activation. Appl. Radiat. Isot. **47**, 723–726 (1996)
19. J.-H. Chao, C.-L. Tseng, Determination of $^{135}$Cs by neutron activation análisis. Nucl. Instrum. Methods Phys. Res. A **372**, 275–279 (1996)
20. A.R. Byrne, L. Benedik, Applications of neutron activation analysis in the determination of natural and man-made radionuclides, including $^{231}$Pa. Czechostovak J. Phys. **49**, 263–270 (1999)
21. S. Landsberger, R. Kapsimalis, Comparison of neutron activation analysis techniques for the determination of uranium concentrations in geological and environmental materials. J. Environ. Radioact. **117**, 41–44 (2013)
22. R. Byrne, L. Benedik, Uranium content of blood, urine hair of exposed and nonexposed persons determined by radiochemical neutron activation analysis, with emphasis on quality control. Sci. Total Environ. **107**, 143–157 (1991)

23. P. Germain, P. Guegueniat, S. May, G. Pinte, Measurement of transuranic elements, chiefly $^{237}$Np (by neutron activation analysis), in the physical and biological compartments of the French Shore of the English channel. J. Environ. Radioact. **5**, 319–331 (1987)
24. R.J. Rosenberg, Non-conventional measurement techniques for the determination of some long-lived radionuclides produced in nuclear fuel a literature survey. J. Radioanal. Nucl. Chem. Art. **171**, 465–482 (1993)

# Radioactive Particle Characterization

# 20

**ABSTRACT**

Characterization of radioactive particles (see Chaps. 4, 5, and 13) is of paramount importance when studying environmental radioactivity, as they rule the behavior and fate of some radionuclides in the environment. Different techniques to determine the morphology, shape, size, density, composition, and chemical form of radioactive particles are described in this chapter.

## 20.1 Radioactive Particle Characterization

The relevance of radioactive particles for environmental radioactivity studies has been established in previous chapters. To isolate and characterize the particles, some complementary techniques are needed. They include not only nuclear but also atomic and molecular analytical techniques. These techniques provide information on relevant aspects that help to understand the environmental behavior of the radionuclides contained in the particle. Features such as particle shape, size, elemental and radionuclide 2D or 3D distribution, surface properties, crystalline or amorphous structures, valence and oxidation states, or radionuclide molecular forms are of high interest. They complement the information derived from radioactivity measurements.

Once the particles are characterized by using these nondestructive techniques, leaching and sequential extraction procedures are very useful to complement the obtained results. An extensive account of these techniques and their applications to real radioactive samples can be found in [1].

Of course, before their characterization, the particles must be isolated.

© The Author(s), under exclusive license to Springer Nature Switzerland AG 2022
M. García-León, *Detecting Environmental Radioactivity*, Graduate Texts
in Physics, https://doi.org/10.1007/978-3-031-09970-0_20

## 20.2    Radioactive Particle Identification and Isolation

The isolation of radioactive particles can be done during sampling or in the laboratory. In Chap. 13, sampling procedures for aerosols, water, soil, or sediment samples are described. Fractionation by filtration or ultrafiltration of water can be performed directly at the site [2]. Cascade impactors classify atmospheric particles by size [2]. Using radiation detectors, it is possible to identify radioactive hot spots in atmospheric or water filters in situ. They give a hint on the presence of radioactive particles. In the case of soils or sediments, a first approximation can also be performed on site with radiation detectors. The main objective is to identify inhomogeneities in the samples that could be the consequence of radioactive particles. Once the first screening is carried out, a more systematic study is necessary, and this is done at the laboratory.

Radioactive particles are usually isolated by the binary splitting method [3]. The sample is divided into two parts, and the radioactivity is measured. The subsamples are in turn divided into two, and so on. The main goal is the isolation of subsamples that contain significantly more radioactivity. The radioactive particles should be there.

### 20.2.1 Radiometric Methods

Radiometric methods can be used to identify particles [1]. The use of portable detectors as GM survey meters serves to define hot spots of radioactivity, although no further information can be obtained from these measurements. Portable Ge $\gamma$-spectrometers that can provide more relevant information about the sample. In some cases, depending on the radionuclide concentration, fission products can be perfectly identified with such devices. Even transuranics can be determined by measuring X-rays or low-energy $\gamma$-emissions. This is the case for the determination of Pu isotopes and $^{241}$Am in radioactive particles taken from weapons testing sites or as a result of accidental releases. NORM radioactive particles can also be identified by in situ $\gamma$-spectrometry since the daughters $^{238}$U, $^{235}$U, and $^{232}$Th emit $\gamma$-radiation.

However, in many cases, the activity concentration, although inhomogeneously distributed, is low or very low. Thus, it is not possible to perform precise measurements in situ. The samples are brought to the laboratory, and Low-Level Counting Techniques must be applied. The subsamples of interest are measured, and some additional splitting is performed to isolate the radioactive particle as much as possible. In Fig. 20.1, we present a subsample $\gamma$-spectrum, showing the Pu X-rays as well as the 59.5 keV $\gamma$-ray from $^{241}$Am. This gives a hint of the likely presence of radioactive particles [4].

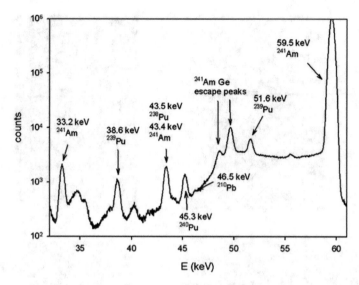

**Fig. 20.1**  γ-spectrum of a radioactive particle showing the presence of $^{239}$Pu and $^{241}$Am. Figure reprinted from [4] with permission of the American Chemical Society

## 20.2.2 Imaging Techniques

A more precise location of the particle requires the use of imaging techniques. Autoradiography can be applied for this purpose. It was already described in Chap. 12. The subsample is covered by a photographic emulsion, and after a long exposure time, the emulsion is developed. In the case of atmospheric or water filters, the sample can be covered directly with the emulsion. The areas containing radioactive particles are expected to produce a more intense blackening of the emulsion. This allows the location of the hot particle in the subsample. In a more recent version of this technique, digital autoradiography, the photographic emulsion is substituted for plates that contain fluorescent materials such as P and Eu [1]. Now, the radiation emitted by the radioactive particles promotes electrons from the valence to the conduction band. However, there are some electron traps within the forbidden band that capture the liberated electrons. The development of the plate is carried out by laser stimulation. The laser wavelength must be selected to provide the trapped electrons with enough energy to leave the trap and return to the conduction band. During this process, light is emitted. Light is captured by a photomultiplier that converts its intensity into an electrical current. A grayscale map of the subsample is generated, where the blackest points correspond to high electron intensities or to high irradiation points. In this way, the black spots identify the presence of radioactive particles. Digital autoradiography provides better spatial resolution than conventional autoradiography. An autoradiography image is shown in Fig. 20.2. In general, the time exposure in autoradiography is long

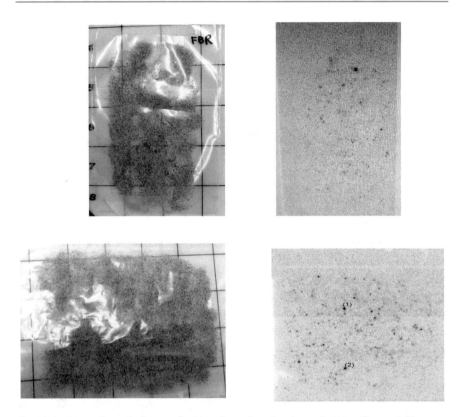

**Fig. 20.2** Autoradiography image of two sand samples taken in southwestern Portugal. The numbers identify the particles. Figure courtesy of Prof. I. Vioque

since the radioactivity concentrations are low. Exposure times close to a week are typical in these experiments [5, 6].

Solid-State Nuclear Track Detectors (SSNTDs) can also be used to locate radioactive particles. They were described in Chap. 12, so no further information is given here on their working principles. There exist real-time imaging devices capable of providing a picture of the subsample and, therefore, locating the position of the particle in real time [7]. They are based on Si or scintillation detectors and work as a gamma camera, an Anger camera [8] after its inventor. Thus, the position is determined by locating the photomultiplier or multichannel plate responsible for the recorded signal. When strip Si detectors are used, the position is determined by locating the semiconductor strip that produces the signal. Spatial resolutions close to $10^2$ μm can be achieved with these devices. Real-time detectors are faster than passive systems. This is their main advantage. However, when dealing with very low-activity concentrations, the background must be taken into account carefully, and the time needed to obtain precise results increases accordingly.

### 20.2.3 Particle Isolation and Manipulation

Once located, the subsample containing the particle is homogeneously distributed onto adhesive carbon tape. This carbon tape is normally used in electron microscopy. The tape prepared in this way is studied with the following techniques.

## 20.3 Radioactive Particle Size, Morphology, and Composition

### 20.3.1 Electron Microscopy

Scanning electron microscopy (SEM) is based on several interaction effects that occur when an electron beam impinges on a sample. In SEM, the sample is irradiated with an electron beam produced by the thermionic effect on a W filament. The electrons are accelerated to energies ranging from 100 keV to 1 meV and driven to the sample by a series of magnetic lenses. The optical elements of the microscope can be used to scan the sample with a spatial resolution below the submicrometer range [1]. The electrons reaching the sample interact with its components and undergo different effects. The most interesting are those presented in Fig. 20.3. The electrons from the beam can be backscattered or removed from the sample surface. In addition, they can produce X-ray fluorescence as a result of the electron vacancies they provoke in the atomic shells in the sample. Backscattered electrons are the result of elastic interactions with the atomic nuclei in the sample. The probability of backscattering occurring strongly depends on Z along a wide range of atomic numbers (see Chap. 7). Therefore, the intensity of the backscattered electrons will inform the atomic number of the sample component. The backscattered electron spectrum is presented in a grayscale map, where the brighter points correspond to higher Z elements. In Fig. 20.4a, a backscattered electron picture of a radioactive particle is given. Brighter points are associated in this case with the presence of U and Pu, while darker areas correspond to the sediment [9].

The electrons removed from the sample are called secondary electrons. They leave the atoms as a consequence of the energy gained from collisions with the impinging electrons. The probability of producing secondary electrons depends mostly on the energy and very little on Z and is more associated with the presence of the atom at the sample surface and the angle between the electron beam and the sample surface. Therefore, they can provide information on the topographical aspects of the sample surface. In practical terms, these secondary electrons cannot be distinguished from the backscattered electrons. For real applications, only those electrons from the sample with energies below 50 keV are considered secondary electrons, and the image is formed only with them [10]. The result is a picture of a radioactive particle, as presented in Fig. 20.4b. A clear image of the morphology of the particles and, more specifically, of its surface can be observed [9].

Many electron microscopes are also equipped with X-ray detectors that can add information on the elemental composition of the irradiated sample point by point.

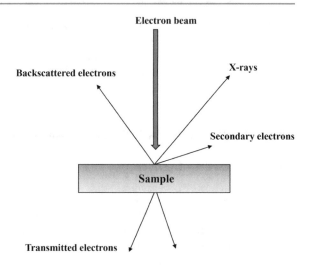

**Fig. 20.3** Interesting interaction mechanisms of e with matter in SEM. X-rays and secondary, transmitted or backscattered electrons provide information on different sample characteristics

The X-rays are produced by fluorescence as a result of the vacancies generated by the electrons in the sample atoms. In this way, an X-ray map can be generated, as presented in Fig. 20.4c, in which U X-rays are colored red and those from Pu are colored blue. The picture corresponds to a radioactive particle, and the X-ray map is overimposed onto a backscattered electron picture of the particle.

Very interestingly, in many cases, a simple inspection of the shape of the X-ray spectrum can give a hint of the mineral contained in the radioactive particle. This is possible because there exists a large catalogue of X-ray spectrum shapes corresponding to typical mineral compositions [11]. An example is given in Fig. 20.5a, where the X-ray spectrum taken from a radioactive particle is clearly identified as an ilmenite mineral whose expected X-ray spectrum is that of Fig. 20.5b [11].

The combination of the three signals gives very complete information on the morphology, size, and sample composition.

## 20.3.2 Computed Tomography (CT)

Computed tomography can provide three-dimensional information on the particle morphology, density, and composition. In this case, the sample is irradiated by an X-ray beam, and the transmitted X-rays are studied [1]. The attenuation of the X-ray beam depends on the density of the irradiated sample area (see Chap. 7). In conventional systems, software can study the differences in the X-ray beam before and after sample irradiation. The obtained information is translated into a grayscale picture in which lighter areas represent high-density zones, while darker areas are associated with low density. This 2D picture can be converted to 3D by rotating the sample and recording consecutive images. This is possible on most CT equipment. The CT device can be used in X-ray fluorescence (XRF) mode, and information on the elemental composition of each sample slice was added to

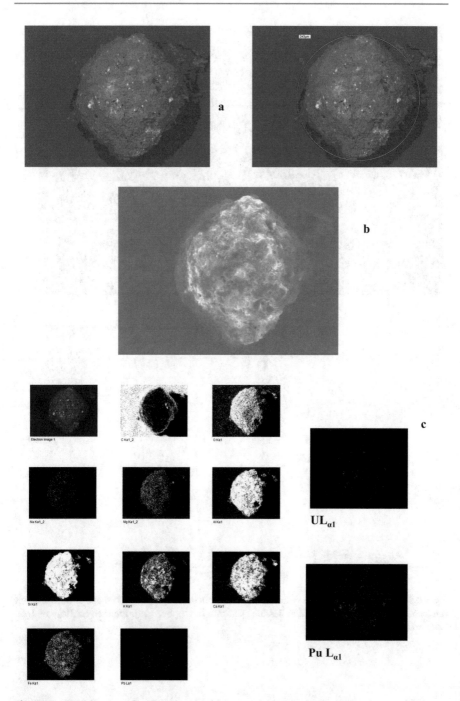

**Fig. 20.4** SEM images of radioactive particles containing U and Pu. Figure **a** corresponds to backscattered electrons, **b** to secondary electrons, and Figure **c** is the X-ray map of the particle according to the elemental composition. Pictures courtesy of Prof. I. Vioque

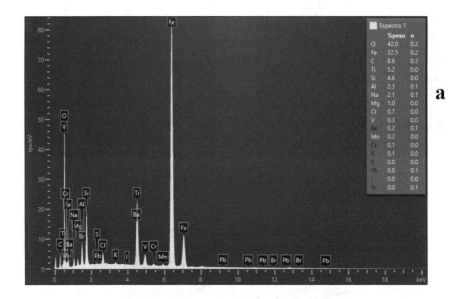

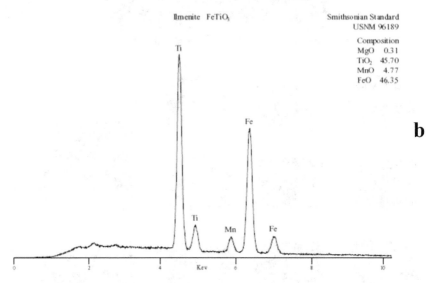

**Fig. 20.5  a** X-ray spectrum of a radioactive particle obtained by SEM. The shape of the spectrum can be identified as ilmenite, which has the spectrum in (**b**). See [11]. Pictures courtesy of J. L. García-León

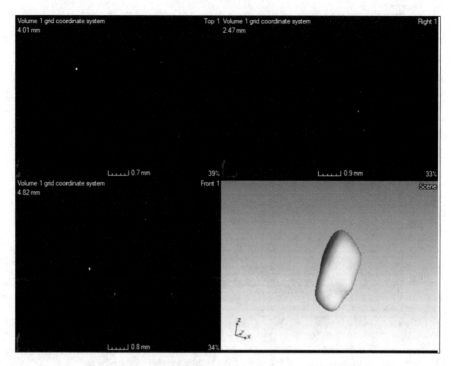

**Fig. 20.6**  CT image of a hot particle. Courtesy of Prof. I. Vioque

the density determination. The micrometer spatial resolution can be reached with CT. The X-ray source in CT experiments can vary from one system to another. In laboratory systems, the X-ray source is similar to that of conventional X-ray tubes [1]. There, a W target is bombarded with an electron beam. The Bremsstrahlung radiation produced in this way lies within the X-ray electromagnetic range and is used to irradiate the sample. Compact CT systems are usually based on this concept. However, synchrotron radiation can also be used (see Chap. 17). The intensity of the radiation source and the spatial resolution provides very sensitive results. Now, the sample must be transported to an accelerator facility. A CT image of a radioactive particle is shown in Fig. 20.6.

### 20.3.3  Nano- and $\mu$-XRF Techniques

X-ray fluorescence (XRF) occurs when electron vacancies are created in the atomic shells. The electrons from the outer shells tend to fill in the vacancy, and an X-ray is emitted during the transition from the outer shell to the inner shell. The X-ray energy corresponds to the energy difference between the atomic levels. The X-ray identifies the element as it depends on Z [12]. The plot in Fig. 20.7 confirms this dependency.

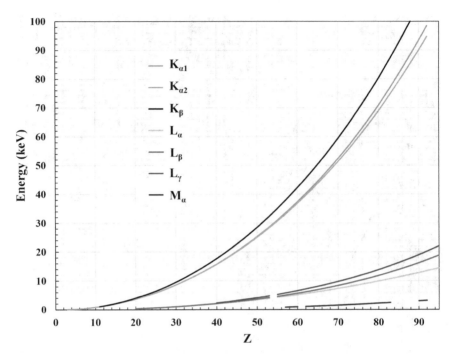

**Fig. 20.7** Energy of the X-ray as a function of the atomic number of the element

There are several mechanisms that create the vacancy and cause X-ray fluorescence. The sample can be irradiated by electromagnetic radiation or charged particles. In this last case, electrons or heavier particles can be used. In electron microscopes, the XRF is produced by irradiating the sample by electrons, as explained before. In the next section, we will describe the production of XRF by protons or α particles.

More conventionally, XRF is produced by irradiating the sample with electromagnetic radiation, either from an X-ray tube or a radioactive sample such as [241]Am that emits low-energy γ-radiation, 59.5 keV. It is sufficient to virtually produce X-ray fluorescence along the periodic system. An XRF spectrum from an environmental sample is shown in Fig. 20.8. The different X-ray peaks identify the elements in the sample. Under well-defined conditions, the net areas under the peaks can give the elemental concentration in the sample.

Usually, the excitation source in XRF laboratory systems is an X-ray tube similar to that used in CT devices. XRF provides information on the elemental composition of the samples at depths higher than those of electrons in SEM. Its combination with CT gives a morphological and elemental composition. In general, XRF instruments in laboratories provide spatial resolutions close to approximately 10 μm. For that, we discuss μ-XRF when referring to XRF laboratory systems [1, 2].

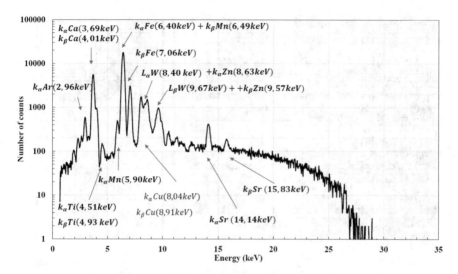

**Fig. 20.8** XRF spectrum of an environmental sample. The elements are identified through the emitted X-rays

XRF can also be produced using synchrotron radiation [12]. In this case, the intensity of the exciting radiation source is high enough to allow for better spatial resolution. In fact, the radiation beam can be focused within $50 \times 50$ nm$^2$ spots. The use of synchrotron radiation has an additional advantage. Indeed, the absence of a Bremsstrahlung background improves the sensitivity of the method. In this case, we discuss nano-XRF [2].

## 20.3.4 Nuclear Microprobes

Proton-Induced X-ray Emission was already described in Chap. 17. The use of protons as an excitation source for XRF is also possible, and this method provides fast and precise results on the elemental composition of a sample. Rutherford backscattering analysis (RBA) was also described in Chap. 17. Both techniques can be used to analyze the composition of radioactive particles. In the so-called nuclear microprobes, the ion beam from the accelerator is conducted through a quadrupole-focusing device placed before the reaction chamber. In this way, the beam area is reduced to a few $\mu$m$^2$ while maintaining an adequate intensity. The sample can be scanned to obtain a composition map with a spatial resolution close to a few $\mu$m$^2$. A typical $\mu$-PIXE map of a radioactive particle is presented in Fig. 20.9a. In Fig. 20.9b, an example of $\mu$-RBS spectra taken at different points of a radioactive particle is given. RBS is very useful for identifying heavy nuclides in low-mass matrices. The thickness of the mass of the Pu + U atom was measured and found to be inhomogeneous within the particle [13, 14].

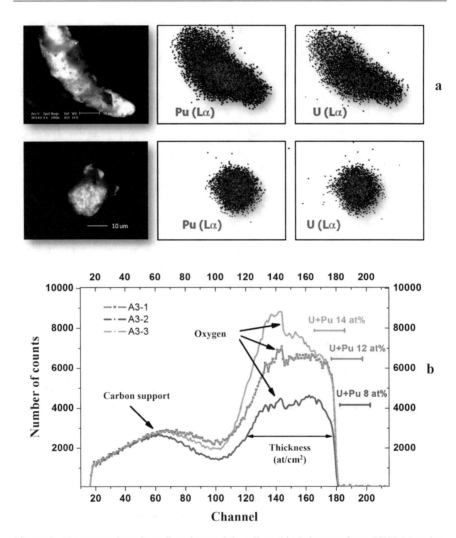

**Fig. 20.9** X-ray mapping of a radioactive particle collected in Palomares by μ-PIXE (**a**) and μ-RBS (**b**) SEM Figures and X-ray maps courtesy of Dr. M. C. Jiménez-Ramos. The μ-RBS Figure reprinted from [13] with permission by Elsevier

## 20.4   Radioactive Particle Characterization

### 20.4.1 X-ray Diffraction (XRD)

X-ray diffraction studies are a very interesting technique that helps to elucidate the morphology of a particle and its crystallographic structure. In XRD, the samples are irradiated with an X-ray source, and the diffracted radiation is studied with an X-ray diffractometer [1]. The application of Bragg laws allows one to calculate the distance among atoms and find the crystal planes, diffraction angles, relative orientations, and consequently the crystalline structure of the sample. The use of synchrotron radiation has contributed to the increase in XRD studies since very intense nano- and $\mu$-focused X-ray beams are available [2]. This permits the analysis of microcrystalline or nanocrystalline structures within the particle.

## 20.4.2 XANES and EXAFS

X-ray Absorption Near the Edge spectrometry is an X-ray Absorption (XAS) technique that accounts for the differences found near the absorption peaks that are related to the chemical state of the element. In a simple XAS experiment, the sample is irradiated by X-rays of different energies. It is expected that the X-ray flux is resonantly absorbed when its energy is exactly the binding energy of one electron in the atomic shell. As shown in Fig. 20.10, a sudden increase in the number of absorbed X-rays occurs for a specific X-ray energy. This peak, called the edge in Atomic Physics, appears at the electron-binding energy. Each element presents characteristic edges at the different shells. In this way, the element could be easily identified by an XAS experiment [15]. The edge position and shape depend on the elemental chemical species. Therefore, it could be possible to distinguish the elemental chemical species in an XAS experiment. XANES is an analytical technique based on the study of the edge shape basically for energies lower, although close, and around the edge. Extended X-ray Absorption Fine Structure (EXAFS), in turn, consists of the same technique based on the analysis of the XAS spectrum at energies higher than the edge.

XANES and EXAFS can provide information on the valence and oxidation states of the elements, coordination number, neighboring atom distance, crystalline form of the sample, and other chemical information [16]. On the other hand, a large variety of cataloged XANES spectra are available in the literature that helps to identify mineral species in radioactive particles [17]. A very important point for XANES and EXAFS is the available X-ray energy beams and their intensity. As the energies become more varied, the detection of chemical species becomes easier. The recent use of synchrotron radiation has boosted these techniques. In addition, in laboratory or synchrotron facilities, it is possible to obtain a focused X-ray beam at the $\mu$m range. Thus, $\mu$-XANES and $\mu$-EXAFS are possible, and the information they provide is available with high spatial resolution. $\mu$-XANES and $\mu$-EXAFS have been successfully applied to Chernobyl radioactive particles [18], revealing different U oxidation states depending on the region where the samples were collected.

## 20.4.3 Electron Spectroscopy Techniques (EELS, STEM-HAADF)

Electron Energy-Loss Spectrometry (EELS) is a technique based on electron beams and is often associated with electron microscopy. The analysis of the energy

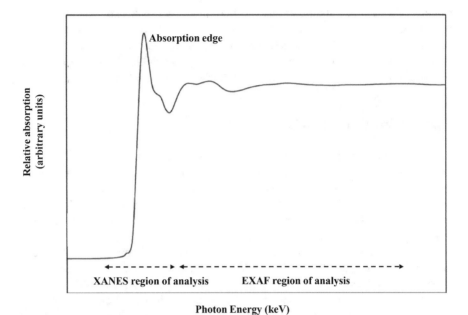

**Fig. 20.10**  General shape of an XAS spectrum

of electron beams after passing through a sample provides much information about its composition and structure. Electrons can lose energy by different mechanisms in their path. They are related to ionization or excitation effects. The energy spectrum of the transmitted electrons shows the energy losses suffered by the beam in its path. When the energy of the transmitted electrons is compared with that of the electron beam, it should be possible to identify a number of effects. Indeed, ionization of atoms or molecules leads to the loss of very specific energies that can be measured at the end of the transmission experiment. This loss identifies the element or chemical species. Specific vibrational modes of a crystal, phonons, can also be excited during the electron pathway within the sample, which produces a loss of a given energy. This can also help identify crystalline structures in the sample. Unfortunately, EELS is restricted to thin samples, since, as we studied in Chap. 7, the range of electrons in matter is relatively short in dense materials. Moreover, high-Z materials rapidly absorb electron energies. For that, EELS is especially useful for low-Z elements. However, because long irradiation times are usually needed, the sample is heated, which may lead to volatilization of some element or chemical species. This poses a clear limitation to this technique. Despite this situation, EELS has been applied to studies of radioactive particle characterization and particle size speciation in aqueous solutions [19].

A very specific technique has been developed in connection to electron microscopy and specifically together with transmission electron microscopy. This is HAADF spectroscopy after high angle annular dark field. In HAADF, the transmitted electrons in an electron microscope are selected by an annular detector that collects those scattered electrons within the defined annulus. This technique is very sensitive to Z changes in the sample, providing a high-contrast image because electrons are chiefly scattered by high-Z components of the sample. This allows us to identify very clearly high-Z elements within a major low-Z background. The spatial resolution of this technique is closely related to the size of the electron beam and consequently to the spatial resolution of the electron microscope. Tenths of nanometers can be obtained, as observed in the literature [20].

**Exercises (Figures in the exercises courtesy of Prof. I. Vioque and Mr. J.-L. García-León)**

1. Autoradiography of a soil sample was performed during a week. The result is collected in Fig. 20.2. Identify the radioactive particles in the picture. How many radioactive particles can you find? The darkness and size of the potential particles vary. Describe the features of the particles that cause these variations.
2. With an electron microscope, we record the backscattered and secondary electron pictures from a radioactive particle.

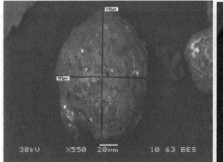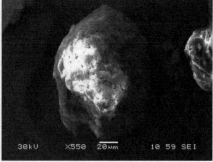

   Identify each of the pictures and explain why. Can you identify high-Z components in any of the pictures? Why?
3. The following X-ray pictures were taken with an electron microscope from two regions of the above-described radioactive particles. One seems to contain U and Pu, while the other does not.

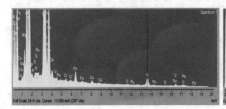

Guess which region corresponds to each of the pictures and explain why.

4. A soil sample was collected from an area contaminated by an unexploded nuclear weapon. A radioactive particle was found in the soil, and a CT image was taken. Explain the differences found between lighter and darker areas and guess their composition.

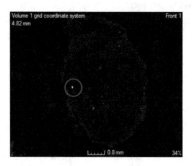

5. In an X-ray fluorescence (XRF) experiment, a $^{241}$Am excitation source is used. The XRF spectrum of a sample known to contain Cu follows.

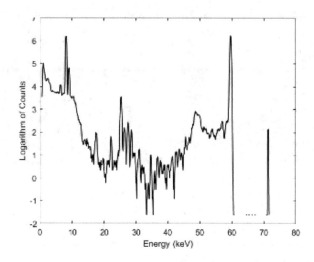

Identify the Cu X-ray peaks. Could you identify a peak from the $^{241}$Am γ-emission?

6. Use the same $^{241}$Am source to obtain the XRF spectrum from a sample comprising U and Pu. Sketch the expected XRF spectrum.

7. Describe the necessary data to obtain the $\frac{U}{Pu}$ composition of the sample.

8. Two radioactive particles containing $^{239}$Pu, $^{241}$Am, and $^{235}$U are sufficiently active to be measured by an extended range energy Ge detector. Sketch the

expected spectra from both particles if particles A and B are known to contain $\frac{Pu}{U}$ concentration ratios close to 2 and 4, respectively, and $\frac{Pu}{Am}$ concentration ratios of approximately 100 and 130, respectively. Use Fig. 11.16 to estimate the counting efficiency.

9. 3 MeV protons are used for a 170° RBS determination of the $^{239}$Pu and $^{235}$U contents in a radioactive particle. Using the Rutherford scattering kinematic factor sketch, the RBS spectrum was obtained for the sample. Suppose that both radionuclides are placed on the surface of the particle.

10. A realistic RBS spectrum is presented in Fig. 20.9. There, the concentrations of Pu and U are distributed in depth and within a matrix made of silicates and other soil components. Explain the shape of the spectrum and compare it with that sketched in Exercise 9.

# References

1. Radioactive particles in the environment: Sources, Particle Characterization and Analytical Techniques, IAEA-TECDOC-1663 (IAEA, 2011)
2. Brit Salbu and Ole Christian Lind, Analytical techniques for characterizing radioactive particles deposited in the environment. J. Environ. Radioact. **211**, 106078 (2020)
3. M.C. Jiménez-Ramos, R. García-Tenorio, I. Vioque, G. Manjón, M. García-León, Presence of plutonium contamination in soils from Palomares (Spain). Environ. Pollut. **142**, 487–492 (2006)
4. M.C. Jiménez-Ramos, S. Hurtado, E. Chamizo, R. García-Tenorio, L. León-Vintró, P.I. Mitchell, $^{239}$Pu, $^{240}$Pu, and $^{241}$Am determination in hot particles by low level Gamma-spectrometry. Environ. Sci. Technol. **44**, 4247–4252 (2010)
5. C.J. Zeissler, R.M. Lindstrom, J.P. McKinley, Radioactive particle analysis by digital autoradiography. J. Radioanal. Nucl. Chem. **248**, 407–412 (2001)
6. R. Ikehara, M. Suetake, T. Komiya, G. Furuki, A. Ochiai, S. Yamasaki, W.R. Bower, G.T.W. Law, T. Ohnuki, B. Grambow, R.C. Ewing, S. Utsunomiya, Novel method of quantifying radioactive Cesium-rich microparticles (CsMPs) in the environment from the Fukushima Daiichi nuclear power plant. Environ. Sci. Technol. **52**, 6390–6398 (2018)
7. M. Eriksson, K. Ljunggrena, C. Hindorf, Plutonium hot particle separation techniques using real-time digital image Systems. Nuc. Instrum. Methods Phys. Res. A **488**, 375–380 (2002)
8. C. Fiorini et al., Present and future Anger cameras based on silicon drift detectors, in *IEEE Nuclear Science Symposium Conference Record* (2007), pp. 2318–2322
9. B. Salbu, V. Kashparov, O.C. Lind, R. García-Tenorio, M.P. Johansen, D.P. Child, P. Roos, C. Sancho, Challenges associated with the behaviour of radioactive particles in the environment. J. Environ. Radioact. **186**, 101–115 (2018)
10. J.I. Goldstein, D.E. Newbury, J.R. Michael, N.W.M. Ritchie, J.H.J. Scott, D.C. Joy, *Scanning Electron Microscopy and X-Ray Microanalysis*, 4th Edn. (Springer, 2018)
11. K.P. Severin, *Energy Dispersive Spectrometry of Common Rock-Forming Minerals* (Springer, 2004)
12. A.C. Thomson (ed.), *X-ray Data Booklet*, 3rd Edn. (Lawrence Berkeley National Laboratory, University of California, 2009)
13. J. García-López, M.C. Jiménez-Ramos, M. García-León, R. García-Tenorio, Characterisation of hot particles remaining in soils from Palomares (Spain) using a nuclear microprobe. Nucl. Instrum. Methods Phys. Res. B **260**, 343–348 (2007)

14. M.C. Jiménez-Ramos, J. García López, M. Eriksson, J. Jernstrom, R. García-Tenorio, PIXE analysis of U and Pu from hot particles: K-lines vs L-lines. Nucl. Instrum. Methods Phys. Res. B **273**, 118–121 (2012)

15. G.S. Henderson, F.M.F. de Groot, B.J.A. Moulton, X-ray absorption near-edge structure (XANES) spectroscopy. Rev. Mineral. Geochem. **78**, 75–138 (2014)

16. O.N. Batuk, D. Conradson, O.N. Aleksandrova, H. Boukhalfa, B.E. Burakov, D.L. Clark, K.R. Czerwinski, A.R. Felmy, J.S. Lezama-Pacheco, S.N. Kalmykov, D.A. Moore, B.F. Myasoedov, D.T. Reed, D.D. Reilly, R.C. Roback, I.E. Vlasova, S.M. Webb, M.P. Wilkerso, Multiscale speciation of U and Pu at Chernobyl, Hanford, Los Alamos, McGuire AFB, Mayak and Rocky Flats. Environ. Sci. Technol. **49**, 6474–6484 (2015)

17. J. Rehr, S. Zabinsky, R. Albers, Ab Initio XAFS and XANES standards. MRS Proc. **307** (1993), http://ixs.iit.edu/database/index.html, and others

18. B. Salbu, T. Krekling, O.C. Lind, D.H. Oughton, M. Drakopoulos, A. Simionovici, I. Snigirevab, A. Snigirev, T. Weitkamp, F. Adams, K. Janssens, V.A. Kashparov, High energy X-ray microscopy for characterisation of fuel particles. Nucl. Instrum. Methods Phys. Res. A **467–468**, 1249–1252 (2001)

19. E.C. Buck, D.A. Moore, K.R. Czerwinski, S.D. Conradson, O.N. Batuk, A.R. Felmy, Nature of nano-sized plutonium particles in soils at the Hanford Site. Radiochim. Acta **102**, 1059–1068 (2014)

20. K.T. Moore, X-ray and electron microscopy of actinide materials. Micron **41**, 336–358 (2010)

# Index

## A

Absorbed dose, 56, 59, 195, 197–199, 201, 203, 206, 216, 376, 377, 379, 380, 389

Absorption coefficient, 135, 229, 232, 244

Accelerator Mass Spectrometry, 121, 331, 383, 418, 456, 521, 522, 525, 527, 531, 540, 547, 548

Activation products, 45, 47, 70, 73, 78, 84–86, 89, 90, 96, 118, 122, 145, 584, 597

Active shielding, 371, 467, 468, 470, 473, 474, 477, 478, 490

Advantage factor, 597

ALARA, 208

Albedo coefficient, 246

Alvarez, 536, 549

Annihilation radiation, 318

Annual Limit of Intake (ALI), 200, 201

Anticoincidence detector, 474, 477

Anti-Compton, 475–477, 588

Antineutrino, 10, 11

Armstrong, 2

Artificial radioactivity, 51, 141, 145, 160, 522

Ashing, 407

Atmospheric tests, 70, 71, 75, 121

Atomic number, 3, 8, 11, 12, 31, 222, 229, 236, 310, 350, 353–355, 367, 371, 468, 469, 473, 611, 616

Attenuation coefficient, 229, 231, 232, 243, 256, 257, 470, 489

Auger, P., 17

Average Aerodynamic Diameter (AMAD), 144

Avogadro number, 21, 25, 236, 448, 449, 580, 583

## B

Backscattered electrons, 611–613

Backscattering edge, 227

Backscattering peak, 319, 334

Bateman equations, 21, 25, 27, 37, 39, 43, 44, 54

Beam Profile Monitors (BPM), 527–529

Becquerel, 6, 7, 24

Beta-spectrometry, 360

Bethe-Bloch equation, 235, 236, 238, 383

Binary splitting method, 608

Binomial distribution, 447, 448

Bioavailability, 53, 105, 106, 144, 149, 432

Blank samples, 435, 436

Bomb pulse, 69, 78, 80, 121, 150, 153, 158, 173, 177

Box model, 109–111

Bragg addition rule, 232, 238, 241

Bragg curves, 243, 248

Bragg-Gray principle, 376

Branching ratio, 16, 29, 31, 389, 390

Bremsstrahlung, 233, 239–241, 247, 249, 256, 268, 282, 309, 317, 319, 332, 371, 432, 469, 615, 617

## C

Carrier, 275–278, 289, 341–346, 348, 350, 362, 407, 421, 422, 511

Cascade impactor, 402, 433, 608

$^{14}$C dating, 3, 61, 63, 331, 445, 467, 478, 485, 486, 549, 561, 563, 570

Channeltrons, 501, 505

Charge collection, 259, 289, 292, 350, 368, 429

Chemical quenching, 328, 423, 425, 426

Cherenkov detectors, 387–389

© The Editor(s) (if applicable) and The Author(s), under exclusive license to Springer Nature Switzerland AG 2022
M. García-León, *Detecting Environmental Radioactivity*, Graduate Texts in Physics, https://doi.org/10.1007/978-3-031-09970-0

Printed in the United States
by Baker & Taylor Publisher Services